# HEAVEN
# AND EARTH

## global warming
## the missing science

## IAN PLIMER

**TAYLOR TRADE TUBLISHING**
lanham • new york • boulder • toronto • plymouth, uk

Published by Taylor Trade Publishing
An imprint of The Rowman & Littlefield Publishing Group, Inc.
4501 Forbes Boulevard, Suite 200, Lanham, Maryland 20706
www.rlpgtrade.com

Distributed by NATIONAL BOOK NETWORK

Library of Congress Preassigned Control Number: 2009930238
ISBN-13: 978-1-58979-472-6 (pbk. : alk. paper)

♾™ The paper used in this publication meets the minimum requirements of American
National Standard for Information Sciences—Permanence of Paper for Printed Library
Materials, ANSI/NISO Z39.48-1992.

Manufactured in the United States of America.

*Dedicated to AGL and CEL (1992 to 2003);*
*White knights against the forces of darkness*

# ABOUT THE AUTHOR

PROFESSOR IAN PLIMER (School of Earth and Environmental Sciences, The University of Adelaide) is Australia's best-known geologist. He is also Emeritus Professor of Earth Sciences at the University of Melbourne. He was Professor and Head at the University of Melbourne (1991–2005) and Professor and Head at the University of Newcastle (1985–1991). He was previously on the staff of the University of New England, the University of New South Wales and Macquarie University. He has published more than 120 scientific papers on geology. This is his seventh book written for the general public, the best known of which are *Telling Lies for God* (Random House), *Milos-Geologic History* (Koan) and *A Short History of Planet Earth* (ABC Books).

He won the Leopold von Buch Plakette (German Geological Society), Clarke Medal (Royal Society of New South Wales), Sir Willis Connolly Medal (Australasian Institute of Mining and Metallurgy), was elected Fellow of the Australian Academy of Technological Sciences and Engineering and was elected Honorary Fellow of the Geological Society of London. In 1995, he was Australian Humanist of the Year and later was awarded the Centenary Medal. He was Managing Editor of *Mineralium Deposita*, president of the SGA, president of IAGOD and president of the Australian Geoscience Council and sat on the Earth Sciences Committee of the Australian Research Council for many years. He is a regular radio and television broadcaster of science to the general public and has received the Eureka Prize for the promotion of science, the Eureka Prize for *A Short History of Planet Earth* and the Michael Daley Prize (now a Eureka Prize) for science broadcasting.

Professor Plimer has spent much of his life in the rough and tumble of the zinc-lead-silver mining town of Broken Hill where an integrated interdisciplinary scientific knowledge intertwined with a healthy dose of scepticism and pragmatism are necessary. At Broken Hill, he is Patron of Lifeline and Patron of the Broken Hill Geocentre. He has worked for North Broken Hill Ltd, is director of CBH Resources Ltd, Ivanhoe Australia Ltd and Kefi Minerals plc and recently had a new Broken Hill mineral, plimerite $ZnFe_4(PO_4)_3(OH)_3$, orthorhombic, named after him in recognition of his contribution to Broken Hill geology. Plimerite is insoluble in alcohol.

# TABLE OF CONTENTS

# Chapter 1

# INTRODUCTION

We are all environmentalists. Some of us underpin our environmentalism with political and romantic idealism, others underpin it with emotion, others have a religious view of the environment, some underpin their environmental view with economic pragmatism and many, like me, try to acquire an integrated scientific understanding of the environment. An integrated scientific view involves a holistic view of the Earth and considers life, ice sheets, oceans, atmosphere, rocks and extraterrestrial phenomena which influence our planet. This is what is attempted in this book. I look at climate over geological, archaeological, historical and modern time. Geology is about time, changes to our environment over time and the evolution of our planet. Geology is the only way to integrate all aspects of the environment. In this book I look at what history tells us about past climate and how the Sun, the Earth, ice, water and air affect climate. In the last chapter, I give some personal views.

Past climate changes, sea level changes and catastrophes are written in stone. Time is a beautiful but misunderstood four-letter word. Most of us can't fathom the huge numbers that geologists and astronomers use, hence most of the community has little knowledge of geology. History and archaeology are rarely integrated with natural geological events. There is little or no geological, archaeological and historical input into discussions about climate change.

It is little wonder then that catastrophist views of the future of the planet fall on fertile pastures. The history of time shows us that depopulation, social disruption, extinctions, disease and catastrophic droughts take place in cold times and life blossoms and economies boom in warm times.

Planet Earth is dynamic. It always changes and evolves. It is currently in an ice age that started 37 million years ago.

## Climate

Climate has always changed. It always has and always will. Sea level has always changed. Ice sheets come and go. Life always changes. Extinctions of life are normal. Planet Earth is dynamic and evolving. Climate changes are cyclical and random. Through the eyes of a geologist, I would be really concerned if there were no change to Earth over time. In the light of large rapid natural climate changes, just how much do humans really change climate?

The Earth's climate is driven by the receipt and redistribution of solar energy. Without this, there would be no life on Earth. Despite well-documented linkages between climate and solar activity, the Sun tends to be brushed aside as the driver of climate on Earth in place of a trace gas (carbon dioxide – $CO_2$), most of which derives from natural processes. The $CO_2$ in the atmosphere is only 0.001% of the total $CO_2$ held in the oceans, surface rocks, air, soils and life.

Although we are in one of the many warm periods[1] between glacial stages in the current ice age, there is a significant amount of ice remaining in the polar regions. Polar ice has been present for less than 20% of geological time, life on Earth for more than 80% of time and liquid water on Earth for 90% of time. Planet Earth is a warm wet volcanic greenhouse planet, which is recovering from glacial times and is naturally warming. Cooling has also occurred in the current interglacial times. Earth has warmed and cooled on all time scales, whether they be geological, archaeological, historical or within our own lifetime. The key questions are: How much of this warming can be attributed to human activity?

If we humans are warming the planet now, how do we explain alternating cool and warm periods during the current post-glacial warming?

Before we can hope to understand present climate change, we must understand how climate has changed in the past. We know that there have been past climate changes which have been extreme and rapid yet we do not understand all the drivers of these past climate changes. Although we know that there are a large number of variables that influence climate, there are probably variables that have not yet been discovered. Some of the known variables have a huge effect on climate, others have a slight effect, but combinations can have an unpredictable effect.

We cannot view planet Earth as a simple scientific experiment where, by changing one variable, we can isolate another variable.

---

[1] These warm periods are called interglacials. Sea level is commonly higher, plant and animal life expand their habitats and the volume of the ice sheets decreases. From now on, the word interglacial will be used for these warm periods.

Calculations on supercomputers, as powerful as they may be, are a far cry from the complexity of the planet Earth, where the atmosphere is influenced by processes that occur deep within the Earth, in the oceans, in the atmosphere, in the Sun and in the cosmos. To reduce modern climate change to one variable ($CO_2$) or, more correctly, a small proportion of one variable (i.e. human-produced $CO_2$) is not science, especially as it requires abandoning all we know about planet Earth, the Sun and the cosmos. Such models fail.

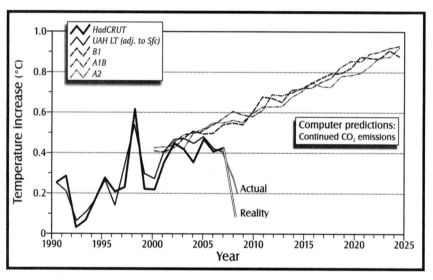

*Figure 1: Five computer predictions of climate made in 2000. These underpin the IPCC predictions and all show that there is no relationship between the predicted future temperature and actual measured temperature over even a short period of time and that there is no relationship between the actual temperature and the atmospheric $CO_2$ content. Computer predictions cannot even predict a decade in advance, let alone 50 years or a century in advance. This diagram shows that the hypothesis that human emissions of $CO_2$ create global warming is invalid.*

The history of temperature change over time is related to the shape of continents, the shape of the sea floor, the pulling apart of the crust, the stitching back together of the crust, the opening and closing of sea ways, changes in the Earth's orbit, changes in solar energy, supernoval eruptions, comet dust, impacts by comets and asteroids, volcanic activity, bacteria, soil formation, sedimentation, ocean currents and the chemistry of air. If we humans, in a fit of ego, think we can change these normal planetary processes, then we need stronger medication.

If we look at the history of $CO_2$ over time, we see the atmospheric $CO_2$ content has been far higher than at present for most of time. Furthermore, atmospheric $CO_2$ follows temperature rise – it does not create a temperature rise. To argue that human emissions of $CO_2$ are forcing global warming requires all the known, and possibly chaotic, mechanisms of natural global warming to be critically analysed and dismissed. This has not even been attempted. To argue that we humans can differentiate between human-induced climate changes and natural climate changes is naïve. To argue that natural climate changes are slow and small is contrary to evidence. The slogan "Stop climate change" is a very public advertisement of absolute total ignorance as it is not cognisant of history, archaeology, geology, astronomy, ocean sciences, atmospheric sciences and the life sciences.

Humans can change the *weather*. The "urban heat island" effect shows that the concentration of roads, concrete, buildings and machinery in towns of more than 1000 inhabitants creates a warmer setting than in a rural setting. In Europe, we see a "winter weekend effect" where cooler wetter weather probably results from human activity. These weather changes do not necessarily mean that humans change *climate*.

## Carbon dioxide and pollution

Pollution shortens your life. However, $CO_2$ is not a pollutant. Global warming and a high $CO_2$ content bring prosperity and lengthen your life. Carbon dioxide is plant food, is necessary for life, and without $CO_2$ there would be no complex life on Earth. In some parts of the world, polluting smogs are common. They currently derive from backyard brickworks, small dirty smelters and furnaces, power stations using high sulphur-high ash coals, forest clearing fires, bush and grass fires and millions of small obsolete and dirty wood, charcoal and coal stoves, heaters, boilers and furnaces. Millions of coal fires caused similar smogs in England up until the 1950s.

The Kyoto Protocol is a treaty to regulate $CO_2$, methane, nitrous and other nitrogen oxides, hydrofluorocarbons, perfluorocarbons and sulphur hexafluoride. It cannot be a treaty to regulate greenhouse gases, because water ($H_2O$) vapour, the main greenhouse gas, is not included. Car exhaust gases consist of harmless gases ($CO_2$, nitrogen, $H_2O$ vapour), pollutants (carbon monoxide, volatile organic compounds, nitric oxide, nitrogen dioxide, sulphur dioxide and PM-10 [very small particulate matter]).[2] A car's catalytic converter converts some 95% of these pollutants into $H_2O$ and

---

[2] Charron, A. and Harrison, R. M. 2003: Primary particle formation from vehicle emissions during exhaust dilution in the roadside atmosphere. *Atmospheric Environment* 37: 4109-4119.

$CO_2$. Smog consists of ozone (formed from the photochemical reaction of nitrogen oxides with hydrocarbons), sulphur dioxide and PM-10.[3] Smog can kill people, plants and animals.[4]

The open combustion of poor quality carbon fuels produces soot, smoke, ash, unburnt fuel and chemicals containing sulphur, chlorine, nitrogen, fluorine and metals. In confined unventilated places, open fires produce poisonous carbon monoxide. At present, China emits more sulphur dioxide than any other country in the world and this chokes people, causes acid rain, damages life and destroys buildings. The "Asian Brown Cloud" covers an area as large as Australia, obscuring the Sun in some polluted Asian cities. It has a profound effect on human health. At times, it drifts right across the Pacific Ocean and covers the Northern Hemisphere. These smogs are not due to $CO_2$, which is invisible. Darker soot falling on snow and ice allows it to absorb more solar energy and may contribute to more rapid melting of snow and ice.

The Western world was bathed in atmospheric pollution half a century ago. The smoke pollution from 1860 to 1960 of London, Manchester and Pittsburgh was far greater than that of Beijing today. Charles Dickens called the notorious pea souper fogs of London "London particular", and Edward I passed a law in 1272 AD trying to get rid of them. London was called the "big smoke" because that's exactly what it was. Little sunshine could penetrate the atmosphere. Children developed rickets from the lack of sunshine, plants and animals died and lung disease was widespread. The smog was commonly so dense that bus drivers could not see the kerb and a passenger had to walk along the edge of the road with a light to lead the way. Trains had difficulty running as drivers could not see the signals, and detonators had to be placed on rail tracks to warn of potential dangers. The Black Fog of 1952, triggered by a temperature inversion over London, reduced visibility to 10 centimetres and 4000 Londoners died of respiratory problems caused by sulphur dioxide. The Clean Air Act of 1956 prevented the use of open fires of coal and wood in big cities. It was reticulated cheap coal-fired and nuclear reactor electricity that stopped pollution in Britain. The same will probably happen in Asia.

The public have rightfully become less tolerant of pollution and much progress has been made to clean up the Western world. Governments, the media and many people are of the view that $CO_2$ is the cause of climate

[3] Shi, J. P., Evans, D. E., Khan, A. A. and Harrison, R. M. 2001: Sources and concentration of nanoparticles (<10 nm diameter) in the urban atmosphere. *Atmospheric Environment* 35: 1193-1202.

[4] Dockery, D. W., Pope, C. A., Xu, X. P., Spengler, J. D., Ware, J. H., Fay, M. E., Ferris, B. G. and Speizer, F. E. 1993: An association between air pollution and mortality in six US cities. *New England Journal of Medicine* 329: 1753-1759.

change, is of human origin and is a pollutant. It would seem to the layman that there is no longer any need for scientific debate about climate change.

There has never been a transparent public debate. The time is now.

## The science of climate

Science is married to evidence derived from observation, measurement and experiment. Evidence is fraught with healthy uncertainties and scientists argue about the methods, accuracy, repeatability and veracity of data collection. If the data can be validated, then this body of new evidence awaits explanation. The explanation is called a scientific theory. This scientific theory must be abandoned or modified if the evidence is not repeatable or if the evidence is not coherent with previously validated evidence. With new evidence, theories are abandoned or refined. A scientific hypothesis tests a concept by the collection and analysis of evidence. Hypotheses are invalidated by just one item of contrary evidence, no matter how much confirming evidence is present. Science progresses by abandoning theories and hypotheses and creating new explanations for validated evidence.

Most scientists are anarchistic, bow to no authority and construct conclusions based on evidence. These conclusions change with more evidence. Science is not dogmatic and the science of any phenomenon is never settled. Matters of science cannot be resolved by authority or consensus. Scientific evidence is unrelated to politics, ideology, popular paradigms, worldviews, fads, ethics, morality, religion and culture. It matters not whether one is from Canada, Chad or Chile, the scientific measurement for the speed of light is about 299,792.5 kilometres per second. If you are Buddhist, Baha'i or Baptist, the speed of light is still about 299,792.5 kilometres per second. If it is dark, the speed of light is still about 299,792.5 kilometres per second.

The level of scientific acceptance of human-induced global warming is misrepresented. Furthermore, the claim by some scientists that the threat of human-induced global warming is 90% certain (or even 99%) is a figure of speech reflecting the speaker's commitment to the belief. It has no mathematical or evidential basis. It is comparable to 100% certainty professed by religious devotees that theirs is the one and only true faith. My experience of dealing with blindingly obvious arguments against creation "science" was that data and logic were treated with anger, rejection and hostility. Scientific arguments were never addressed. With some rabid environmentalists, human-induced global warming has evolved into a similar religious belief system. This, I argue in the last chapter of this book, is an urban atheistic religion disconnected from Nature and it evolved to fill a yawning spiritual vacuum in the Western world. Contrary scientific data and conclusions are greeted with anger, rejection and hostility. As more contrary data is aired, the defence of the indefensible produces grimmer and grimmer future climate scenarios. The scientific arguments are not addressed. These are the characteristics of a fundamentalist religion.

Aristotle established a general principle of scientific enquiry: "First we must seek the fact, then seek to explain."[5]

The scientific method is now popularly conceptualised that the science on global warming is settled as a process where authorities balance volumes of opinions. That's it. A phenomenon is now scientifically proven because various authorities and some scientists say so. Evidence now no longer matters. And any contrary work published in peer-reviewed journals is just ignored. We are told that the science on human-induced global warming is settled. We have not advanced much since Galileo's spot of bother on 22 June 1633.

Climate science lacks scientific discipline. Studies of the Earth's atmosphere tell us nothing about future climate. An understanding of climate requires an amalgamation of astronomy, solar physics, geology, geochronology, geochemistry, sedimentology, tectonics, palaeontology, palaeoecology, glaciology, climatology, meteorology, oceanography, ecology, archaeology and history. This is what is attempted in this book.

At times, primary scientific evidence is manipulated and simplified by computer models. The extensive reliance by global warmers on computer models impresses those with little scientific training. However, the significant manipulation of the source data and the lack of use of many known variables create uncertain outputs. Furthermore, scientific data yet to be discovered cannot be used in a model. It is very easy for the modeller to produce the predestined outcome before the model can be run. This is a common flaw of mathematical modelling. A model is not real. Models are not evidence. Models with simulations, projections and predictions prove nothing. All a model shows is something about the model itself and the modellers, normally their limitations. As the Talmud states: "We do not see things as they are. We see them as we are."

Data collection in science is derived from observation, measurement and experiment, not from modelling. We can't make Nature conform to virtual computer models. Climate catastrophes regularly occur and, no matter what models are used, they will still occur. Collection of new scientific data by observation, measurement and experiment is now out of fashion. It is now the height of fashion to use someone else's data in the virtual world of a

---

[5] *Posterior Analytics* II, 1, 89b, 29-31.

computer model. The problem is that the virtual world is just not related to the real world, especially as Nature is so fickle.

Observations in Nature differ markedly from the results generated by nearly two dozen climate models. These climate models exaggerate the effects of human $CO_2$ emissions into the atmosphere because few of the natural variables are considered.

What actually has been omitted from these models to reach conclusions that are not in accord with observations in Nature?

What are the natural mechanisms that control climate?

How well can they be predicted?

Insurance modellers did not factor in two Boeing 767 jets destroying the World Trade Center and financial modellers did not predict the global financial crisis commencing in 2008. Natural systems are far more complex and it is naïve to think that a model can predict future events on the Earth. To try to predict the future based on just one variable ($CO_2$) in extraordinarily complex natural systems is folly. Modern global temperature trends are doing their best to show us that $CO_2$ is not a driver of climate.

The hypothesis that human emissions of $CO_2$ can create global warming can be tested by measurement. This is how science works. Temperature measurements using ground-based thermometers, balloon-mounted radiosondes and satellite-mounted microwave sensing units all show that no warming has occurred since 1998. Once the urban heat island effect with ground thermometers and the 1998 El Niño are considered, there has been little warming since 1979. During that time atmospheric $CO_2$ has increased. Climate models using increasing $CO_2$ predict simultaneous and intense warming in both polar areas yet this has not happened in modern or ancient times. The test of the hypothesis above shows that there is no relationship between measured temperature and $CO_2$ emissions. The hypothesis fails.

Many questions are asked and answered in this book.

Are the speed and extent of modern climate change unprecedented?

Is dangerous warming occurring?

Is the temperature range observed in the 20th Century outside the range of normal variability?

Do volcanoes change climate?

Do wobbles in the Earth's orbit change climate?

Have past climate changes driven extinction?

Do human emissions of $CO_2$ create sea level rise?

Will the seas become acidic?

Does sea level rise kill coral atolls?

Are humans forcing changes in ocean currents?

Do higher sea temperatures cause more hurricanes?

Is global warming melting the polar ice caps and alpine valley glaciers?

Does the Sun influence the Earth's climate?

Do extraterrestrial forces influence the Earth's climate?

In the past, what has stopped temperature rising and rising and rising until the Earth became uninhabitable?

Why do human emissions of $CO_2$ continue to increase yet, since 1998, temperature has decreased?

Why do Antarctic temperatures decrease when Greenland temperatures increase and *vice versa* despite the almost uniform spread of $CO_2$ over the globe?

Are temperature and $CO_2$ measurements reliable?

How does the greenhouse effect work?

Where does $CO_2$ come from?

Where does $CO_2$ go?

One thing I have learned from more than 40 years in science is that surprises abound. When I look at the history of planet Earth, no surprise can surprise me any longer. The impossible happens.

Political decisions based on the statement "the science is settled" have a guaranteed short life.

## Why the Intergovernmental Panel on Climate Change (IPCC)?

In 1827, Josephe Fourier suggested that the Earth's atmosphere traps heat radiated by the Sun. In 1860, John Tyndall reported that it is only the greenhouse gases in the atmosphere that have this property. Water vapour contributed to 95% of the "greenhouse effect" followed by $CO_2$ (3.62%), nitrous oxide (0.95%), methane (0.36%) and others (0.07%).

In 1896, the Swedish chemist Svante Arrhenius tried to calculate the effect of $CO_2$ that was being added to the atmosphere by the burning of fossil fuel. He calculated that if atmospheric $CO_2$ were to double, the temperature would rise by 5°C. He was wrong.

Because of warming in the 1920s and 1930s, in 1938 the English meteorologist Guy Challender suggested the temperature rise might be due to the release of $CO_2$ into the atmosphere from human activities. Challender argued that the increased $CO_2$ would be good for agriculture. He was right. He also argued that $CO_2$ might dampen the effects of the next inevitable ice age.[6] As soon as Challender had argued that $CO_2$ was good for humans, the climate started to cool (1940–1976).

---

[6] Challender, G. S. 1938: The artificial production of carbon dioxide and its influence on climate. *Royal Meteorological Society Quarterly Journal* 64: 223-240.

On 24 June 1974, *Time* magazine warned that we would suffer from a new ice age, as did *Newsweek* (28 April 1975) and *National Geographic*[7] in 1976. On 3 April 2006, *Time* was at it again. This issue had a special report on global warming. The effects of the predicted global cooling on humans in 1974 were exactly the same as the effects of global warming on humans in 2006. Climatologist Stephen Schneider co-authored a book in 1977 warning us of the horrors of a new ice age.[8] He now warns us of the horrors of global warming. In Lowell Ponte's 1975 book on global cooling,[9] he states:

> Global cooling presents humankind with the most important social, political, and adaptive challenge we have had to deal with for 110,000 years. Your stake in the decisions we make concerning it is of ultimate importance: the survival of ourselves, our children, our species.

We only need to substitute the word "warming" for "cooling" and we have the identical alarmism, 30 years later. Meanwhile, the climate stubbornly refuses to co-operate with computer models and the writers of alarmist popular articles and books.

During the Cold War, there was considerable concern that the nuclear arsenal of the USSR and USA was enough to eliminate all humans on Earth many times. Environmental groups were then anti-nuclear groups. After the end of the Cold War, it was difficult for environmental groups to attract attention. Many were looking for a new global problem to tackle. The establishment by the UN of the Intergovernmental Panel on Climate Change (IPCC) in 1988 gave an opportunity to make global warming the main theme of environmental groups. This theme had the ability to attract public interest. And it did. Crispin Tickell, a lobbyist for the planned IPCC was the UK's permanent representative on the UN. Earlier he had published a book[10] on the dangers of global cooling, now he was warning of the dangers of global warming.

Under the auspices of the UN's World Meteorological Organization and the UN Environment Program, the IPCC was established. As global warming was a hot topic, it was discussed in 1989 by the US Senate Committee on Science, Technology and Space chaired by Senator Al Gore. The Committee heard submissions by Dr Roger Revelle (who had taught Gore at Harvard)

---

[7] Matthews, S. W. 1976: What's happening to our climate? *National Geographic* 150:5, 576-615.
[8] Schneider, S. and Mesirow, L. E. 1977: *The genesis strategy: Climate and global survival.* First Delta.
[9] Ponte, L. 1975: *The cooling. Has the next ice age already begun? Can we survive it?* Prentice Hall.
[10] Tickell, C. 1977: *Climate change and world affairs.* University Press of America.

and Dr James Hansen (Director, Goddard Institute for Space Studies). Hansen claimed that the warm summer of 1988 was due to global warming. The summer of 1988 was one of drought, fires were consuming Yellowstone National Park and the city of Washington sweltered. The Senate hearing room was hot and stuffy. Hansen announced at the Senate hearing, "with a high degree of confidence", that global warming had arrived.[11] Evidence urging caution and that the science was controversial was dismissed.[12]

The Climate Action Network was formed, the media went into years of brouhaha about global warming and fellow travellers boarded the bandwagon at every opportunity. The cause became fashionable, especially amongst climate experts such as Robert Redford, Barbra Streisand, Meryl Streep and numerous other show business folk. Gore went from strength to strength, those that had other scientific views were attacked and, in the New York Times, Gore even compared "true believers", such as himself, to Galileo who bravely stood up for the truth against the blind orthodoxy of the time.

The IPCC gathered many climatologists, meteorologists, environmentalists and political activists and published several voluminous publications, the first of which was in 1990. These reports comprised a three-part scientific report under the IPCC's directed headings. Three working groups had authors who contributed to a series of chapters under the guidance of lead authors and a lead chapter author. These people are touted as the 2500 scientific experts who constitute a consensus.

In the 1996 report on the impact of global warming on health, one contributing author was an expert on the effectiveness of motorcycle helmets. That author had also written on the health effects of mobile phones. Other authors were environmental activists, one of whom had written on the health effects of mercury poisoning from land mines. If a land mine explodes, the last thing one thinks about is the health effects of mercury poisoning. In the 2007 report, the health effects of global warming were expertly dealt with by two lead authors, one of whom was a hygienist and another a specialist in coprolites (fossil faeces). Those who drove the publication of the chapters on the health effects of global warming had no formal expertise in the chapters' subject material, especially tropical diseases.[13] In fact, the expert opinions of tropical disease scientists were ignored by the other lead authors with no

---

[11] Kerr, R. A. 2007: Pushing the scary side of global warming. Science 316: 1412-1415.

[12] Lindzen, R. S. 1992: Global warming: The origin and nature of the alleged scientific consensus. Proceedings of the OPEC Seminar on the Environment, 13-15 April 1992.

[13] Reiter, P., 2005: Written evidence to the House of Lords Select Committee on Economic Affairs, The economics of climate change, Vol. II: Evidence (2005).

experience in the field.

The second stage of the IPCC process is that the draft *Summary for Policymakers* is submitted to governments, each of which can insist upon changes. These changes are made behind closed doors, the scientists who wrote three bulky volumes have no avenue for objection to political changes, and the final draft of the *Summary* forms the basis for a negotiating process between a few lead scientists and politicians. This is not the process of science and is not a peer review process. It is the process of politics. The IPCC head, Rajendra Pachauri, showed his cards with the release of the 2007 *Summary for Policymakers*: "I hope that this will shock the governments so much that they take action."

The IPCC process is related to environmental activism, politics and opportunism. It is unrelated to science.

Although it is commonly cited that 2500 scientists wrote the IPCC's Fourth Assessment report, a head count shows that there were 1656 authors and many of them were authors of many parts of the Report. Some of them used their given name in one part, used an initial in another and used an abbreviation in another. Furthermore, if we investigate the biographies of the 2500 "climate scientists", we find that many are not even scientists. To claim that this group of 2500 people represents the world's top scientists is untrue. It seems that of the 1190 separate individuals who wrote the scientific part of the report, many were not scientists but were political and environmental activists.

This is why the *Summary* is significantly different in key areas from the main scientific report. The *Summary* is the most widely read, publicised and quoted part of various IPCC Reports. The *Summary for Policymakers*, using the expert "scientific" opinion in the chapter on the health effects of global warming, was able to categorically state: "climate change is likely to have wide-ranging and mostly adverse effects on human life with significant loss of life".

The *Summary* predicted that 60% of humans were vulnerable to malaria, leading to an additional 50–80 million cases per year. This was contrary to expert opinion ignored by the IPCC.

The IPCC Reports gave the global warming campaign enormous momentum. The 1992 Earth Summit in Rio de Janeiro attracted 20,000 environmental activists from all over the world and politicians from 170 countries. God knows what this cost and how much $CO_2$ was added to the atmosphere by those who volunteered to have our interests at heart. Gore was the hero; his book *Earth in the Balance* had been published, and all this helped his vice-presidential nomination.

IPCC Reports became bolder. The 1996 *Summary for Policymakers* claimed that: "the balance of evidence suggests that there is discernible human influence on global climate".

This statement, derived from a scientific chapter (Chapter 8), sent the media into overdrive with alarming headlines, environmentalists increased their pressure on governments and the public was convinced that the august body of IPCC scientists had given a considered consensus opinion. What was not known then was that after the authors of Chapter 8 had signed off, a lead author had added the statement above about "discernible human effect" and deleted passages in Chapter 8 that stated:[14,15,16]

> None of the studies cited above has shown clear evidence that we can attribute the observed changes to the specific cause of increases in greenhouse gases [and]
>
> No study to date has positively attributed all or part (of the climate change observed) to (man-made) causes [and]
>
> Any claims of positive detection and attribution of significant climate change are likely to remain controversial until uncertainties in the total natural variability of the climate system are reduced [and] When will an anthropogenic effect on climate be identified? It is not surprising that the best answer to this question is 'We do not know'.

The lead author then added references to his own work which showed warming from 1943 to 1970.[17] However, when a full set of data from 1905 to after 1970 was analysed by others, no warming was seen.[18] The *Wall Street Journal* had a blistering editorial[19] exposing the IPCC process ("Cover-up in the Greenhouse") and a strident article[20] by the former president of the National Academy of Sciences entitled "Major Deception on Global Warming".

---

[14] De Freitas, C. R. 2002: Are observed changes in the concentration of carbon dioxide in the atmosphere really dangerous? *Bulletin of Canadian Petroleum Geology* 50: 297-327.

[15] Singer, S. F. and Avery D. T. 2007: *Unstoppable global warming: Every 1,500 years.* Rowman and Littlefield.

[16] Booker, C. and North, R. 2007: *Scared to death: From BSE to global warming – how scares are costing us the Earth.* Continuum.

[17] Santer, B., Taylor, K. E., Wigley, T. M. L., Joghns, T. C., Karoly, D. J., Mitchell, J. F. B., Oort, A. H., Penner, J. E., Ramaswamy, V., Schwarzkopf, M. D., Stouffer, R. J. and Tett, S. 1996: A search for human influences on the thermal structure of the atmosphere. *Nature* 382: 39-46.

[18] Michaels, P. J. and Knappenburger, P. C. 1996: Human effects on global climate. *Nature* 384: 522-523.

[19] *Wall Street Journal,* 11th June 1996.

[20] *Wall Street Journal,* 12th June 1996.

The IPCC continued to claim a consensus of scientists despite the fact that the UN *Climate Change Bulletin* in 1996 reported that only 10% of 400 American, Canadian and German climate researchers expressed strong agreement that they are "certain that global warming is a process already underway". Some 48% of those surveyed stated they did not have faith in global climate forecast models.[21] This finding was confirmed in 1997 by a survey of climatologists employed by 50 states in the USA.[22] Nevertheless, the IPCC and the media still claimed consensus as support for their findings.

One of the persistent problems that the IPCC faced was the Little Ice Age (1280–1850 AD) and the Medieval Warming (900–1300 AD). Evidence from a great diversity of sources showed that during the Medieval Warming, the global temperature was a few degrees higher than today. This created a problem for the IPCC because there were no major $CO_2$ emitting industries at that time. The solution was simple and elegant – change history. By creating *ex nihilo* a "hockey stick" graph that showed that the Little Ice Age and the Medieval Warming did not exist and that temperature started to rise dramatically in the early 20th Century, clearly a result of industrialisation. In the 2001 version of the IPCC's report, the "hockey stick" was used as proof that we were all doomed to fry and that it was all our fault. It was highlighted on the first page of the *Summary for Policymakers* and was shown another four times in the 2001 *Summary for Policymakers*. The IPCC's intent was clear. The media went into a frenzy, governments had even more pressure from environmental lobby groups and the global warming juggernaut increased in momentum. It took eight years to show that the "hockey stick" was fraud.[23] The IPCC, without explanation, quietly withdrew the "hockey stick" from the *Summary for Policymakers* in subsequent publications and had it buried in a scientific chapter of the 2007 report.[24]

Many policy makers, environmental groups and the media conclude that an IPCC *Summary for Policymakers* is actually the consensus view of a large number of scientists. It is not. It is the consensus of governments with a great diversity of agendas. At times, the *Summary for Policymakers* has been underpinned by fraud, undetected by environmental agitators, journalists or the public.

[21] Parry, M., Arnell, N., Hulme, M., Nicholls, R. and Livermore, M. 1998: Adapting to the inevitable. *Nature* 395: doi: 10.1038/27316.

[22] Bray, D. and von Storch, H. 1999: Climate science: An empirical example of postnormal science. *Bulletin of the American Meteorological Society* 80: 439-456.

[23] Holland, D. 2007: Bias and concealment in the IPCC process: The "hockey stick" affair and its implications. *Energy and environment* 18: 951-983.

[24] Chapter 6, as one of the reconstructions of past climate.

The IPCC is clearly an ascientific political organisation in which environmental activists and government representatives are setting the agenda for a variety of reasons including boosting trade, encouraging protectionism, adding costs to competitors and pushing their own sovereign barrow.

## Climate change

Climate change can be summarised:

(a) The Earth's climate has always changed with cycles of warming and cooling long before humans appeared on Earth. Numerous overlapping cycles range from 143 million years to 11.1 years. These cycles can be greatly affected by sporadic unpredictable processes such as volcanoes.

(b) Measured global warming in the modern world has been insignificant in comparison with these natural cycles.

(c) Although man-made increases in atmospheric $CO_2$ may theoretically make some contribution to temperature rise, such links have not been proven and there is abundant evidence to the contrary.

(d) Contrary to nearly two dozen different computer models, temperature has not increased in the last decade despite an accelerated input of $CO_2$ into the atmosphere by human activities.

(e) Other factors such as major Earth processes, variable solar activity, solar wind and cosmic rays appear to have a far more significant factor on the Earth's climate than previously thought. The IPCC has not demonstrated that the Sun was not to blame for recent warmings and coolings.

(f) Humans have adapted to live at sea level, at altitude, on ice sheets, in the tropics and in deserts. As in the past, humans will again adapt to any future coolings or warmings.

| Climate Changes | |
| --- | --- |
| Pleistocene ice age | 110,000 - 14,700bp |
| Bölling | 14,700 - 13,900bp |
| Older Dryas | 13,900 - 13,600bp |
| Allerød | 13,600 - 12,900bp |
| Younger Dryas | 12,900 - 11,600bp |
| Holocene Warming a | 11,600 - 8,500bp |
| Egyptian Cooling | 8,500 - 8,000bp |
| Holocene Warming b | 8,000 - 5,600bp |
| Akkadian Cooling | 5,600 - 3,500bp |
| Minoan Warming | 3,500 - 3,200bp |
| Bronze Age Cooling | 3,200 - 2,500bp |
| Roman Warming | 500BC - 535AD |
| Dark Ages | 535AD - 900AD |
| Medieval Warming | 900AD - 1300AD |
| Little Ice Age | 1300AD - 1850AD |
| Modern Warming | 1850AD - ......... |
| | bp = years before the present |

*Figure 2: Alternating warming and cooling cycles of climate since the last interglacial. On all scales, climate is cyclical and the numerous previous warming events occurred before industrialisation. In a few decades time, we will be able to determine whether the Late 20th Century Warming (Modern Warming) finished with the 1998 El Niño.*

The warm climate of Greenland 1000 years ago allowed the growing of grain, sheep and cattle. This warm climate could not have resulted from human emissions of $CO_2$. A few hundred years later, the bitterly cold weather of the Little Ice Age could not have derived from a decrease in human emissions of $CO_2$. There must be other causes of warming and cooling. How can we know that the slight warming since 1850 is due to humans adding $CO_2$ to the atmosphere? Furthermore, there have also been coolings since 1850. There must be other global-scale natural processes at work and the question must be asked: Does atmospheric $CO_2$ have anything at all to do with climate?

If $CO_2$ derived from modern industrialisation is the culprit for global warming, then why did the global temperature increase from 1918 to 1940, decrease from 1940 to 1976, increase from 1976 to 1998 and decrease from 1998 to the present? Throughout this period, humans were adding increasing amounts of $CO_2$ to the atmosphere. The IPCC does not explain the temperature variations in the 20th Century. There was alarm in the 1970s that the decreasing temperature was heralding another ice age. This was an important lesson from which nothing was learned. After 1976, temperature started to rise and again there was alarm, this time that there was going to be a period of global warming. Then temperatures started to fall after 1998. There is now silence. It is not possible to make computer model forecasts of climate change for the year 2040, 2100 or 2300 based on a few decades of data.

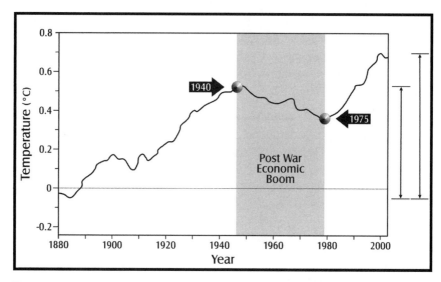

*Figure 3: Thermometer temperature measurements in the 20th Century showing both cooling and warming. The cooling was during and after World War II industrialisation emitted increasingly large amounts of $CO_2$ into the atmosphere. The 20th Century, like any other time period was one of both warming and cooling.*

If we change the time scale and look at the last 6 million years, for 3 million years it was warmer than now. For the other 3 million years there was an increase in the magnitude of high-frequency warm and cold cycles. During the last three warm cycles, it was 5°C warmer than now. Past natural climate changes have been partly cyclical and partly unpredictable and have nothing to do with human additions of $CO_2$ to the atmosphere .

Why are slight temperature changes in our lifetime related to humans adding $CO_2$ to the atmosphere whereas past slight and large climate changes cannot possibly be related to industrialisation?

There is no problem with global warming. It stopped in 1998. The last two years of global cooling have erased nearly thirty years of temperature increase.

The year of 2008 was an exceptionally cold year.[25] By the end of January 2008, blizzards and cold temperatures in China had killed 60 people, millions lost electricity service, nearly a million buildings were damaged, airports were closed and Hong Kong had the second longest cold spell since 1885. In February 2008, cold weather in Vietnam destroyed 40% of the rice crop and killed 33,000 head of livestock. In Mumbai (India), the lowest temperature for 40 years was recorded. In the USA, International Falls (Minnesota) set a new record (-40°C) breaking the old record (-37°C) set in 1967. In Reading (Pennsylvania), the temperature stayed below -40°C for six consecutive days for the first time since the 18th Century. Alaskan glaciers grew. On October 29 2008, the USA beat or tied 115 low-temperature records for that date. Alaska, which was unusually warm in 2007, recorded -32°C for that night, beating the previous low by 2°C.

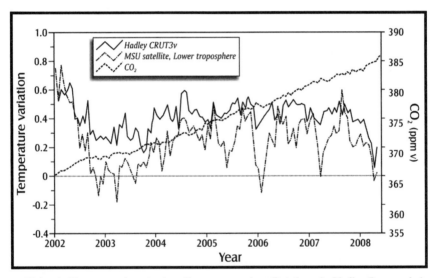

*Figure 4: Temperature determinations (thermometer and satellite) from the Hadley Centre and the University of Alabama (Huntsville) showing a decrease in global temperature in the early 21st Century. By contrast, atmospheric $CO_2$ is increasing thereby showing no relationship between global temperature and $CO_2$. This diagram shows that the hypothesis that human emissions of $CO_2$ create global warming is invalid.*

In the first week of December 2008, blizzards closed roads and schools across northern England and Scotland. Large parts of the UK were blanketed with snow for the third time in the 2008-09 winter. At the same time the UK government's Committee on Climate Change issued its first report on how Britain is to handle the terrifying threat of runaway global warming. Nature

[25] http://www.washingtontimes.com/news/2008/dec/10/global-warming/freeze/

certainly has a keen sense of humour.

With 24-hour news, we always have weather and climate disasters somewhere in the world in real time. Some place on Earth will always be breaking a local, regional or national record for temperature or precipitation.

## A retreat to the good old days

We humans on Earth have never had it better. Compared with 100, 200 or 500 years ago, we Westerners live longer, have a better diet, are wealthier, are healthier and the elderly have far better health care.[26,27,28,29] Even in the Third World, times are far better than before, despite population increase. The daily food intake has increased by 38% since the 1960s to 2666 calories per person per day despite the fact that the population in those countries has increased by 83%. Food prices have decreased by 75%. Improved agricultural production and freer trade have guaranteed that a smaller proportion of the Third World starves. In the 1970s, 16% of people in the Third World subsisted on less than $1 per day. Now it is 6%. Those living on $2 per day dropped from 39% to 18%. Life expectancy has also improved. In China, it has increased from 41 years in the 1950s to 71 today; in India it has increased from 39 to 63. In both countries the lifespan of 2 billion people has been almost doubled. In 1900, the average life expectancy globally was 31 years. Now it is 67 years. There is a long way to go, but never in human history have so many people been liberated from the clutches of starvation and poverty.[30]

Many environmentalists have a romantic view of the past and do not acknowledge the enormous struggle to stay alive in past times when unemployment, famines, disease, high child mortality, shortened lives and bitter cold dominated everyday life.

Don't give me the environmental romantic view about the good old days. They were not.

---

[26] Goklany, I. M. 2007: *The improving state of the world: Why we are living longer, healthier, more comfortable lives on a cleaner planet.* Cato Institute.

[27] Moore, S. and Simon, J. L. 2000: *It's getting better all the time: 100 greatest trends of the last 100 years.* Cato Institute.

[28] Simon, J. L. 1993: *Population matters: People, resources, environment and immigration.* Transaction Publishers.

[29] Lomborg, B. 1998: *The skeptical environmentalist: Measuring the real state of the world.* Cambridge University Press.

[30] Goklany, I. M. 2007: *The improving state of the world: Why we are living longer, healthier, more comfortable lives on a cleaner planet.* Cato Institute.

## Acknowledgements

This book evolved from a dinner in London with Andrew Wright (Cobbetts LLP), Kerry Stevenson (Apollo Global) and three young lawyers from Cobbetts LLP (Nia Bryan, Chris Tang, Fiona Twigg). The three young lawyers all supported the popular paradigm that human emissions of $CO_2$ are changing the climate. Although these three had more than adequate intellectual material to destroy the popular paradigm, they had neither the scientific knowledge nor the scientific training to pull it apart stitch-by-stitch. This was done at dinner and stimulated me to write a scientific book on climate change for Nia, Chris and Fiona and those out there with an open mind wanting to know more about how the planet works. The mind is like a parachute, it only works when it is open.

Numerous conferences have allowed the ideas in this book to be aired and criticised: Sydney Mining Club, Quadrant Dinner, YPO, Minesite Forum, Australian Skeptics, Paydirt, Federal and Supreme Court Judges, WA Pastoralists and Graziers Association, Reinsurance Rendezvous, NZ Insurance Council, Excellence in Mining and Exploration, Commonwealth Club, Australasian Institute of Mining and Metallurgy, AMEC, Royal United Services Institute of SA, Australia-Israel Chamber of Commerce.

Discussions, criticism and comments were solicited from Nick Badham, Bob Besley, Colin Brooks, Bob Carter, Kate Hartley, Tim Hartley, John Holland, John Nethery, Cliff Ollier, Rowl Twidale, Jim Wall and Peter Whellum. These critics gave huge amounts of time and effort as readers and critics and, in places, left coffee and red wine stains on the manuscript. Without dispassionate criticism by these scientists and non-scientists, it would have been nigh on impossible to publish such a book, and I have a huge debt of gratitude to those above who worked under difficult time pressures. They made numerous suggestions on style and content. The editor, George Thomas, was the necessary pedant and the publisher Anthony Cappello (Connor Court Publishing, Ballan, Victoria) and the publisher of the United Kingdom edition, Niam Attallah (Quadrant Books, Mayfair, London) both gave me all the encouragement needed. Notwithstanding, any errors in this book are purely my responsibility. My colleagues at the University of Adelaide suffered long periods of my isolation, a necessary condition for the writing of a book. Andy Lidgard kindly converted fly scratchings into line diagrams, again under huge time pressures. Why he has put up with me for over two decades is one of life's mysteries.

Upstairs, my wife Maja writes her book on the incredible story of Charles Rasp, the discoverer of the Broken Hill silver-lead-zinc orebody, and downstairs I wrote *Heaven and Earth*. She is my greatest fan and also my greatest critic. Ideas and draft chapters bounced up and down the stairs, life is joyous and highly irregular and we wouldn't have it any other way. It is not easy to live with someone writing a book.

Heartfelt thanks are to three very busy people, Lord Lawson of Blaby[31], Professor Geoffrey Blainey[32] and Dr Václav Klaus[33], who were kind enough to give the time to write cover notes.

[31] After a number of years in journalism, including as Editor of *The Spectator* from 1966 to 1970, Nigel Lawson became conservative MP in 1974. He served in the Thatcher government from 1979 to 1989 as Financial Secretary to the Treasury, Secretary of State for Energy, and, from 1983, Chancellor of the Exchequer. He entered the House of Lords in 1992, and is a member of the Lords' Select Committee on Economic Affairs which, in 2005 produced a substantial report on *The Economics of Climate Change*. He is author of *A view from No.11: Memoirs of a Tory Radical*, *The Power Game*, *The Nigel Lawson Diet Book* and *An Appeal to Reason: A cool look at global warming*.

[32] Professor Geoffrey Blainey AC is Australia's best-known historian. He has written 36 books, including *The Tyranny of Distance*, *Triumph of the Nomads*, *A Short History of the 20th Century*, the best selling *A Short History of the World* and *Sea of Dangers: Captain Cook and his rivals*. He held Chairs in history and economic history at The University of Melbourne for 21 years. He was delegate to the 1998 Constitutional Convention and chaired many Commonwealth Government bodies including the Australia Council, the Literature Board, the Australia-China Council and the National Council for the Centenary of Federation.

[33] Dr Václav Klaus is President of the European Union (2009) and the second President of the Czech Republic. He was Prime Minister (1992-1997), Minister of Finance (1989-1992), Chairman of the Chamber of Deputies (1998-2002) and an economist who studied in Prague, Italy and the USA. He has had various positions in the Institute of Economics in the Czechoslovak Academy of Sciences and the Czechoslovak State Bank. He has published over 20 books on social matters, politics and economics. He received the Julian L. Somin award from the Competitive Enterprise Institute, has been awarded numerous doctorates and prizes from international bodies and is author of *Blue planet in green shackles*.

# Chapter 2

## HISTORY

Question: Are the speed and amount of modern climate change unprecedented?
Answer: No.

Question: Is dangerous warming occurring?
Answer: No.

Question: Is the temperature range observed in the 20th Century outside the range of normal variability?
Answer: No.

*During the last interglacial period, sea level was 6 metres higher than today. Air temperature was anything from 2°C to 6°C warmer. The ice sheets retreated but did not completely melt. Alpine valley glaciers retreated. Vegetation and animal habitats changed. Trees advanced up slope and to higher latitudes and there was no extinction of life. Life on Earth thrived and there were fewer cold snaps. There was no industry emitting $CO_2$ at that time so this warming can only be natural.*

*The last glaciation started 116,000 years ago. Ice sheets, glaciers and sea ice expanded. Temperature and sea level fell. Some plants and animals moved and those that could not adapt became extinct. During the last ice age there were short periods of warmth followed by rapid collapses into bitterly cold conditions. Humans lived at the edge and were lucky to survive the last glaciation.*

*The last glaciation finished 14,000 years ago. There was rapid global warming and rapid sea level rise followed. Sea level rose at least 130 metres at the rate of 1 centimetre per year. Trees migrated up slope and to higher latitudes, animals migrated and humans thrived.*

*The climate rapidly plunged into very cold periods from 12,900 to 11,500 years ago and 8500 to 8000 years ago. These changes stressed life on Earth, changed plant and animal distribution, and led to the expansion of ice sheets and alpine valley glaciers. During a warm period 6000 years ago, sea level was 2 metres higher than at present. Temperature was also higher.*

*It only took years to decades to change from a warm to a cold climate. Fluctuating warm and cold conditions continued during a period of thousands of years characterised by warmth. A 300-year drought started in 2200 BC. This led to the collapse of empires and depopulation.*

*In the Roman Warming from 250 BC to 450 AD, temperature was at least 2°C higher than today. It was a period of global warming. Population increased, there was excess wealth and warm climate agriculture could be undertaken in areas at much higher latitudes and altitudes than now. Forests expanded. This warming could not be due to human emissions of $CO_2$.*

*The Dark Ages followed. This was a bitterly cold period of crop failure, famine, disease, war, depopulation, expansion of ice and increased wind. During the Dark Ages there was great social disruption and murderous climate refugee gangs wandered Europe looking for food. Civilisations such as the Mayans collapsed.*

*The Medieval Warming (900–1300 AD) was a wonderful time for life on Earth. Ice sheets, glaciers and sea ice contracted, enabling sea exploration and settlement at high latitudes. Grain crops, cattle, sheep, farms and villages were established on Greenland which was at least 6°C warmer than today. Although there was a cold period of 40 years in the Medieval Warming, crop failures and famine were rare. The population increased and there was enough food to feed the additional tens of millions of people. Excess wealth created over generations was used to build cathedrals, monasteries and universities. The Medieval Warming was global. This warming could not possibly be due to human emissions of $CO_2$.*

*The Little Ice Age started in the late 13th Century with a decrease in solar activity. The Little Ice Age was characterised by rapidly fluctuating climate and extraordinarily cold periods during solar inactivity (1280–1340, 1450–1540, 1645–1715 and 1795–1825). It was very cold. It was a global climate change. There was crop failure, famine, disease, war and depopulation. There was social disruption, and hungry wandering climate refugees resorted to cannibalism. Food prices increased at times of weak solar activity. The Vikings on Greenland died out. It was not a good time to live.*

*The Little Ice Age ended in 1850 and since then there has been a warming trend with cooler periods (1940–1976 and 1998–present). History, archaeology and geology show that we currently live in an interglacial and variable climate.*

*The slight changes that we can observe with modern instrumentation are very small.*

*Both the rates and magnitude of climate change are less than changes over the last 1000, 10,000 or 100,000 years. Global warming has brought excesses of food and wealth, social stability and a rapid diversification of life on Earth.*

*History and archaeology show us that global cooling results in drought, social disruption, climate refugees, famine, disease, war, depopulation, collapse of civilisations and extinctions of plants and animals. Great civilisations prospered in warm times. We live in the best times that humans have ever had on planet Earth.*

*We are the only generation of humans to fear warm times! Global warming makes us richer and healthier.*

## Our changing climates

Climate always changes. That's what climate does. This is no surprise, as planet Earth is dynamic and evolving. Without looking into the past, it is impossible to predict climate changes or to understand modern climate. In the Upper Palaeolithic some 35,000 to 30,000 years ago during the grip of a glaciation[34], there were only a few hundred thousand humans on Earth. The glaciation ended about 14,000 years ago and by 10,000 years ago the global population had grown to about 5 million.[35] With the coming of agriculture and growth of civilisations, the global population rose to 100–150 million people some 2500 years ago.[36] During the last glaciation (118,000 – 14,000 years ago), temperature and the rate of temperature change varied wildly. In the modern and previous interglacials, temperature variation and the rate of temperature change was small.

History shows us that climate rules our lives. For example, the subsistence crises in the Northern Hemisphere in the 17th and 18th Centuries resulted from the Little Ice Age. The wet weather and resultant bad harvests in 1697 brought disaster to the farming communities. In Finland in 1697, the famine killed one third of the population.[37] We are currently enjoying an interglacial within a glacial period that has already lasted tens of millions of years.[38] During cold phases of the glaciation alpine snow level in Europe was some

---

[34] An ice age is when ice covers polar and high altitude areas.

[35] Burroughs, W. J., 2005: *Climate change in prehistory. The end of the reign of chaos.* Cambridge University Press.

[36] Kremer, M., 1993: Population growth and technological change: One million BC to 1990. *Quarterly Journal of Economics* 108, 681-716.

[37] Burroughs, W. J., 1997: *Does the weather really matter?* Cambridge University Press.

[38] Pleistocene glaciation, sometimes called the Quaternary glaciation.

1400 metres lower than now in the Würm (50,000 years ago), Riss (150,000 years ago), Mindel (470,000 years ago) and Günz (650,000 years ago) events. Blizzards raged and there was the ominous presence of ice sheets and alpine valley glaciers which waxed and waned and scoured the surface of soil and vegetation.

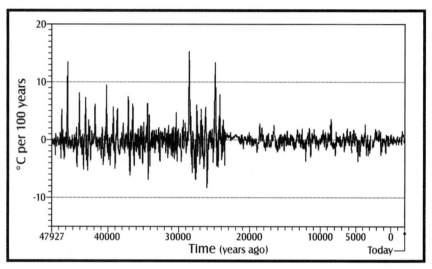

*Figure 5: The amount of temperature and rate of temperature change over the last 50,000 years showing wild swings in temperature during glaciation and far more stable temperatures during the current interglacial. Measured modern temperature changes are well within variability.*

Do these natural changes affect humans? Do these changes lead to extinctions of life on Earth? Do they result in human depopulation or adaptation? Are they rapid? Are the changes large? There is overwhelming evidence to show that cold climate results in a decrease in human population and warm climate results in prosperity. For example, during the last glaciation there was a warm period 32,000 to 28,000 years ago. It was at this time that there was a great migration of Cro-Magnon man in Europe. Warmer periods during and after the glaciation 20,000, 14,000 and 11,000 years ago were coincidental with other waves of human migration, such as that from Asia to North America about 12,000 years ago.

Previous generations of humanity enjoyed good weather and suffered bad weather. Although climate cycles were recognised by the Egyptians who depended on the productivity of the Nile River for thousands of years, more commonly those who lived through historic climate changes were not aware of climate cycles.[39]

---

[39] Hassan, F. A. 1998: Climatic change, Nile floods and civilisation. *Nature and Resources* 34: 34-40.

Nonetheless, the ancient Greek philosopher Plato (427–347 BC) argued in *Timaeus* that global warming occurs at regular intervals, often leading to great floods. In *Critias*, Plato showed that such floods cause soil erosion and his student, Aristotle (382–322 BC), recorded evidence for climate change in *Meteorologica*. He noted that during the Trojan War, Argos was marshy and not arable, whereas during Mycenaean times, the land was temperate and fertile. Theophrastus (374–287 BC), in turn a student of Aristotle, followed the tradition with *De ventis* and observed that Crete's mountains had previously produced fruit and grain whereas at the time he wrote, the winters were more severe and had more snow falls. In *De causis plantarum*, Theophrastus also noted that the Greek city of Larissus once had plentiful olive trees but falling temperatures had killed them.

Roman scribes such as Columella (in *De re rustica*) noted that areas once too cold for olives and grapes could in later times produce olive harvests and wine. Many places in central Europe and the UK have place and street names suggesting that wine grapes and olives were once grown there when the climate was warmer.

In Medieval and later times, people started recording climate-related phenomena such as the dates when plants blossomed annually, the dates and volume of harvests, population records and the advances and retreats of glaciers. Even though changes were recorded, they were not co-ordinated and understood because they were buried in monastery records, financial accounts and harvest records of estates, records of taxes, legal papers, government reports, harbour and shipping reports, changed sea routes, ship logs, local reports on road conditions and freezing of rivers, commodity prices and miscellaneous writings of scholars and adventurers.

We can acquire factual hard data on past climate from ice. However, we need not rely solely on this because there is a wealth of information in sea floor and lake sediments, tree rings, bogs, peat, pollen and historical records that show there were very significant rapid climate changes in the past. Previous warming trends occurred well before industrialisation which cannot be related to human emissions of $CO_2$ and must be natural. Knowledge of these previous warmings is validated from science and history. If human emissions of $CO_2$ have forced warming in the late 20th Century, then those making such a claim need to show that this warming is above and beyond natural warming. This has not been done.

The last 2.67 million years (Ma)[40] has been a time of cyclical icehouse and greenhouse conditions. During the last 730,000 years there have been ten

---

[40] The abbreviation Ma, million years ago or million years, will now be used in the rest of the book.

major glaciations separated by interglacials. Records of these glaciations are best seen in deep sea sediment cores and are present in all oceans.[41] Climate change was global. Full glaciation and full deglaciation appear to be related to heat exchange processes between ice and ocean and not to $CO_2$. Previous interglacials have been up to 5°C warmer in polar regions than the current interglacial.

Wobbles in the Earth's orbit have produced repeated cycles of 90,000 years of cold climate followed by 10,000 years of warmth. Before 1 Ma, there were 41,000-year cycles. For at least the last million years, a 1500-year warm-cold cycle[42] has coincided with a 1500-year solar cycle. These 1500-year cycles were commonly marked by changes in temperature of up to 4°C and these were superimposed over the longer icehouse (glacial) and greenhouse (interglacial) cycles. Other solar cycles on the scale of 210 years, 80 to 90 years, 33 years, 22 years and 11 years also cause climate change in varying degrees.[43] During cool times, climate has been very variable whereas during warm times, climate has been far steadier.

At times, events such as massive volcanic eruptions (e.g. Toba 74,000 years ago, Krakatoa 735 AD, Rabaul 736 AD, Tambora 1815 AD, Krakatoa 1883 AD) and cometary impacts (736 AD) fill the atmosphere with dust and sulphuric acid which reflects solar radiation and results in rapid cooling. The emission of sulphurous gases from volcanoes, even very small volcanoes, can change climate.

Melting ice sheets deposit debris which dams valleys that then fill with melt water. When melt water dams burst, huge floods scour the landscape and flood the oceans with cool water. This cool fresh water lies above warmer dense saline water in the oceans and the climate suddenly becomes cool until the waters have mixed. Ocean currents transfer heat around the world and are driven by wind, orbital, solar and water salinity changes. This has happened many times in the past and is still happening today.

## The last great warming

The last interglacial, about 125,000 years ago, was a short warm period between two longer colder periods.[44] On a more detailed scale, the last

[41] Emiliani, C. 1978: The cause of ice ages. *Earth and Planetary Science Letters* 37: 349-354.

[42] Dansgaard-Oeschger cycles.

[43] Blaauw, M., Blaauw, M., van Geel, B. & van der Plicht, J. 2004: Solar forcing of climatic change during the mid-Holocene: indications from raised bogs in the Netherlands. *The Holocene* 14: 35-44.

[44] Kukla, G. J., Bender, M. L., de Beaulieu, J.-L., Bond, G., Broecker, W. S., Cleveringa, P., Gavin, J. E., Herbert, T. H., Imbrie, J., Jouzel, J., Kelgwin, L. D., Knudsen, K.-L., McManus, J. F., Merkt, J., Muhs, D. R., Müller, H., Poore, R. Z., Porter, S. C., Seret, G., Shackleton, N. J., Turner, C., Tzedakis, P. C. and Winograd, I. J. 2002: Last interglacial climates. *Quaternary Research* 58: 2-13.

glaciation ended with sudden warming about 14,700 years ago,[45] after which it suddenly became cold again some 13,000 years ago[46] after which the modern interglacial started at about 11,700 years ago.[47] During the last interglacial, the climate was warmer than today.[48,49] The temperature was up to 6°C warmer at the poles and 2°C warmer at the equator.[50] This interglacial[51] spanned from 130,000 to 116,000 years ago with the peak 125,000 years ago.[52] Global ice volume was low and sea level was 4 to 6 metres higher than now.[53] Tree lines followed the glacier retreats and expanded to high latitude areas and to mountains.[54] Much low-lying land became covered by warm shallow seas. Indeed, many flat coastal plains were directly caused by the deposition of suspended clay and silt in a shallow marine setting. Sea surface temperature rose and this correlates with a time of raised coral reefs.[55] It also correlates with polar temperature calculated from ice cores.[56,57] This warm interglacial was worldwide. During the interglacial, land that was once covered by thick ice sheets started to rebound and rise, only to be

---

[45] Late Saalian warming (MIS6).

[46] Kattegat Stadial; Heinrich Event 1.

[47] Beets, D. J., Beets, C. J. and Cleveringa, P. 2005: Age and climate of the late Saalian and early Eemian in the type-area, Amsterdam basin, The Netherlands. *Quaternary Science Reviews* 25: 876-885.

[48] Kukla, G. J. 2000: The last interglacial. *Science* 287: 987-988.

[49] Otto-Bliesner, B. L., Marshall, S. J., Overpeck, J. T., Miller, G. H. and Hu, A. 2006: Simulating Arctic climate warmth and icefield retreat in the last interglaciation. *Science* 311: 1751-1753.

[50] Cuffey, K. M. and Marshall, S. J. 2000: Substantial contribution to sea-level rise during the last interglacial from the Greenland ice sheet. *Nature* 404: 591-594.

[51] Called the Eemian Interglacial (Netherlands), Sangamon Interglacial (North America), Ipswichian Interglacial (UK) and Riss-Würm Interglacial (European Alps). It occurred at Marine Isotope Stage 5e and, at 116,000 years ago, at the end of the Eemian, Marine Isotope Stage 5d commenced.

[52] Schokker, J., Cleveringa, P. and Murray, A. S. 2004: Palaeoenvironmental reconstruction and OSL dating of terrestrial Eemian deposits in the southeastern Netherlands. *Journal of Quaternary Science* 19: 193-202.

[53] Cuffey, K. M. and Marshall, S. J. 2000: Substantial contribution to sea-level rise during the last interglacial from the Greenland ice sheet. *Nature* 404: 591-594.

[54] Muhs, D. R., Ager, T. R. and Begét. J. E. 2001: Vegetation and paleoclimate of the last interglacial period, Central Alaska. *Quaternary Science Reviews* 20: 41-61.

[55] Henderson, G. M., Slowey, N. C. and Fleischer, M. Q. 2001: U-Th dating of carbonate platform and slope sediments. *Geochimica et Cosmochimica Acta* 65: 2757-2770.

[56] Oxygen 16 is preferentially evaporated from the oceans and falls in snow. This depletes the oceans in $O^{16}$ (light oxygen) and they become relatively enriched in oxygen 18 (heavy oxygen). Ocean life therefore become relatively enriched in $O^{18}$. The proportion of $O^{18}$ to $O^{16}$ in ice and fossil of shells of floating animals can be used to calculate surface water temperature and the air temperature at the time snow fell. The use of oxygen chemistry in this book is a reference to the use of $O^{18}/O^{16}$ to determine ancient temperature.

[57] Mangerud, J., Dokken, T., Hebbeln, D., Heggen, B., Ingólfsson, O., Landvik, J. Y., Mejdahl, V., Svendsen, J. I. and Vorren, T. O. 1998: Fluctuations of the Svalbar-Berents sea ice sheet during the last 150,000 years. *Quaternary Science Reviews* 17: 11-42.

covered by ice and depressed again a few thousand years later by ice sheets.[58] During the interglacial, high latitude tundra was replaced with trees, and thick forests again covered continental Europe,[59,60,61] the UK[62] and elsewhere in the Northern Hemisphere. Not only was it warmer in northern Europe, Greece was also warm and warmer than at present, showing that the warming was not just restricted to high latitudes.[63] This warming was global, sea level in Western Australia was at least 3 metres higher than at present and coral reefs thrived between 128,000 and 121,000 years ago in areas where water temperature is now far too cool for coral.[64] During this time modern man, *Homo sapiens*, evolved in East Africa.

During the last interglacial, when temperature and sea level were higher than now, the polar ice sheets did not completely melt.[65] Polar bears did not become extinct. There was no runaway greenhouse effect or "tipping point". Human populations adapted. Climate did what climate always does: change. This 14,000-year interglacial warm period was not driven by high atmospheric $CO_2$. Orbital and solar mechanisms were the driving forces.

## The last big freeze

The last glaciation[66] started 116,000 years ago. Over the last 100,000 years we have had both climate change and climate variability. Evidence suggests that the shift from interglacial to glacial conditions occurred in only 400 years. Snowlines throughout the world were 900 metres lower than today.[67] Air temperature at the glaciers was some 5°C cooler than today and the tropical

---

[58] Beets, D. J., Beets, C. J. & Cleveringa, P. 2005: Age and climate of the late Saalian and early Eemian in the type area, Amsterdam basin, The Netherlands. *Quaternary Science Reviews* 25: 876-885.

[59] De Beaulieu, J.-L and Reille, M. 1992: The last climatic cycle at La Grande Pile (Vosges, France). A new pollen profile. *Quaternary Science Reviews* 11: 431-438.

[60] Aaby, B. and Tauber, H. 1995: Eemian climate and pollen. *Nature* 376: 27-28.

[61] Caspers, G., Freund, H., Merkt, J. and Müller, H. 2002: The Eemian interglaciation in northwestern Germany. *Quaternary Research* 58: 49-52.

[62] Gascoyne, M., Currant, A. P. and Lord, T. C. 1981: Ipswichian fauna of Victoria Cave and the marine palaeoclimate record. *Nature* 338: 309-313.

[63] Frogley, M. R., Tzedakis, P. C. and Heaton, T. H. E. 1999: Climate variability in Northwest Greece during the last Interglacial. *Science* 285: 1886-1889.

[64] Stirling, C. H., Esat, T. M., Lambeck, K. and McCulloch, M. T. 1998: Timing and duration of the Last Interglacial: evidence for a restricted interval of widespread coral reef growth. *Earth and Planetary Science Letters* 160: 745-762.

[65] Winograd, I. J., Landwehr, J. R., Ludwig, K. R., Coplen, T. P. and Riggs, A. C. 1997: Duration and structure of the past four interglaciations. *Quaternary Research* 48: 141-154.

[66] The glaciation is known as the Wisconsin (North America) and Würm (Europe).

[67] Porter, S. C. 2000: Snowline depression in the tropics during the last glaciation. *Quaternary Science Reviews* 20: 1067.

sea surface temperature was 3°C cooler.[68] Open vegetation replaced thick forests. The forests retreated to lower latitude areas, ice sheets expanded to lower altitudes and latitudes and forest lands such as the Amazon gave way to grasslands. Forests in Europe disappeared abruptly 107,000 years ago and cold water invaded the central North Atlantic Ocean.[69] Evaporated water fell as snow, accumulated in ice sheets, and was not recycled back to the oceans, resulting in a lowering of sea level. The ice sheets waxed and waned, as did alpine valley glaciers. With less water falling as rain on vegetation, forest lands retreated to lower latitudes. Dune sands and sea spray were deposited over large areas of Africa, Australia, Asia and the Americas. The climate was not only cold, it was windier and drier. As the glaciation commenced, sea level dropped and extended rivers cut new ravines at least 10 metres deep into previously submerged coastal plains.[70]

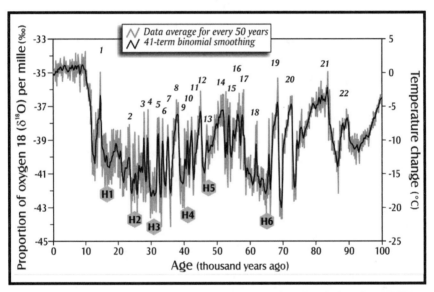

*Figure 6: Sea surface temperature proxy using oxygen isotopes from Greenland ice sheet (GISP2 ice core). Sea surface temperature varied rapidly by up to 20°C with more than 20 Dansgaard/Oeschger warming events. Ice sheets shed armadas of icebergs (Heinrich Events H1 to H6) during glaciation and there was great temperature instability in the last glaciation. During the current interglacial, sea surface temperature was higher and there was far less temperature variation.[71]*

[68] Broecker, W. S. 2001: Was the Medieval Warm Period global? *Science* 291: 1497-1499.

[69] Knudsen, K.-L., Seidenkrantz, M.-S. and Kristensen, P. 2002: Last interglacial and early glacial circulation in the northern North Atlantic Ocean. *Quaternary Research* 58: 22-26.

[70] Törnqvist, T. E., Wallinga, J., Murray, A. S., de Wolf, H., Cleveringa, P. and de Gans, W. 2000: Response of the Rhine-Meuse system (west-central Netherlands) to the last Quaternary glacio-eustatic cycles; a first assessment. *Global and Planetary Change* 27: 89-111.

[71] World Data Center for Paleoclimatology, Boulder, Colorado.

During a glaciation, there are great variations in air and sea surface temperature, ice volume and sea level. Temperature reconstructions from the shells of floating animals provide a sea surface temperature proxy.

Some 74,000 years ago it became intensely cold after the Indonesian volcano of Toba filled the atmosphere of both hemispheres with dust and sulphuric acid aerosols. This dust reflected heat and light.[72] There was a brief respite between 60,000 and 55,000 years ago when it became slightly warmer[73] and glaciers started to retreat. Then it cooled again to the zenith of the last ice age at 21,000 to 17,000 years ago. Areas not covered by ice were windy cold deserts. Sea level was at least 130 metres lower than now. During the peak of the last glaciation some 20,000 years ago, lake sediments in Africa show that there was aridity, lake levels were low and the winds were stronger.[74]

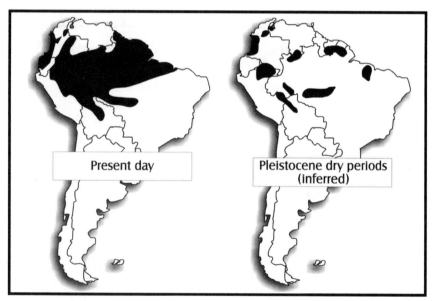

*Figure 7: Present day rainforest distribution and distribution of Amazon forest during the peak of the last big freeze showing that forests are dynamic and come and go.*[75]

[72] Rampino, M. R. and Self, S. 1992: Volcanic winter and accelerated glaciation following the Toba super eruption. *Nature* 359: 50-52.

[73] Mikkelsen, N. and Kuijpers, A. 2001: *The climate system and climate variations.* Geological Survey of Denmark and Greenland.

[74] Filippi, M. L. and Talbot, M. R. 2005: The palaeolimnology of northern Lake Malawi over the last 25 ka based on the elemental and stable isotopic composition of sedimentary organic material. *Quaternary Science Reviews* 24: 1303-1328.

[75] From: Simberloff, D. 1998: Flagships, umbrellas, and keystones: Is single-species management passé in the landscape era. *Biological Conservation* 83: 247-257.

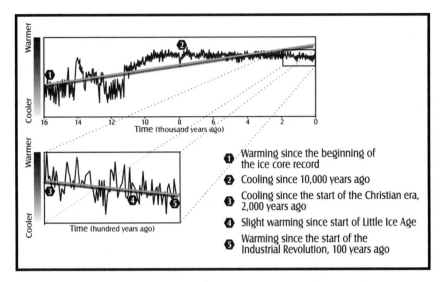

*Figure 8: The post-glacial warming showing the change from wildly fluctuating temperature to the stable temperatures of the current interglacial. Any modern climate change is well within historical variability and, because there are no extraordinary changes, it is concluded that either $CO_2$ has had no effect on climate or that modern climate changes are so small that the effect of $CO_2$ cannot be recognised.*

During this glaciation, humans existed at the edge of ice sheets, in mountain terrains, on the coastal plains and next to water. Although there were at least three hominid species at the start of the last ice age, by the end of the ice age, only one species had survived.

That was us, *Homo sapiens*, and we almost didn't make it.

## The end of the freeze

The Earth is normally free of ice.

After the last glaciation, planet Earth did not just steadily warm. Climate fluctuated wildly[76] and cyclically.[77]   The deglaciation that followed this glaciation was dramatically interrupted by cooling a number of times, the most intense episode of which was the Younger Dryas.

Our story in this section starts at the end of the last glaciation, only 14,000 years ago. Since that time, planet Earth has been recovering from the

[76] Wohlfarth, B., Schwark, L., Bennike, O., Filimonova, L., Tarasov, P., Bjorkmanj, L., Brunnberg, L., Demidov, I. and Possnert, G. 2004: Unstable early-Holocene climatic and environmental conditions in northwestern Russia derived from a multidisciplinary study of a lake-sediment sequence from Pichozero, southeastern Russian Karelia. *The Holocene* 14: 732-746.

[77] Müller, U. C., Klotz, S., Geyh, M. A., Pross, J. and Bond, G. C. 2005: Cyclic climate fluctuations during the last interglacial in central Europe. *Geology* 33: 449-452.

last glaciation, and more recently, from the Little Ice Age (1300–1850 AD), which marked a cold fluctuation within the current interglacial. The Earth became relatively free of ice after the glaciation and it remains so. Greenland temperatures have risen by more than 20°C in the 20,000 years since the height of the last glaciation.[78,79] Once the glaciation started to lose its grip, climate instability took over. Conditions were generally warmer and wetter but with extreme and rapid climate changes. Sea level has risen by at least 130 metres over the last 14,000 years, an average rate of over a centimetre a year. At times sea level rose quickly, at other times slowly, sometimes it did not rise at all and at other times it fell. The rate of post-glacial temperature increase and sea level rise were far faster than any such change in the 20th Century which is estimated at less than 0.5°C and 1.5 millimetres per year. During the period 14,500–12,900 years ago,[80] a 20-metre sea level rise occurred at an average of 1.25 centimetres per year, almost ten times the rate of the 20th Century rise.

Alaska and Siberia were joined by a coastal plain which a subsequent sea level rise flooded to create the Bering Strait. Similarly, a river flowed between England and France and post-glacial sea level rise formed the English Channel. The same rise fragmented England and Scotland from Ireland and formed the Shetlands and Orkneys as islands. Corsica and Sardinia were split apart by rising water, Sicily was separated from Italy and many other islands were greatly reduced in area. Japan was split from the Asian mainland by rising sea level. A land bridge between the Mediterranean and Black Seas was breached by the sea level rise creating the Bosphorus and the Black Sea. Papua New Guinea, mainland Australia and Tasmania were one landmass until sea level rise created the Torres and Bass Straits. The Gulf of Carpentaria changed from a huge inland lake to a shallow sea. Coastal areas that were lowlands were inundated as the sea level rose and coastal populations moved inland. Plants and animals were stranded on islands. In many areas (e.g. Milos Island, Greece), the stranding of animals led to dwarfism.[81]

---

[78] Mayewski, P. A., Meeker, L. D., Twickler, M. S., Whitlow, S. I., Yang, Q. and Prentice, M. 1997: Major features and forcing of high latitude Northern Hemisphere atmospheric circulation over the last 110,000 years. *Journal of Geophysical Research* 102: 26,345-26,366.

[79] Mayewski, P.A., Meeker, L. D., Whillow, S., Twickler, M. S., Morrison, M. C., Grootes, P. M., Bond, G. C., Alleys, R. B., Meese, D. A., Gow, A. J., Taylor, K. C., Ram, M. and Wumkes, M. 1994: Changes in atmospheric circulation and ocean ice cover over the North Atlantic during the last 41,000. *Science* 263, 1747–1751.

[80] The 14,500-12,900 BP Bölling-Allerød interstadial.

[81] Plimer, I. R. 2001: *Milos- geologic history*. Koan.

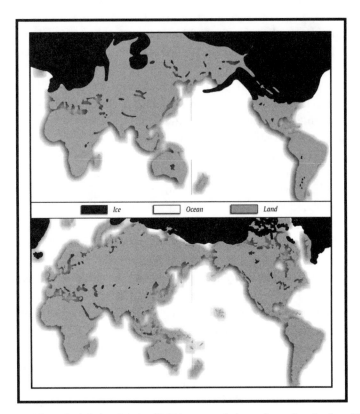

*Figure 9: The peak of the last glaciation 20,000 years ago had extensive northern ice sheets (black) and connection of land masses because sea level was 130 metres lower that at present (upper diagram). Ice sheet melting resulted in a sea level rise of 130 metres giving the current geography (lower diagram).*

Within the last post-glacial warming, a short sharp cooling from 13,900 to 13,600 years ago intervened[82] and then warming continued until 12,900 years ago when another short-lived cold period commenced.[83] I stress that this intense cooling event was only 1500 years after the Earth started to recover from a major ice age.

This very intense cold period from 12,900 to 11,500 years ago is called the Younger Dryas.[84,85] It was a brief bitterly cold period that lasted for about 1300 years that saw a return to the glaciation from which the Northern Hemisphere had just escaped. Parts of Greenland were 15°C colder than

---

[82] The 13,900-13,600 BP Older Dryas.

[83] The 13,600-12,900 BP Allerød stadial.

[84] The Younger Dryas is named after a tundra flower *Dryas octopetala*.

[85] The 12,900-11,500 BP Younger Dryas, also known as the Nahanagan Stadial (Ireland) and the Loch Lomond Stadial (UK).

now, in England fossil beetles show that the temperature dropped to -5°C and ice fields and glaciers formed. Lake sediments in Germany show that in the Younger Dryas the wind strength increased due to an abrupt change in the North Atlantic westerlies.[86] Glaciers surged, ice broke off to form icebergs and armadas of ice that drifted south to lower latitudes. During the Younger Dryas, changing ocean currents resulted in changes to the distribution of heat.[87]

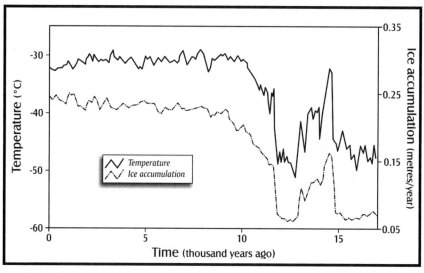

*Figure 10: Extreme variations in temperature towards the end of the last glaciation with the sudden cooling in the Younger Dryas, even more rapid warming after the Younger Dryas and temperature stabilisation during the current interglacial. The warmer more humid times resulted in higher precipitation and greater ice accumulation.*

The change from warmth to the bitter cold of the Younger Dryas took less than 100 years and maybe only a decade. No climate change of this size, rapidity and extent has occurred since the Younger Dryas. It was possibly caused by a change in the ice sheet configuration which saw dammed glacial melt waters suddenly cascade into the North Atlantic Ocean.[88] One such

[86] Brauer, A., Haug, G. H., Dulski, P., Sigman, D. M. and Negendank, J. F. W. 2008: An abrupt wind shift in western Europe at the onset of the Younger Dryas cold period. *Nature Geoscience* 1: 520-523.

[87] Tarasov, L. and Peltier, W. R. 2005: Arctic freshwater forcing of the Younger Dryas cold reversal. *Nature* 435: 662-665

[88] Nesje, A., Dahl, S. O. and Bakke, J. 2004: Were abrupt late glacial and early-Holocene climatic changes in northwest Europe linked to freshwater outbursts to the North Atlantic and Arctic Oceans? *The Holocene* 14: 299-310.

event was the breaching of the glacial debris dam wall of the huge Lake Agassiz in North America, which released large volumes of cold fresh water into the North Atlantic Ocean.[89] The sudden release of melt waters changed the landscape. Massive sheet erosion occurred, river systems were rerouted by floodwaters, glaciers collapsed and ocean circulation changed. This flood water was less dense than ocean water, and it froze and reduced ocean water circulation until the deeper saline and surface fresh waters were again thoroughly mixed. The warm Gulf Stream was covered by very cold water. It took over a thousand years for the waters to mix.

There may be an astronomical reason for the sudden start to the Younger Dryas. Tree rings show a steep rise in the concentration of carbon 14 ($C^{14}$) at the beginning of the Younger Dryas. This increase in $C^{14}$ can only occur with an increased bombardment of the surface of the Earth by cosmic rays, decreased solar activity, or both.[90] This is validated by deep sea[91] and ice core[92,93] measurements of increases in $C^{14}$, beryllium 10 ($Be^{10}$) and chlorine 36 ($Cl^{36}$), all of which form by cosmic ray bombardment of the atmosphere and Earth.

However, this origin for the Younger Dryas does not explain why a similar cooling in the Southern Hemisphere preceded it, and nor does it explain why the Younger Dryas came to a sudden end at 11,530 ± 50 (GRIP ice core, Greenland[94]), 11,530 ± 40–60 (Kråkanes Lake, Norway[95]), 11,570

---

[89] Broecker, W. S., Kennett, J. P., Flower, B. P., Teller, J. T., Trumbore, S., Bonan, G. and Wolfli, W. 1989: Routing of meltwater from the Laurentide Ice Sheet during the Younger Dryas cold episode. *Nature* 341: 318-321.

[90] Muscheler, R., Kromer, B., Björck, S., Svensson, A., Friedrich, M., Kaiser, K. F. and Southon, J. 2008: Tree rings and ice cores reveal $^{14}C$ calibration uncertainties during the Younger Dryas. *Nature Geoscience* 1: 263-267.

[91] Muscheler, R., Beer, J., Wagner, G. and Finkel, R. C. 2000: Changes in deep-water formation during the Younger Dryas cold period inferred from a comparison of $^{10}Be$ and $^{14}C$ records. *Nature* 408: 567-570.

[92] Yiou, F., Raisbeck, G. M., Baumgartner, S., Beer, J., Hammer, C., Johnsen, S., Jouzel, J., Kubik, P. W., Lestringuez, J., Stievenard, M., Souter, M. and Yiou, P. 1997: Beryllium 10 in the Greenland Ice Core Project ice core at Summit, Greenland. *Journal of Geophysical Research* 102: 26783-26794.

[93] Muscheler, R., Beer, J., Kubic, P. W. and Synal, H.-A. 2005: Geomagnetic field intensity during the last 60,000 years based on $^{10}Be$ & $^{36}Cl$ from the Summit ice cores and $^{14}C$. *Quaternary Science Reviews* 24: doi:10.1029/2005JA011500.

[94] Grootes, P. M., Stuiver, M., White, J. W. C., Johnsen, S. and Jouzel, J. 1993: Comparison of oxygen isotope records from GISP 2 and GRIP Greenland ice cores. *Nature* 366: 552-554.

[95] Gulliksen, S., Birks, H. H., Possnert, G. and Mangerud, J. 1998: A calendar age estimate of the Younger Dryas-Holocene boundary at Kråkenes, western Norway. *The Holocene* 8: 249-259.

(Cariaco Basin core, Venezuela[96]), 11,570 (tree rings, Germany[97]) and 11,640 ± 280 (GISP2 ice core, Greenland[98]) years ago. Another suggestion for the Younger Dryas is that a meteorite impact could have breached Lake Agassiz and stressed the American fauna to extinction. Evidence for an impact 12,900 years ago comes from sedimentary strata that contain chemical fingerprints, microdiamonds and glassy spheres from vaporised and molten rock.

The abrupt cooling in the Younger Dryas led to the replacement of forest in Scandinavia with glacial tundra, the advance of glaciers and increased snowfall, high winds removing dust from the deserts of Asia into the atmosphere, deposition of wind blown sands[99] in Asia and a prolonged drought in the Eastern Mediterranean. A south-north flowing current in the North Atlantic[100] was reduced, there were warmer sea surface temperatures in the equatorial South Atlantic Ocean, the African summer monsoonal winds were reduced and West Africa suffered drought.[101]

Cooling in the North Atlantic Ocean may have preceded the Younger Dryas by a few hundred years. In Antarctica, a cold period appears to have started at about 14,000 years ago, 1000 years before the Younger Dryas. Sudden cooling in South America started at about the same time as the Younger Dryas and ended suddenly. The Sajama (Bolivia) ice core shows that cooling may have started in the Southern Hemisphere 14,000 years ago, 1100 years earlier than in the Northern Hemisphere.[102] At that time, for example, the Amazon Basin had far less vegetation than now and comprised grasslands with copses of trees.

The start of the Younger Dryas coincided with sudden widespread changes in animal life, changes in Palaeolithic cultural development in the Americas, the extinction of the American megafauna (including mammoths,

[96] Hughen, K. A., Overpeck, J. T., Peterson, L. C. and Anderson, R. F. 1996: The nature of varved sedimentation in the Cariaco Basin, Venezuela, and its palaeoclimatic significance. *Geological Society of London Special Publications* 116: 171-183.

[97] Hughen, K. A., Overpeck, J. T., Lehman, S. J., Kashgarian, M., Southon, J., Peterson, L. C., Alley, R. and Sigman, D. M. 1998: Deglacial changes in ocean circulation from an extended radiocarbon calibration. *Nature* 391: 65-68.

[98] Grootes, P. M., Stuiver, M., White, J. W. C., Johnsen, S. and Jouzel, J. 1993: Comparison of oxygen isotope records from the GISP2 and GRIP Greenland ice cores. *Nature* 336, 552-554.

[99] Loess.

[100] Atlantic meridional overturning circulation.

[101] Chang, P., Zhang, R., Hazeleger, W., Wen, C., Wan, X., Ji, L., Haarsma, R. J., Breugem, W.-P. and Seidel, H. 2008: Ocean link between abrupt changes in the North Atlantic Ocean and the African monsoon. *Nature Geoscience* 1: 444-448.

[102] Thompson, L. G., Mosely-Thompson, E. and Henderson, K. A. 2000: Ice-core palaeoclimate records in tropical South America since the last glacial maximum. *Journal of Quaternary Science* 15: 377-394.

mastodons, horses and ground sloths[103]) and the termination of the Clovis and other Palaeolithic cultures.[104] Extinction of the megafauna may have been assisted by over-hunting by the Clovis people, a disease pandemic or the effects of a meteorite impact.[105,106]

It appears that the end of the Younger Dryas took place over 40–50 years in three different steps, each of about five years duration. Other data indicates a warming of 7°C in only a few years, half of the warming taking place in a 15-year period. Such a warming rate is far higher than even the most alarmist catastrophic warming suggested by models of human-induced global warming. The warming was to temperatures similar to those today. Humans, plants and animals adapted to the rapid intense warming and rapid sea level rise after the Younger Dryas. This post-Younger Dryas warm period lasted for 2600 years.

Rapid accumulation of snow in Greenland occurred immediately after the Younger Dryas as warming allowed air to hold more moisture.[107] For every 1°C rise in the sea surface temperature, air can hold another 7% of the evaporated moisture. After the Younger Dryas, ice sheets retreated, forests rapidly expanded and sea level rose. Trees replaced grass, and grass replaced desert.[108]

The end of the Younger Dryas heralded the start of the Holocene, the time of modern humans. This warm period after the Younger Dryas was from 11,500 to 8900 years ago.[109] Rapid expansion of forests occurred in the Northern Hemisphere,[110] despite land clearing for agriculture. The change in the regional climate stimulated the upslope movement of the tree line higher

---

[103] Haynes, G. 2002: The catastrophic extinction of North American mammoths and mastodonts. *World Archaeology* 33: 391-416.

[104] Barnosky, A. D., Koch, P. L., Feranec, R. S., Wing, S. L. and Shabel, A. B. 2004: Assessing the causes of Late Pleistocene extinctions on the continents. *Science* 306: 70-75.

[105] Alroy, J. 2001: A multispecies overkill simulation of end-Pleistocene megafaunal mass extinction. *Science* 292: 1893-1896.

[106] Grayson, D. K. and Meltzer, D. J. 2002: Clovis hunting and large mammal extinction: A critical review of the evidence. *Journal of World Prehistory* 16: 313-359.

[107] Alley, R. B., Meese, D., Shuman, C. A., Gow, A. J., Taylor, K., Ram, M., Waddington, E. D. and Mayewski, P. A. 1993: Abrupt increase in Greenland snow accumulation at the end of the Younger Dryas event. *Nature* 362, 527-529.

[108] Emeis, K.-C. and Dawson, A. G. 2003: Holocene palaeoclimate records over Europe and the North Atlantic. *The Holocene* 13: 305-309.

[109] The 11,500-10.190 BP warming is the Preboreal and the 10,190-8,900 BP warming is the Boreal.

[110] Gobet, E., Tinner, W., Bigler, C., Hochuli, P. A. and Ammann, B. 2005: Early-Holocene afforestation processes in the lower subalpine belt of the Central Swiss Alps as inferred from macrofossil and pollen records. *The Holocene* 15: 672-686.

and the retreat of glaciers.[111] Vegetation studies of peat, pollen and spores from lake sediments[112,113] show that, once again, the climatic and sea level changes were widespread and cyclical.[114,115,116] Summer monsoonal rainfall patterns changed from one continent to another[117], river flow patterns changed[118] and some regions fluctuated between drought with some vegetation to ample rainfall with lush vegetation.[119] Even though the Holocene was warm, at many times it was punctuated by short sharp cold periods. The Holocene warming was global.[120] Lake levels rose in cold times due to decreased evaporation and increased precipitation, and fell in warm times in response to changes in climate.[121] Ice melting in warmer times decreased the salinity of seas, while in colder times the seas were more saline.[122] Cold periods had high winds and the resultant drought, dust storms, sand dunes and desertification.[123] The fluctuating post-glacial climates can even be measured from insects trapped

[111] Emeis, K.-C. and Dawson, A. G. 2003: Holocene palaeoclimate records over Europe and the North Atlantic. *The Holocene* 13: 305-309.

[112] Feurdean, A. 2005: Holocene forest dynamics in northwestern Roumania. *The Holocene* 15: 435-446.

[113] Chen, F.-H., Cheng, B., Zhao, Y., Zhu, Y. and Madson, D. B. 2006: Holocene environmental change inferred from high-resolution pollen record, Lake Zhueze, arid China. *The Holocene* 16: 675-684.

[114] Hu, F. S., Kaufman, D., Yoneji, S., Nelson, D., Shemesh, A., Huang, Y., Tian, J., Bond, G., Clegg, B. and Brown, T. 2003: Cyclic variation and solar forcing of Holocene climate in the Alaskan subarctic. *Science* 301: 1890-1893.

[115] Yu, Z., Campbell, I. D., Campbell, C., Vitt, D. H., Bond, G. C. and Apps, M. J. 2003: Carbon sequestration in western Canadian peat highly sensitive to Holocene wet-dry cycles at millennial timescales. *The Holocene* 13: 801-808.

[116] Yu, S.-Y. 2003: Centennial-scale cycles in middle Holocene sea level along the southeastern Baltic Coast. *Bulletin of the Geological Society of America* 115: 1404-1409.

[117] Maher, B. A. and Hu, M. 2006: A high-resolution record of Holocene rainfall variations from the western Chinese Loess plateau: antiphase behaviour of the African/Indian and East Asian summer monsoons. *The Holocene* 16: 309-319.

[118] Macklin, M. G., Johnstone, E. and Lewin, J. 2005: Pervasive and long-term forcing of Holocene river instability and flooding in Great Britain by centennial-scale climate change. *The Holocene* 15: 937-943.

[119] Feng, Z.-D., An, C. B. and Wang, H. B. 2006: Holocene climatic and environmental changes in the arid and semi-arid areas of China: a review. *The Holocene* 16: 119-130.

[120] Enzel, Y., Ely, L. L., Mishra, S., Ramesh, R., Amit, R., Lazar, B., Rajaguru, S. N., Baker, V. R. and Sandler, A. 1999: High-resolution Holocene environmental changes in the Thar Desert, northwestern India. *Science* 284: 125-128.

[121] Magny, M., Begeot, C., Guiot, J., Marguet, A. and Billaud, Y. 2003: Reconstruction and palaeoclimatic interpretation of mid-Holocene vegetation and lake-level changes at Saint-Jorioz, Lake Annecy, French Pre-Alps. *The Holocene* 13: 265-275.

[122] Emeis, K.-C., Struck, U., Blanz, T., Kohly, A. and Voss, M. 2003: Salinity changes in the central Baltic Sea (NW Europe) over the last 10,000 years. *The Holocene* 13: 411-421.

[123] Miao, X., Mason, J. A., Swinehart, J. B., Loope, D. B., Manson, P. R., Goble, R. J. and Liu, X. 2007: A 10,000 year record of dune activity, dust storms, and severe drought in the central Great Plains. *Geology* 35: 119-122.

in stalagmites.[124] Stalagmites are especially good temperature proxies as the formation of each layer can be dated by carbon or uranium-thorium dating techniques and caves are buffered from the more rapid environmental changes that occur outside.

Another period of cooling from 8900 to 8500 years ago may have been caused by other meltwater dam collapses.[125] This was a widespread and perhaps a global series of events.[126] Meltwater dams in far northern Canada collapsed 8500 years ago, which was followed by a 500-year period of intense cold windy glacial climate,[127,128] rather similar to the Younger Dryas. Greenland ice cores show that there was a cold period from about 8400 to 8000 years ago.[129] There was more dust and sea spray preserved in the ice core layers at that time, accompanied by a decrease in the methane content of the atmosphere.[130] Again, a cold freshwater layer covered the warmer saline Gulf Stream waters so that less heat was brought by ocean currents to higher latitudes. Again, there were great ice armadas in the oceans around 8300 years ago,[131] as was the case in the Younger Dryas.

The post-glacial climate variability, especially during the Younger Dryas, possibly prompted development of agriculture and animal husbandry in the Levant at that time.[132] Even though climate varied greatly during post-glacial times, solar variability influenced climate and therefore the development of human culture.[133] Populations built villages rather than continuing to subsist as nomadic hunters and gatherers. In the cold period from 8500 to

---

[124] Polyak, V. J., Cokendolpher, J. C., Norton, R. A. and Asmerom, Y. 2001: Wetter and cooler late Holocene climate in the southwestern United States from mites preserved in stalagmites. *Geology* 29: 643-646.

[125] Barber, D. C., Dyke, A., Hillaire-Marcel, C., Jennings, A. E., Andrews, J. T., Kerwin, M. W., Bilodeau, G., McNeely, R., Southon, J., Morehead, M. D. and Gagnon, J.-M. 1999: Forcing of the cold event 8,200 years ago by catastrophic draining of Laurentide lakes. *Nature* 400: 344-348.

[126] Douglass, D. C., Singer, B. S., Kaplan, M. R., Ackert, R. P., Mickelson, D. M. and Caffee, M. W. 2005: Evidence of early Holocene glacial advances in southern South America from cosmogenic surface-exposure dating. *Geology* 33: 237-240.

[127] Andrews, J. T. and Dunhill, G. 2004: Early to mid-Holocene Atlantic water influx and deglacial meltwater events, Beaufort Sea slope, Arctic Ocean. *Quaternary Research* 61: 14-21.

[128] Lauriol, B. and Gray, J. T. 1987: The decay and disappearance of the Late Wisconsin ice sheet in the Ungava Peninsula, Northern Quebec, Canada. *Arctic & Alpine Research* 19: 109-126.

[129] Rohling, E. J. and Pälike, H. 2005: Centennial-scale climate cooling with a sudden event around 8,200 years ago. *Nature* 434: 975-979.

[130] Alley, R. B., Mayewski, P. A., Sowers, T., Stuiver, M., Taylor, K. C. and Clark, P. U. 1997: Holocene climate instability; a prominent widespread event 8200 yr ago. *Geology* 25: 483-486.

[131] Clarke, G., Leverington, D., Teller, J. and Dyke, A. 2003: Superlakes, megafloods, and abrupt climate change. *Science* 301: 922-923.

[132] Feynman, J. and Ruzmaikin, A. 2007: Climate stability and the development of agricultural societies. *Climate Change* 84: 295-311.

[133] Feynman, J. 2007: Has solar variability caused climate change that affected human culture. *Advances in Space Research* 40: 1173-1180.

8000 years ago, highland populations moved to lower altitudes. At that time, sea level was about 3 metres lower than at present. The tree line descended and vegetation in the Alps of Europe responded to cooler times.[134] People from the Anatolian Highlands deserted their villages[135] and moved into the grasslands of the 160,000-square-kilometre Black Sea basin.[136] Almost a quarter of the Black Sea basin is flat, lying 100 metres beneath the present sea level. These rich plains were ideal grasslands for stock. Wheat was grown, villages were established and the big rivers (Don, Dnieper, Danube) fed two large freshwater lakes which have since been inundated by the Black Sea.

In the Northern Hemisphere, warming started again from about 8100 years ago[137] and continued to about 4030 years ago.[138] It was unusually warm in Greenland and, at this time, the greatest summer melting of ice over the last 10,000 years occurred. By contrast, it was especially cold in Antarctica. This suggests a polar seesaw (i.e. Antarctic climate anomaly) with warmth flipping from one hemisphere to the other.[139] During the Holocene maximum, the Arctic was up to 3°C warmer than today. Lake fossils and pollen in Iceland show that the temperature at sea level some 8000 years ago was 1.5°C warmer than now and possibly 2–3°C higher than the 1961–1990 average. This warmth in Iceland was a little later than similar warmth in Greenland and east Arctic Canada.[140] This was a global event, also recorded in cave stalagmites of New Zealand where the temperature was at least 2.3°C warmer.[141]

Rock paintings show that during the earlier Holocene the Sahara had herds of animals and people. It is, of course, now a dry unpopulated desert. The paintings, in a variety of styles that must represent thousands of years of artistic activity, show that the Sahara was wet enough to support herds

---

[134] Koffler, W., Krapf, V., Oberhuber, W. and Bortenschlager, S. 2005: Vegetation responses to the 8200 cal. BP cold event and to long-term climatic changes in the Eastern Alps: possible influence of solar activity and North Atlantic freshwater pulses. *The Holocene* 15: 779-788.

[135] Esin, U. 1996: *Aşikli, ten thousand years ago: A habitation model from central Anatolia*. Tarih Vakfi, Istanbul,

[136] Wilson, I. 2001: *Before the flood*. Orion.

[137] Viau, A. E., Gajewski, K., Fines, P., Atkinson, D. E. and Sawada, M. C. 2002: Widespread evidence of 1500 yr climatic variability in North America during the past 14,000 yr. *Geology* 30: 455-458.

[138] Atlantic I warming 8,100-6,700 years BP and Atlantic II warming 6,700-4,030 years BP.

[139] Shackleton, N. 2001: Climate change across the hemispheres. *Science* 291: 58-59.

[140] Caseldine, C., Langdon, P. and Holmes, N. 2006: Early Holocene climate variability and the timing and extent of the Holocene thermal maximum (HTM) in northern Iceland. *Quaternary Science Reviews* 25: 2314-2331.

[141] Williams, P. W., Marshall, A., Ford, D. C. and Jenkinson, A. V. 1999: Palaeoclimatic interpretation of stable isotope data from Holocene speleotherms of the Waitomo district, North Island, New Zealand. *The Holocene* 9: 649-657.

of giraffes, hippopotami and elephants somewhat like the Serengeti today. The herd paintings showing the lushest Sahara are dated at about 6000 years ago.[142] What is now the Western Desert of Egypt were grasslands served by monsoons between 7000 and 4300 BC.[143] The area was seasonally occupied by nomadic cattle herders. A compilation of some 500 C[14] dates and sediment features from lakes, soils, wadis and archaeological sites in Egypt and Northern Sudan shows that the area was warm, wet and hospitable from 7000 to 4000 BC. At this time, the global sea level was about 2 metres higher than at present[144] and stayed at that level until about 3000 years ago.[145,146] This was possibly the peak of the interglacial.

In Palaeolithic and Neolithic times, extreme climate variability and prolonged drought caused the collapse of civilisations. When the Western Desert became arid in 4300 BC as a result of changes in the monsoon strength, the area was depopulated. Concurrently, the Nile dwellers began to worship cattle and create monumental architecture.[147] The same drought was seen in the Sahel.[148] The area underwent desertification, which fostered technological innovation, migration and settlement elsewhere and the further development of agrarian communities in complex cultures.[149,150]

When a prolonged drought began in northern Mesopotamia in 4200 BC, the region was largely abandoned and population migrated south where there was irrigation-based agriculture. There was a similar collapse of Neolithic cultures around Central China at the same time, with drought in the north and flooding in the south.[151] This followed a warm wet period (6000–5800 BC) when tropical plants were abundant in China. Human settlement disappeared from the Yangtze Delta from 5240–3320 BC, as a result of flooding and a

---

[142] Lhote, H., 1958: *The search for the Tassili frescoes*. E. P. Dutton.

[143] Post boreal warming.

[144] Woodroffe, S. A. and Horton, B. P. 2005: Holocene sea-level changes in the Indo-Pacific. *Journal of Asian Earth Sciences* 25: 24-39.

[145] Sloss, C. R., Murray-Wallace, C. V. and Jones, B. G. 2007: Holocene sea-level change on the southeast coast of Australia: a review. *The Holocene* 17: 999-1014.

[146] von der Borch, C. C., Bolton, B. and Warren, J. K. 1977: Environmental setting and microstructure of subfossil lithified stromatolites associated with evaporites, Marion Lake, South Australia. *Sedimentology* 24: 693-708.

[147] Dalfes, H., Kukla, G. and Weiss, H. 1997: *Third Millennium BC Climate Change and Old World Collapse*. NATO ASI Series 1, Volume 49, Springer-Verlag.

[148] Dumont, H. J. 1978: Neolithic hyperarid period preceded the present climate of the Central Sahel. *Nature* 274: 356-358.

[149] Nicoll, K. 2004: Recent environmental change and prehistoric human activity in Egypt and Northern Sudan. *Quaternary Science Reviews* 23: 561-580.

[150] Nicoll, K. 2000: Radiocarbon chronologies for prehistoric human occupation and hydroclimatic change in Egypt and Northern Sudan. *Geoarchaeology* 16: 47-64.

[151] Wenxiang, W. and Tungsheng, L. 2004: Possible role of the "Holocene Event 3" on the collapse of Neolithic cultures around the Central Plain of China. *Quaternary International* 117: 153-166.

high water table during cold wet times.[152]   The population returned to the delta as sea levels retreated and developed in the period 2410–1250 BC. A 200-year cooling event started at 2200 BC and humans migrated from mountains to low-lying deltas.[153]

In the mid-Holocene (6000–1000 BC), profound cultural changes occurred that also were possibly driven by rapid climate change.[154]   In the summer of 2003 AD, the retreat of the ice at Schnidejoch (Switzerland) revealed a 4700-year-old archer's quiver. The Schnidejoch must have been a short unfrozen route across the Alps around 2700 BC. Subsequent work has shown that there were four periods during the last 5000 years when the Schnidejoch was warmer than today.[155]   It was also warmer in the Southern Hemisphere where old beach terraces show that from 5660 to 4040 years ago, sea level was about 1.7 metres higher than at present.[156]   Other events or monuments from this time include the first temple mounds of Peru, the first pyramids of Egypt, settled agrarian societies worldwide and the rise and fall of civilisations in the Near East. The El Niño-Southern Oscillation was absent from 6900 to 3800 BC[157] despite warmer sea surface temperatures than today[158] and warm tropical water as far south as 10°S.[159] Andean ice core, lake sediments from Ecuador, Chile and the Galapagos Islands show increased variability in rainfall after 3800 BC. Pollen from Australia shows transition to an El Niño-Southern Oscillation[160] dominated climate system with greater variability about 4000 years ago.[161]   A similar transition to modern conditions

[152] Yu, S., Zhu, C., Song, J. and Qu, W. 2008: Role of climate in the rise and fall of Neolithic cultures on the Yangtze Delta. *Boreas* 29: 157-165.

[153] Chen, Z., Wang, Z. and Schneiderman, J. 2005: Holocene climate fluctuations in the Yangtze delta of eastern China and the Neolithic response. *The Holocene* 15: 915-924.

[154] Sandweiss, D. H., Maasch, K. A. and Anderson, D. G. 1999: Transitions in the Mid-Holocene. *Science* 283: 499-500.

[155] Svensmark, H. and Calder, N. 2007: *The chilling stars – A new theory of climate change.* Icon Books.

[156] Beaman, R., Larcombe, P. and Carter, R. M. 1994: New evidence for the Holocene sea-level high from the Inner Shelf, Central Great Barrier Reef, Australia. *Journal of Sedimentary Research* 64a: doi:10.1036/D4267EF1-2B26-11D7-8648000102C1865d

[157] Sandweiss, D. H., Richardson, J. B. III, Reitz, E. J., Rollins, H. B. and Maarsch, K. A. 1996: Geoarchaeological evidence from Peru for a 5000 years B.P. onset of El Niño. *Science* 273: 1531-1533.

[158] Gagan, M. K., Ayliffe, L. K., Hopley, D., Cali, J. A., Mortimer, G. E., Chappell, J., McCulloch, M. T. and Head, M. J. 1998: Temperature and surface-ocean water balance of the Mid-Holocene tropical Western Pacific. *Science* 279: 1014-1018.

[159] Gagan, M. K., Ayliffe, L. K., Beck, J. W., Cole, J. E., Druffel, E. R. M., Dunbar, R. B. and Schrag, D. P. 2000: New views on tropical paleoclimates from corals. *Quaternary Science Reviews* 19: 45-64.

[160] The El Niño-Southern Oscillation is dealt with more fully in Chapter 6: Water.

[161] Schulmeister, J. and Lees, B. G. 1995: Pollen evidence from tropical Australia for the onset of the ENSO-dominated climate at c. 4000 BP. *The Holocene* 5: 10-18.

at about 3800 BC is seen in the northwest Pacific.[162]

Changing climates change ecosystems. This is well documented in the scientific literature. For example, over the last 6000 years, lake sediment studies show that the Sahara changed from a green warm wet environment to a desert about 2700 years ago.[163] This is in accord with the evidence from archaeology,[164] geology,[165] fossil pollen[166,167] and deposition of Saharan dust in the Atlantic Ocean sea floor sediments.[168] This was not the end of the world, but simply meant that one drier ecosystem replaced another wetter one. The silica shells of floating animals, organic carbon and nitrogen from lake sediments in southwestern Alaska also show cyclic variations similar to those found in the North Atlantic Ocean, offshore West Africa and Greenland.[169]

Climate changed at about 3800 BC, from the previous 400 years of dry cold times that led to the collapse of Saharan and Mesopotamian civilisations to warmer wetter times. Temperature was about 2°C warmer than today. As the Saharan and Arabian deserts became wetter, humans moved into these areas for hunting, herding and some agriculture.[170] The rest of Africa was also warm.[171] Cultures adapted to fluctuating climate. Changes in diet, agriculture, pottery and tools coincided with this event which may have been

[162] Lutaenko, K. A. 1993: Climatic optimum during the Holocene and the distribution of warm water mollusks in the Sea of Japan. *Palaeogeography, Palaeoclimatology, Palaeoecology* 102: 273-281.

[163] Kröpelin, S., Verschuren, D., Lézine, A.-M., Eggermont, H., Cocquyt, C., Francus, P., Cazet, J-P., Fagot, M., Rumes, B., Russell, J. M., Darius, F., Conley, D. J., Schuster, M., von Suchodoletz, H. and Engstrom, D. R. 2008: Climate-driven ecosystem succession in the Sahara: the past 6000 years. *Science* 320: 765-768

[164] Kuper, R. and Kröpelin, S. 2006: Climate-controlled Holocene occupation in the Sahara: Motor of Africa's evolution. *Science* 313: 8803-807

[165] Verschuren, D., Briffa, K., Hoelzmann, P., Barber, K., Barker, P., Scott, L., Snowball, I, Roberts, N. and Battarbee, R. 2004: Holocene climate variability in Europe and Africa: a PAGES PEP III time-stream synthesis. In: *Past climate variability through Europe and Africa* (eds R. W. Battarbee, Gasse, F., Stickley, C. E.) Kluwer, 567-582

[166] Salzman, U., Hoelzmann and Morczinek, I. 2002: Late Quaternary climate and vegetation of the Sudanian Zone of Northeast Nigeria. *Quaternary Research* 58: 73-83

[167] Waller, M. P., Street-Perrott, F. A. and Wang, H. 2007: Holocene vegetation history of the Sahel: pollen, sedimentological and geochemical data from Jikariya Lake, north-eastern Nigeria. *Journal of Biogeography* 34: 1575-1590.

[168] deMenocal, P. B. 2001: Cultural responses to climate change over the last 1000 years. *Science* 289: 270-277

[169] Hu, F. S., Kaufman, D., Yoneji, S., Nelson, D., Shemesh, A., Huang, Y., Tian, J., Bond, G., Clegg, B. and Brown, T. 2003: Cyclic variation and solar forcing of Holocene climate in the Alaskan subarctic. *Science* 301: 1890-1893.

[170] Malek, J. 2000: The Old Kingdom (c. 2686-2160 BC). *The Oxford History of Ancient Egypt.* Oxford University Press.

[171] Nguetsop, V. F., Sevant-Vildary, S. and Servant, M. 2004: Late Holocene climatic changes in West Africa, a high resolution diatom record from Equatorial Cameroon. *Quaternary Science Reviews* 23: 561-609.

initially driven by the return of El Niño.

After this warming, the inevitable happened and it became cold again, peaking around 3600 to 3300 BC. Alpine glaciations hit Europe, cattle grazing ended in what is now the Sahara, and Saharan populations dispersed into West Africa and Egypt. Pollen records in Greek lake sediments show that massive deforestation occurred in the early Bronze Age (3500–3100 BC)[172], probably as a response to the cold climate.[173,174] Grazing and farming on steep slopes led to catastrophic erosion that produced barren landscapes which are still a feature of much of modern Greece.[175]

Greek mythology makes reference to deforestation, flooding, siltation of irrigation channels, salination and the collapse of the Sumerian city-states.[176] Written records dating back to 5000 years ago describe declining crop yields and decreasing production of wheat relative to the more salt-tolerant barley. Patches of soil turned white,[177] suggesting salt accumulation on the surface of agricultural lands. The drier conditions made it impossible to flush salt from fields.

This cool dry period was again followed by a warm period. Culture thrived during this warm period among a population living off the bounty of the Nile River.[178] The lower Nile was one of the cradles of civilisation, totally dependent upon floods from the upper Nile, just as Egypt is today. Even today, some 300 million people depend upon the Nile River for food. The Nile had monstrous floods in the periods 14,700 to 13,100, 9700 to 9000, 7900 to 7600, 6300 and 3200 to 2800 years ago. The Egyptian Old Kingdom (2350–2200 BC) thrived on the banks of the Nile during warm times. This was short-lived because, inevitably, subsequent climate changes occurred.

The next cool period around 2200 BC was accompanied by a catastrophic drought. A profound cooling commenced in the Northern Hemisphere 4030 years ago.[179] This was also recorded in the Southern Hemisphere. Cave

---

[172] Use of terms such as Bronze Age and Iron Age are terms are descriptive time periods in the Northern Hemisphere (especially continental and Mediterranean Europe and the UK).

[173] Atherden, M., Hall, J. and Wright, J. C. 1993: A pollen diagram from the northeast Peloponnese, Greece: implications for vegetation history and archaeology. *The Holocene* 3: 351-356.

[174] Jahns, S. 1993: On the Holocene vegetation history of the Argive Plain (Peloponnese, southern Greece). *Vegetation History and Archaeobotany* 2: 187-203.

[175] Van Andel, T. H., Zangger, E. and Demitrack, A. 1990: Land use and soil erosion in prehistoric and historical Greece. *Journal of Field Archaeology* 17: 379-396.

[176] Carpenter, R. 1966: *Discontinuity in Greek civilisation*. Cambridge University Press.

[177] Jacobsen, T. and Adams, R. M. 1958: Salt and silt in ancient Mesopotamian agriculture: Progressive changes in soil salinity and sedimentation contributed to the breakup of past civilizations. *Science* 128: 1251-1258.

[178] The Egyptian (Menes) 1st United Kingdom Warm Period from 2,686 to 2,160 BC.

[179] The 4,030-2,850 years BP Suboreal cooling.

deposits from New Zealand show air temperature cooled at least 1.5°C.[180] This was a solar-driven global mega-drought which led to widespread famine.[181] Empires collapsed. The period from 2200 to 1900 BC was a dark age with Dynasty VI Egypt ending in anarchy, the disintegration of the Akkadian Empire, destruction of Byblos and other sites in Syria, and, further afield, the destruction of Troy and the decline of Indus Valley civilisations (e.g. Harappan).[182] The Indo-European peoples from northern Europe migrated to southern Europe, Greece, southern Russia, Turkey, Iran, India and Xinjiang (northwest China).[183] Glaciers advanced in the Swiss Alps.

The Akkadian Empire ruled a region from the headwaters of the Tigris and Euphrates Rivers to the Persian Gulf. Increased aridity and wind on the Habur Plains of Syria heralded the onset of this 300-year savage drought. Wind-blown dust is most common in periods of cool climate when wind and drought are more prevalent.[184] The northern Mesopotamian civilisation depended upon regular rains and, after four centuries of urban life at Tell Leilan, the city was abandoned. This led to the collapse of the Akkadian Empire. The synchronous collapse of civilisations in adjacent regions shows that the abrupt climate change was extensive.[185] Although civilisations collapsed during cooling and desertification, in some places humans were able to adapt to climate change, especially warming.[186]

Wind-blown sediments deposited in the Gulf of Oman show that starting at about 2562 ± 125 BC, there was a 300-year period of intense windiness. Such winds are characteristic of aridity and the wind-blown dust was mostly derived from northern Mesopotamia.[187] The coldness and desertification at this time may have been caused by the weakening of Northern Hemisphere

---

[180] Williams, P. W., Marshall, A., Ford, D. C. and Jenkinson, A. V. 1999: Palaeoclimatic interpretation of stable isotope data from Holocene speleotherms of the Waitomo district, North Island, New Zealand. *The Holocene* 9: 649-657.

[181] Bell, B. 1975: Climate and the history of Egypt: the Middle Kingdom. *American Journal of Archaeology* 79: 223-269.

[182] Bell, B. 1971: The Dark Ages in ancient history. *American Journal of Archaeology* 75: 1-26.

[183] Hsu, K. J. 2000: *Climate and people: A theory of history*. Orell Fussli.

[184] deMenocal, P., Ortiz, J., Guilderson, T. and Sarthein, M. 2000: Coherent high- and low-latitude climate variability during the Holocene Warm Period. *Science* 288: 2198-2202.

[185] Weiss, H., Courty, M.-A., Wetterstrom, W., Guichard, F., Senior, L., Meadow, R. and Curnow, A. 1993: The genesis and collapse of third millennium North Mesopotamian civilisation. *Science* 261: 995-1004.

[186] Smit, B. and Wandel, J. 2006: Adaptation, adaptive capacity and vulnerability. *Global Environmental Change* 16: 282-292.

[187] Cullen, H. M., deMenocal, P. B., Hemming, S., Hemming, G., Brown, F. H., Guilderson, T. Sirocko, F. 2000: Climate change and the collapse of the Akkadian empire: Evidence from the deep sea. *Geology* 28: 379-382.

ocean currents.[188] Ice cores from the Mt Kilimanjaro glacier show abrupt climate changes including a 300-year drought at about 2000 BC.[189] Stalagmites from a cave in West Virginia provide a detailed record of climate in North America over the last 7000 years. Cave deposits are undisturbed. They give a better climate record because burrowing animals can disturb ocean and lake sediments which are commonly used to track past climates. The cave deposits of West Virginia show that when the Earth received less solar radiation every 1500 years, the Atlantic Ocean cooled, icebergs increased and rainfall decreased. This led to long droughts, especially between 4300 and 2200 years ago.[190] Clearly the great drought of Mesopotamia was more widespread than just in the Middle East. Climate change is a powerful causal agent for the evolution of civilisation. Global cooling is generally associated with a collapse of civilisation whereas global warming is associated with great advances in civilisation.[191]

Several millennia later, a similar fate destroyed the Mayan civilisation of Central America, which also depended upon seasonal rainfall for irrigation. This culture collapsed during a very severe drought in 899–900 AD. The Mayans, like the Akkadians, were not able to sustain their prosperous civilisations in a prolonged drought. History shows us that global warming gives us nothing to fear. If we need to fear something, then the best candidate is a global mega-drought associated with cooling and driven by solar activity. It's happened before and it will happen again.

In the Indus Valley in what is now Pakistan and northwest India between 4500 and 3500 years ago, cities such as Harappa and Mohenjo-dar lived by cultivating grain, cotton, melons and dates. Harappa just disappeared off the map. As no record of war exists, the most logical conclusion is that a climate change led to the abandonment of settlements. Although the area occupied by the Harappans is now arid, fossils, pollen and lake sediments tell a different history.[192] The area was lush with sedge, grass, mimosa and

[188] Shaowu, W., Tianjun, Z., Jingning, C., Jinhong, Z., Zhihui, X. and Daoyi, G. 2004: Abrupt climate change around 4 ka BP: Role of the thermohaline circulation as indicated by a GCM experiment. *Advances in Atmospheric Sciences* 21: 291-295.

[189] Thompson, L. G., Mosely-Thompson, E., Davis, E., Henderson, K. A., Brecher, H. H., Zagorodnov, V. S., Mashiotta, T. A., Lin, P.-N., Mikhalenko, V. N., Hardy, D. R. and Beer, J. 2002: Kilimanjaro ice core records: Evidence of Holocene climate change in tropical Africa. *Science* 298: 589-593.

[190] Williams, A. N., Rowe, H., Springer, G., Cheng, H. and Edwards, L. R. 2005: Interpreting trace-metal and stable isotopic results from a Holocene stalagmite from Buckeye Creek cave, West Virginia. *GSA Salt Lake City AGM*, October 16-19, 2005; paper131-33.

[191] deMenocal, P. 2001: Cultural responses to climate change during the Late Holocene. *Science* 292: 667-673.

[192] Mandella, M. and Fuller, D. Q. 2006: Palaeoecology and the Harappan civilisation of South Asia: a reconsideration. *Quaternary Science Reviews* 25: 1283-1301.

jamun trees. Jamun needs a rainfall of at least 50 cm per annum, twice the modern rainfall in this area.[193] Lake sediments show the area was very wet from 3000–1800 BC and then it was dry until 500 BC.

Sediment layers and $C^{14}$ dating of coastal plain marine sediments near Karachi (Pakistan) provide clues about rainfall in the hinterland. Rainfall decreased from 2000 to 1500 BC coincidental with increasing aridity in the Near East and Middle East, as documented by decreasing Nile River runoff and influx to lakes from Turkey through to northwestern India. These lake sediment records show cyclical drought periods from 200 BC to 100 AD, around 1000 AD and from 1300 to 1600 AD (Late Middle Ages).[194]

Rapid climate change was not only recorded in the Northern Hemisphere. Increased runoff in wet times and a decrease in dry times is a feature of many South American lakes.[195,196,197] Similar climate changes have also been recorded in South Africa,[198] showing that climate changes may have been global.

An intense period of desertification starting at about 2000 BC was also recorded from China. This was certainly global because it has been recorded in North Europe, Mediterranean Europe, the northern Middle East, northern East Asia, East Africa, the Middle East, the Indian Peninsula, the Americas and the Yellow River Valley.[199]

After this intense 300-year drought that caused so much havoc in the old world, a warm period of the Bronze Age prevailed from 1470 to 1300 BC.[200] People migrated northward into Scandinavia and reclaimed farmland with growing seasons that were at the time probably the longest for two millennia.

[193] Shaw, J., Sutcliffe, J., Lloyd-Smith, L., Schwenninger, J.-L. and Chauhan, M. S. 2007: Dates and pollen sequences from the Sanchi Dams. *Asian Perspectives* 46: 166-201.

[194] Rad, U. von, Schaaf, M., Michels, K. H., Schultz, H., Berger, W. H. and Sirocko, F. 1999: A 5,000-years record of climate change from the oxygen minimum zone off Pakistan, Northeastern Arabian Sea. *Quaternary Research* 51: 39-53.

[195] Chepstow-Lusty, A. J., Bennett, K. D., Fjeldsa, J., Kendall, A., Galiano, W. and Tupayachi Herrara, A. 1998: Tracing 4,000 years of environmental history in the Cuzco area, Peru, from the pollen record. *Mountain Research and Development* 18: 159-172.

[196] Valero-Garces, B. L., Delgado-Huertas, A., Ratto, N., Navas, A. and Edwards, L. 2000: Paleohydrology of Andean saline lakes from sedimentological and isotopic records, Northwestern Argentina. *Journal of Paleolimnology* 24: 343-359.

[197] Iriondo, M. 1999: Climatic changes in the South American plains: Records of a continent-scale oscillation. *Quaternary International* 57: 93-112.

[198] Holmgren, K., Tyson, P. D., Moberg, A. and Svanered, O. 2001: A preliminary 3,000-year regional temperature reconstruction for South Africa: Research letter. *South African Journal of Science* 97: 49-51.

[199] Drysdale, R., Zanchetta, G., Hellstrom, J., Maas, R., Fallick, A., Pickiett, M., Cartwright, I. and Piccini, L. 2006: Late Holocene drought responsible for the collapse of Old World civilisations is recorded in an Italian cave flowstone. *Geology* 34: 101-104.

[200] Hempel, L. 1987: The "Mediterraneanization" of the climate in the Mediterranean countries – a cause of unstable ecobudget. *GeoJournal* 14: 163-173.

The Assyrian Empire, the Hittite Kingdom, the Shang Dynasty in China and the Middle Egyptian Empire flourished.[201] This warming, in Minoan times, led to a thriving of culture and a growth of empires. The Minoan empire, greatly weakened by the eruption of Santorini, was displaced by the Mycenaean empire. This was the most favourable warming period of the Holocene Maximum.

The Bronze Age came to an end with the Centuries of Darkness. Between 1300 and 500 BC, another period of global cooling prevailed.[202] Glaciers in Alaska, Utah, Scandinavia and Patagonia advanced again and armadas of icebergs again appeared in the oceans. This global cooling may have been a factor in coincident mass migrations, invasions and wars. The Hittite empire in Anatolia went into decline in 1200 BC and disappeared soon after. The Mycenaean civilisation fell at the expense of the rise of the Assyrian, Phoenician and Greek civilisations. Records from Troy show that it was cold, with famine around 1259 to 1241 BC and no recovery until 800 BC. About that time Egypt went into a prolonged decline while Babylonia and Assyria were also weak for most of the period from 1100–1000 BC.[203] At this time, the Jews made their exodus from Egypt when the Nile was consistently running low.

Massive flooding affected the Nile around 800 BC. The Egyptians recorded cooling from 750 to 450 BC. The river was sourced from an area that was becoming increasingly cool and dry. Sediments from the main river source, Lake Victoria in East Africa, show that the level of the lake progressively declined. In response, dams and canals were built to retain as much floodwater as possible for Egyptian agriculture. Sea level was lower and, in response, the Sweetwater Canal was built in Egypt for irrigation but the canal filled with silt.[204] North Africa had a cool dry windy period from 600 to 200 BC. Grasslands of the Sahara and Arabian deserts retreated, and humans migrated.

At the transition from the Bronze Age to the Iron Age (about 800 BC), the climate was miserable. The early Roman authors wrote of a frozen Tiber River and of snow lying on the ground for a long time. Such events don't happen in Italy today. Archaeological sites in West Friesland (the Netherlands) show prolonged periods of wet cold weather, the ground sodden prompting

[201] Perry, C. A. and Hsu, K. J. 2000: Geophysical, archaeological, and historical evidence support a solar-output model for climate change. *Proceedings of the National Academy of Sciences* doi: 10.1073/pnas.230423297.

[202] The Sub-Atlantic cooling.

[203] Bell, B. 1971: The Dark Ages in ancient history. *American Journal of Archaeology* 75: 1-26.

[204] Oliver, J. E. 1973: *Climate and man's environment.* Wiley, New York.

migration from low-lying areas to higher drier country.[205] Massive armadas of icebergs drifted to lower latitudes, melted and dropped coarse-grained boulders and sand onto the sea floor muds. This cooling was not restricted to Europe and the Middle East. It was probably global. A stalagmite from a cave in the Makapansgat Valley of South Africa records cold periods from 800–200 BC.[206]

During the 850 BC Bronze Age-Iron Age transition in western Britain, a period of substantial cooling, pollen studies show that there was not widespread abandonment of settlements, but upland agricultural areas were abandoned in preference to more intense farming in lowland areas. No nearby catastrophic volcanic events impacted on land use.[207] Farming and grazing on uplands and lowlands followed climate and climate-induced vegetation changes.[208] Farming activities adapted by moving to lowlands[209] and some areas were abandoned in the UK[210] and Europe.[211]

Ice cores record only slight variations in Holocene climate. However, other methods of reconstructing climate and recorded history tell a different story. Peat bogs in northwest Europe and from South America show that there was a rapid global cooling around 800 BC.[212] The timing, nature and global synchronicity suggest that it was driven by solar activity, possibly amplified by oceanic circulation changes.

----

[205] Van Geel, B. and Renssen, H. 1998: Abrupt climate change around 2,650 BP in northwest Europe: Evidence for climatic teleconnections and a tentative explanation. In: *Water, Environment and Society in Times of Climatic Change* (eds Issa, A. S. and Brown, N.), 1-21, Kluwer.

[206] Holmgren, K., Tyson, P. D., Moberg, A. and Svanered, O. 2001: A preliminary 3,000-year regional temperature reconstruction for South Africa: Research letter. *South African Journal of Science* 97, 49-51.

[207] Dark, P. 2006: Climate deterioration and land-use change in the first millennium BC: perspectives from the British palynological record. *Journal of Archaeological Science* 33: 1381-1395.

[208] Fyfe, R. M., Brück, J., Johnston, R., Lewis, H., Roland T. P. and Wickstead, H. 2008: Historical context and chronology of Bronze Age land enclosure on Dartmoor, UK. *Journal of Archaeological Sciences* 35: 2250-2261.

[209] Tipping, R., Davies, A., McCulloch, R. and Tisdall, E. 2008: Response to late Bronze Age climate change of farming communities in north east Scotland. *Journal of Archaeological Sciences* 35: 2379-2386.

[210] Amesbury, M. J., Carman, D. J., Fyfe, R. M., Langdon, P. G. and West, S. 2008: Bronze Age upland settlement decline in southwest England: testing the climate change hypothesis. *Journal of Archaeological Science* 35: 87-98.

[211] Dark, P. 2006: Climate deterioration and land-use change in the first millennium BC: perspectives from the British palynological record. *Journal of Archaeological Sciences* 33: 1381-1395.

[212] Chambers, F. M., Mauquoy, D., Brain, S. A., Blaauw, M. and Daniell, J. R. G. 2007: Globally synchronous climate change 2800 years ago: Proxy data from peat in South America. *Earth and Planetary Science Letters* 253: 439-444.

## The Roman Warming (250 BC—450 AD)

Warming started about 250 BC and was enjoyed by the Greeks and Romans. The Romans had it easy. Although the Empire started in a cool period, grapes were grown in Rome in 150 BC. By the 1st Century BC, Roman scribes record little snow and ice and that vineyards and olive groves extended northwards in Italy.[213] At the peak of the Roman warming, olive trees grew in the Rhine Valley of Germany. The location of vineyards is a good climate proxy. Citrus trees and grapes were grown in England as far north as Hadrian's Wall and most of Europe enjoyed a Mediterranean climate. This suggests a very rapid warming. It was also wetter. Temperatures in the Roman Warming were 2 to 6°C warmer than today. Sea level was slightly lower than today despite the fact that times were warmer[214] suggesting that land movements associated with the collision of Africa with Europe influenced local sea level. Roman clothing also shows it was warmer than today.

Tropical rains in Africa caused huge flooding of the Nile and many of the great buildings were inundated. These changes in rainfall, river flow and lake levels were widespread.[215,216] By 300 AD, the global climate was far warmer than at present.[217] Weather records of southern Italy kept by Ptolemy in the 2nd Century AD show that rain fell all year round whereas now the region enjoys only winter rain. Furthermore, because of regular and higher rainfall, North Africa became a breadbasket for the Carthaginians and later the Romans. It is now mostly desert. Not only did the Romans enjoy a warmer wetter climate, Central America was wetter than now and Central Asia responded to the warmer wetter times with a strong population increase around 300 AD.[218]

Pollen studies show that vegetation thrived and a good example is the long-term data from Spain. A warm wet climate existed in Spain 5000 years

[213] Allen, H. W. 1961: *The history of wine.* Faber & Faber, London.

[214] Lambeck, K., Anzidei, M., Antonioli, F., Benini, A. and Esposito, A. 2004: Sea level in Roman time in the Central Mediterranean and implications for recent change. *Earth and Planetary Sciences* 224: 563-575.

[215] Laird, K. R., Cumming, B. F., Wunsum, S., Rusak, J. A., Odlesby, R. J., Fritz, F. C. and Leavitt, P. R. 2003: Lake sediments record large-scale shifts in moisture regimes across the northern prairies of north America during the past two millennia. *Proceedings of the National Academy of Sciences* 100: 2483-2488.

[216] Lebreiro, S. M., Frances, G., Abrantes, F. F. G., Diz, P., Bartels-Jonsdottir, H. B., Stroynowski, Z. N., Gil, I. M., Pena, L. D., Rodrigues, T., Jones, P. D., Nombela, M. A., Alejo, I., Briffa, K. R., Harris I. and Grimalt, J. O. 2006: Climate change and coastal hydrographic response along the Atlantic Iberian margin (Tagus Prodelta and Muros Ria) during the last two millennia. *The Holocene* 16: 1003-1015.

[217] Lamb, H. H. 1977: *Climate, history and the future.* Methuen, London.

[218] Claiborne, R. 1970: *Climate, man and history.* Norton, New York.

ago. Land was cleared to plant crops.[219]   In northern Spain (Galicia), there was a major climate-driven reduction of forests at about 975 BC. A low pollen count from 1400 to 1860 AD reflects the Little Ice Age with the minimum pollen influx at 1700 AD, at the time of the Maunder Minimum (1645–1750 AD). Maximum pollen influx at 250 BC–450 AD reflects the Roman Warming and increased pollen influx from 950–1400 AD reflects the Medieval Warming.[220]   Pollen analysis is so accurate that it records the 20th Century introduction of Australian *Eucalyptus* trees.

In the crater lake of an extinct volcano, Mt Kenya, drill cores recording the period from 2250 BC to 750 AD showed a distinct high temperature period from 350 BC to 450 AD. This was the Roman Warming reflected as a warmer climate in equatorial Africa.[221] The researchers correlated this with data from Swedish Lapland, the northeastern St Elias Mountains in Alaska and the Canadian Yukon, which indicates that the Roman Warming was global. It was also warmer in Antarctica in Roman times. Skin and hair from mummified elephant seals preserved on Holocene raised beaches in Antarctica tell a story. Mummified seals dating from the Roman Warming and Medieval Warming were found well south of the modern breeding and moulting grounds. In the intervening Dark Ages, it was cold but not cold enough to drive the seals from Antarctica.[222]

The good weather during the Roman Warming meant that crop failures and famine became a rarity. There was an excess of food, population increased and the great Roman construction projects were undertaken using the excess labour and wealth. England had at least 5.5 million people, all of whom could be fed. It was not until the Medieval Warming (900 to 1280 AD) and the late 16th Century that England again had a population exceeding 5.5 million. The Dark Ages quickly depopulated areas that had thrived in the Roman Warming.

---

[219] Badal, E., Bernabeu, J. and Vernet, J. L. 1994: Vegetation changes and human action from the Neolithic to the Bronze Age (7000-4000 B.P.) in Alicante, Spain, based on charcoal analyses. *Vegetation History and Archaeobotany* 3: 155-166.

[220] Desprat, S., Sánchez Goñi, M. F. and Loutre, M.-F. 2003: Revealing climatic variability of the last three millennia in northwestern Iberia using pollen influx data. *Earth and Planetary Science Letters* 2134: 63-78.

[221] Rietti-Shati, M., Shemesh, A. and Karlen, W. 1998: A 3,000-year climate record from biogenic silica oxygen isotopes in equatorial high-altitude lake. *Science* 281: 980-982.

[222] Hall, B. L., Hoelzel, A. R., Baroni, C., Denton, G. H., le Boeuf, B. J., Overturf, B. and Töpf, A. L. 2006: Holocene elephant seal distribution implies warmer-than-present climate in the Ross Sea. *Proceedings of the National Academy of Sciences* 103: 10213-10217.

## The Dark Ages (535–900 AD)

The Dark Ages were a terrible time to be alive. Sudden cooling took place in 535 and 536 AD and the Earth plunged into the Dark Ages until about 900 AD.[223] It was cold, there were famine, war, change of empires and the stressed humans succumbed to the plague.

Around 540 AD, trees almost stopped growing.[224] Flooded bog oaks and timber from this time have very narrow growth rings. This was a global event because it is also recorded in tree rings from Ireland, England, Siberia, North America and South America. Snow fell in Mediterranean Europe and coastal China and there were savage storms in Scandinavia and South America. The sky was dim, there were meteor and comet swarms, flooding was common and, after the famines of the late 530s, the plague attacked Europe between 542 AD and 545 AD. The weather of Constantinople was described by Procopius:

> The Sun gave forth its light without brightness, like the Moon during the whole year and it seemed exceedingly like the Sun in eclipse, for the beams it shed were not clear, nor such as it is accustomed to shed.

A similar account was given from a more southern city by John of Ephesus:[225]

> The Sun became dark and its darkness lasted for 19 months. Each day it shone for about four hours, and still this light was only a feeble shadow ... the fruits did not ripen and the wine tasted like sour grapes.

From even further south, John the Lydian[226] wrote in De Ostentis: "The Sun became dim for nearly the whole year ... the fruits were killed at an unseasonable time."

The Black Sea froze in 800, 801 and 829 AD. Ice formed on the Nile River. Such freezing has not happened since then. It was very cold.

Long bitter droughts in Europe between 300 AD and 800 AD led to population displacement (the *Völkerwanderungen* or migrating wandering people), social tensions and famine. The lack of sunlight and drought caused crop failure, and famine followed.[227] Weakened populations and new groups of *Völkerwanderungen* with no resistance fell prey to the bubonic plague.[228]

---

[223] Keys, D. 1999: *Catastrophe: An investigation into the origins of the modern world*. Ballantine.

[224] Baillie, M. G. L. 1994: Dendrochronology raises questions about the AD 536 dust-veil event. *The Holocene* 4: 212-217.

[225] Keys, D. 1999: *Catastrophe: An investigation into the origins of the modern world*. Ballantine.

[226] Maas, M. 1992: *John Lydus and the Roman past: Antiquarianism and politics in the age of Justinian*. Routledge, London.

[227] Mango, C. 1980: *Byzantium, the Empire of New Rome*. Scribner, New York.

[228] An infection from the bacterium *Yersinia pestis*.

Rats and their fleas spread the plague although at that time it was suspected that sailors spread the disease. Sailors were constrained to their ships, rats and their fleas jumped ship, trade stopped and economies collapsed while the plague just continued to spread. The plague, Justinian's Plague, killed about 200,000 inhabitants of the Byzantine Empire, then killed about one third of the inhabitants of Eastern Europe and about half those of Western Europe. By 590 AD when the plague had run its course, some 25 million people had been killed. Plague pandemics did not take place again until the 14th Century which was, not coincidentally, another time of global cooling, famine and social disruption.

Peat in blanket mires may appear wet and dirty but it hides a treasure trove of information about past vegetation and climate. Five sites in the UK and additional information from Scandinavia show wetter cooler times during the Dark Ages.[229] In Scandinavia, studies of tree rings and algae microfossils indicate a very cold period around 500 AD and this was coincidental with farmlands retreating to lower altitudes.[230] The same data shows warmer times (Medieval Warming) from 700 to 1200 AD.

Glaciers in Pacific North America expanded and tree lines decreased in altitude.[231] Studies of lake fauna in Mexico show that after a long period of warm wet conditions, there was a dry period in which there were exceptionally arid events at 862, 986 and 1051 AD. This was a period of frequent intense drought in Central America which coincided with the collapse of the Mayan civilisation. Although the exact date of the Mayan collapse is not known, it was in the period 750–900 AD.[232] At about 1064 AD, warm wet conditions returned. By then, it was too late for the Mayans.[233] Another dry period commenced at 1391 AD, reflecting the cool dry times of the Little Ice Age.[234] The droughts were cyclical, corresponding with a solar cycle.[235,236] The Mayan civilisation grew during times of global warming and was killed

[229] Blackford, J. J. and Chambers, F. M. 1991: Proxy records of climate from blanket mires: evidence for a Dark Age (1400 BP) climatic deterioration in the British Isles. *The Holocene* 1: 53-67.

[230] Berglund, B. E. 2003: Human impact and climate changes. *Quaternary International* 105: 7-12.

[231] Reyes, A. V., Wiles, G. C., Smith, D. J., Barclay, D. J., Allen, S., Jackson, S., Larocque, S., Laxton, S., Lewis, D., Calkin, P. E and Clague, J. J. 2006: Expansion of alpine glaciers in Pacific North America in the first millennium A.D. *Geology* 34: 57-60.

[232] Hodell, D. A., Curtis, J. H. and Brenner, M. 1995: Possible role of climate in the collapse of Classic Maya civilisation. *Nature* 375: 391-394.

[233] Peterson, L. C. and Haug, G. H. 2005: Climate and the collapse of Maya civilization. *American Scientist* 93: 322-329.

[234] Curtis, J. H., Hodell, D. A. & Brenner, M. 1996: Climate variability on the Yucatan Peninsula (Mexico) during the past 3500 years, & the implications for Maya cultural evolution. *Quaternary Research* 46: 37-47.

[235] The 210-year DeVries-Suess solar cycle.

[236] Hodell, D. A., Brenner, M., Curtis, J. H. and Guilderson, T. 2001: Solar forcing of drought frequency in the Maya lowlands. *Science* 292: 1367-1370.

off during global cooling. There are also suggestions that the collapse of the Mayan civilisation resulted from changes in diet[237] but, because the Mayan civilisation had no international trade, the changes in diet most likely reflect the changes in climate anyway.

The Dark Ages was a global cold period. For example, coastal sediments in Venezuela show that there was very little runoff water at that time, suggesting a prolonged drought. This is the same drought that caused the collapse of Mayan cities in Central America.[238] Studies of peat bogs in the Tibetan Plateau also recorded the Dark Ages. In addition, three severely cold events within the Dark Ages were noted.

## The Medieval Warming (900–1300 AD)

The Dark Ages ended as quickly as they began as the world became warm again. The Medieval Warming from 900 to 1280 AD was followed by two decades of very changeable weather as the Medieval Warming changed to the ensuing Little Ice Age. In the Medieval Warming, it was far warmer than the present and warming was widespread.[239] The Medieval Warming was not all beer and skittles, because there was a cold period from 1040 to 1080 AD when the Sun was very inactive (Oort Minimum).

The Medieval Warming is just one of the many warm periods that Earth has enjoyed. The much warmer and longer Holocene Climate Optimum lasted between 7000 BC and 3000 BC. (The Late 20th Century Warming was somewhat cooler than the Medieval and Roman Warmings.) Although it was warm, there were the inevitable intermittent periods of bad weather. However, on balance summers were longer and warmer, crops were plentiful and there were few serious famines. Kings and landowners prospered and peasants rarely went hungry. The amount of land devoted to agriculture increased and fields crept up to higher altitudes where farming had previously not taken place. In the Northern Hemisphere, crops that enjoy warmth were farmed further and further north.

Europe was warm, rainfall was higher, the climate was stable and agricultural productivity was very high. There was excess food, excess labour and excess wealth. There was prosperity and there were funds to fight the Crusades. Cultivation was higher in the mountains than it had ever been, and tree ring studies in California suggest that North America was also

---

[237] Wright, L. E. and White, C. D. 1996: Human biology in the Classic Maya collapse: Evidence from paleopathology and paleodiet. *Journal of World Prehistory* 10: 147-198.

[238] Haug, G. H., Günther, D., Peterson, L. C., Sigman, D. M., Hughen, K. A. and Aeschlimann, B. 2003: Climate and collapse of Maya civilisation. *Science* 299: 1731-1735.

[239] Grove, J. M. 1988: *The Little Ice Age*. Methuen.

enjoying the warm times.[240] Excess food in Europe led to a 50% increase in population. Although grain could not be effectively stored from rats and other pests, the regular reliable harvests created stability and certainty.

In Europe, cities grew, transport networks were established and excess labour was employed in construction of the great monasteries, cathedrals and universities.[241] The great architectural structures took generations to build, showing that there was sustained generational wealth. Universities were established to train young men for the priesthood. The modern secular universities owe their origin to the Medieval Warming. Any visitor to Europe can see the results of the boom in cathedral building, a direct result of the great prosperity that warm times brought. In Europe, new cities were built and the population increased from 30 million to 80 million. At the same time, the thousands of temples at Angkor Wat in southeast Asia were built. In China, these warmer conditions led to a doubling of the population in 100 years. The Medieval Warming was the zenith of Muslim imperialism, culture and science. Enough food was available to feed more people and travellers could venture great distances.

Economies boomed. In colder times during the Dark Ages, the economy was organised around self-sufficient estates which grew their own food, flax and wool, wove their own clothing and had little or no trade with other estates or internationally. The Medieval Warming was a time of excess when luxury goods such as spices from Oriental caravans, sugar from Cyprus and glass from Venice could be acquired through trade. Overseas and coastal trade was easier because of the decline in the frequency and intensity of high winds and fierce storms. More sunshine meant that, even if there was heavy rain, roads dried quickly, allowing more reliable traffic between farms and towns. Mountain passes were open for longer, allowing a longer season of trade. The warm climate allowed the emergence of European trade fairs. In 1000 AD there were more than 70 mints in market towns coining German silver for purchase of wool and fish. In return, the English purchased wine, furs, cloth and slaves.[242]

Agriculture thrived high in the Alps of Europe. At the Grosser Aletsch Glacier, a larchwood aqueduct was constructed to supply water to an alpine village around 1200 AD. It was destroyed by glacier advance in 1240 and had to be totally rerouted in 1370 during the Little Ice Age after further

---

[240] Broecker, W. S. 2001: Was the Medieval Warm Period global? *Science* 291: 1497-1499.

[241] Gimpel, J. 1961: *The cathedral builders*. Grove Press, New York.

[242] Lacey, R. and Danziger, D. 1999: *The year 1000: What life was like at the turn of the First Millennium*. Little Brown, Boston.

glacial advance.[243] It was finally destroyed in the peak of the Little Ice Age. The numerous glacial retreats and advances in the Alps are a proxy for air temperature.[244,245] By measuring glacial advances and retreats in New Zealand, a correlation of 1500-year cycles can be made.[246] In the North Atlantic, ice packs retreated northward (especially in summer) and the incidence of severe storms decreased.

The Vikings, who were already great mariners, sailed north and west and established settlements in Greenland, Iceland and North America, such as L'Anse aux Meadows. The ice-free North Atlantic meant that the Vikings could travel and they called Newfoundland "Vinland" because of the vineyards there. Cattle, sheep and barley were grown in Greenland, tree roots could penetrate soil that was once tundra, fishing for cod and seals took place on ice-free seas, burials could be undertaken in soils that were not frozen, villages were established and the Pope sent a bishop to Greenland to care for his Norse flock. The Vikings also traded as far south as Persia and the Arab states.[247]

The Doomsday Book of England shows where grapes were grown, in places where no grapes could now be cultivated for wine production. England, now a cool damp place, was warmer and drier in the Medieval Warming. England thrived and its population grew from 1.4 million to 5.5 million. France's population tripled to 18 million.

Vineyards in Germany were up to 780 metres above sea level whereas today the maximum altitude is 560 metres above sea level. Temperature usually decreases by 0.6 to 0.7°C per 100 metres of altitude gained, so the average mean temperature must have been 1.0 to 1.4°C warmer than now.[248] Settlements, land clearing and farming in valleys and slopes spread 100 to 200 metres higher in altitude in Norway, again suggesting that summer

---

[243] Holzhauser, H. 1997: Fluctuations in the Grosser Aletsch Glacier and the Gorner Glacier during the last 3,200 years: new results. *Paläoklimaforschung* 24: 35-58.

[244] Hormes, A., Schlüchter, C. and Stocker, T. F. 1998: Minimal extension phases of Unteraarglacier (Swiss Alps) during the Holocene based on $^{14}C$ analysis of wood. *Radiocarbon* 40: 809-817.

[245] Hormes, A., Müller, B. U. and Schlüchter, C. 2001: The Alps with little ice: Evidence for eight Holocene phases of reduced glacier extent in the Central Swiss Alps. *The Holocene* 11: 255-265.

[246] Hormes, A., Preusser, F., Denton, G., Hajdas, I., Weiss, D., Stocker, T. F. and Schlüchter, C. 2003: Radiocarbon and luminescence dating of overbank deposits in outwash sediments of the Last Glacial Maximum in North Westland, New Zealand. *New Zealand Journal of Geology and Geophysics* 46: 95-106.

[247] Thachuk, R. D. 1983: The Little Ice Age. *Origins* 10: 51-65.

[248] Arendes, C. 2007: Joseph H. Reichholf: Eine kurze Naturgeschichte des letzen Jahrtausends. Frankfurt a.M. *Zeitschrift für Geschichtswissenschaft* 56: 1-462.

temperatures were 1°C higher than now.[249] Tree lines moved upslope in the Medieval Warming and the stumps and roots are still preserved above the current tree line in many alpine areas. Stumps and logs of *Larix sibirica* 30 metres above the current tree line in the Polar Urals have been dated and show that at 1000 AD the tree line was higher than now.[250] This tree line receded around 1350 AD, indicating the effects of the following Little Ice Age.

Copper, gold and emeralds were mined at high altitude in the Alps of Europe during both the Roman Warming and the Medieval Warming.[251] These mines were covered by ice and abandoned during the Dark Ages and were again covered by ice during the Little Ice Age. Some of these areas have just been exposed by glacier retreat in the Late 20th Century Warming.[252]

Sediments from Lake Neufchatel in Switzerland show an abrupt temperature decrease of 1.5°C at the end of the Medieval Warming. These sediments also show that mean annual temperatures in the Medieval Warming were higher than at present.[253] In the Baltic Sea, the Medieval Warming allowed tropical and sub-tropical marine plankton to survive. Despite the Late 20th Century Warming, they have not returned because the Baltic Sea is still colder than it was in the Medieval Warming. At about 1200 AD, the warm water micro-organisms were replaced by cold water organisms. The faunal replacement reflects the onset of the Little Ice Age.[254]

Boreholes give accurate temperature histories for about 1000 years into the past because rock conducts past surface temperatures downward only slowly. In the Northern Hemisphere, borehole data shows the Medieval Warming and a cooling of about 2°C from the Medieval Warming to the Little Ice Age.[255] A study of 6000 boreholes on all continents has shown that temperature in the Medieval Warming was warmer than today and that the

[249] Fagan, B. 2000: *The Little Ice Age: How climatic change made history 1300-1850*. Basic Books, New York

[250] Esper, J. and Schweingruber, F. H. 2004: Large scale tree line changes recorded in Siberia. *Geophysical Research Letters* 31, 10.1029/2003GLO019178.

[251] Pohl, W. 1984: Metallogenetic evolution of the East Alpine Proterozoic basement. *International Journal of Earth Sciences* 73: 131-147.

[252] Hyde, W. W. 1935: The Alps in history. *Proceedings of the American Philosophical Society* 75: 431-442.

[253] Filippi, F. L., Lambert, P., Hunziker, J., Kübler, B. and Bernasconi, S. 1999: Climatic and anthropogenic influence on stable isotope record from bulk carbonates and Ostracodes in Lake Neufchatel, Switzerland, during the last two millennia. *Journal of Palaeolimnology* 21: 19-34.

[254] Andren, E., Andren, T. and Sohlenius, G. 2000: The Holocene history of the southwestern Baltic Sea as reflected in a sediment core from the Bornholm Basin. *Boreas* 29: 233-250.

[255] Steig, E. J., Brook, E. J., White, J. W. C., Sucher, C. M., Bender, M. L., Lehman, S. J., Morse, D. L., Waddington, E. D. and Clow, G. D. 1998: Synchronous climate changes in Antarctica and the North Atlantic. *Science* 282: 92-95.

temperature fell 0.2 to 0.7 °C during the Little Ice Age.[256]

Advance of tree lines, retreat of glaciers, reduced lake-catchment erosion (i.e. fewer storms) and temperature proxies show that between 700 and 1200 AD, Scandinavia was warm.[257] Other data from Scandinavia shows a cool period from 500 AD to 700 AD (i.e. the Dark Ages) with 660 AD a very cold year. An overall warm period from 720 AD to 1360 AD (Medieval Warming) had exceptionally warm times in the 10th, 11th, 12th Centuries. There was also a warm period in the early 15th Century. After 1430 it was cold (i.e. Little Ice Age).

The Medieval Warming also affected the eastern Mediterranean.[258] Lake Van in eastern Turkey had a high water level[259] as did lakes in the Sahara Desert.[260] The Dead Sea[261] and the Sea of Galilee[262] were also full and rainfall in the Nile headwaters was excessive.[263] The Nile has two headwaters. The Blue Nile comes from Ethiopia. This mainly contributes silt. The White Nile rises in Lake Victoria and contributes most of the water. The Nile River has a flooding and sedimentation record that integrates two widely separated areas. A 1300-year tree ring record from Pakistan shows that the warmest decades occurred between 800 AD and 1000 AD in the Medieval Warming and the coldest between 1500 and 1700 in the Little Ice Age.[264]

The spread of villages in southern Africa occurred during the warm wet periods of the Early Iron Age (650–300 BC). Another warm wet period extended from 900–1290 AD, the Medieval Warming. The appearance of many villages and the rise of Great Zimbabwe coincided with the beginning

---

[256] Huang, S., Pollack, H. N. and Shen, P. Y. 1997: Late Quaternary temperature change seen in worldwide continental heat flow measurements. *Geophysical Research Letters* 24: 1947-1950.

[257] Berglund, B. E. 2003: Human impacts & climate changes. *Quaternary International* 105: 7-12.

[258] Schilman, B., Bar-Matthews, M., Almogi-Labin, A. and Luz, B. 2001: Global climate instability reflected by Eastern Mediterranean marine records during the Late Holocene. *Palaeogeography, Palaeoclimatology, Palaeoecology* 176: 157-176.

[259] Schoell, M. 1978: Oxygen isotope analysis of authigenic carbonates from Lake Van sediments and their possible bearing on the climate of the past 10,000 years. In: *Geology of Lake Van, Kurtman* (ed. E. T. Degens), 92-97. MTA.

[260] Nicholson, S. E. 1980: Saharan climates in historic times. In: *The Sahara and the Nile* (ed. M. A. J. Williams). Balkema.

[261] Issar, A. S. 1989: Climatic changes in Israel during historical times and their impact on hydrological, pedalogical, and socioeconomic systems. In: *Palaeoclimatology and Palaeometeorology: Modern and past patterns of global atmospheric transport* (ed. M. Leinen and M. Sarnthein), 535-541.

[262] Frumkin A., Magaritz, M., Carmi, I. and Zak, I. 1991: The Holocene climatic record of the salt caves of Mount Sedom, Israel. *The Holocene* 1: 191-200.

[263] Hassan, F. 1981: Historical Nile floods and their implications for climatic change. *Science* 212: 1142-1145.

[264] Esper, J., Schweingruber, F. H. and Winiger, M. 2002: 1,300 years of climate history for West Central Asia inferred from tree rings. *The Holocene* 12: 267-277.

of the dry Little Ice Age, while a warm pulse in the 15th and 16th Centuries created the conditions for mixed farming in the highveld. Another warm and wet period at the end of the 18th Century contributed to the spread of maize, increased populations and more military action.[265]

In East Africa, an 1100-year rainfall record using sediments, fossil diatoms and numbers of species of midges shows alternating dry and wet conditions. Off the coast of West Africa, sea surface temperature dropped. Onshore, the landmass was drier for centuries during cool periods and there were massive rainfalls that created lakes in the Sahara Desert in warm periods. However, other work shows that in greenhouse times, such as the Medieval Warming, East Africa was drier. In the Little Ice Age a colder wetter period was frequently interrupted by droughts.[266] Over the past millennium, equatorial East Africa alternated between dry and wet conditions. In the Medieval Warming, East Africa was drier than today whereas in the Little Ice Age, East Africa was far wetter than now. However, the Little Ice Age had three prolonged dry periods. Lake Naivasha in Kenya had a higher rainfall than now. The writings of Arab travellers in North Africa indicate that the rainfall was higher than in the Little Ice Age and today. In South Africa, a stalagmite from a cave in the Makapansgat Valley showed a warm period from 1000 to 1300 AD.[267]

China flourished in the Medieval Warming. Palace records, official histories, year books, gazettes and diaries record the arrival and departure of migratory birds; the distribution of plants, bamboo groves and fruit orchards; the patterns of elephant migrations; the flowering times of plants; and major floods and droughts.[268,269,270] The growing seasons were longer and more reliable and citrus orchards moved north, only to move south once the Little Ice Age commenced.[271] China enjoyed the Holocene Climate Optimum (8000 BC to 3000 BC), the Roman Warming and the Medieval Warming and, on the

---

[265] Huffman, T. N. 1996: Archaeological evidence for climatic change during the last 2000 years in southern Africa. *Quaternary International* 33: 55-60.

[266] Verschuren, D., Laird, K. R. and Cumming, B. F. 2000: Rainfall and drought in equatorial East Africa during the past 1100 years. *Nature* 403: 410-444.

[267] Tyson, D., Karlen, W., Holmgren, K. and Heiss, G. A. 2000: The Little Ice Age and Medieval Warming in South Africa. *South African Journal of Science* 96: 121-126.

[268] Ko Chen, C. 1973: A preliminary study on the climatic fluctuations during the last 5,000 years in China. *Scientia Sinica* 16: 483-486.

[269] Saho Wu, W. and Zong Ci, Z. 1981: Droughts and floods in China. In: *Climate and history: Studies in past climates and their impact on man.* (eds T. M. L. Wigley et al.). Cambridge University Press.

[270] Zhang, J. and Crowley, T. J. 1989: Historical climate records in China and reconstruction of past climates. *Journal of Climate* 2: 830-849.

[271] Deer, Z. 1994: Evidence for the existence of the Medieval Warm Period in China. *Climatic Change* 26: 289-297.

basis of pollen studies, China was at least 2 to 3°C warmer than now.[272]

Such warming periods created great wealth in China. Wealth had been rising from 200 BC to its peak at 1100 AD with the greatest increases in the Han period (206 BC to 220 AD) and the Northern Sung Dynasty (961 AD to 1127 AD).[273] These two warm periods in China coincided with the Roman Warming and the Medieval Warming elsewhere. China's temperature history has been reconstructed for the last 2000 years from ice cores, lake sediments, peat bogs, tree rings and historic documents.[274] The 2nd and 3rd Centuries AD, towards the end of the Roman Warming, were the warmest. It was also warm from 800 AD to 1400 AD during the Medieval Warming, cold in the Little Ice Age from 1400 to 1920 and then warm again after 1920 during the Late 20th Century Warming. Cave stalagmites from China show a strong warming from 700 to 1000, corresponding to the Medieval Warming, and a cooling from 1500 to 1800 when the air temperature was 1.2°C cooler that at present.[275]

Similar good times were enjoyed by the Japanese as official records on weather, floods, droughts, heavy snows, long rains and mild winters show.[276] It was warm from the 10th Century to the 14th Century, as in Europe. Official records allowed a detailed analysis which showed that relatively hot conditions continued until the 8th Century, then cool conditions appeared for a short time in the late 9th Century. Warm conditions existed from the 10th Century to the early 15th Century and late in the 15th Century cooling commenced. It became very cold at the beginning of the 17th Century. The carbon chemistry of Japanese cedar records the Dark Ages, the Medieval Warming and the Little Ice Age, again showing that these climate changes were widespread.[277]

North America also thrived in the Medieval Warming. Increased rainfall

---

[272] Feng, Z. 1993: Temporal and spatial variations in climate in China during the last 10,000 Yrs. *The Holocene* 3: 174-180.

[273] Caho, K. 1986: *Man and land in China: An economic analysis.* Stanford University Press.

[274] Yang, B., Braeuning, A., Johnson, K. R. and Yafeng, S. 2002: General characteristics of temperature variation in China during the last two millennia. *Geophysical Research Letters* 29: 1029/2001GL014485.

[275] Zhibang, M., Hongchun, L. I., Ming, X. I. A., Tehlung, P., Zicheng, P. and Zhaofeng, Z. 2003: Palaeotemperature changes over the past 3,000 years in Eastern Beijing, China: A reconstruction based on Mg/Sr records in a stalagmite. *Chinese Science Bulletin* 48: 395-400.

[276] Tagami, Y. 1993: Climate change reconstructed from historical data in Japan. *Proceedings of the International Symposium on Global Change, International Geosphere-Biosphere Program,* 720-729.

[277] Kitagawa, H. and Matsumoto, K. 1995: Climatic implications of $^{13}$C variations in a Japanese cedar (*Cryptomeria japonica*) during the last two millennia. *Geophysical Research Letters* 22: 2155-2158.

cut channels in the Great Plains[278] and Alaska warmed quickly.[279] Vegetation studies in northern Quebec show a cold period (760 to 860 AD) coinciding with the Dark Ages, a warming from 860 to 1000 AD reflecting the Medieval Warming and severe cold from 1025 to 1400 reflecting the Little Ice Age.[280] A similar study in southern Ontario showed forest changes at the end of the Medieval Warming. Warmth-loving beech trees were replaced by the cold-tolerant oak and then later by the cold-loving pine. The change from the Medieval Warming to the Little Ice Age resulted in deforestation and a loss of 30% of the mass of the forests. The Ontario forests have still not recovered from the Little Ice Age and have not returned to the diversity and productivity of the Medieval Warming.[281] In the United States, the Medieval Warming is detected by studies of moisture records of lodgepole pines at Lake Tenaya in the high Sierra Nevada Mountains.[282] Lake levels have changed with the release of water from melting snow. A similar pattern was recognised at Mono Lake and the Walker River in California.[283]

The Anasazi Indians' agriculture and culture spread in the early part of the Medieval Warming as rains were more consistent. Tree rings from Sand Canyon show low rainfall in 1125 to 1180 and 1270 to 1274 and a 24-year drought late in the 13th Century. These led to food scarcities, internal conflict, the building of fortress-like cliff dwellings and the eventual sacking of the fortresses in the Little Ice Age.[284] By 1400, the maize crop failures had driven the Anasazi from their cliff dwellings and to extinction. In the Sierra Nevadas of California, living and dead trees provide a 3000-year record of tree line changes.[285] Dense forests grew above the current tree line in the Roman Warming and from 400 AD to 1000 AD. The tree line moved rapidly down slope between 1000 and 1400 and continued to move down slope,

---

[278] Daniels, J. M. and Knox, J. C. 2005: Alluvial stratigraphic evidence for channel incisions during the Medieval Warm Period on the central Great Plains, USA. *The Holocene 15*: 736-747.

[279] Hu, F. S., Ito, E., E., Brown, T. A., Curry, B. B. and Engstrom, D. R. 2001: Pronounced climatic variations in Alaska during the last two millennia. *Proceedings of the National Academy of Sciences* 98: 10,552-10,556.

[280] Arseneault, D. and Payette, S. 1997: Reconstruction of millennial forest dynamic from tree remains in a subarctic tree line peatland. *Ecology* 78: 1873-1883.

[281] Campbell, I. D. and McAndrews, J. H. 1993: Forest disequilibrium caused by rapid Little Ice Age cooling. *Nature* 366: 336-338.

[282] Stine, S. 1998: Medieval climate anomaly in the Americas. In: *Water, Environment and Society in Times of Climatic Change* (eds Issar, A. S. and Brown, N.), 43-67. Kluwer.

[283] Stine, S. 1994: Extreme and persistent drought in California and Patagonia during mediaeval time. *Nature* 369: 546-549.

[284] Fagan, B. 1999: *Floods, famines and emperors: El Niño and the fate of civilisations.* Basic Books, New York.

[285] Graumlich, L. J. 2000: Global change in wilderness areas: Disentangling natural and anthropogenic changes. *U.S. Department of Agriculture Forest Service Proceedings* RMRS-P-15, Vol 3.

albeit more slowly, from 1500 to 1900. The current tree line has not changed since 1900. In southern Ontario, pollen studies show that beech trees in the Medieval Warming were replaced by oaks in the Little Ice Age. Beech trees enjoy warm conditions whereas oaks are tolerant of cold conditions.[286] In southern Alberta (Canada), lake sediments show the increased runoff during the Medieval Warming and the decreased runoff reflecting the drier conditions of the Little Ice Age.[287]

The Southern Hemisphere also experienced the Dark Ages, Medieval Warming and Little Ice Age. In Argentina, the carbon chemistry of prehistoric villages shows that villagers clustered in the lower valleys during the Dark Ages. Villages moved upslope to altitudes as high as 4300 metres in the Central Peruvian Andes during the Medieval Warming to capitalise on the stable warmer climate. In 1320, villagers moved back down slope as the colder unstable Little Ice Age commenced.[288] The compilation of flood reports, sailors' handbooks and folk records show that central Argentina had more rain during the Medieval Warming than now and that temperatures were up to 2.5°C higher than now.[289]

Vegetation in South America also felt the effects of climate change, especially the Little Ice Age. Pollen from lake sediments in Peru provides a 4000-year record of climate. The Roman Warming could be seen and rainfall declined during the Dark Ages. In the Medieval Warming, increased pollen indicated warmer temperatures, more plants and greater plant diversity, followed by a pollen decline in the Little Ice Age.[290] Elsewhere in South America, lake sediments from a high volcanic plateau showed that climate and rainfall changed quickly, with the Little Ice Age being a prominent feature.[291]

In the South Pacific, during the Roman Warming there was island-hopping migration by Polynesians. Easter Island was settled about 400 AD. Great blocks of stone were carved there between 1000 and 1350 during the

---

[286] Campbell, I. D. and McAndrews, J. H. 1993: Forest disequilibrium caused by rapid Little Ice Age cooling. *Nature* 366: 336-338.

[287] Campbell, C. 1998: Late Holocene lake sedimentology and climate change in Southern Alberta, Canada. *Quaternary Research* 49: 96-101.

[288] Cioccale, M. A. 1999: Climatic fluctuations in the Central Region of Argentina in the last 1000 years. *Quaternary International* 62: 35-47.

[289] Iriondo, M. 1999: Climatic changes in the South American plains: Records of a continent-scale oscillation. *Quaternary International* 57-58: 93-112.

[290] Chepstow-Lusty, A. J. , Frogley, M. R., Bauer, B. S., Bush, M. B. and Herrera, A. T. 2003: A late Holocene record of arid events from the Cuzco region, Peru. *Journal of Quaternary Science* 18: 491-502.

[291] Valero-Garces, V. L., Delgado-Huertas, A., Ratto, N., Navas, A. and Edwards, L. 2000: Palaeohydrology of Andean saline lakes from sedimentological and isotopic records, northwestern Argentina. *Journal of Palaeolimnology* 24: 343-359.

time of plenty in the Medieval Warming. In 1350 famine set in and the wet tropical island became a dry cool desert island during the Little Ice Age. By 1600, Easter Islanders had resorted to cannibalism and the population greatly declined.[292] The Tokelau, Society, Austral, Marshall and Marquesas Islands and Tonga and Fiji were all settled in the Roman Warming. New Zealand was first settled in the Medieval Warming[293] when South Pacific island populations thrived.[294]

An analysis of the physical evidence from 112 studies of the Medieval Warming in Greenland, Europe, Russia, USA, China, Japan, Africa, Chile, Argentina, Peru, Australia and Antarctica[295] showed that the Medieval Warming was recorded. The Medieval Warming can also be measured in sea floor sediments in the North Atlantic, the South Atlantic near Antarctica, the central and southern Indian Ocean and the Central and Western Pacific Ocean.

There were no $CO_2$ emitting industries in the Medieval Warming. This natural warming event was greater than the Late 20th Century Warming, which we are told is due to human emissions of $CO_2$.

## The Little Ice Age (1280–1850 AD)

The Medieval Climate Optimum ended rapidly with the Little Ice Age, starting in 1303 AD. This major climate change took only 23 years. It led to famine, depopulation, war and disease.[296] The Little Ice Age started when the Sun again became lazy. The Wolf Minimum (1280 to 1340 AD) was a time when there were few sunspots, and the lack of solar activity resulted in increased cloudiness. The planet became cold. The Little Ice Age had a number of intense periods when the Sun emitted less energy. These were the Spörer Minimum (1450–1540 AD), the Maunder Minimum (1645–1715 AD) and the Dalton Minimum (1795–1825 AD).[297] The Maunder Minimum was the most bitterly cold time of the Little Ice Age. Times of feast suddenly changed to times of famine.[298] The Little Ice Age was not a good time to be

---

[292] McGall, G. 1995: *Pacific Islands Yearbook*. Fiji Times.

[293] Houghton, P. 1996: *People of the great ocean: aspects of human biology of the early Pacific.* Cambridge University Press.

[294] Nunn, P. 2007: *Climate, environment and society in the Pacific during the last millennium.* Developments in Earth and Environmental Sciences 6: Elsevier.

[295] Soon, W. and Baliunas, S. 2003: Reconstructing climatic and environmental changes of the past 1,000 years: A reappraisal. *Energy and Environment* 14: 233-296.

[296] Lamb, H. H. 1982: *Climate, history and the modern world*. Routledge.

[297] Ribes, J. C. and Nesme-Ribes, E. 1993: The solar sunspot cycle in the Maunder minimum AD 1645 to AD 1715. *Astronomy and Astrophysics* 276: 549-563.

[298] Ladurie, L. and Ladurie, E. 1971: *Times of feast, times of famine*. Noonday Press (translated by B. Bray).

alive on planet Earth.[299]

The Little Ice Age was not really an ice age. In reality, it was a cool interval within the current interglacial. What made the Little Ice Age particularly difficult was that there had been hundreds of years of warmth in the Medieval Warming and the increased population was supported by subsistence farming. Subsistence farming was later replaced in Britain by specialist farming to support city populations. The Northern Hemisphere had adapted to warm times and was not prepared for the sudden onset of cold times. This created an environmental catastrophe. There was massive depopulation. This catastrophe was global. Pacific Island populations were greatly reduced at the beginning of the Little Ice Age.[300] Other parts of the world were cold and dry, especially during the Spörer and Maunder Minima.[301] Not only was it cold during the Little Ice Age, but there were rapid fluctuations in temperature and precipitation. During the Maunder Minimum, a year of record cold temperatures (1683–1684) was followed by a year of record heat (1685–1686). Change to glacial climate is characterised by drastic changes in temperature, storminess and precipitation without warming. These changes were local, global and rapid. They had a profound effect on human society.[302]

We have a reliable picture of the extremely cold periods during the Little Ice Age from the weather records. Private diaries, ships' logs, accounts of military campaigns and similar sources give descriptions of the wind directions, wind speed, cloud formations and other weather indicators. Precisely dated annals, chronicles, audited accounts, agricultural records and tax ledgers provide indirect information, particularly on extreme weather events. Records of wine grape harvests, salt harvest from evaporation pans and grain prices are a good proxy for temperature, rainfall and wind. For example, the price of grain was higher in periods of weak solar activity (Maunder Minimum 1645–1715 AD and Dalton Minimum 1775–1825) when Europe was extraordinarily cold. Additional evidence from debris left behind by glaciers, lake and ocean muds, pollen and insects in mud, tree rings, coral growth structures, ice core analysis, boreholes, archaeological

---

[299] Fagan, B. 2000: *The Little Ice Age: how climatic change made history 1300-1850 AD*. Basic Books.

[300] Nunn, P. D. 2000: Environmental catastrophe in the Pacific Islands around A.D. 1300. *Geoarchaeology* 15: 715-740.

[301] Touchan, R., Akkemik, U., Hughes, M. K. and Erkan, N. 2007: May-June precipitation reconstruction of southwestern Anatolia, Turkey during the last 900 years from tree rings. *Quaternary Research* 68: 196-202.

[302] Nunn, P. D. and Britton, J. M. R. 2001: Human-environment relationships in the Pacific Islands around A.D. 1300. *Environment and History* 7: 3-22.

site investigations and historical records can all be used to reconstruct the conditions during the Little Ice Age. It was not a pretty sight.[303]

The cold climate and glacier expansion in the Little Ice Age are documented from all continents and on major islands from New Zealand in the South Pacific Ocean to Svalbard in the Arctic Sea.[304] The Little Ice Age was not a single, uniformly cold episode. There were warm and exceptionally cold periods and distinct variations in climate and glacier activity took place on a regional basis. In Europe[305] and North America[306], at least six phases of glacier expansion occurred. These were separated by warm periods.[307,308]

Corals in the Florida Straits reveal variations in $C^{14}$ during the Little Ice Age,[309] which shows that the Earth's atmosphere was being bombarded by additional cosmic rays in the coldest time of the Little Ice Age. The coral carbon chemistry shows that the Maunder Minimum in Florida was at the same time as the Maunder Minimum in Europe. The effects of the Little Ice Age must have been widespread.

Studies of lichen are able to give a window into the extent of ice in the Little Ice Age in Iceland.[310] Four glaciers were studied and these show that the maximum ice extent was in the mid 19th Century and that there is a relationship between the mass of glacial ice and mean summer temperature.

Elsewhere in North America, the forests also responded to the extreme cold of the Little Ice Age.[311] The foxtail pine and western juniper of the southern Sierra Nevada Mountains show that it was warmer than current times from 1100 to 1375 AD and colder from 1450 to 1850. Tree rings in the long-lived bristle cone pines on the California-Nevada border show that some trees are 5500 years old. From 800 AD to the present, the hundred-year averages of temperature correlate statistically with the temperatures derived from central England.[312]

[303] Tuchman, B. W. 1979: *A distant mirror: the calamitous 14th Century*. Penguin.

[304] Grove, J. M. 1988: *The Little Ice Age*. Methuen.

[305] Svendsen, J. I. and Mangerud, J. 1997: Holocene glacial and climatic variations on Spitsbergen, Svalbard. *The Holocene* 7: 45-57.

[306] Luckman, B. H., Holdsworth, G. and Osborn, G. D. 1993: Neoglacial fluctuations in the Canadian Rockies. *Quaternary Research* 39: 144-153.

[307] Magnusson, M. 1987: *Iceland saga*. Bodley Head.

[308] Nesje, A. and Dahl, S. O. 2000: *Glaciers and environmental change*. Arnold.

[309] Druffel, E. M. 1982: Banded corals: Changes in oceanic carbon-14 during the Little Ice Age. *Science* 218: 13-19.

[310] Caseldine, C. J. 1985: The extent of some glaciers in northern Iceland during the Little Ice Age and the nature of recent deglaciation. *The Geographical Journal* 151: 215-227.

[311] Graumlich, I. D. 1993: A 1,000-year record of temperature and precipitation in the Sierra Nevada. *Quaternary Research* 39: 249-255.

[312] LaMarche, V. C. 1974: Palaeoclimatic interferences from long tree ring records. *Science* 183: 1043-1048.

History and knowledge of modern times show that the world was a different place in the Little Ice Age. In the second half of the 17th Century, the French army used frozen rivers as thoroughfares to invade the Netherlands, while New Yorkers walked from Manhattan to Staten Island. Sea ice surrounded Iceland, trapping the population and causing famine. This was not the first time this had happened. In the period 1420 to 1570, Vikings on Greenland lost their livestock, agriculture and lives in the first phase of the Little Ice Age. There is a view that the chronic shortages of grain and bread in the late 1700s in France due to poor climate led to the social discontent that fuelled the French Revolution.

The Little Ice Age had two cold phases and included four intense cold periods at times of reduced sunspot activity. This has been validated by measuring heavy and light oxygen isotopes in cave stalagmites from Ireland. The stalagmites also identified the Medieval Warming, the Dark Ages and the Roman Warming.[313] Glaciers advanced and retreated in the Little Ice Age. During glacial advance, European alpine villages were destroyed and forests were flattened. The northeast Pacific region of Alaska shows evidence of two major glacial advances that destroyed forests. Glaciers stabilised after advancing, some retreated slightly and the glacial fluctuations were on a decadal scale.[314] This also shows that the Little Ice Age was not restricted to Europe.

During the first phase (1280–1550 AD) of the Little Ice Age, the climate was far more variable than in the Medieval Warming or the second phase. The extreme variability brought warm and very dry summers in some years and very cold wet summers in other years. Storm frequency in the North Sea and the English Channel increased.[315] There were Arctic winters, stinking hot summers, major droughts, torrential rains and floods, long winters and long summers. In high latitudes, the Little Ice Age was heralded by the growth of the ice sheets in Greenland in the early 13th Century. Ice then covered much of Iceland, Scandinavia and northern Europe and landslides, avalanches and floods were far more common.[316]

---

[313] McDermott, F., Mattey, D. P. and Hawkesworth, C. 2001: Centennial-scale Holocene climate variability revealed by high-resolution speleotherm $\partial^{18}O$ record from SW Ireland. *Science* 294: 1328-1331.

[314] Wiles, G. C., Barclay, D. J. and Calkin, P. E. 1999: Tree-ring-dated 'Little Ice Age' histories of maritime glaciers from western Prince William Sound, Alaska. *The Holocene* 9: 163-173.

[315] Luterbacher, J., Rickli, R., Xoplaki, E., Tinguely, C., Beck, C., Pfister, C. and Wanner, H. 2001: The Late Maunder Minimum (1675-1715) – A key period for studying decadal scale climatic change in Europe. *Climatic Change* 49: 441-462.

[316] Grove, J. M. 1972: The incidence of landslides, avalanches and floods in western Norway during the Little Ice Age. *Arctic and Alpine Research* 4: 131-138.

The Gulf Stream, which helps to bring warm weather to much of the North Atlantic region, was significantly weakened during the Little Ice Age.[317] It is suggested that from 1200 to 1850 AD, the Gulf Stream, a vast pattern of currents that carries warm surface waters from the tropical Atlantic northeast towards Europe, decreased in flow by some 10%, thereby transporting less heat to Europe.[318] Foraminifer fossils from sediment cores show that there was a southward shift of the zone of tropical rain that feeds fresh water into the Atlantic. This rain produces a less dense top layer of water that bolsters the surface current flowing north.

In Eastern Europe, pronounced variability in the weather appeared in the 12th Century. In Western Europe, the 14th Century was very wet, especially between 1313 and 1321. In 1315, crops failed. It was wet and cold. Torrential rains removed topsoil. Harvests very commonly failed in the shortened growing season. Areas at altitude and high latitude that had been fertile fields during the Medieval Warming were abandoned because of the lower temperature and increased wetness. Landslides in alpine areas became more common[319] and, together with advancing glaciers, destroyed many villages. With crop failure, famine, the bubonic plague and the collapse of society, the feudal system of Europe started to fall apart. The stressed human population of the Northern Hemisphere was attacked by the plague in 1347. The depopulation was so intense that it took 250 years for the Northern Hemisphere population to return to the levels of 1280. The plague, the Black Death, was the midwife to modern Europe. It appeared in the Dark Ages when it was cold and again appeared in the Little Ice Age. The Little Ice Age probably was the forcing mechanism for the plague. The rapid spread of the plague was assisted by huge segments of the population who had left the fields and lived in cramped quarters in towns.[320]

Marine life migrated as the sea ice advanced. The cod fields that had served the Vikings well retreated, as cod have a limited tolerance to low temperature and suffer kidney failure at temperatures of less than 2°C. In Norway, Greenland and Iceland, the abundant supplies of fish that were an essential source of protein disappeared. Sweden and Finland also had an

[317] Lund, D. C., Stieglitz, J. and Curry, W. B. 2005: Gulf Stream density structure and transport during the last millennium. *EOS* 86: 52.

[318] Lund, D. C., Lynch-Stieglitz, J. and Curry, W. B. 2006: Gulf Stream density, structure and transport during the last millennium. *Nature* 444: 601-644.

[319] Dapples, F., Oswald, D., Raetzo, H., Lardelli, T. and Zwahlen, P. 2003: New records of Holocene landslide activity in the Western and Eastern Swiss Alps: Implications for climate and vegetation changes. *Ecologae Geologicae Helvetiae* 96: 1-9.

[320] Cantor, N. F. 2002: *In the wake of the plague: The black death and the world it made.* Harper Perennial.

expansion of ice and a loss of agricultural land[321] and tax records show that there was destocking and economic collapse in the highlands of Norway.[322] Some five years after Christopher Columbus "discovered" North America, Basque fishing boats were catching cod off the east coast of Canada. They had probably been fishing these waters for centuries before the New World was discovered. In an especially cold period in the mid-late 1600s (Maunder Minimum), the abundant cod disappeared from the waters around the Faeroe Islands.

Land abandonment, crop failure and soil losses were catastrophic because 90% of the population were subsistence farm families who needed enough grain to see them through winter and enough spare grain to sow for the following year's crop. Both the quantity and quality of harvests were vital for survival. Grain rotted in the fields and sometimes couldn't be planted at all. Crop failure led to famine, famine led to disease and death, famine led to a breakdown in society and even cannibalism. Gangs of desperately hungry peasants roamed the countryside searching for food. The harvesting and storage of wet grain, especially rye, stimulated ergot fungus which ruined grain stockpiles. Hungry people ate mouldy grain which contained fungal toxins. This led to ergotism (St Anthony's Fire) which causes convulsions, hallucinations, mass hysteria and death. In extreme cases, internal poisoning from fungal toxins leads to gangrene causing victims' limbs to fall off.[323] Woodcuts from the 14th Century show St Anthony surrounded by detached hands and feet. The ghastly weather was a clear sign to some Christians that Satan was dominating the Earth. Witches were blamed and thousands were burned because it was well known then that witches caused continuous crop failures.[324]

The typical northern European dwelling was a small room with an earthen floor, no insulation, no glass in the windows and a leaky thatched roof. People sat around a central fire on low stools to avoid smoke. Wood was scarce because it required ownership of a forest and metal cutting tools. The overcrowding, dampness, persistently damp clothing, malnutrition, poor sanitation and scarcity of heating fuel were an ideal environment for disease.

---

[321] Pettersson, O. 1914: Climate variations in historic and prehistoric times. *Svenska Hydrografis Biology Konnor Skriften* 5.

[322] Grove, J. and Battagal, A. 1990: Tax records from western Norway as an index of Little Ice Age environmental and economic deterioration. *Climate Change* 5: 265-282.

[323] Garn, S. M. and Leonard, W. R. 1989: What did our ancestors eat? *Nutrition Reviews* 47: 337-345.

[324] Behringer, W. Climate change and witch-hunting. The impact of the Little Ice Age on mentalities. History Department, The University of York http://www.york.ac.uk/depts/hist/staff/wmb1.

Epidemics of typhoid, spread by lice, were more common in winter because malnourished people huddled together in huts to share body warmth and fires. Colds turned to pneumonia. Tuberculosis thrived in crowded areas, as did typhoid, diphtheria and whooping cough. Disease in the Little Ice Age resulted in massive depopulation. And, while huddled around the smoky fire, what did the Europeans talk about? The weather, of course.

The cold weather led to inventiveness. Glass windows were a response to these times. They kept out the cold and still allowed a view of the world. The Dutch thrived as fish migrated from high latitudes into Dutch waters. The repeated violent storms of the 16th and 17th Centuries led the Dutch to develop the technology to reclaim low-lying land from the sea.[325]

The trade and travel of the Medieval Warming ended. At times, the seas were stormier and the waves were higher. There was increased sea ice. On the land, roads became impassable bogs, mountain passes were closed for long periods and trade fairs were a thing of the past. By the 1340s, the sea route between Iceland and Greenland had to follow longer, more southerly routes to avoid ice and treacherous weather.[326] In the 13th and 14th Centuries, fierce storms devastated large tracts of the lowlands of north Germany, Holland and Denmark.[327] More than 100,000 people were killed.

Desperation set in among the Norse colonies in Greenland, where there was a shorter growing season and less grass for cattle and sheep, sea ice prevented seal boats from sailing and there was no timber for fires. The freezing climate and lack of adequate diet had a severe effect on Greenlanders. In Osterbygd, there were some 225 deserted farms in 1500. Skeletons in the graveyards showed that the average height of the Greenlanders decreased by at least 12 cm over the first 200-year period of the Little Ice Age. A study of the chemistry of Viking teeth shows that between 1100 and 1400 there was a 1.5°C drop in temperature.[328,329] Examination of Viking cemeteries showed that with time, the graves became shallower as the permafrost returned.[330] Greenland and Antarctica became stormier and windier at the start of the Little Ice Age, as shown by the increase in sea spray in ice cores.[331] This was

[325] Wolff, W. J. 2006: Netherlands – Wetlands. *Hydrobiologia* 265: 1-14.

[326] Krogh, K. J. 1967: *Viking Greenland*. National Museum of Denmark.

[327] Hebbeln, D., Scheurle, C. and Lamy, F. 2003: Depositional history of Helgoland mud area, German Bight, North Sea. *Geo-Marine Letters* 23: 81-90.

[328] Monastersky, R., 1994: Viking teeth recount sad Greenland tale. *Science News* 19: 310.

[329] Fricke, H. C., O'Neil, J. R. and Lynnerup, N. 1995: Oxygen isotope composition of human tooth enamel from medieval Greenland: Linking climate and society. *Geology* 23: 869-872.

[330] Jones, J. G. 1968: *A history of the Vikings*. Oxford University Press.

[331] Kreutz, K. J., Mayewski, P. A., Meeker, L. D., Twickler, M. S., Whitlow, S. I. and Pittalwala, I. I. 1997: Bipolar changes in atmospheric circulation during the Little Ice Age. *Science* 277: 1294-1296.

the death knell for the Vikings, who had greater difficulty in escaping from Greenland through the pack ice and in the stormier windier seas. Most did not escape and the colony of Greenland, once promoted by Eric the Red in the Medieval Warming, was no longer populated by the immigrants. Only the Inuit people survived.

The second phase of the Little Ice Age (1550–1850 AD) was even colder and more variable. In the middle of the 16th Century a very rapid change occurred. This change is reflected in vegetation. An upland blanket of peat in southern Scotland provides a vegetation and climate record over the last 5500 years and shows 210-year cycles of alternating wet-cool and warm climate with the coldest wettest time in the Little Ice Age during the Spörer Minimum (1450–1540 AD).[332] This coincides with a solar cycle of 210 years in length, the DeVries-Suess Cycle.

The first half of the 16th Century in Europe appears to have been much warmer than the previous 150 years, which had seen a steady decline in temperatures after the Medieval Warm Period. During this early 16th Century warmth, people were able to bathe in the Rhine River in January. A brief warm period in the 1500s allowed the return of ships to Greenland, only to find that the stranded Viking population had starved and frozen to death.

However, this early 16th Century warmth was not to last and a rapid cooling occurred. The winter of 1564–1565 was long and bitter. It heralded many similar winters which brought hardship and social unrest throughout Europe. The next 150 to 200 years was the zenith of the Little Ice Age and temperatures were lower than any other period since the last major ice age.[333] The impact of this sudden cooling in the middle of the 16th Century was widespread.[334] Glaciers advanced rapidly in Greenland, Iceland, Scandinavia and the European Alps. Large areas of high latitude and alpine land were abandoned, snowfall was much higher and snow lay on the ground for many months longer than it does today. Many springs and summers were very cold and wet. Seasons became more variable between years and groups of years. European farmers tried to adapt by changing cropping practices for the shortened, less reliable growing season but there were many years of famine. Violent storms created havoc, flooding and loss of life with some areas along the Danish, German and Dutch coasts lost permanently to the sea. Contemporary painters recorded these scenes. Pieter Brueghel the Elder (1525–1569) started a snow scene genre which included biblical scenes, such

[332] Chambers, F. M., Barber, K. E., Maddy, D. and Brew, J. 1997: A 5500-year proxy-climate and vegetation record from blanket mire at Talla Moss, Borders, Scotland. *The Holocene* 7: 391-399.

[333] Grove, J. M. 1988: *Little ice age*. Routledge Keegan and Paul.

[334] Lamb, H. H., 1982: *Climate, history and the modern world*. Routledge.

as the *Adoration of the Magi* in a snowstorm!

Rapid climate changes were also recorded in Africa. In Ethiopia and Mauritania, permanent snow was reported on mountain peaks at levels where it does not occur today. The Niger River flooded Timbuktu at least 13 times yet there are no records of similar flooding before or since the mid 16th Century. In North America it was a similar story and European settlers reported exceptionally severe winters. From 1607–1608, Lake Superior's ice persisted until mid summer.

If the air is cold, the ground beneath our feet cools. Periods of extreme cold coincided with the Sun's weakest output of energy. Examination of temperature indicators in boreholes in Australia has given a 500-year record of temperature.[335] The 17th Century was the coolest in this 500-year period, with warming in the 19th and 20th Centuries. The warming of Australia over the past five centuries is only about half that experienced by the continents of the Northern Hemisphere in the same period. This geothermal reconstruction agrees with tree ring evidence from Tasmania and New Zealand. Because Australia had no snow cover in the Little Ice Age, its borehole data is far more accurate than borehole data from areas in the Northern Hemisphere. More importantly, the Australian and South Pacific[336] data shows that the Little Ice Age was global. This is contrary to the suggestion that the Little Ice Age was restricted to the Northern Hemisphere and was caused by a weakening in the Gulf Stream.[337] Stalagmites in a cave in the Makapansgat Valley of South Africa show that the region was 1°C cooler from 1300 to 1800 AD. The lowest temperatures recorded in South Africa were in the Maunder and Spörer Minima.[338] Again it is clear that the Little Ice Age was global and not regional.

The Maunder Minimum (1645–1715 AD) was bitterly cold. In China from 1654 to 1676, orange groves that had existed for centuries in Kiangsi Province were abandoned. Cool climate oak forests appeared in Mauritania, suggesting that south of the Sahara Desert it was far cooler and wetter than now, and the water level in Lake Chad was about 4 metres higher than now. In 1676, the artist Abraham Hondius painted hunters pursuing a fox across the frozen Thames River of England. Ice fairs were held on the Thames, the

---

[335] Pollack, H. N., Huang, S. and Smerdon, J. E. 2006: Five centuries of climate change in Australia: The view from underground. *Journal of Quaternary Science* 21: 701-706.

[336] Nunn, P. 2007: *Climate, environment and society in the Pacific during the last millennium.* Developments in Earth and Environmental Sciences 6, Elsevier.

[337] Lund, D. C. 2006: Gulf Stream density structure and transport during the past millennium. *Nature* 444, 601-604.

[338] Tyson, D., Karlen, W., Holmgren, K. and Heiss, G. A. 2001: The Little Ice Age and Medieval Warming in South Africa. *South African Journal of Science* 96: 121-126.

last of them in 1813–1814 towards the end of the Little Ice Age. In 1684, the coast of the English Channel had a 5 km belt of ice. It was so cold in 1695 that ice blocked all the coast of Iceland in January and stayed for most of the year. Cod fishing was not possible, there was little hay and the only source of food was sheep and cattle.[339] One observer in Switzerland in the early 1600s reported that a glacier was advancing daily as far as one could shoot a musket. Between 1695 and 1728, Orkney islanders in far northern Scotland saw Inuits paddling kayaks. One kayaker went as far south as the River Don near Aberdeen. Arctic sea ice had pushed these hunters, seals and cod south.

Famine in Europe killed millions between 1690 and 1700 AD and these were followed by famines in 1725 and 1816.[340] With the eruption of Tambora in Indonesia on 10 April 1815, the situation was only exacerbated. The explosion was heard 850 km away and the top 1400 metres of the volcano was blasted into the air leaving a crater 6 km across and 1 km deep. The blast was equivalent to 60,000 Hiroshima-sized bombs. Tambora launched more than seven times the number of ash particles into the atmosphere than the more famous Krakatoa eruption of 1883. The Indonesian islands were plunged into darkness for two days. Most crops were destroyed by the ash fall and tsunamis. Nearly 10,000 islanders on Tambora were engulfed in ash, all vegetation on Lombok and Bali died and the epidemics and famine in the months following killed more than 80,000 islanders. Contemporary Chinese records show that at Hainan Island, 2000 km north of Tambora, the Sun disappeared. The combination of low temperatures, excessive rainfall and unseasonal frosts played havoc with subsistence farming. China experienced an exceptionally cold and stormy winter in 1816–1817 with disastrous crop failures.

The large volume of fine particles of dust filled the atmosphere in both hemispheres, reflected light and reflected heat, as did a monstrous volume of sulphuric acid droplets. Ash was trapped in the Greenland ice sheet. It was at this time that landscape artists such as J.M.W. Turner painted brilliant sunsets and stormy seas. The winter of 1815–1816 was known as "the year without a summer". Three long cold periods ravaged Canada and the New England region of the USA. The first, in June, killed most crops. The second, in July, killed replanted crops and the third, in August, killed corn, potatoes, beans and vines. Severe cold and crop failures ranged from North America to the Ottoman Empire in the Middle East, into parts of North Africa and

[339] Thoroddsen, P. 1908-1922: *The climate of Iceland through one thousand years*. Karysmannaliofn, Reykjavik (Vol 1, 1916-1917; Vol 2, 1908-1922).

[340] Ladurie, L. and Laduries, E. 1971: *Times of feast, times of famine*. Noonday Press, New York.

across Europe. Typhus epidemics followed the crop failures and the bubonic plague again appeared. The Rev. Ezra Styles, the president of Yale University, started keeping daily temperatures from 1779. The June 1816 measurements were the coldest thermometer measurement ever recorded in Connecticut, some 2.6°C lower than the 1780–1968 long-term mean, and 1816 was the coldest year on record in the USA.

Europe was still recovering from the disruption brought on by the Napoleonic Wars, which took place during a period of cool wet years.[341] The cold years of 1816–1817 created a food crisis and widespread unrest, especially in France. This drove immigration from Europe to the USA and American farmers from northern to warmer latitudes. Average temperatures in the UK were 2°C lower and it rained or snowed almost every day. Prices on the London Grain Exchange skyrocketed. Crop failures in Bengal in 1816 resulted in famine which triggered a major outbreak of cholera. This spread from Bengal and was the world's first cholera pandemic. It reached northwestern Europe, Russia and the eastern USA in the summer of 1832. At Lake Geneva, the poet Lord Byron and his guests Mary and Percy Shelley used the gloomy summer of 1816 to write. Mary Shelley wrote *Frankenstein* and Byron wrote *Darkness*.[342] Tambora gives a pretty bleak picture into the effect of an equatorial volcanic eruption or a small asteroidal impact. It also provides a window into the effects of global cooling.

A tree ring study shows that growth in high-altitude European forests slowed because of the cold in the Little Ice Age. Elsewhere forests became stressed and were replaced by tundra.[343] Between 1625 and 1720 in the Maunder Minimum, trees showed exceptionally narrow growth rings producing dense and strong wood. These properties may have enhanced the quality of the violins made by Stradivarius, who produced his most famous instruments between 1700 and 1720.[344]

---

[341] The Dalton Minimum.

[342] I had a dream, which was not all a dream,
The bright star was extinguish'd, and the stars
Did wander darking in the eternal space,
Rayless, and pathless, and the icy earth
Swung blind and blackening in the moonless air;
Morn came and went – and came, and brought no day,
And men forgot their passions in the dread
Of this their desolation; and all hearts
Were chill'd into a selfish prayer for light.

[343] Campbell, I. D. and McAndrews, J. H. 1993: Forest disequilibrium caused by rapid Little Ice Age cooling. *Nature* 366: 336-338.

[344] Burckle, L. and Grissino-Mayer, H. 2003: Stradivari, violins, tree rings, and the Maunder Minimum: a hypothesis. *Dendrochronologicia* 21: 41-45.

Low clouds occur during cold times. In a remarkable study of clouds by a meteorologist who studied more than 6000 landscape paintings in galleries in Europe and North America painted from 1400 to 1967, a statistical analysis showed a slow increase in cloudiness between the early 15th and mid 16th Centuries (i.e. Spörer Minimum). Low clouds increased after 1550 and declined after 1850 (end of Little Ice Age). Summer in the 18th and 19th Centuries showed that 50 to 75% of the summer sky was covered with cloud[345] (Maunder and Dalton Minima). Although artists surely took some licence with their subjects, none of the British paintings viewed showed a clear sky whereas some 12% of Mediterranean paintings showed a completely clear sky. The data showed an increase in cloudiness between 1400 and 1550 and then an abrupt further increase (more than 50%), especially in the abundance of low clouds. Cloudiness peaked in the 17th Century (Maunder Minimum).

Landslides form in colder times when the process of freezing and thawing moves unconsolidated material and in spring when unconsolidated material is destabilised by meltwater acting as a lubricant. Most major landslides in the Swiss Alps occurred in a cold period before the Roman Warming, the Dark Ages and the Little Ice Age.[346]

The most recent cool period, the Little Ice Age, and preceding climate changes are well recorded in ocean floor sediments off West Africa, Greenland ice cores, Swiss landslides, seafloor sediments from the North Atlantic Ocean, seafloor sediments from the Arabian Sea, cave stalagmites from Germany and Ireland, sea surface temperatures and plankton in the Sulu Sea.[347] For example, the Sargasso Sea shows that sea surface temperature was about 1°C lower than today in the Little Ice Age and about 1°C warmer than now in the Medieval Warming.[348] Life in lakes, fjords[349] and oceans[350] adapted to the colder climate of the Little Ice Age and the rapid climate variations.

[345] Neuberger, H. 1970: Climate in art. *Weather* 25: 46-56.

[346] Dapples, F., Oswald, D., Raetzo, H., Lardelli, T. and Zwahlen, P. 2003: New record of Holocene landslide activity in the western and eastern Swiss Alps: Implication of climate and vegetation changes. *Ecologae Geologicae Helvetiae* 96: 1-9.

[347] Rosenthal, Y., Oppo, D. W. and Linsley, B. K. 2003: The amplitude and phasing of climate change during the last deglaciation in the Sulu Sea, western equatorial Pacific. *Geophysical Research Letters* 30: doi:10.1029/2002GL016612.

[348] Keigwin, L. D. 1996: The Little Ice Age and Medieval Warm Period in the Sargasso Sea. *Science* 274: 1503-1508.

[349] Jensen, K. G., Kuijpers, A., Koc, N. and Heinemeier, J. 2004: Diatom evidence of hydrographic changes and ice conditions in Igaliku Fjord, South Greenland, during the past 1500 years. *The Holocene* 14: 152-164.

[350] Lassen, S. J., Kuijpers, A., Kunzendorf, H., Hoffman-Wieck, G., Mikkelsen, N. and Konradi, P. 2004: Late-Holocene Atlantic bottom-water variability in Igaliku Fjord, South Greenland, reconstructed from foraminiferal faunas. *The Holocene* 14: 165-171.

Sediments from volcanic crater lakes in Uganda show that the Little Ice Age in Africa had a number of cool and warm periods[351] and periods of intense rainfall and drought.[352] While Europe was cold and wet, tropical central Africa was cold and dry. Tropical South America also enjoyed a fluctuating and cooler drier climate in the Little Ice Age.[353]

While in the grip of the Little Ice Age, there were 10 eruptions from the Laki craters (Iceland) between June 1783 and February 1784. Not only was Europe covered in a dry sulphuric acid fog as a result, there was additional cooling and damage to vegetation.[354,355,356] These eruptions created a cascade of events that led to record low levels of water in the Nile River in Africa. Unusual temperature and rainfall patterns peaked in 1783 causing below normal rainfall in most of the Nile source areas. Europe was colder, tree rings in Alaska and Siberia show stunted growth, and a lack of monsoons led to a reduction in cloud cover of the Sahel of Africa, the southern Arabian Peninsula and India.[357]

A study of Vermetid reefs off the Sicilian coast in the Mediterranean documented the history of sea surface temperature over the last few centuries.[358] The Maunder Minimum and the Late 20th Century Warming could be seen as a chemical signature in the reefs.

During the Little Ice Age, Greenland was very cold and Antarctica was relatively warm. Between 1550 and 1700, melting of Greenland ice was common with 8% of the years experiencing melting and elevated summer

[351] Russell, J. M., Verschuren, D. and Eggermont, H. 2007: Spatial complexity of 'Little Ice Age' climate in East Africa: sedimentary records from two crater lake basins in western Uganda. *The Holocene* 17: 183-193.

[352] Russell, J. M. and Johnson, T. C. 2007: Little Ice Age drought in equatorial Africa: Intertropical Convergence Zone migrations and El Niño-Southern Oscillation variations. *Geology* 35: 21-24.

[353] Polisaar, P. J., Abbott, M. B., Wolfe, A. P., Bezada, M., Rull, V. and Bradley, R. S. 2006: Solar modulation of Little Ice Age climate in the tropical Andes. *Proceedings of the National Academy of Sciences* 103: 8,937-8,942.

[354] Grattan, J. and Charman, D. J. 1994: Non-climatic factors and the environmental impact of volcanic volatiles: implications of the Laki fissure eruption of AD 1783. *The Holocene* 4: 101-106.

[355] Jacoby, G. C., Workman, K. W. and D'Arrigo, R. D. 1999: Laki eruption in 1783, tree rings, and disaster for northwest Alaska Inuit. *Quaternary Science Reviews* 18: 1365-1371.

[356] Thordarson, T. and Self, S. 1999: Atmospheric and environmental effects of the 1783-1784 Laki eruption: A review and reassessment. *Journal of Geophysical Research* 108: doi:10.1029/2001JD002042

[357] Oman, L., Robcock, A., Stenchivov, G. L. and Thordarson, T. 2006: High-latitude eruptions cast shadow over the African monsoon and the flow of the Nile. *Geophysical Research Letters* 33: L18711, doi:10.1029/2006GL027665.

[358] Silenzi, S., Antonioli, F. and Chemello, R. 2005: A new marker for sea surface temperature trend during the last centuries in temperate areas: Vermetid reef. *Global and Planetary Change* 40: 105-114.

temperatures.[359] Although the Little Ice Age was global, the Antarctic Anomaly was apparent. Notwithstanding, the Little Ice Age was felt in the Southern Hemisphere. The transition from the Medieval Warming to the Little Ice Age at around 1300 in the South Pacific created social upheaval[360] which was followed by famine, migration and depopulation.[361]

Ships' logbooks provide a daily detailed insight on the weather and climate of the Little Ice Age.[362,363,364] A study of more than 6000 logbooks ranging from Nelson's *Victory* and Cook's *Endeavour* to the humblest frigate provide a contemporary unified record of wind force, direction, precipitation and notes about the weather. Most records also show air pressure, air temperature and sea surface temperature. These logs show that the decades of the 1680s and 1690s were the coldest for 1000 years and that there was a surge in summer storms.[365] This was in the Maunder Minimum (1645–1715 AD).

The popular belief was that hurricanes form in the east Atlantic Ocean and track westwards. It was a surprise when, in 2005, Hurricane Vince moved northeast and hit southern Spain and Portugal. For many, this was proof of unusual weather conditions derived from human emissions of $CO_2$. However, the same happened in 1842, well before industrial emissions of $CO_2$. The 100,000 Royal Navy logbooks from 1670 to 1850 are a goldmine of information as are the 900 logbooks of the East India Company covering 1780 to 1840. These can provide a global perspective (e.g. Robert FitzRoy's two expeditions in HMS *Beagle* in the 1820s and 1830s with Charles Darwin), information on the Arctic, East and West Indies and the Mediterranean Sea (Horatio Nelson's voyages) and information on the Pacific Ocean (Captain James Cook in the 1760s and 1770s).

The Little Ice Age was savage. It was global. Conditions changed from pleasant warmth to bitter cold in just two decades. This dramatic climate

[359] Joughun, I., Das, S. B., King, M. A., Smith, B. E., Howat, I. M. and Moon, T. 2008: Seasonal speedup along the western flank of the Greenland ice sheet. *Science* 320: 781-783.

[360] Nunn, P. D. 2007: *Climate, environment and society in the Pacific during the last millennium.* Developments in Earth and Environmental Sciences 6, Elsevier.

[361] Nunn, P. D. 2008: *Vanished islands and hidden continents of the Pacific.* University of Hawaii Press.

[362] Wheeler, D. 2001: The weather of the European Atlantic seaboard during October, 1805: An exercise in historical climatology. *Climate Change* 48: 361-365.

[363] Woodruff, S. D., Diaz, H. F., Worley, S. J., Reynolds, R. W. and Lubker, S. J. 2005: Early ship observational data and ICOADS. *Climate Change* 73: 169-194.

[364] García-Herrera, R., Können, G. P., Wheeler, D. A., Prieto, M. R., Jones, P. D. and Koek, F. B. 2005: CLIWOC: A climatological database for the world's oceans 1750-1854. *Climate Change* 73: 1-12.

[365] Wheeler, D. and Suarez-Dominguez, J. 2006: Climatic reconstructions for the northeast Atlantic region AD 1685-1700: a new source of evidence from naval logbooks. *The Holocene* 16: 39-49.

change was initiated by changes in solar activity, and the coldest periods in the Little Ice Age were when the Sun was relatively inactive (see Chapter 3).

The Little Ice Age brought famine, disease, death, depopulation, war and social disintegration. The previous cooling, the Dark Ages, did the same. Over the last 1000 years in Europe, there is a correlation between violent conflict, cold weather and precipitation.[366,367] Cold times bring violence, war, depopulation and human misery.

## The Late 20th Century Warming  (1850 AD to the present)

The Earth is recovering from the Little Ice Age. The Late 20th Century Warming has just finished. An analysis of 102 scientific studies of the Late 20th Century Warming showed that 78% of the studies found that earlier periods lasting at least 50 years were warmer than any period in the 20th Century. Three studies stated that the 20th Century was the warmest century and four studies rated the early part of the 20th Century, before humans released much $CO_2$ into the air, as the warmest part of the 20th Century. The Late 20th Century Warming was nothing unusual.

As with previous climate changes, the Late 20th Century Warming has not been a period of steady warming. As with previous climate changes, the evidence shows that the primary driver is a change in solar activity (see Chapter 3). There was warming from 1850–1940, cooling from 1940–1976, warming from 1976–1998 and cooling since 1998.

The evidence from history, archaeology and science is overwhelming. It shows substantial changes in climate over the last 130,000 years. Many of these changes are cyclical and coincidental with solar cycles. These changes are rapid. The evidence from geology is also overwhelming. Since the explosion of multicellular life (at 542 Ma), there were times when Earth was far colder and far warmer than now. It also includes times when atmospheric $CO_2$ was far higher than now. Records of climate sensitivity are based on measurements over a few decades to thousands of years ago when $CO_2$ levels and temperature were similar to or lower than today. In the past, rises in atmospheric temperature, accompanied by a rise in atmospheric $CO_2$, may have increased the rate of weathering of silicate minerals at the Earth's surface. Increased weathering increases $H_2O$ and $CO_2$ consumption by minerals, thereby removing $CO_2$ from the atmosphere (negative feedback).

---

[366] Zhang, D. D., Brecke, P., Lee, H. F., He, Y.-Q. and Zhang, J. 2007: Global climate change, war, and population decline in recent history. *Proceedings of the National Academy of Sciences* 4: 19,214-19,219.

[367] Tol, R. S. J. 2008: Climate change and violent conflict in Europe over the last millennium. http://ideas.repec.org/p/sgc/wpaper/154.html

Regardless of whether climate is viewed from a historical, archaeological or geological time perspective, climate sensitivity greater than 1.5°C has been a robust feature of the Earth's climate system over the last 420 million years.[368]

If it is acknowledged that there have been rapid large climate changes before industrialisation, then the human production of $CO_2$ cannot be the major driver for climate change. The evidence is overwhelming that another mechanism or combination of mechanisms drives climate change, such as variations in solar activity, cosmic ray input, orbit and terrestrial processes. This being the case, then the whole purpose of the IPCC ceases to exist.

One infamous scientific paper[369] attempted to rewrite the climate record, leaving out the awkward Little Ice Age and Medieval Warming, and attributing $CO_2$ from industry as the only driving force of climate change.

## The long tale of the lone pine[370]

The Roman Warming, the Dark Ages, the Medieval Warming and the Little Ice Age invalidate all arguments supporting human-induced global warming. This is because climates far warmer than the Late 20th Century Warming existed before industrialisation and human emissions of $CO_2$. The notion that climate change is tied only to human activity with known atmospheric and ocean feedbacks is a simple and erroneous explanation of modern and ancient climates. To argue that modern climate is driven by slight changes in a trace gas in the atmosphere ($CO_2$) requires many non-scientific leaps of faith. History commonly is rewritten for nefarious reasons and this is exactly what happened with the climate cycles over the last 2000 years.[371]

In the IPCC Second Assessment *Summary for Policymakers* in 1996, a diagram showing the past 1000 years of Earth temperatures from tree rings, ice cores and thermometers showed the Medieval Warming, the Little Ice Age and the Late 20th Century Warming. Although we might argue about integrating proxies for temperature (tree rings, ice cores) with thermometer measurements (and all the limitations of these measurements), the diagram showed us what we knew from history.

Five years later, the IPCC's Climate Change Report presented a totally different diagram for the past 1000 years of Earth temperatures. The

---

[368] Royer, D. L., Berner, R. A. and Park, J. 2007: Climate sensitivity constrained by $CO_2$ concentrations over the past 420 million years. *Nature* 446: 530-532.

[369] Mann, M. E., Bradley, R. S. and Hughes, M. K. 1998: Global-scale temperature patterns and climate forcing over the past six centuries. *Nature* 392: 779-787.

[370] For the complete tenacious exposure of the whole "hockey stick" debacle, a series of postings on www.climateaudit.org is illuminating.

[371] Bradley, R. S. 2003: Climate in Medieval time. *Science* 302: 404-405.

Medieval Warming and the Little Ice Age had been expunged from history and a significant warming from 1910 onwards was the highlight of the diagram. The diagram, which looked like a hockey stick, had a great visual impact and the implication was that the runaway temperature rise in the 20th Century was due to industrialisation. The next IPCC Report omitted the "hockey stick" without explanation.

This "hockey stick" diagram derives from a 1998 study by Michael Mann and colleagues.[372] Mann was a recent graduate and this study brought him fame, fortune and great prestige. At an early age, he became editor of a major scientific journal[373] and became an IPCC lead author. Mann was a hero, and was besieged by the media. It is worthwhile to evaluate the study critically. For the period from 1000 to 1980. Mann mainly used tree ring data as a temperature proxy. This assumes that in warmer times trees grow more vigorously than in colder times. To this were added thermometer measurements, mainly from urban areas, to cover the period after 1908. The expunging of the Medieval Warming and the Little Ice Age was contrary to thousands of historical records and scientific studies, yet these were reduced to insignificance in the Mann study. The *IPCC Climate Change 2001 Report* (Section 2.3.3 "Was there a 'Little Ice Age' and 'Medieval Warm Period'?") restricted the Medieval Warming and Little Ice Age to very slight changes that occurred only in the North Atlantic region and suggested that they were due to slightly altered patterns of atmospheric circulation.

The IPCC uncritically accepted the "hockey stick" and rejected, without explanation, the thousands of scientific studies on the Medieval Warming and the Little Ice Age. The Medieval Warming had been dismissed. It was a nuisance. The impact of the Little Ice Age was greatly diminished and was attributed to "constricted" global circulation. I am sure that the millions of people who died of cold or cold-induced famine in the Little Ice Age would be relieved. The IPCC then claimed that there was unprecedented warming in the 20th Century. The "hockey stick" showed it. The IPCC used the "hockey stick" on the first page of the *Summary for Policymakers* and displayed it four more times, in some places occupying half a page. Clearly the IPCC endorsed the "hockey stick". In 2000, Mann's "hockey stick" was featured in a US government report.[374]

---

[372] Mann, M. E., Bradley, R. S. and Hughes, M. K. 1998: Global-scale temperature patterns and climate forcing over the past six centuries. *Nature* 392: 779-787.

[373] *The Journal of Climate.*

[374] US National Assessment of the Potential Consequences of Climate Variability and Change, 2000.

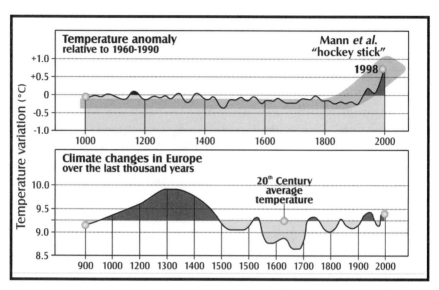

*Figure 11: The Mann et al. "hockey stick" (upper diagram) which neither records the Medieval Warming (900–1280 AD) nor the Little Ice Age (1300–1850 AD) and yet shows an abrupt 20ᵗʰ Century warming. By contrast, the temperature history of the last 1000 years derived from hundreds of studies shows the Medieval Warming, the intense cold of the Little Ice Age (with short-lived warm pulses) and the Late 20th Century Warming. The Mann et al. "hockey stick" was not in accord with hundreds of previous validated studies yet failed to show why it differed. This "hockey stick" was the icon of the IPCC.*

The methodology of science is such that new data and the resulting conclusions are critically analysed, repeated, refined or rejected. This "hockey stick" graphic was contrary to conclusions derived from thousands of studies using boreholes in ice, lakes, rivers and oceans, glacial deposits, flood deposits, sea level data, soils, volcanoes, wind blown sand, isotopes, pollen, peat, fossils, cave deposits, agriculture and contemporary records. When extraordinary conclusions are made, there needs to be extraordinary data in support.

This is exactly what happened with the Mann study. It was demolished on the basis of statistics.[375] Two Canadians, Steven McIntyre and Ross McKitrick, requested the original data from Mann that underpinned his study. This was like extracting teeth. After much bluster, stonewalling and hiding behind the veil of confidentiality, the data was provided in dribs and drabs. The original data set provided for validation and repeatability, a normal process of science, was incomplete. Because US federal funds had been used to support Mann's study, by law the data had to be made available.

---

[375] McIntyre, S. and McKitrick, R. 2003: Corrections to the Mann et al. "Proxy data base and Northern Hemisphere temperature series, 1998". *Energy and Environment* 14: 751-771.

In other jurisdictions, it may not be possible to obtain the primary data for government-supported research.

It seemed clear that no reviewer of the Mann *et al.* paper in *Nature* had requested the original data upon which the paper was based, for otherwise *Nature* would not have published a paper using such incomplete data. This is not the place to speculate on whether this was a lapse in editorial standards or whether *Nature* was following another agenda. However, extraordinary conclusions and the dismissal of thousands of previous scientific studies on the Medieval Warming and Little Ice Age should have stimulated reviewers and editors of *Nature* to view the primary data and calculations as a normal part of scientific due diligence.

McIntyre and McKitrick found that the Mann data did not produce the claimed results:

> due to collation errors, unjustifiable truncation or extrapolation of source data, obsolete data, geographical location errors, incorrect calculation of principal components and other quality control defects.

The IPCC used the Mann diagram in 2001 as the central tool to show that human-induced global warming started in the 20th Century. It is clear that Mann's data used to construct the "hockey stick" was meaningless, that adequate due diligence was not undertaken by the authors, reviewers and editors.

Using Mann's own data, McIntyre and McKitrick showed that the warming in the early 15th Century exceeds any warming of the 20th Century. McIntyre and McKitrick showed that the Mann study gave great weight to the 20th Century tree ring data from the Sierra Nevada Mountains of California. This data was collected by others and was not compared with thermometer measurements that existed for that area but was compared with thermometer measurements from urban areas. The trees used were ancient, slow-growing, high-altitude bristlecone pine trees. Such trees can live up to 5000 years, hence are ideal for climate studies. The trees showed a growth spurt after 1910. The research on the bristlecone trees[376] was used by Mann to show that temperature started to increase in 1910. However, the original paper used by Mann not only demonstrated that the tree ring temperature proxy must be used with caution, but also that the bristlecone tree ring data showing a post-1910 growth spurt could not be explained by local or regional

---

[376] Graybill, D. A. and Idso, S. B. 1993: Detecting the aerial fertilisation effect of atmospheric $CO_2$ enrichment in tree ring chronologies. *Global Biogeochemical Cycles* 7: 81-95.

temperature changes. This was ignored by Mann.

The explanation for the growth spurt was that the bristlecone pine grows at the limit of moisture and fertility at altitude and hence shows strong responses to $CO_2$ fertilisation. This was the point of the bristlecone study. This point could not have been missed by Mann, because it is the title of the paper from which Mann derived the critical data to show the post-1910 temperature rise.

Mann *et al.*[377] issued a "correction" later which admitted that their proxy data contained some errors but "none of these errors affect our previously published results". This means that Mann was quite happy to publish work that he had either not checked or he knew was wrong. Mann was unable and unprepared to argue against the statistics of McIntyre and McKitrick and dogmatically stated that he was correct. He did not address the issue that bristlecone pine growth, his principal data set for his "hockey stick", was unrelated to temperature.

The "hockey stick" graphic used by the IPCC sent a very misleading message to the public. Furthermore, the 1996 IPCC report showed the Medieval Warming and the Little Ice Age. Mann's "hockey stick" was used in the IPCC's 2001 report[378] and the Medieval Warming and Little Ice Age were expunged from the record of modern climates. In the next IPCC report, the Medieval Warming and Little Ice Age mysteriously reappeared.

This suggests that the IPCC knew that the "hockey stick" was invalid. This is a withering condemnation of the IPCC. The "hockey stick" was used as the backdrop for announcements about human-induced climate change,[379] it is still used by Al Gore, and it is still used in talks, on websites and in publications by those claiming that the world is getting warmer due to human activities. Were any of those people who view this graphic told that the data before 1421 AD was based on just one lonely alpine pine tree?

Mann had not released all his data and calculation methods to McIntyre and McKitrick, and was reported in public as stating that he would not be intimidated into disclosing the algorithm by which he obtained his results. This attracted the interest of the US House Energy and Commerce Committee.[380] Its members read the McIntyre and McKitrick articles and became concerned about allegations that Mann had withheld adverse statistical results and that

---

[377] Mann, M., Bradley, R. S. and Hughes, M. K. 2004: Corrigendum: Global-scale temperature patterns and climate forcing over the past six centuries. *Nature* 430: 105.

[378] IPCC Climate Change 2001: The Scientific Basis; Figure 2.20.

[379] It was the first image in the *U.S. National Assessment of the Potential Consequences of Climate Variability* (2000).

[380] http://energycommerce.house.gov/108/home/07142006_Wegman_fact_sheet.pdf

his results depended upon bristlecone pine ring widths, well known to be a questionable measure of temperature. In June 2005, they sent questions to Mann and his co-authors about verification statistics and bristlecone pines, asked Mann for the algorithm he used, and asked *pro forma* questions about federal funds used in their research. This caused a storm with allegations of intimidation. Various learned societies, none of which had been offended by Mann's public refusal to provide full disclosure, were outraged that a House committee (representing the taxpayers who had paid for the results) should be trying to find out how Mann derived his results.

A turf war started. The House Science Committee felt its jurisdiction had been impinged upon. After a few months of battles, the House Science Committee asked the National Academy of Sciences (NAS) to evaluate criticism of Mann's work and to assess the larger issue of historical climate data reconstructions. The NAS agreed but only under terms that precluded a direct investigation of the issues that prompted the original dispute – whether Mann *et al.* had withheld adverse results and whether the data and methodological information necessary for replication were available.

In the March 2006 hearings of the NAS, no claim of McIntyre and McKitrick was refuted. However, the NAS issued a press release[381] on 22 June 2006 stating:

> There is sufficient evidence from tree rings, boreholes, retreating glaciers, and other 'proxies' of past surface temperatures to say with a high level of confidence that the last few decades of the 20th Century were warmer than any comparable period in the last 400 years.

What was not said was that planet Earth was in the grip of the Little Ice Age 400 years ago and that it is no surprise that temperature increased when the Little Ice Age ended. The NAS statement could be very misleading to the general public. The report accurately suggested that temperature measurements with instruments date back only 150 years and hence other proxies such as coral, ocean and lake sediments, ice cores, cave deposits, and documentary sources such as cave paintings must be used. The report also states that the globally averaged warming of 0.6°C is reflected in proxies but does not mention that during the last 150 years, the global average temperature has both risen and fallen. What was also not said is that proxies must be used with great caution. As with any method of science, there are limitations. Tree rings are very commonly used to indicate past climates. However, when

---

[381] www.nationalacademies.org/oninews/newsitem.aspx?RecordID=11676

modern trees are measured there is a huge variability of responses to climate change. Fossil tree ring studies are probably inadequate in distinguishing a palaeoclimate signal from the background of variability.[382] Great caution needs to be expressed in the use of tree ring proxies unless there are associated well-constrained palaeoecological studies. This caution was not exercised by Mann *et al.* nor was there caution expressed by the NAS.

However, the devil lies in the detail and the NAS press release is very different from the body of the report. The report concedes that almost every criticism of the Mann work is well founded and does not agree with the Mann statistics for temperature changes over decades and especially individual years. The NAS committee concludes that it is plausible that:

> the Northern Hemisphere was warmer during the last few decades of the 20th Century than during any comparable period over the preceding millennium.

Again, the devil lies in the detail and the studies used to reach this conclusion suffer from the same methodological and data problems of Mann, which were conceded by the NAS committee. In the political heat, it would not have been politically possible for the NAS committee to state that the Mann *et al.* papers were fraudulent, wrong or biased. This would have unstitched the IPCC. However, the detailed NAS report shows extensive criticism of the methodology of Mann and states:

> Some of these criticisms are more relevant than others, but taken together, they are an important aspect of a more general finding of this committee, which is that uncertainties of the published reconstructions have been underestimated.

The House Energy and Commerce Committee appointed an eminent team of statisticians led by Dr Edward Wegman to investigate.[383] The conclusions of the Wegman investigation were confirmed by another independent statistical analysis of Mann's data.[384] Wegman's committee had

---

[382] Falcon-Lang, H. J. 2006: Global climate analysis of growth rings in woods, and its implications for deep-time palaeoclimate studies. *Paleobiology* 31: 434-444.

[383] US Congress House Committee on Energy and Commerce, http://republicans.energycommerce.house.gov/108/home/07142006_Wegman_Report.pdf

[384] North, G. R. 2006: NRC, 2006: *Committee on surface temperature reconstructions for the last 2,000 years*. National Research Council, National Academies Press.

some interesting statements about the Mann *et al.* publication:[385]

> It is important to note the isolation of the palaeoclimate community; even though they rely heavily on statistical methods they do not seem to be interacting with the statistical community. Additionally, we judge that the sharing of research materials, data and results was haphazardly and grudgingly done. In this case we judge that there was too much reliance on peer review,

---

[385] Executive summary:

The Chairman of the Committee on Energy and Commerce as well as the Chairman of the Subcommittee on Oversight and Investigations have been interested in an independent verification of the critiques of Mann *et al.* (1998, 1999) [MBH98, MBH99] by McIntyre and McKitrick (2003, 2005a, 2005b) [MM03, MM05a, MM05b] as well as the related implications in the assessment. The conclusions from MBH98, MBH99 were featured in the *Intergovernmental Panel on Climate Change* report entitled *Climate Change 2001: The Scientific Basis*. This report concerns the rise in global temperatures, specifically during the 1990s. The MBH98 and MBH99 papers are focused on paleoclimate temperature reconstruction and conclusions therein focus on what appear to be a rapid rise in global temperature during the 1990s when compared with temperatures of the previous millennium. These conclusions generated a highly polarised debate over the policy implications of MBH98, MBH99 for the nature of global climate change, and whether or not anthropogenic actions are the source. This committee, composed of Edward J. Wegman (George Mason University), David W. Scott (Rice University) and Yasmin H. Said (The John Hopkins University), has reviewed the work of both articles, as well as a network of journal articles that are related either by authors or subject matter, and has come to several conclusions and recommendations. This Ad Hoc Committee has worked pro bono, has received no compensation, and has no financial interest in the outcome of the report.

Recommendations

*Recommendation 1*: When massive amounts of public monies and human lives are at stake, academic work should have a more intense level of scrutiny and review. It is especially the case that authors of policy-related documents like the *IPCC report, Climate Change 2001: The Scientific Basis*, should not be the same people as those that constructed the academic papers.

*Recommendation 2*: We believe that federally funded research agencies should develop a more comprehensive and concise policy on disclosure. All of us writing this report have been federally funded. Our experience with funding agencies has been that they do not in general articulate clear guidelines to the investigators as to what must be disclosed.

Federally funded work including code should be made available to other researchers upon reasonable request, especially if the intellectual property has no commercial value. Some consideration should be granted to data collectors to have exclusive use of their data for one or two years, prior to publication. But data collected under federal support should be made publicly available. (As federal agencies such as NASA do routinely).

*Recommendation 3*: With clinical trials for drugs and devices to be approved for human use by the FDA, review and consultations with statisticians is expected. Indeed, it is standard practice to include statisticians in the application-for-approval process. We judge this to be good policy when public health and also when substantial amounts of monies are involved, for example, when there are major policy decisions to be made based on statistical assessments. In such cases, evaluation by statisticians should be standard practice. This evaluation phase should be a mandatory part of all grant applications and funded accordingly.

*Recommendation 4:* Emphasis should be placed on the Federal funding of research related to fundamental understanding of the mechanisms of climate change. Funding should focus on interdisciplinary teams and avoid narrowly focused discipline research.

which was not necessarily independent. Moreover, the work has been sufficiently politicised that this community can hardly reassess their public positions without losing credibility. Overall, our committee believes that Dr Mann's assessments that the decade of the 1990s was the hottest decade of the millennium and that 1998 was the hottest year of the millennium cannot be supported by his analysis.[386]

It appears that the science of Mann is poorly communicated.

The papers of Mann *et al.* in themselves are written in a confusing manner, making it difficult for the reader to discern the actual methodology and what uncertainty is actually associated with these reconstructions. Vague terms such as 'moderate certainty' (Mann *et al.* 1999) give no guidelines to the reader as to how such conclusions should be weighed. While the works do not have supplementary websites, they rely heavily on the reader's ability to piece together the work and methodology from raw data. This is especially unsettling when the findings of these works are said to have global impact, yet only a small population could truly understand them. Thus, it is no surprise that Mann *et al.* claim a misunderstanding of their work by McIntyre and McKitrick.

and

In their works, Mann *et al.* describe the possible causes of global climate change in terms of atmospheric forcings, such as anthropogenic, volcanic or solar forcings. Another questionable aspect of these works is that linear relationships are assumed in all forcing-climate relationships. This is a significantly simplified model for something as complex as the earth's climate, which most likely has complicated non-linear cyclical processes on a multi-centennial scale that we do not yet understand. Mann *et al.* also infer that since there is a partial correlation between global mean temperatures in the 20th century and $CO_2$ concentration, greenhouse-gas forcing is the dominant external forcing of the climate system. Osborn and Briffa make a similar statement, where they casually note that evidence for warming also occurs at a period where $CO_2$ concentrations are high. A common phrase among statisticians is correlation does not imply causation. Making conclusive statements without specific findings with

---

[386] www.climateaudit.org and http://scienceandpublicpolicy.rg/monckton/what_hockey-stick.html

regard to atmospheric forcings suggests a lack of scientific rigor and possibly an agenda.

and

Specifically, global warming and its potentially negative consequences have been central concerns of both governments and individuals. The 'hockey stick' reconstruction of temperature graphic dramatically illustrated the global warming issue and was adopted by the IPCC and many governments as the poster graphic. The graphic's prominence together with the fact that it is based on incorrect use of PCA puts Dr Mann and his co-authors in a difficult face-saving problem.

The network analysis of Mann and 42 other authors by Wegman's statisticians shows diagrammatically how they formed a closed coterie, who not only co-authored but also refereed each other's publications. This phenomenon is, of course, not new, but has never been so powerful in world affairs.

The report finds that:

a.  Mann *et al.* misused certain statistical methods in their studies which inappropriately produce "hockey stick" shapes in the temperature history.

b.  The claim that the 1990s were the warmest decade of the millennium could not be substantiated.

c.  The cycle of the Medieval Warm Period and the Little Ice Age disappeared from Mann *et al.* analysis thereby making it possible to make the claim about the hottest decade.

d.  A social network analysis revealed that the small community of palaeoclimate researchers appear to review each other's work, and reuse many of the same data sets, which calls into question the independence of peer review and temperature reconstructions.

e.  It is clear that many of the proxies are re-used in most of the papers. It is not surprising that the papers would obtain similar results and so cannot claim to be independent verifications.

f.  Although the researchers rely heavily on statistical methods, they do not seem to be interacting with the statistical community. The public policy implications of this debate are financially staggering and yet apparently no independent statistical independent expertise was sought or used.

g.  Authors of policy-related science assessments should not assess their own work. It is especially the case that authors of policy-related documents like the IPCC report, *Climate Change 2001: The Scientific Basis* should not be the same people that constructed the academic papers. Policy-related climate science should have a more intense level of scrutiny and review involving statisticians.

h. Federal research should involve interdisciplinary teams to avoid narrowly focused discipline research.

i. Federal research should emphasise fundamental understanding of the mechanisms of climate change and should focus on interdisciplinary teams to avoid narrowly focused discipline research.

j. While the palaeoclimate reconstruction has gathered much publicity because it reinforces a policy agenda, it does not provide insight and understanding of the physical methods of climate change.

The Chairman of the NAS committee was later asked at the US Senate House Energy and Commerce hearings whether or not the NAS agreed with Wegman's harsh criticisms.

*Chairman Barton:* Dr North, do you dispute the conclusions or the methodology of Dr Wegman's report?

*Dr North:* No we don't. We don't disagree with their criticism. In fact, pretty much the same thing is said in our report.

*Dr Bloomfield:* Our committee reviewed the methodology used by Dr Mann and his co-workers and we felt that some of the choices were inappropriate. We had much the same misgivings about his work that was documented at much greater length by Dr Wegman.

Mann claims that the NAS panel vindicated him.

In many fields of science, this would have been considered as fraud. In many fields of endeavour, Mann would have been struck off the list of practitioners. In the field of climate studies, he was thrashed in public with a feather and still gainfully practises his art. Mann should be grateful for being dealt with in such a gentle manner, given his rather thuggish behaviour in trying to prevent valid criticism being published. I'm sure St Peter will judge Mann accordingly!

A dispassionate reading of Dr Steve McIntyre's exposure of Mann shows the systematically dishonest manner in which the "hockey stick" graph was used to show that it was far warmer today than in the Medieval Warming. This was adopted as the poster child for climate panic by the IPCC in 2001 and retained in the 2007 report despite having been demolished in the scientific literature. The original work of McIntyre and McKitrick[387] showing that Mann *et al.* were, at best, misleading has been expanded[388,389] and independently validated by many others.[390,391,392,393,394,395] After reading the history of the "hockey stick",[396] no one could ever again trust the IPCC or the scientists and environmental extremists who author the climate assessments. The IPCC has encouraged a collapse of rigour, objectivity and honesty that were once the hallmarks of the scientific community. McKitrick stated[397] that had the IPCC undertaken the kind of rigorous review that they boast of:

> they would have discovered that there was an error in a routine calculation step (principal component analysis) that falsely identified a hockey stick shape as the dominant pattern in the data. The flawed computer program can even pull out spurious hockey stick shapes from lists of trendless random numbers.

The modern media barrage has conditioned us to think that we are approaching an unprecedented catastrophic warming and that we humans can actually change climate.

Those who claim the Earth is suffering human-induced global warming cite NASA's Goddard Institute of Space Studies (GISS) as an authority to support their beliefs. The GISS director[398] claimed that nine of the ten

---

[387] McIntyre, S. and McKitrick, R. 2003: Corrections to the Mann et al., "Proxy data base and Northern Hemisphere temperature series, 1998". *Energy and Environment* 14: 751-771.

[388] McIntyre, S. and McKitrick, R. 2005a: The M&M critique of the MB1198 Northern Hemisphere Climate Index: Update and implications. *Energy and Environment* 16: 69-100.

[389] McIntyre, S. and McKitrick, R. 2005b: Hockey sticks, principal components and spurious significance. *Geophysical Research Letters* 32: L03710, doi: 10.1029/2004GL021750.

[390] Von Storch, H., Zarita, E., Jones, J. M., Dimitriev, Y., Gonza'lez Rouco, F. and Tett, S. F. B. 2004: Reconstructing past climate from noisy data. *Science* 306: 679-682.

[391] Bürger, G. and Cubasch, U. 2005: Are multiproxy climate reconstructions robust? *Geophysical Research Letters* 32: L23711, doi: 10.1029/2005GL024155.

[392] Von Storch, H. and Zarita, E. 2005: Comment on "Hockey stick principal components, and spurious significance" by S. McIntyre and R. McKitrick. *Geophysical Research Letters* 32: L20701, doi: 10.1029/2005GL022753.

[393] Bürger, G., Fast, I and Cubasch, U. 2006: Climate reconstructions by regression – 32 variations on a theme. *Tellus* 58A: 227-235.

[394] Huybers, Y. 2005: Comment on "Hockey stick principal components, and spurious significance" by S. McIntyre and R. McKitrick. *Geophysical Research Letters* 32: L20705, doi: 10.1029/2005GL023395.

[395] Crok, M. 2005: Kyoto Protocol based on flawed statistics. *Natuurwetenskap Techniek* 2: 20-31.

[396] www.climateaudit.org and http://scienceandpublicpolicy.rg/monckton/what_hockey-stick.html

[397] R. McKitrick, in evidence given to the House of Lords Select Committee on Economic Affairs, *The Economics of Climate Change, Volume II: Evidence* (2005).

[398] Hansen, J., Ruedy, R., Glascoe, J. and Sato, M. 1999: GISS analysis of surface temperature change. *Journal of Geophysical Research* 104: 30,997-31,022.

warmest years in history have occurred since 1995, with the warmest being 1998. This was accompanied by a huge media fanfare. When NASA had to reverse its position on the basis of the work undertaken by Toronto-based statistician Steve McIntyre,[399] there was no fanfare. NASA now states that the top four years of high temperatures are from the 1930s (1934, 1931, 1938 and 1939). The warmest year was 1934, The years 1998, 1921, 2006, 1999 and 1953 were also warm. Several previously allegedly warm years (2000, 2002, 2003 and 2004) are now cool years. Similarly, the UK's Meteorological Office has now confirmed a fall in average global temperatures since 1998, despite a 25% increase in the burning of coal, oil and natural gas which produced voluminous $CO_2$ additions to the atmosphere. These facts are only uncomfortable if history is ignored, including the last ten years which provided the best data ever collected.

Mother Nature does not obey computer models and ideology. Declarations that a particular year was the warmest ever are nothing more than calculating an average value of temperatures recorded at measuring stations. Such calculations can be misleading since the distribution of observation points over land and ocean is uneven and there are large areas of the Earth that have few measurements.

Data from the 3000 scientific robots in the world's oceans shows that there has been a slight cooling over the past five years. While we are getting hot and bothered about a possible global warming, we are ignoring the announcements by Nature of the next inevitable global cooling.[400] It has happened before, it will happen again. Quickly. And all we can do is to adapt, as we have done in the past.

The 2008 draft *Global Change Impacts in the United States* report (co-chaired by Thomas R. Karl, Jerry Melillo and Thomas C. Peterson) states: "Historical climate and weather patterns are no longer an adequate guide to the future."

History cannot be rewritten just because it does not fit a computer model with a pre-ordained conclusion.

[399] Holland, D. 2007: Bias and concealment in the IPCC process: The "hockey-stick" affair and its implications. *Energy and Environment* 18: 951-983.

[400] Kerr, R. 1999: The Little Ice Age: Only the latest big chill. *Science* 269: 1431-1433.

# Chapter 3

# THE SUN

Question: Does the Sun influence the Earth's climate?
Answer: Yes.

*T*he Sun is the primary driving force of climate. The Sun provides Earth
with a staggering amount of energy. It drives weather, ocean currents and
evaporation and provides the energy for life on Earth. The Sun also prevents the
oceans freezing or boiling.

Global warming occurs on other planets and moons in our Solar System.
It cannot be related to human emissions of $CO_2$ on Earth. Planets orbiting
stars outside our Solar System also show global warming resulting from orbital
variations and changes in energy emitted from the parent star.

Very slight changes in solar energy output have a profound effect on the
Earth's climate. An energetic Sun blasts away cosmic radiation, there is less
low-level cloud, and the planet reflects less energy back into space. The surface
of the Earth warms. A weak Sun allows cosmic radiation to form low-level
clouds which reflect energy back into space and the surface of Earth cools.
This phenomenon has been calculated and validated by experiments and
observations. Clouds are the engine of weather. The Earth also has a variable
flux of galactic cosmic rays.

Candidates for climate drivers are the variable Sun (solar driver), planetary
perturbations (Milankovitch forcing) and variable cosmic ray flux (cosmic ray
forcing). The effects of greenhouse gases in the atmosphere piggy-backs on the
principal drivers of climate and may amplify changes.

The solar driving and cosmic ray forcing of climate are seen globally on
geological, archaeological, historical and modern time scales. The solar cycles
of 11, 22, 87, 210 and 1500 years have been detected in ice sheets, ice melting,
floods, droughts, lake sediments, deep sea sediments, cave deposits, boreholes,
tree rings, pollen, peat and floating organisms in both the Northern and

*Southern Hemisphere. There is no relationship between atmospheric $CO_2$ and temperature over time.*

*The 23 climate models of the IPCC ignore or minimise the role of the Sun. All the models failed to predict cooling in the early 21st Century. None of the models predicted El Niño-La Niña events which transfer huge amounts of energy around the planet's surface.*

## The bringer of life, heat and cold

There is a big incandescent thermonuclear reactor in the sky that emits huge amounts of energy to the Earth. The Sun provides enough energy to power the great oceanic and atmospheric currents, the cycle of evaporation and condensation that brings fresh water inland and drives river flow, hurricanes and tornadoes that destroy our natural and built landscape. The Sun provides the energy for photosynthesis. The Sun is the bringer of life to Earth. If the Sun were more energetic, the oceans would boil. If the Sun were less energetic, they would freeze and all life on Earth would be destroyed. The Sun drives weather on our planet and drives climate. Changes in the Sun's output result in changes on Earth on time scales less than a human lifetime.

Every second, the Sun delivers to Earth the total amount of energy released by an earthquake of Richter magnitude 8. The amount of energy humans use annually is delivered from the Sun to the Earth in one hour. The known recoverable resource of oil contains the energy that the Sun delivers to the Earth in 36 hours.

We are all very much aware that it is hotter in the blazing sunshine than when there is cloud cover. We also know that in the humid tropics, summer air temperature is much lower than in deserts at higher latitudes where the air is dry. The maximum temperatures recorded on Earth are in deserts at mid latitudes, not in the tropics. We also know that a humid winter tropical night is far warmer than a winter night in the dry desert at the same latitude. How many of us have camped out in the desert, frying during the day and freezing at night because of the dry air and lack of cloud cover? It is clear that clouds and humidity combined with the energy emitted by the Sun affect air temperature. Clouds reflect radiated heat from the Sun back into space. This causes the planet to cool. It was thought that clouds were caused by climate change but measurements, calculations and experiments now show that cosmic radiation forms clouds.[401,402] Clouds are one of the major drivers of climate change.

---

[401] Svensmark, H., Pedersen, J. O. P., Marsh, N. D., Enghoff, M. B. and Uggerhøf, U. I. 2007: Experimental evidence for the role of ions in particle nucleation under atmospheric conditions. *Proceedings of the Royal Society Journal A: Mathematical, Physical and Engineering Sciences* 463: 385-396.

[402] Svensmark, H. and Calder, N. 2007: *The chilling stars: A new theory on climate change*. Icon.

However, there are other external influences that affect climate on the Earth. These are the passage of the Solar System through our galaxy. This results in Earth acquiring variable amounts of space junk (mainly dust) and being bombarded by variable amounts of galactic cosmic rays from past supernova explosions.

## Dirt in the air

There is an extraordinary amount of dust in the air. Each breath we take contains 50 million particles. Particles are more abundant in city, coastal and urban areas and less abundant in polar, marine and desert areas. Most particles are from human activities although, in some areas, emissions from plants and the oceans dominate. Volcanoes and soils also deliver particles to the atmosphere. Many particles derive from elsewhere in our Solar System.[403]

Planet Earth is not a spaceship. Material from space is being constantly added to Earth. Outer space may have a far greater influence on Earth processes than we intuitively think. Measurements from high-flying aircraft show that some 40,000 tonnes of extraterrestrial dust drops to Earth each year.[404] This dust carries minerals, amino acids (the building blocks of life) and chemicals that can only form from blasting by cosmic rays. Besides bringing dust, comets also bring an unknown amount of water and $CO_2$ to Earth.[405] Ice sheets, lake sediments and deep-sea sediments are ideal places to measure the input of extraterrestrial dust over time. This dust also picks up atoms of helium 3 ($He^3$), an isotope that can only form in outer space, and the comings and goings of supernova, comets and meteors can be tracked by $He^3$ in the Earth's ocean and lake sediments.[406,407,408,409] Some of the original Solar System $He^3$ leaks out from deep in the Earth, mainly at mid ocean ridges, and could also be absorbed onto atmospheric dust particles. Dust particles coming from space to Earth are about one hundredth the diameter of a

[403] Buseck, P. R. and Adachi, K. 2008: Nanoparticles in the atmosphere. *Elements* 4: 389-394.

[404] Parkin, I. W. and Tilles, D. 1968: Influx measurement of extraterrestrial material. *Science* 159: 936-946.

[405] Hut, P. 1987: Comet showers as a cause of mass extinctions. *Nature* 329: 118-125.

[406] O'Sullivan, D., Zhou, D. and Flood, E. 2001: Investigation of cosmic rays and their secondaries at aircraft altitudes. *Radiation Measurements* 34: 277-280.

[407] Ozima, M., Takayanagi, M., Zashu, S. and Amari, S. 1984: High $^3He/^4He$ ratio in ocean sediments. *Nature* 311: 448-450.

[408] Matsuda, J., Murota, M. and Nagao, K. 1989: Investigation of high He-3/He-4 ratio in deep sea sediments. *Antarctic Meteorites* XIV: 139-141.

[409] Farley, K. A., Montanari, A., Shoemaker, E. M. and Shoemaker, C. S. 1998: Geochemical evidence for a comet shower in the Late Eocene. *Science* 280: 1250-1253.

human hair. Planet Earth has been increasing in mass since the beginning of time due to the addition of dust, meteors and comets.[410]

Since 1992, the Ulysses spacecraft has been monitoring the stream of stardust flowing through our Solar System. A dust grain takes 20 years to traverse the Solar System. The DUST experiment on board Ulysses shows that the stream of stardust is highly affected by the Sun's magnetic field. Stardust increases during solar maxima. The DUST experiment on Ulysses showed a three-fold increase during the last solar maximum.[411] Collision of these high-speed dust particles with asteroids, meteors and comets creates more dust particles.

Another source of dust is from within the Solar System. The Earth rotates around the Sun inside the zodiac dust cloud. This cloud is between the Sun and the main belt of asteroids, located between Mars and Jupiter. The inner planets, asteroids and meteor fluxes pass through the zodiac cloud. There are more than 2.5 million comets in the Solar System and these are a major source of the dust.[412]

Why worry about a few specks of dust? Dust particles (as well as water droplets and ice) in the atmosphere reflect solar radiation back into space. Dust particles are also the condensation centres for water droplets. Dust particles from large volcanic eruptions (e.g. Krakatoa, 1883; Vesuvius, 1906, 1944; Agung, 1963; Pinatubo, 1991) have a short residence time in the atmosphere of only a few years. After a large explosive volcanic eruption, there can be significant cooling for a few years after ejection of material up to 40 km into the atmosphere.[413] Volcanic eruptions are sporadic. However, they not only add dust to the atmosphere but also add sulphate aerosols which can also cool the planet.[414]

Extraterrestrial dust falls all the time, albeit at a varying rate depending upon the position of the Solar System within the galactic arm, the strength of the solar wind and cometary fragmentation. There is a correlation between atmospheric dust in both Greenland and Antarctic ice over the last 420,000

[410] Cloud, P. 1968: Atmospheric and hydrostatic evolution of the primitive Earth. *Science* 160: 1135-1143.

[411] Landgraf, M., Kruger, H., Altobelli, N. and Grun, E.. 2003: Penetration of the heliosphere by the interstellar dust stream during solar maximum. *Journal of Geophysical Research* 108: 8030

[412] Gorbanov, Y. M. and Knyaskova, E. F. 2003: Young meteorite swarms near the Sun. I. Statistical relationship of meteors with families of short perihelion comets. *Astronomichesky vestnik* 37: 555-568 (in Russian).

[413] Lamb, H. H. 1970: Volcanic dust in the atmosphere; with a chronology and assessment of its meteorological significance. *Philosophical Transactions of the Royal Society of London. Series A, Mathematical and Physical Sciences* 266: 425-533.

[414] Handler, P. 1989: The effect of volcanic aerosols on global climate. *Journal of Volcanology and Geothermal Research* 37: 233-249.

years with cyclical changes every 100,000 years.[415,416] More recent studies show an 800,000-year correlation between dust and cold climate.[417] Tree ring studies in the Arctic of Fennoscandia and Siberia show that there is a correlation between the variation in stardust in the Solar System, solar activity and tree growth.[418] The 22-year cycle of solar magnetic field reversals allows a window of opportunity for extra stardust to penetrate the solar defences and there is increasing rain of stardust onto Earth.

Ice sheets are good places to construct the history of dust. Even the Little Ice Age (1280–1850 AD) can be detected in both Greenland and Antarctic ice cores. There was far less terrestrial dust and sea spray in the Medieval Warming (900–1300 AD) than in the Little Ice Age. Ice cores show that the more dust in the atmosphere, the colder the climate and glacial conditions have over 50 times as much dust in the atmosphere as interglacial times.[419]

During periods of glaciation, there is less water vapour in air. Hence, there is less rainfall and less vegetation. This results in more desertification and more dust. In the past, global warming has not produced more desertification. It is global cooling that creates desertification. The 20th Century dust deposition in Antarctic ice was higher than in the 19th Century. The dust content is still more than 30 times that from glacial times.[420] Because of the short residence time of terrestrial dust in the atmosphere, terrestrial volcanic dust from one eruption could not influence climate over longer periods of time. However, if there are numerous eruptions from many volcanoes at almost the same

[415] Fuhrer, K., Wolff, E. W. and Johnsen, S. J. 1999: Timescales for dust variability in the Greenland Ice Core Project (GRIP) in the last 100,000 years. *Journal of Geophysical Research* 104: 31043-31052.

[416] Petit, J. R., Jouzel, J., Raynaud, D., Barkov, N. I., Barnola, J. M., Basile, I., Bender, M., Chappalez, J., Davisk, M., Delaygue, G., Delmotte, M., Kotlyakov, V. M., Legrand, M., Lipenkov, V. Y., Lorius, C., Pépin, L., Ritz, C., Saltzmark, E. and Stievenard, M. 1999: Climate and atmospheric history of the past 420,000 years from the Vostok ice core, Antarctica. *Nature* 399: 429-436.

[417] Lambert, F., Delmonte, B., Petit, J. R., Bigler, M., Kaufmann, P. R., Hutterli, M. A., Stocker, T. F., Ruth, U., Steffensen, J. P. and Maggi, V. 2008: Dust-climate couplings over the past 800,000 years from the EPICA Dome C ice core. *Nature* 452: 616-619.

[418] Kastakina, E., Shumilov, O., Lukina, N. V., Krapiec, M. and Jacoby, G. 2007: Interstellar dust: a significant driver of climate change? *Dendrochronologica* 24: 131-135.

[419] Petit, J. R., Jouzel, J., Raynaud, D., Barkov, N. I., Barnola, J. M., Basile, I., Bender, M., Chappalez, J., Davisk, M., Delaygue, G., Delmotte, M., Kotlyakov, V. M., Legrand, M., Lipenkov, V. Y., Lorius, C., Pépin, L., Ritz, C., Saltzmark, E. and Stievenard, M. 1999: Climate and atmospheric history of the past 420,000 years from the Vostok ice core, Antarctica. *Nature* 399: 429-436.

[420] McConnell, J. R., Aristarain, A. J., Banta, J. R., Edwards, P. R. and Simöes, J. C. 2007: 20th Century doubling in dust archived in an Antarctic Peninsula ice core parallels climate change and desertification in South America. *Proceedings of the National Academy of Sciences* 104: 5743-5748.

time, this could trigger a climate change. The 1960s had 21 great eruptions, including three in 1963. In the 1980s, there were 15 great eruptions, including three in 1983. These were not enough to trigger climate change.

The constant flux of extraterrestrial dust into the atmosphere is a likely candidate for cooling the Earth. Not only would this dust come from the zodiac dust cloud but the entry of meteors and comets into the Earth's upper atmosphere also adds dust as these extraterrestrial visitors "burn" into dust. Cycles of extraterrestrial dust at 194, 64, 32 and 21 years correspond to solar cycles suggesting input of dust during cosmic ray bombardment.[421]

The Solar System rotates around the galaxy centre and crosses galactic arms. The amount of material inside the galactic arm is more than outside the galactic arm and the gravitational influence of this matter increases the influence of comets on the Solar System.

## Snowballs and spiral arms

To understand modern climate, we must understand ancient climate. And the biggest climate changes in the history of the Earth were during the Neoproterozoic some 750 to 635 Ma.

The fundamental questions for the Neoproterozoic glaciation is: How did planet Earth become glaciated at sea level and at the equator, then have a very hot interglacial period with sea temperature at +40°C and an associated sea level rise of at least 600 metres, then drop into another glaciation and then have another hot interglacial? There is great discussion in the scientific literature about the triggers and effects of the Neoproterozoic glaciation.[422,423,424,425,426,427,428,429]

[421] Reach, W. T. 1988: Zodiacal emission. I- Dust near the earth's orbit. *Astrophysical Journal* 335: 468-485.

[422] Oglesby, R. J. and Ogg, J. G. 1988: The effect of large fluctuations in obliquity on climates of the Late Proterozoic. *Paleoclimates* 24: 293-316.

[423] Rubincam, D. P. 1995: Has climate changed the Earth's tilt? *Paleoceanography* 10: 365-372.

[424] Schmidt, P. W. and Williams, G. W. 1995: The Neoproterozoic climate paradox: Equatorial palaeolatitude for Marinoan glaciation near sea level in South Australia. *Earth and Planetary Science Letters* 134: 107-124.

[425] Veevers, J. J. 1990: Tectonic-climate supercycle in the billion year plate-tectonic eon: Permian Pangean icehouse alternates with Cretaceous dispersed continents greenhouse. *Sedimentary Geology* 68: 1-68.

[426] Williams, G. E. 1975: Late Precambrian glacial climate and the Earth's obliquity. *Geological Magazine* 112: 441-465.

[427] Williams, G. E. 1993: History of the Earth's obliquity. *Earth Science Reviews* 24: 1-45.

[428] Young, G. M. 1991: The geological record of glaciation: Relevance to climatic history of Earth. *Geoscience Canada* 18: 100-108.

[429] Hoffman, P. F., Kaufman, A. J., Halverson, G. P. and Schrag, D. P. 1998: A Neoproterozoic snowball Earth. *Science* 281: 1342-1346.

A recent suggestion for the Neoproterozoic glaciation takes a galactic view. Four major arms or segments of arms in our galaxy cross the path of the Solar System in its 26 km per second passage through the Milky Way. During the Neoproterozoic glaciation, the Solar System passed through the Sagittarius-Carina Arm of the Milky Way. This would have produced great variations in Solar System dust and variations in the input of cloud-forming cosmic rays.

It was not until the 1950s that measurements of hydrogen gas in space showed that our galaxy, the Milky Way, is a spiral galaxy. Gravitational forces between stars generate waves of dense and less dense matter, which create the spiral, and the spiral rotates around the centre of the Milky Way. The density waves disrupt the interstellar gas producing dense clouds from which stars form. These short-lived bright blue stars along the arms of the galaxy are unstable and form supernovae which, after explosion, emit cosmic rays.

## Galactic time travel

Over the last 545 Ma, the Solar System has crossed spiral galactic arms four times. The Israeli astronomer Nir J. Shaviv[430] proposed that a large cosmic ray flux should occur from passages of the Solar System through the Milky Way's spiral arms that harbour most of the star formation activity. Such passages occur in cycles of $143 \pm 10$ million years.[431] The Canadian isotope geochemist Jan Veizer[432] independently noted climate variations on cycles of $135 \pm 9$ million years in the chemical evolution of seawater. He later suggested that over the last 545 Ma, there has been no relationship between atmospheric $CO_2$ and temperature.[433] In fact, over the last 545 Ma, the atmospheric $CO_2$ content has been up to 25 times greater than now. There has been neither runaway greenhouse nor extinctions due to this high atmospheric $CO_2$. For the last 545 Ma, the atmospheric $CO_2$ content has been decreasing, and it is now at an all-time low because the Earth has been efficiently sequestering $CO_2$ into sedimentary rocks.[434]

---

[430] Shaviv, N. J. 2002: Cosmic ray diffusion from the galactic spiral arms, iron meteorites, and a possible climate connection. *Physical Review Letters* 89: 51-102.

[431] Shaviv, N. J. 2002: The spiral structure of the Milky Way, cosmic rays, and ice age epochs on Earth. *New Astronomy* 8: 39-77.

[432] Veizer, J., Ala, D., Axmy, K., Bruckschen, P., Buhl, D., Bruhn, F., Carden, G. A. F., Diener, A., Ebneth, S., Goddéris, Y., Jasper, T., Korte, C., Pawellek, F., Podlaha, O. G. and Strauss, H. 1999: $^{87}Sr/^{86}Sr$, $\partial^{13}C$ and $\partial^{18}O$ evolution of Phanerozoic seawater. *Chemical Geology* 16: 158-188.

[433] Veizer, J., Godderis, Y. and François, L. M. 2000: Evidence for decoupling of atmospheric $CO_2$ and global climate during the Phanerozoic eon. *Nature* 408: 698-701.

[434] Flower, B. P. 1999: Warming without $CO_2$. *Nature* 399: 313-314.

When Shaviv and Veizer, two scientists in different fields and from different parts of the world, came together they showed that at least 66% of the variation in temperature over the last 545 Ma was due to cosmic ray variations as the Solar System passed through the spiral arms.[435] Integration of various disciplines of science commonly creates surprises. Such surprises always show that in any area of science, the science is never settled.

During the Neoproterozoic glaciation (750–635 Ma), the Solar System was in the Sagittarius-Carina Arm. During the Ordovician-Silurian glaciation (450–420 Ma), there was an encounter with the Perseus Arm. Glaciation at 300–260 Ma in the Permo-Carboniferous (Norma Arm) and at 151–132 Ma in the Jurassic to early Cretaceous[436] Period (Scutum-Crux Arm) can also be related to encounters with spiral galactic arms. The early Cretaceous glaciation was mild, probably due to a high atmospheric $CO_2$ content[437] and a quick crossing of the Scutum-Crux Arm. Cooling in the Miocene (Sagittarius-Carina Arm) was quickly followed by the Pleistocene glaciation (Orion Spur). There is a suggestion that the onset of the most intense period of the Pleistocene glaciation at 2.75 Ma was from cosmic rays emitted from a nearby supernova.[438] The question remains: Is the crossing of a galactic arm enough to produce glaciation? Are other factors needed?

The Solar System has been in Gould's Belt for the past few million years. This is where the Sun and the Earth have been bombarded by cosmic rays from short-lived massive exploding stars. The Sun oscillates about the galactic mid-plane and the coolest phase, every 34 million years, corresponds with a crossing of the mid-plane where cosmic rays are most intense.

This raises fascinating questions about the origin of tropical glaciations at sea level at about 2400–2100 Ma (Huronian Glaciation) and 750 Ma (Cryogenian or Neoproterozoic Glaciation). After each of these glaciations, there was an increase in atmospheric oxygen and rapid evolution of life. At the same time as the Huronian Glaciation, there was a mini-starburst some 2400–2000 Ma. There was then a billion-year period when the rate of star formation decreased greatly. During this time, there was no glaciation. Even

---

[435] Shaviv, N. J. and Veizer, J. 2003: Celestial driver of Phanerozoic climate. *GSA Today* 13: 4-10.

[436] Alley, N. F. and Frakes, L. A. 2003: First known Cretaceous glaciation: Livingston Tillite Member of the Cadna-powie Formation, South Australia. *Australian Journal of Earth Sciences* 50: 139-144.

[437] Royer, D. L., Berner, R. A., Montañez, I. P., Tabor, N. J. and Beerling, D. J. 2004: $CO_2$ as a primary driver of Phanerozoic climate. *GSA Today* 14: 4-10.

[438] Knie, K., Korschinek, G., Faestermann, T., Dorfi, E. A., Rugel, G. and Wallner, A. 2004: $^{60}$Fe anomaly in deep-sea manganese crust and implications for a nearby supernova source. *Physical Review Letters* 93: 171103-171107.

though spiral arms may have been visited, there were not enough cosmic rays derived from recently exploded supernovae to create icehouse conditions. There was a very high star formation rate in the Milky Way, the highest since the Earth was formed,[439] peaking at about 750 Ma. At this time, the Earth commenced the Neoproterozoic glaciation,[440,441] the most intense glaciation of all time. The Neoproterozoic glaciation some 750–635 Ma may have rendered the Earth into a snowball, or at least a slush ball.[442]

This model creates a problem. When both the Earth and the Sun were young at 4000 Ma, the Earth was not frozen. The Sun emitted 25% less energy than now, hence the Earth's surface temperature should have frozen all surface water. But there is abundant evidence to show that the Earth had liquid water, possibly as far back as 4400 Ma. Sedimentary rocks in Greenland show that there was certainly liquid water at 3800 Ma.[443] This paradox, the faint Sun paradox, is explained by the Earth having a $CO_2$-dominant atmosphere and a high heat flow.[444,445,446] However, evidence from the carbon chemistry of life and sediments shows that this was not the case.[447] An alternative view is that planet Earth had fewer clouds at that period of time thereby producing warming, especially in the oceans.[448,449] This means that there was little or no cloud on early Earth.[450]

In the very distant future, the Sun and the Earth will again pass through the Perseus Arm and, further in the future, the Norma, Scutum-Crux and

---

[439] Marcos, R. de la Fuente and Marcos, C. de la Fuente 2004: On the recent star formation history of the Milky Way disc. *New Astronomy* 9: 475-502.

[440] Rocha-Pinto, H. J., Flynn, C., Scale, J., Hänninen, J., Maciel, W. J. and Hensler, G. 2004: Chemical enrichment and star formation in the Milky Way disc. III. Chemodynamical constraints. *Astronomy and Astrophysics* 423: 517-535.

[441] Shaviv, N. J. and Veizer, J. 2003: Celestial driver of Phanerozoic climate? *GSA Today* 13: 4-10.

[442] Hoffman, P. F., Kaufman, A. J., Halverson, G. P. and Schrag, D. P. 1998: A Neoproterozoic snowball Earth. *Science* 281: 1342-1346.

[443] Cloud, P. 1988: *Oasis in Space. Earth history from the beginning*. W. W. Norton and Co.

[444] Kaufman, A. J. 1997: An ice age in the tropics. *Nature* 386: 227-228.

[445] Crowley, T. J. and Baum, S. K. 1993: Effect of decreased solar luminosity on late Precambrian ice extent. *Journal of Geophysical Research* 98: 16723-16732.

[446] Kasting, J. F. and Ackerman, T. P. 1986: Climatic consequences of very high carbon dioxide levels in the Earth's early atmosphere. *Science* 234: 1383-1385.

[447] Veizer, J. 2005: Celestial climate driver: A perspective from four billion years of carbon cycle. *Geoscience Canada* 32: 13-28

[448] Rossow, W. B., Henderson-Sellers, A. and Weinreich, S. K. 1982: Cloud feed-back: A stabilising effect for the early earth? *Science* 217: 1245-1247.

[449] Shaviv, N. J. 2002: Cosmic ray diffusion from the galactic spiral arms, iron meteorites, and a possible climatic connection. *Physical Review Letters* 89: 51102-51106.

[450] Marsh, N. and Svensmark, H. 2003: Galactic cosmic ray and El Niño-Southern Oscillation trends in International Satellite Cloud Climatology Project D2 low-cloud properties. *Journal of Geophysical Research* 108: doi:10.1029/2001JD001264.

Sagittarius-Carina Arms. There will again be ice ages.

Don't wait up.

## Galactic bullets

The Earth is constantly bombarded with extraterrestrial atomic bullets. Most of these are deflected by the Sun's magnetic shield, the solar wind and the Earth's magnetic field. However, some atomic bullets penetrate all these defences. If we ignore the role of the Sun and the cosmos in our study of the Earth, then we simplify the interrelated dynamic processes on our planet to the point of absurdity.[451]

The hypothesis that the global warming of the past century is man-made is based on the results of computer models in which the main drivers of climate are not adequately considered. This is clear from the RSS AMSU satellite measurements that show that global temperature has not increased since 1998, despite the continuing massive increase in $CO_2$ emissions. Balloon measurements also show this decrease. The most important climate driver (besides solar luminosity) comes from the interplay of the Earth's orbit with solar activity, interplanetary magnetic field strength, cosmic ray intensity and cloud cover in the Earth's atmosphere. Many of these phenomena are generated by galactic density waves affecting the Sun's core.

Whenever the Solar System passes a relatively dense cloud of interstellar gas during its galactic travels, the magnetic field around the Sun is reduced in size from way outside the Solar System to near Jupiter. This reduction to almost a quarter of its diameter results in a doubling of the amount of cosmic radiation hitting the Solar System. At present, the Solar System is in a region where there is little interstellar gas and a large dense cloud of interstellar gas will not be encountered for another million years.

When cosmic radiation hits nitrogen in the atmosphere, it forms the short-lived carbon isotope ($C^{14}$).[452] The highest rate of $C^{14}$ production takes place at altitudes of 9 to 15 km and at high geomagnetic latitudes. The $C^{14}$ readily mixes and becomes evenly distributed throughout the atmosphere and reacts with oxygen to form radioactive carbon dioxide. When cosmic rays hit dust particles and sea spray in the atmosphere, the traces of boron in the dust

---

[451] Scherer, K., Fichtner, H., Bormann, T., Beer, J., Desorgher, L., Flükiger, E., Fahr, H.-J., Ferreira, S. E. S., Langner, U. W., Potgieter, M. S., Heber, B., Masarik, J., Shaviv, N. and Veizer, J. 2007: Interstellar-terrestrial relations; Variable cosmic environments, the dynamic heliosphere and their imprints on terrestrial archives and climate. *Space Science Reviews* 127: 327- 465.

[452] When cosmic rays enter the atmosphere, they undergo various transformations, including the production of neutrons. The neutrons react with nitrogen: $n + N^{14} = C^{14} + H^1$. Carbon 14 has a half life of 5,730 years.

and sea spray are converted to a short-lived isotope of beryllium ($Be^{10}$). $Be^{10}$ produced in the atmosphere by cosmic rays readily attaches to the aerosols and ends up in polar snow. It is an indicator of the variations in the cosmic ray flux. The decrease in $Be^{10}$ since 1900 reflects the decrease in the cosmic ray flux over this period and this, in turn, is related to solar flux. A measurement of $C^{14}$ in fossils, tree rings, bogs and peat and $Be^{10}$ in muds and dust in ice gives an indication of the intensity of past cosmic ray bombardment. More $C^{14}$ and $Be^{10}$ are formed when the Sun is inactive and unable to blast away cosmic rays. More clouds are formed in these times of solar inactivity and the planet is cooler. There is a very tight correlation between $C^{14}$ and $Be^{10}$ in a range of materials and global temperatures, suggesting that solar activity has been a driver of past surface temperature on Earth.[453] The $C^{14}$ and $Be^{10}$ measurements on materials up to tens of thousands of years old have shown a variable Sun, cycles of solar energy output and cycles of cooling and warming.

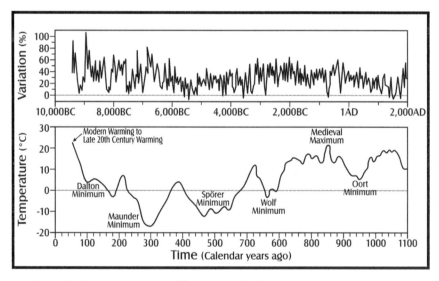

*Figure 12: Plot of the variation in $C^{14}$ over the last 12,000 years (upper diagram) showing that either the Sun's energy changes or the flux of cosmic rays striking the Earth changes. The last 1000 years shows that when it was exceptionally cold in the Little Ice Age (lower diagram), there was little sunspot activity (Oort, Wolf, Spörer, Maunder and Dalton Minima) and warmer times (Medieval and modern times) were characterised by many sunspots.*

[453] Parker, G. 1999: The sunny side of global warming. *Nature* 399: 416.

However, these radioactive elements may be changed by processes such as variations in the Earth's magnetic field and hence a number of reconstructions of past solar energy output using $C^{14}$ and $Be^{10}$ differ. There is a loose correlation between climate and the Earth's magnetic field, mainly driven by wobbles in the Earth's orbit changing the behaviour of the Earth's core. This in turn changes the Earth's magnetic field.[454] Although this is a speculative climate forcing mechanism, a change in the Earth's magnetic field changes the amount of solar and cosmic material that can enter the Earth, hence the Earth's magnetic field can indirectly affect climate.[455] Since atomic bomb testing began in 1945, there have been more than 2200 atomic tests which have added $C^{14}$ to the atmosphere, life, soils and rocks.

Problems with $C^{14}$ and $Be^{10}$ measurement can be reconciled by looking into space. Meteors buzz around in space before they fall to Earth. When in space, they are bombarded by cosmic rays which create new radioactive isotopes in the meteor. Titanium 44 ($Ti^{44}$) is one of these isotopes. The beauty of this technique is that the $Ti^{44}$ formed in space is unaffected by processes that occurred on Earth. By measuring the amount of $Ti^{44}$ in meteorites that have fallen to Earth over the last 240 years, the cosmic ray activity was plotted against solar activity and it was shown that the Sun's activity has increased over the last 100 years.[456]

Cosmic rays form electrically charged particles when they hit the atmosphere. These particles attract water molecules from the air which clump together until they condense as clouds. The cosmic ray assistance in the forming of clouds is over decadal, centennial and millennial time scales whereas, over longer time intervals, the changing galactic environment of the Solar System has dramatic climate effects.[457] A good example of this can be seen in a Wilson cloud chamber where cosmic rays form cloud traces (as do particles from radioactive breakdown).

The number of cosmic particles hitting Earth changes with the changing magnetic activity of the Sun. The Sun operates as the cosmic gatekeeper

[454] Mursula, K. and Zieger, B. 2001: Long-term north-south asymmetry in solar wind speed inferred from geomagnetic activity: A new type of century-scale solar oscillation? *Geophysical Research Letters* 28: 95-98.

[455] Courtillot, V., Gallet, Y., Mouël, J.-L., Fluteau, F. and Genevey, G. 2006: Are there connections between the Earth's magnetic field and climate? *Earth and Planetary Science Letters* 253: 328-339.

[456] Usoskin, I. G., Solanki, S. K., Taricco, C., Bhandari, N. and Kovaltsov, G. A. 2006: Long term solar activity reconstructions: direct test by cosmogenic ⁴⁴Ti in meteorites. *Astronomy and Astrophysics* 457: L25-28.

[457] Svensmark, H. 2007: Astronomy and geophysics cosmoclimatology, a new theory emerges. *Astronomy and Geophysics* 48: 1.18-1.24, doi:10.10.1111/j.1468-4004.2007.48118.x.

blasting away cosmic rays with solar particles and providing a shielding magnetic field through which most cosmic rays cannot penetrate. During periods of high solar activity, the cosmic particles are blasted away, less low-level clouds form and there is global warming.

## The engine of weather

Clouds reflect 60% of the Sun's radiation. A change of just 1% in cloudiness of planet Earth could account for all of the 20th Century warming. However, IPCC computers don't do clouds. Fine particles in the atmosphere are the nuclei for condensation of water vapour into the water droplets that form low-level clouds. Wind-blown dust (especially if it is rich in clay) from continents, meteoritic and cometary dust and volcanic dust block the input of energy to Earth. This can affect climate. However, these dust particles are normally too large to act as nuclei for water droplets and have little effect on the formation of clouds at low altitudes (i.e. less than 3 km altitude).

Sulphuric acid aerosols formed from dimethyl sulphide released from micro-organisms in the oceans affect low-level cloud formation over a very large area because the oceans cover some 70% of the Earth's surface. Dimethyl sulphide reacts with water and sunlight to produce sulphuric acid droplets. This reaction is accelerated by lightning in a process called ion seeding.[458] Sea spray provides very small grains of sodium chloride from storm waves, especially in winter at latitudes 40–60°. Sporadic volcanic eruptions, hot gas vents and hot springs release large amounts of sulphurous gases into the atmosphere.[459] Various sulphur gases in the atmosphere combine with water and form sulphuric acid droplets. These droplets are also the nuclei for low-level cloud formation.

Updrafts in cumulus clouds carry the clouds' water droplets to colder regions of the atmosphere. They freeze to form snow and hail. At high latitudes, water vapour droplets can form ice and these we see in high cirrus clouds. These nuclei of ice are the sites for further water condensation. Because there are billions of cloud condensation nuclei, many meteorologists conclude that there are enough cloud-forming nuclei such that cosmic rays are not necessary for cloud formation. In a similar way, charged atoms and molecules from burnt aircraft fuel assist in the formation of nuclei that form

[458] Raes, F., Janssens, A. and van Dingenen, R. 1986: The role of ion-induced aerosol formation in the lower atmosphere. *Journal of Aerosol Science* 17: 466-470.

[459] Ackermann, T. P. 1988: Aerosols in climate modelling. In: *Aerosols and climate* (eds Hobbs, P. V. and McCormick, M. P.). Deepack, 335-348.

the cloud condensation trails of aircraft.[460] Cosmic rays produce ions which form cloud condensation nuclei and hence clouds. Electrically charged ions induce water molecules to nucleate even when there are not enough sulphuric acid nuclei to stimulate water droplet formation.

The Danish National Space Centre[461,462] has shown that cosmic radiation, which derives from high-speed atomic particles originating in exploded stars far away in the galaxy, plays a significant role in promoting cloud formation. The causal mechanism whereby cosmic rays facilitate the formation of clouds in the Earth's atmosphere has been experimentally determined.[463] If this mechanism of forming clouds by cosmic radiation is valid, it should be able to be measured. And it has. The correlation between the Earth's magnetic field and rainfall in the tropics can only be explained if cosmic rays have an influence on the formation of clouds.[464]

The observed variation of 3–4% of global cloud cover during Solar Cycle 22 was strongly correlated with the cosmic ray flux.[465] This, in turn, is related to solar activity and the Earth's magnetic field.[466] The weaker the solar activity, the more cosmic rays strike Earth and the more abundant is cloud cover in the lower atmosphere.[467,468] The effect is greater at high latitudes because of the shielding effect on the Earth's magnetic field on high-energy charged particles.[469] We must also be mindful of the fact that the Earth's magnetic field constantly changes.

[460] Yu, K. Kim, C. S., Adachi, M. and Okuyama, K. 2005: An experimental study of ion-induced nucleation using a drift tube ion mobility spectrometer/mass spectrometer and a cluster-differential mobility analyzer/Faraday cup electrometer. *Journal of Aerosol Science* 36: 1036-1049.

[461] http://www.spacecenter.dk/publications/press-releases/getting-closer-to-the-cosmic-connection-to-climate

[462] Svensmark, H., Pedersen, J. A. P., Marsh, N. D. and Enghoff, M. B. 2006: Experimental evidence for the role of ions in particle nucleation under atmospheric conditions. *Proceedings of the Royal Society Journal A: Mathematical, Physical and Engineering Sciences* 10: 1098.

[463] Barlow, A. K. and Latham, J. 1983: A laboratory study of the scavenging of sub-micron aerosol by charged raindrops. *Quarterly Journal of the Royal Meteorological Society* 109: 763-770.

[464] Knudsen, M. F. and Riisager, P. 2009: Is there a link between Earth's magnetic field and low-latitude precipitation? *Geology* 37: 71-74.

[465] Svensmark, H. and Friis-Christensen, E. 1997: Variation of cosmic ray flux and global cloud cover – a missing link in solar-climate relationships. *Journal of Atmospheric and Solar-Terrestrial Physics* 59: 1225-1232.

[466] Anderson, R. Y. 1992: Possible connection between surface winds, solar activity and the Earth's magnetic field. *Nature* 358: 51-53.

[467] Dickinson, R. 1975: Solar variability and the lower atmosphere. *Bulletin of the American Meteorological Society* 56: 1240-1248.

[468] Labitzke, K. and van Loon, H. 1993: Some recent studies of probable connections between solar and atmospheric variability. *Annals of Geophysics* 11: 1084-1094.

[469] Ohring, G. and Clapp, P. F. 1980: The effect of changes in cloud amount on the net radiation at the top of the atmosphere. *Journal of Atmospheric Science* 37: 447-454.

The relationship between cosmic ray flux and cloud cover is the link between solar activity,[470,471] especially the solar cycle length,[472,473] stratospheric oscillations[474] and global temperature.[475] Many have shown the link between solar activity, low-level clouds and climate.[476,477,478,479]

Climate change occurs over time scales greater than one solar cycle. Furthermore, not all short-term variation in cloud cover is driven by cosmic rays. No solar physicist has ever suggested this. Cloud cover within the solar cycle is driven by far more powerful forces, namely the ocean lag effect. As solar activity increases, the atmosphere will warm much more quickly than the ocean. This means less cloud. As solar activity decreases, the opposite occurs. Cloud cover varies because of local temperature and humidity. El Niño events create an uplift of warm moist air in the central Pacific resulting in much cloud. La Niña events create rising moist air further to the west. All this creates grief for the UN's Intergovernmental Panel on Climate Change (IPCC) computer modellers. Computer models can only process the information that is fed into them.

The IPCC has almost two dozen climate models. All are more sensitive in their cloud feedback than estimates of cloud feedback measured in real climate systems.[480] There is not enough cloud observation data to measure

[470] Zhang, Q., Soon, W. H., Baliunas, G. W., Lockwood, G. W., Skiff, V. A. and Radick, R. R. 1994: A method of determining possible brightness variations of the Sun in past centuries from observations of solar-type stars. *Astrophysical Journal* 427: L111-L114.

[471] Tinsley, B. A. 1994: Solar wind mechanism suggested for weather and climate change. *EOS* 75: 369.

[472] Friis-Christensen, E. and Lassen, K. 1991: Length of solar cycle: an indication of solar activity associated with climate. *Science* 254: 698-700.

[473] Lassen, K. and Friis-Christensen, E. 1995: Variability of the solar cycle length during the last five centuries and the apparent association with terrestrial climate. *Journal of Atmospheric and Solar-Terrestrial Physics* 57: 835-845.

[474] Pudovkin, M. and Veretenenko, S. 1995: Cloudiness decreases with decreases of galactic cosmic rays. *Journal of Atmospheric and Solar-Terrestrial Physics* 57: 1349-1355.

[475] Ney, E. R. 1959: Cosmic radiation and the weather. *Nature* 183: 451-452.

[476] Udelhofen, P. M. and Cess, R. D. 2001: Cloud cover variations over the United States: An influence of cosmic rays or solar variability? *Geophysical Research Letters* 28: 2617-2620.

[477] Kristjánsson, J. E., Staple, A. and Kristiansen, J. 2002: A new look at possible connections between solar activity, clouds and climate. *Geophysical Research Letters* 29: doi:10.1029/2002GJ015646.

[478] Pallé, E., Butler, C. J. and O'Brien, K. 2004: The possible connection between ionization in the atmosphere by cosmic rays and low level clouds. *Journal of Atmospheric and Solar-Terrestrial Physics* 66: 1779-1790.

[479] Ososkin, I. G., Marsh, N., Kovaltsov, G. A., Mursula, K. and Gladysheva, O. G. 2004: Latitudinal dependence of low cloud amount on cosmic ray induced ionization. *Geophysical Research Letters* 31: doi:10.1029/2004GL019507.

[480] Forster, P. M. and Gregory, J. M. 2006: The climate sensitivity and its components diagnosed from Earth budget radiation data. *Journal of Climate* 19: 39-52.

any long-term changes in cloudiness with an accuracy of less than 1%. This means that we really do not know how much of the 20th Century warming is natural and, if there are large errors on our observational estimates of feedback, then it is quite possible that the real climate system cannot be modelled. Furthermore, not one of the IPCC models predicted that there would be cooling after 1998. If the models cannot make testable predictions a few years in advance, then they are totally useless for the prediction of climate change hundreds of years into the future.

The IPCC theoretical models have a fixed number of climate forcing and feedback mechanisms. This allows the total effect of solar influences to be diminished. The problem is that such models dismiss unknown feedback mechanisms and alternative solar effects on climate such as UV energy changes in production and loss of ozone and variations in solar wind that affect the formation of clouds.[481] While these factors remain unknown, they will be poorly modelled or not used at all (which is the approach taken by the IPCC). Solar changes are far greater than solar assumptions made in climate models[482,483,484,485,486] and solar changes trigger temperature changes. Changes in temperature trigger changes in the amount of greenhouse gases in the atmosphere, a process seen in Antarctic ice core.[487]

## That big ball of heat in the sky

The Sun formed on the collapsed core of a supernova. We might have learned at school that the Sun is composed of hydrogen and helium. Suspicions that the Sun is not that simple have been around for more than 30 years. Isotopes

[481] Pap, J. M. and Fox, P. 2004: Solar variability and its effects on climate. *Geophysical Monograph Series Volume* 141.

[482] Stevens, M. J. and North, G. R. 1996: Detection of climate response to the solar cycle. *Journal of Atmospheric Sciences* 53: 2594-2609.

[483] Houghton, J. T., Ding, Y., Griggs, D. J., Noguer, M., van der Linden, P. J., Dai, X., Maskell, K. and Johnson, C. A. 2001: *Climate change 2001: The scientific basis.* Cambridge University Press.

[484] Douglass, D. H. and Clader, B. D. 2002: Climate sensitivity of the Earth to solar irradiance. *Geophysical Research Letters* 29: doi:10/1029/2002GL015345.

[485] Hansen, J., Ruedy, R., Sato, M., Imhoff, M., Lawrence, W., Easterling, D., Peterson, T. and Karl, T. 2001: A closer look at United States and global surface temperature change. *Journal of Geophysical Research* 106: 23947-23963.

[486] Foukal, P., North, G. and Wigley, T. 2004: A stellar view on solar variations and climate. *Science* 306: 68-69.

[487] Petit, J. R., Jouzel, J., Raynaud, D., Barkov, N. I., Barnola, J.-M., Basile, I., Bender, M., Chappellaz, J., Davis, M., Delaygue, G., Delmotte, M., Kotlyakov, V. M., Zlegrand, M., Lipenkov, V. Y., Lorious, C., Pépin, L., Ritz, C., Saltzman, E. and Stievenard, M. 1999: Climate and atmospheric history of the past 420,000 years from the Vostok ice core, Antarctica. *Nature* 399: 429-436.

of oxygen, magnesium, xenon and nitrogen in the Sun, solar wind, planets, meteorites, solar flares and the Moon suggest that the Sun consists mostly of the same elements (oxygen, iron, magnesium, calcium, sulphur, nickel and strontium) as ordinary meteorites and the rocky planets.[488,489,490,491,492,493] There is recent visual evidence of rigid iron-rich structures below the Sun's fluid outer zone.[494] Elements such as iron, oxygen, nickel, sulphur and silicon are only made in the deep interior of supernovae. The Sun is actually a pulsating star[495] and this creates variable energy output. We are continually finding supernovae debris in our Solar System and the view is now that it is this material that collapsed to form the Sun.[496,497,498] The Sun is recycled stardust.

Present knowledge shows that the Sun is composed of ordinary elements derived from the debris of an earlier star roughly in the same position as the present Sun. It is a pulsar, dynamic, and emits a variable amount of energy. This variation in energy emitted has not been great enough to allow the Earth's oceans to freeze or to fry life on Earth.

## Angry solar emissions

The NASA solar storm warnings suggest that we are heading for a time of low solar activity with fewer solar flares and radiation storms. That's the good news. If there are lunar and Martian missions during this time, the chances of computer, communication and navigation equipment being disabled by

[488] Newman, M. J. and Rood, R. T. 1972: Implications of solar evolution for the Earth's early atmosphere. *Science* 198: 1035-1037.

[489] Esat, T. M., Brownlee, D. E., Papanastassiou, D. A. and Wasserburg, G. J. 1977: Magnesium isotopic composition of interplanetary dust particles. *Geophysical Research Letters* 4: 190-197.

[490] Manual, O. K., Sabu, D. D., Lewis, R. S., Sirnivasan, B. and Anders, E. 1977: Strange xenon, extinct superheavy elements and the solar neutrino puzzle. *Science* 195: 208-210.

[491] Huss, G. R. 2006: Solar system: when the dust unsettles. *Nature* 440: 751-752.

[492] Manuel, O., Kamat, S. A. and Mozina, M. 2005: Isotopes tell origin and operation of Sun. First Crisis in Cosmology Conference, Moncao, Portugal, 23-25 June 2005. *Astrophysics* arXiv:astro-ph/0510001v1

[493] Ireland, T. R., Holden, P., Norman, M. D. and Clark, J. 2006: Isotopic enhancements of $^{17}O$ and $^{18}O$ from solar wind particles in the lunar regolith. *Nature* 444: 776.

[494] http://www.thesurfaceofthesun.com/index.html

[495] Toth, P. 1977: Is the Sun a pulsar? *Nature* 270: 159-160.

[496] Ballad, R. V., Oliver, L. L., Downing, R. G. and Manual, O. K. 1979: Isotopes of tellurium, xenon and krypton in Allende meteorite retain record of nucleosynthesis. *Nature* 277: 615-620.

[497] Manuel, O., Plees, M., Singh, Y. and Myers, W. A. 2005: Nuclear systematics: Part IV: Neutron-capture cross sections and solar abundance. *Journal of Radioanalytical and Nuclear Chemistry* 266: 159-163.

[498] Manuel, O., Mozina, M. and Ratcliffe, H. 2005: On the cosmic nuclear cycle and the similarity of nuclei and stars. *Journal of Fusion Energy* 25: 107-114.

a restless Sun are lessened. However, Solar Cycle 25 will be weak, peaking around 2022. We earthlings may enjoy more clouds and cooler times.

The bad news is that fewer cosmic rays will be swept away by the solar wind. Space travel will be more dangerous. The constraint on space travel to Mars and other distant planets is that once astronauts are outside the Earth's protective magnetic shield, the greatly intensified bombardment by cosmic radiation over long periods of space travel cause an increased risk of cancer, cataracts and other maladies. Cosmic rays penetrate metal, so increasing the weight of spacecraft using metal shielding will not solve the problem. If astronauts leave the Sun's protective magnetic shield to explore the far reaches of the Solar System, it would be a one-way trip.

Solar storms are common. The Sun emits a relentless current of charged particles, the solar wind. These warp the Earth's magnetic field. There are also vast balls of ionised gas that are released by the Sun as solar flares. These are best seen during a total eclipse. Minor solar flares occur once or twice each decade. Solar activity had its first effect on the modern world when at 6.30 pm on 28 August 1859, the telegraph lines out of Boston failed. Elsewhere, electrical equipment burst into flames. The Northern Lights were seen as far south as the Bahamas. Mayhem was caused by a huge solar flare with its associated electromagnetic blast. A contemporary report by Stuart Clark recorded:

> In September of 1859, the entire Earth was engulfed in a cloud of seething gas, and a blood-red aurora erupted across from the poles to the tropics. Around the world, telegraph systems crashed, machines burst into flames, and electric shocks rendered operators unconscious. Compasses and other sensitive instruments reeled as if struck by a massive magnetic fist. For the first time, people began to suspect that the Earth was not isolated from the rest of the Universe.

A solar flare in 1989 knocked out the electricity grid in northern Quebec. If the Sun had produced a superflare 10,000 times more energetic than the 1989 flare, the ice on Jupiter's moons would have melted and all life on Earth would have been fried. All amateur radio operators, satellite technicians, pilots and astronomers keep a watchful eye on solar variations. We are fortunate that the Sun has been extremely stable over a long period. Its very slight variations appear to drive climate.[499] We can thank our lucky star for long-

---

[499] Kuhn, J. R., Libbrecht, K. G. and Dickie, R. H. 1988: The surface temperature of the Sun and changes in the solar constant. *Science* 242: 908-911.

term temperature stability, life on Earth, weather and climate.

There is a correlation between sudden bursts of solar activity, expressed as solar flares, and climate.[500] An angry Sun produces auroras and the Northern Lights become more active and creep south. The long-term observational record of the number of auroras in the Northern Hemisphere is a measurement of solar flare activity and the number of auroras per decade shows a close correlation with climate.[501] The hydrogen and helium degassed from the Earth's core and mantle are lost into space. However, there is an inflow of atomic bullets of hydrogen. The varying atomic hydrogen inflow into the Earth's atmosphere from the Sun and space (hydrogen forcing) is a highly speculative climate forcing mechanism.[502]

The Sun also emits a variable amount of ultra-violet (UV) energy. UV energy affects ozone in the stratosphere. However, ozone in the stratosphere is also controlled by solar cycles,[503] volcanic emissions,[504] ozone depleting substances[505] and climate change.[506] Variations in ultra-violet radiation are 0.5–0.8%. Variations have been measured, deduced from models and determined from the changes in the UV-absorbing compounds in the spore wall of club mosses.[507] This variation is cyclical (22 years and 80–90 years).[508] This has

[500] Scafetta, N. and West, B. J. 2003: Solar flare intermittency and the Earth's temperature anomalies. *Physical Review Letters* 90: 248701-248705.

[501] Ruzmaikin, A., Feynman, J. and Yung, Y. L. 2006: Is solar variability reflected in the Nile River? *Journal of Geophysical Research* 111: D21114, doi:10.1029/2006JD007462.

[502] Scherer, K., Fichtner, H., Bormann, T., Beer, J., Desorgher, L., Flükiger, E., Fahr, H.-J., Ferreira, S. E. S., Langner, U. W., Potgieter, M. S., Heber, B., Masarik, J., Shaviv, N. and Veizer, J. 2007: Interstellar-terrestrial relations; Variable cosmic environments, the dynamic heliosphere and their imprints on terrestrial archives and climate. *Space Science Reviews* 127: 327-465.

[503] Rozema, J., van Geel, B., Bjorn, I. O. , Lean, J. and Madronich, S. 2002: Paleoclimate: Toward solving the UV puzzle. *Science* 296: 1621-1622.

[504] Tabazadeh, A., Drdla, K., Schoeberl, M. R., Hamill, P. and Toon, O. B. 2002: Arctic 'ozone hole' in a cold volcanic stratosphere. *Proceedings of the National Academy of Sciences* 99: 2609-2612.

[505] Farman, J. C., Gardiner, B. G. and Shanklin, J. D. 1985: Large losses of total ozone in Antarctica reveal seasonal $ClO_x/NO_x$ interaction. *Nature* 315: 207-210.

[506] Goutail, F. , Pommereau, J.-P., Lefèvre, F., van Rozendael, M., Andersen, S. B., Kåstad Høiskar, B.-A., Dorokhov, V., Kyrö, E., Chipperfield, M. P. and Feng, W. 2005: Early un usual ozone loss during the Arctic winter 2002/2003 compared to other winters. *Atmospheric Chemistry and Physics* 5: 665-677.

[507] Lomax, B. H., Fraser, W. T., Sephton, M. A., Callaghan, T. V., Self, S., Harfoot, M., Pyle, J. A., Wellman, C. H. and Beerling, D. J. 2008: Plant spore walls as a record of long-term changes in ultra-violet-B radiation. *Nature Geoscience* 1: 592-596.

[508] Lohmann, G., Rimbu, N. and Dima, M. 2004: Climate signature of solar irradiance variations: analysis of long-term instrumental, historical and proxy data. *International Journal of Climatology* 24: 1045-1056.

an impact on the amount of ozone produced.[509] There is a link between the amount of solar energy hitting the atmosphere and upper atmosphere wind currents,[510] and the amount of UV energy changes the amount of sulphur compounds that move between the atmosphere and the ocean.[511] This, in turn, affects the number of cloud condensation nuclei, the reflection of solar energy back into space and, as a result, the sea surface temperature. The variability in the UV radiation from the Sun not only impacts on ozone production (and the "hole" in the ozone layer) but also can affect global temperature.[512]

When next the Sun becomes exceptionally angry, grid electricity, radio, television, internet, telephone, satellite and navigation systems will collapse. We might then sit around under the lights of an aurora and talk to each other.

The first conversational topic will be that the Sun is the meaning of life.

## Inner turbulence of the Sun

Solar physicists use a standard solar model to calculate many of the properties of the Sun's interior. These calculations suggest a core (~25% solar radius) where nuclear fusion occurs, a surrounding radiative zone (~70% solar radius) and an enveloping convection zone (~5% solar radius) where heat makes its way to the surface by convective flow. Solar instability can be generated by rotating plasma in the presence of a magnetic field. This gives rise to thermal fluctuations and deviations from the standard solar model.[513] Constraints on the size of the core magnetic field are such that there is a magnetic field[514] which displays considerable fluctuations.[515]

---

[509] Krivova, N. A., Solanki, S. K. and Floyd, L. 2006: Reconstruction of solar UV irradiance in cycle 23. *Astronomy and Astrophysics* 452: 631-639.

[510] Balachandran, N. K., Rind, D., Lonergan, P. and Shindell, D. T. 1999: Effects of solar cycle variability on the lower stratosphere and troposphere. *Journal of Geophysical Research* 104: 27,321-27,339.

[511] Larsen, S. H. 2005: Solar variability, dimethyl sulphide, clouds and climate. *Global Biogeochemical Cycles* 19: GB1014,doi:10.1029/2004GB002333.

[512] Krivova, N. A., Solanki, S. K. and Flotd, L. 2006: Reconstruction of solar UV irradiance in cycle 23. *Astronomy and Astrophysics* 452: 631-639.

[513] Grandpierre, A. and Agoston, G. 2005: On the onset of thermal metastabilities in the solar core. *Astrophysics and Space Science* 298: 537-552.

[514] Gough, D. O. and MacIntyre, M. E. 1998: Inevitability of a magnetic field in the Sun's interior. *Nature* 394: 755-757.

[515] Friedland, A. and Gruzinov, A. 2004: Bounds on the magnetic fields in the radiative zone of the Sun. *Astrophysics Journal* 601: 570.

The Sun is a magnetic plasma diffuser that selectively moves light elements like hydrogen and helium and the lighter isotopes of other elements to its surface. Hydrogen ions, generated by emission and decay of neutrons at the core, are accelerated upward by deep magnetic fields, thus acting as a carrier gas that maintains separation of lighter from heavier components in the Sun. Neutron emissions from the centre of the Sun trigger a series of reactions that generate solar luminosity, solar neutrinos, outpourings of the neutron decay product (hydrogen) in the solar wind and separation of heavy atoms from lighter atoms. This process also takes place in many other stars.[516]

The Sun's great internal conveyor belt is a massive circulating current of hot plasma within the Sun. It has a northern and southern branch. Each takes some 40 years to perform a complete circuit and the turning of the belt controls the sunspot cycle. Normally, the conveyor belt moves at walking pace, about 1 metre per second, and it has been at this speed since the 19th Century. In recent years it has slowed to 0.75 metres per second in the northern branch and 0.35 metres per second in the southern branch. The belt plunges some 200,000 km beneath the Sun's surface and is observed as sunspot activity. Sunspots are magnetic knots that bubble up from the base of the conveyor belt, eventually popping through to the surface of the Sun. They are surface expressions of the twisting and untwisting of the magnetic field between the Sun's zones of radiation and convection. Sunspots drift from mid latitudes to the equator. This is caused by the motion of the conveyor belt and this drift is caused by the speed of the conveyor belt. Because the belt controls sunspot activity, the speed of the belt foretells the intensity of sunspot activity some two decades into the future. A slow conveyor belt movement in the Sun means lower solar activity.

The GOLF (Global Oscillation at Low Frequency) instrument on the SOHO (Solar and Heliospheric Observatory) is thought to have detected gravitational oscillations within the core of the Sun. If this is the case, then the Sun's core rotates faster than its surface.[517] The Solar System comprises the Sun and the orbiting bodies, of which the four major planets (Jupiter, Saturn, Uranus and Neptune) are the most important. All bodies of the Solar System, including the Sun, orbit around the centre of mass of the Solar System. The distance between the Sun and the centre of the Solar System varies, which creates a gravitational wobble in the Sun's orbit.[518] The alternating grouping

---

[516] Manual, O., Kamat, S. A. and Mozina, M. 2007: The Sun is a plasma diffuser that sorts atoms by mass. *Astrophysics* 654: 650-664.

[517] http://www.alphagalileo.org/index.cfm?fuseaction=readrelease&releaseaid-520213

[518] Juckett, D. A. 2003: Solar activity cycles, north/south asymmetries, and differential rotation associated with spin-orbit variations. *Solar Physics* 191: 201-206.

and dispersion of the four major planets occurs at regular intervals. The large outer planets (Jupiter, Saturn, Uranus and Neptune) make sure that the Sun is not the centre of gravity of the Solar System and the Sun rotates around the centre of gravity of the Solar System about every 11.1 years.[519,520] This spin-orbit coupling means that the solar plasma circulates every 11.1 years, producing an 11.1-year solar cycle.[521]

At times, the Sun is up to 1 million kilometres from the centre of gravity and at other times it almost coincides with the centre of gravity. This leads to great variations in turbulence inside the Sun. It was not until we could make observations from outside the Earth's atmosphere using satellites that we could measure that the Sun's energy output varied. The variation is only by fractions of a per cent but the Sun is large and emits a huge amount of energy. These slight solar variations have a huge effect on the Earth.

When the Sun's position trails that of the centre of the Solar System, the Sun accelerates. This happens during the first rotation of the Solar System and has a duration of 11 years. The Sun continues rotating about the centre of mass of the Solar System but its galactic velocity decreases as it returns to the trailing position. This 22-year solar cycle (Hale Cycle) is then repeated. Both the acceleration and deceleration process result in an increase in sunspot numbers, while the intervening sunspot minima occur when the Sun is in the trailing and leading positions.[522]

## Shock, horror. The solar constant is not constant

The solar constant, the energy coming from the Sun, is not a constant.[523] Satellite measurements since 1979 and sunspot activity show that the solar constant is not constant. The IPCC claims that the variation in the solar constant is less than 0.1% and concludes that it has no impact on climate compared to the effect of $CO_2$. This claim is misleading because the 0.1% variation does not refer to the complete difference between the maximum and

---

[519] de Jager, C. and Versteegh, G. J. M. 2005: Do planetary motions drive solar variability. *Solar Physics* 229: 175-179.

[520] Wilson, I. R. G., Carter, B. D. and Waite, I. A. 2008: Does a spin-orbit coupling between the Sun and the Jovian planets govern the solar cycle? *Publications of the Astronomical Society of Australia* 25: 85-93.

[521] Shirley, H. 2006: Axial rotation, orbital revolution and solar spin-orbit coupling. *Monograph Note of the Royal Astronomical Society* 368: 280-282.

[522] Javaraih, J., Bertello, L. and Ulrich, R. K. 2005: Long-term variations in solar differential rotation and sunspot activity. *Solar Physics* 232: 25-40.

[523] Kuhn, J. R., Libbrecht, K. G. and Dickie, R. H. 1988: The surface temperature of the Sun and changes in the solar constant. *Science* 242: 908-911.

minimum.[524,525,526] If this is considered, the variation is 0.22%.[527] Changes over long periods may be three to five times the measured variation. Furthermore, a very slight variation in any complex multicomponent system can have a profound effect. The seven-year solar constant variation of 0.22% equates to a surface temperature variation on Earth of 0.45°C and, combined with the effects of urban warming of at least 0.1°C, the total surface temperature effect is at least 0.55°C. Any temperature increase at the Earth's surface may be purely due to solar changes.

Total solar irradiance reconstructions for the period 1900–1980[528] and two different satellite composites (ACRIM and PMOD) that measured total solar irradiance[529] can be used to show 20th Century climate histories. By these methods it was calculated that the Sun contributed to some 46–49% of global warming of the Earth[530] and, considering that there are uncertainties of 20–30%, the Sun may have been responsible for as much as 60% of the 20th Century temperature increase.

The climate modelling community has vastly underestimated the role of the Sun. The energy balance models they use produce estimates of solar-induced warming over this period that are two to ten times lower than was actually found.[531,532]

If we can measure slight variations in the solar constant from satellites,

---

[524] Fröhlich, C. 1995: Variations in total solar irradiance. In *Solar output and climate during the Holocene*. (ed. B. Frenzel). Gustav Fischer Verlag, 125-127.

[525] Hansen, J. E., Lacis, A. A. and Ruedy, R. A. 1990: Comparison of solar and other influences on long-term climate. In: *Climate impact of solar variability* (eds Schatten, K. H. and Arking, A.). NASA, 142.

[526] Hoyt, D. V. and Schatten, K. H. 1997: *The role of the sun in climate change*. Oxford University Press.

[527] Satellite measurements show that the solar constant is 1367 W/m² and 0.22% of this energy is 3 W/m². The solar constant is the amount of energy that reaches the edge of the Earth's atmosphere. Some 30% of this energy is reflected and the irradiated area of the Earth's surface is one quarter of this surface. Hence there is only 239 W/m² available to heat the atmosphere. Consequently the variation of 3 W/m² has only a climate effect of 0.53 W/m². How this affects global temperature depends on which global circulation computer model is used to calculate climate sensitivity. The mean of the range of 0.3 to 1.4°C/W/m² is 0.85°C/W/m² which yields a temperature effect of 0.45°C.

[528] Lean, J., Beer, J. and Bradley, R. 1995: Reconstruction of solar irradiance since 1610: implications for climate change. *Geophysical Research Letters* 22: 3195-3198.

[529] Wilson, R. C. and Mordvinov, A. V. 2003: Secular total solar irradiance trend during solar cycles 21-23. *Geophysical Research Letters* 30: 10.1029/2002GL016038.

[530] Scarfetta, N. and West, B. J. 2006: Phenomenological solar contribution to the 1900-2000 global surface warming. *Geophysical Research Letters* 33: 10.1029/2005GL025539.

[531] Pap, J. M. and Fox, P. 2004: Solar variability and its effects on climate. *Geophysical Monograph Series Volume* 141.

[532] Stevens, M. J. and North, G. R. 1996: Detection of climate response to the solar cycle. *Journal of Atmospheric Sciences* 53: 2594-2609.

can the results of these very small variations be measured on Earth? The answer is a resounding yes. These variations have been traced back in time[533],[534] using the material dropped by icebergs on the bottom of the North Atlantic Ocean. Some of this material is from sample sites thousands of kilometres apart, suggesting that the cycles were global. The amount of material dropped increased every 1500 years as icebergs floated further south into the Atlantic Ocean during temporary cold periods.

The irregular orbit of the Sun about the centre of mass of the Solar System is driven by the combined angular momentum of the giant outer planets. Thus, widely variable solar activity is the electromagnetic outcome of delays in the response to the Sun's irregular orbit. The Sun's most obvious magnetic features are sunspots. They are magnetic fields that rip through the Sun's surface. A magnetically active Sun boosts the number of sunspots, indicating that vast amounts of energy are being released from deep in the Sun. Typically, sunspots flare up and settle down in cycles of 11.1 years. Over the last 50 years we have been living in a period of abnormally high solar activity. Sunspots die out about every 200 years as solar activity diminishes (DeVries-Suess Cycle). When the sunspot activity collapses, the Earth cools dramatically to a Grand Minimum, a phenomenon that has occurred many times over the last 10,000 years. Because of the lag between sunspot activity and the Sun's great conveyor belt,[535] most astronomers now predict the return of a quieter Sun.[536] The decreased solar activity would result in increased cosmic radiation attacking the Earth, resulting in increased cloudiness. Low-level clouds reflect the Sun's energy back into space, resulting in cooling of the Earth.

There does not have to be a great change in the Sun's output of energy to have a profound effect on the Earth's climate. If the Sun emitted only 1 to 1.5 watts per square metre less than now, we would be in conditions similar to a very cold period of the Little Ice Age (Maunder Minimum).[537] Sunspots

[533] Bond, G., Kromer, B., Beer, J., Muscheler, R., Evans, M. N., Showers, W., Hoffmann, S., Lotti-Bond, R., Hajdas, I. and Bonani, G. 2001: Persistent solar influence on North Atlantic climate during the Holocene. *Science* 294: 2130-2136.

[534] Kerr, R. 2001: A variable Sun paces millennial climate. *Science* 294: 1431-1433.

[535] Temmer, M., Veronig, A. and Hansimeier, A. 2003: Does solar flare activity lag behind sunspot activity? *Solar Physics* 215: 111-126.

[536] Dikpati, M., De Toma, G., Gilman, P. A., Argue, C. N. and White, O. R. 2004: Diagnostics of polar field reversal in solar cycle 23 using a flux transport dynamo model. *Astrophysics Journal* 601: 1136-1151.

[537] Tapping, K. F., Boteler, D., Crouch, A., Charbonneau, P., Manson, A. and Paquette, H., 2006: Modelling solar magnetic flux and irradiance during and since the Maunder Minimum. *Solar Physics* 246: 309-326.

have already started to weaken,[538] especially during Solar Cycle 23.[539] At 27 September 2008, there had been 200 days without sunspots. Trends in solar activity are such that some astronomers are predicting that there will be a rapid decline in sunspot activity starting with Solar Cycle 24. If this trend continues, it is suggested that there could be another Maunder Minimum coincidental with reduced sunspot activity.[540]

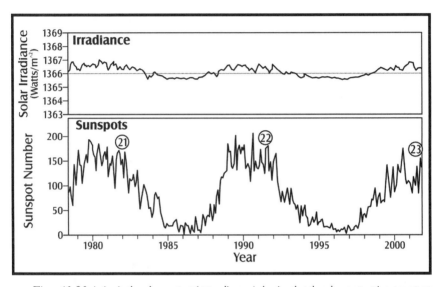

*Figure 13: Variation in the solar constant (upper diagram) showing that the solar constant is not constant. This solar constant variation can be correlated with the numbers of sunspots (lower diagram). Sunspot Cycles 21, 22 and 23 shown on lower diagram.*

Some astronomers are suggesting that the decreased sunspot activity will produce a cooler climate by 2030 AD.[541] Others are predicting that Earth will have another very cold time rather like the Dalton (1795–1825 AD) or Maunder Minima (1645–1715 AD).[542] These were times of low temperatures, increased precipitation and increased social conflict.[543]

---

[538] Livingston, W. 2004: Sunspots observed to physically weaken in 2000-2001. *Solar Physics* 207: 41-45.

[539] Penn, M. J. and Livingston, W. 2006: Temporal changes in sunspot umbral magnetic fields and temperatures. *The Astrophysical Journal* 649: L45-L48, doi: 10.1086/508345.

[540] Schatten, K. H. and Tobiska, W. K. 2003: Solar activity heading for a Maunder Minimum. *Abstract, 24th Solar Physics Division Meeting*, Laurel, MD.

[541] De Jager, C. 2005: Solar forcing of climate 1: Solar variability. *Space Science Reviews* 120: 197-241.

[542] Landscheidt, T. 2003: New Little Ice Age instead of global warming? *Energy and Environment* 14: 327-350.

[543] Ruzmaikin, A. 2007: Effect of solar variability on the Earth's climate patterns. *Advances in Space Research*: doi:10.1016/j.asr.2007.01.076

Contrary to the popular belief that we humans are creating global warming, there is a scientific opinion from astronomers at the Pulkovo Observatory in Russia[544,545] that the Earth will be facing a slow decrease in temperatures in 2012–2015. The gradually falling amounts of solar energy will be expected to reach their minimum level by 2040, inevitably leading to a deep freeze around 2055–2060 AD. This is a view supported by other astronomers.[546,547,548,549]

## Blemishes on her beauty

Sunspots have been known for more than 2000 years. The first recorded observations were in East Asia. Once telescopes became the tool of astronomers in the 17th Century, there was more intense observation of sunspots. In Danzig (now Gdansk) in 1647, Johannes Hevelius (1611–1687) plotted the movement of sunspots eastwards and towards the Sun's equator. In 1801, the astronomer William Herschel (1738–1822) correlated the annual number of sunspots with the price of grain in London recorded in the 1776 work by Adam Smith, *The Wealth of Nations*. In many ways, this relationship between sunspots and grain yields still drives agriculture.

British astronomers some 150 years ago published a well-documented linkage between sunspot activity and famine in India. In *Cycles of Drought and Good Seasons in South Africa*, published in 1889, D.E. Hutchins showed that in South Africa, there was a synchronous link between sunspot activity, temperature, rainfall and river flow. It was only in 1843 that the cyclical nature of sunspots was first measured by Heinrich Schwabe (1789–1875). There were speculations in the period 1843–1851 by Schwabe and Rudolf Wolf that the 11.1-year cycle of sunspots may be related to the orbital period of Jupiter (11.86 years).

---

[544] Abdussamatov, H. I. 2005: On long-term variations of the total irradiance and on probable changes in temperature in the Sun's core. *Kinematics and Physics of Celestial Bodies* 21: 471-477 (In Russian).

[545] Abdussamatov, H. I. 2006: On long-term variations of the total irradiance and decrease of global temperature of the Earth after a maximum of XXIV cycle of activity and irradiance. *Bulletin of Crimea Observatory* 103: 122-127 (In Russian).

[546] Lin, Z-S and Xian, S. 2007: Multi-scale analysis of global temperature changes and trend of a drop in temperature in the next 20 years. *Meteorology and Atmospheric Physics* 95: 115-121.

[547] Nagovitsyn, Y. A. 2001: Solar activity during the last two millennia: Solar patrol in ancient and medieval China. *Geomagnetism and Aeronomy* 41: 680-688.

[548] Hathaway, D. H. and Wilson, R. M. 2004: What the sunspot record tells us about space climate. *Solar Physics* 224: 5-19.

[549] Svalgaard, L., Cliver, E. W. and Kamide, Y. 2005: Sunspot cycle 24: Smallest cycle in 100 years? *Geophysical Research Letters* 32: L01104, doi:1010.1029/2004GL021664.

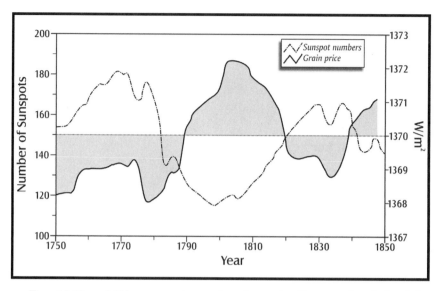

*Figure 14: Times of high sunspot numbers are times of prosperity with excess grain and relatively low grain prices whereas times of low sunspot activity are times of crop failure and relatively high grain prices.*[550]

Sunspots, some 37,000 km in diameter, appear as dark spots within the outermost layer of the Sun. The outermost layer is about 400 km deep and provides most of the solar radiation. At the inner boundary of this layer, it is about 6000°C and on the outside it is about 4200°C.[551] The temperature within sunspots is about 4600°C. There is a strong radial magnetic field within a sunspot and the direction of the magnetic field reverses in alternate years within the major sunspots in a group. Magnetic field lines emerge from one sunspot and enter another spot. There are more sunspots during periods of increased magnetic activity. When there is increased magnetic activity, more highly charged particles are emitted from the Sun's surface, more UV energy is also emitted and the Sun is brighter. In years with a large number of sunspots, polar auroras on Earth are far more common and can extend to latitudes as low as 40°.

There is a 25-month fluctuation of sunspots and, superimposed on the 11-year cycle (Schwabe Cycle) and 22.2-year cycle (Hale Cycle), are other solar cycles (33-year Bruckner Cycle; 87-year Gleissberg Cycle, 210-year DeVries-

[550] Berner, U. and Streif, H. 2001: *Klimafakten. Der Rückblick – Ein Schlüssel für die Zukunft.* E. Schweizerbart'sche Verlagsbuchhandlung.
[551] Kuhn, J. R., Libbrecht, K. G. and Dickie, R. H. 1988: The surface temperature of the Sun and changes in the solar constant. *Science* 242: 908-911.

Suess Cycle and the 1500 ± 500 year Dansgaard-Oeschger cycle). The main cycles that have driven past climate change on Earth are the Schwabe, Hale, Gleissberg and Dansgaard-Oeschger cycles. There are no reasons to suppose anything different for the future.

The length of the sunspot cycle is from 9.8 to 12.0 years with the maximum length of sunspot cycles occurring in 1770, 1845 and 1940. Sunspot Cycle 23 was short, 10.0 years rather than the average of 11.1 years. During times of high solar activity, such as 2000 AD, the Sun is about 0.07% brighter.[552] Both Sunspot Cycles 22 and 23 had two peaks and in late 2006-early 2007, there was a sunspot minimum.[553,554] This is consistent with a warmer than usual climate.[555] Sunspot Cycle 24 is upon us, and now the Sun will be quieter, there will be fewer sunspots[556] and we will be facing cooler times.[557] There were a few sunspots in early 2008 with Sunspot Cycle 24 characteristics. By September 2008, there had been 200 consecutive spotless days. The last time there was a cluster of spotless months was in the early 1800s. This was a very cold period (Dalton Minimum).

The cyclical variation in the length and number of sunspots is not unique to the Sun. There is an increasing body of information to show that the Sun and other stars spend about 25% of their time with no sunspots.

The sunspot cycle length is an inadequate measure of solar activity variability. It is the totality of solar activity that is important (i.e. irradiation, matter, electromagnetic and gravitational fields).[558] The 11-year cycle varies in amplitude and length. The length of the solar cycle is probably a surrogate measure of irradiation. The Armagh Observatory in Northern Ireland has been keeping records since 1795. It is cooler whenever the Sun's sunspot cycle is long and it is warmer whenever the Sun's sunspot cycle is shorter.[559,560]

---

[552] Scafetta, N. and West, B. J. 2006: Phenomenological solar signature in 400 years of reconstructed Northern Hemisphere temperature record. *Geophysical Research Letters* 33: L17718, doi:10.1029/2006GL027142.

[553] Atac, T. and Ozgus, A. 2006: Overview of the solar activity during cycle 23. *Solar Physics* 233: 139-153.

[554] Ishov, V. N. 2005: Properties of the current 23rd solar-activity cycle. *Solar System Research* 39: 453-461.

[555] Friis-Christensen, E. and Lassen, K. 1991: Length of the solar cycle – an indicator of solar activity closely associated with climate. *Science* 254: 5032, 698-700.

[556] Clilverd, M. A., Clarke, E., Ulrich, T., Rishbeth, H. and Jarvis, M. J. 2007: Predicting Solar Cycle 24 and beyond. *Space Weather* 4: S09005, doi.10.1029/2005SW000207.

[557] Svalgaard, L., Cliver, E. W. and Kamide, Y 2005: Sunspot cycle 24: Smallest cycle in 100 years? *Geophysical Research Letters* 32: L01104, doi:10.1029/2004GL021664.

[558] Hoyt, D. V. 1979: Variations in sunspot structure and climate. *Climate Change* 2: 79-92.

[559] Wilson, R. M. 1997: Evidence of solar-cycle forcing and similar variables in the Armagh Observatory temperature record 1844-1999. *Journal of Geophysical Research* 103: D19, 11.

[560] Palle, B. E. and Butler, C. J. 2001: Sunshine records from Ireland, cloud factors and their link to solar activity and cosmic rays. *International Journal of Climate* 21: 709-729.

There has been steadily increasing geomagnetic activity in the Sun since the beginning of the 20th Century.[561] The solar wind mediates the transfer of angular momentum from the Sun to the Earth, this creates a lag and varies the rate of the rotation of the Earth. The rate of rotation of the Earth is measured as the length of the day.

Variation in solar activity creates a delay in the Earth's rotation and this cycle in the Earth's rotation rate in turn forces surface temperature variations in the order of 0.022°C for each millisecond change in the length of the day.[562] The 22-year Hale solar magnetic cycle mirrors length of the day measurements. The length of the day is at a maximum (i.e. when the Earth rotates the slowest) in maximum positive solar polarity (i.e. close to a sunspot minimum between an even and odd numbered sunspot cycle) and the length of the day is at a minimum during maximum negative polarity.[563] After cooling from 1940 to 1975, globally averaged temperature rose 0.3°C in 1976. Sea surface temperature also rose. This reordering of the ocean and atmosphere heat balance coincided with a change in the rate of change of the length of the day.[564] Temperature then started to increase and stopped increasing 22 years later in 1998.

The correlation between solar activity and Northern Hemisphere land temperatures was an even better correlation if the length of the solar cycle is used to represent the Sun's variability rather than the number of sunspots. There is a strong correlation with the global cooling that has occurred since the Medieval Warming (900–1300 AD) including the Maunder Minimum (1645–1715 AD) and the Dalton Minimum (1795–1825 AD). These periods of intense cooling during the Little Ice Age were when there were few sunspots or when sunspots disappeared altogether.

Each 11-year solar cycle is numbered. Counting only started with a somewhat nondescript cycle that peaked in 1760 AD, and sunspot activity before 1760 can be deduced from measurements of $C^{14}$ and $Be^{10}$ in ice, mud, wood, stalactites and fossils. There was a slight cooling event within the Medieval Warming (Oort Minimum, 1040–1080 AD) and, in the Little Ice Age, there were further sunspot minima at 1450–1540 AD (Spörer Minimum) and 1280–1340 AD (Wolf Minimum). The coldest periods of the Little Ice Age (early 14th to late 19th Century) coincided with the Wolf, Spörer, Maunder

[561] Georgieva, K. and Kirov, B. 2007: Long term changes in solar meridional circulation as the cause for the long-term changes in the correlation between solar and geomagnetic activity. *Annales Geophysicae* arXiv:physics/0703187v1.

[562] Duhau, S. 2006: Solar activity, Earth's rotation rate and climate variations in the secular and semi-secular time scales. *Physics and Chemistry of the Earth* 31: 99-108.

[563] Georgieva, K. J. 2006: Solar dynamics and solar-terrestrial influences. In: *Space science: New research* (ed. Maravell, N.) Nova Science Pub. Inc.

[564] Plimer, I. R. 2001: *A short history of planet Earth*. ABC Books.

and Dalton Minima.

The Wolf Minimum heralded the end of the Medieval Warming and the beginning of the 600-year Little Ice Age. It took only 23 years to change from a warm climate to a cool climate. Furthermore, the correlation of sunspot numbers with Northern Hemisphere mean temperature since 1861 and the even better correlation with the length of the solar cycle and temperature shows that the Sun is fundamental for global temperature. This has been long known[565] and, with increasing solar observation[566,567] and studies on cosmic radiation, the Sun has now emerged as the major driver of climate changes[568] such as the Medieval Warming, the Little Ice Age and the Late 20th Century Warming. The $C^{14}$ measurements from 8000-year-old bristlecone pine trees show 18 sunspot minima over the last 7800 years.[569] Sunspot minima and the associated cold climate are the norm and not a feature just of the Little Ice Age.

If variations in solar activity influence climate on Earth, then they also should influence climate elsewhere in the Solar System. They have. In 1998, the Hubble telescope showed that a moon of Neptune (Triton) had warmed since it was visited by the Explorer space probe in 1989.[570] In 2002, it was shown that air pressure on Pluto had tripled in 14 years, indicating a 2°C rise.[571] Pluto's atmosphere became denser.[572] In 2003, NASA's Odyssey mission reported that there was evidence of global warming on Mars and in 2005,[573] NASA reported that the ice caps on Mars' South Pole had diminished for three consecutive years.[574] In 2006, the Hubble telescope showed a new red storm spot on Jupiter and a temperature rise of 1°C.[575] For hundreds of years, changes in Mars have been observed. The reflection of solar energy of Mars changes over decadal cycles and is unrelated to dust storms. Up to 53 million square kilometres of the Martian surface changes in

---

[565] Eddy, J. A. 1981: Climate and the role of the Sun. In: *Climate and history* (eds Rotberg I. and Rabb, T. K.). Princeton University Press.

[566] Willson, R. C. 1997: Total solar irradiance trend during solar cycles 21 and 22. *Science* 277: 1963-1965.

[567] Friis-Christensen, E. and Lassen, K. 1991: Length of the solar cycle, an indication of solar activity closely associated with climate. *Science* 254: 698-700.

[568] Lane, L. J., Nichols, M. H. and Osborn, H. B. 1994: Time series analysis of global change data. *Environmental Pollution* 83: 63-68.

[569] Sonnet, C. P. and Suess, H. E. 1984: Correlation of bristlecone pine ring widths with atmospheric $^{14}$C variations: a climate-Sun relation. *Nature* 307: 141-143.

[570] MIT News Office, 24th June 1998.

[571] *Global warming on Pluto puzzles scientists,* www.space.com, 9th October 2002.

[572] ABC News, 26th July 2006.

[573] NASA's Jet Propulsion Laboratory, Pasadena, http://mars.jpl.nasa.gov/odyssey/newsroon, 8th December 2003

[574] *National Geographic News,* 27th February 2007.

[575] *USA Today,* 4th May 2006.

brightness by 10% or more. It is not exactly known how these changes affect the environment on Mars. However, it appears that they bring about large-scale weather changes and recent climate changes.[576] Mars has warmed by 0.65°C between the 1970s and the 1990s, similar to the Earth's rise of 0.7°C over the last century.

Mars, Triton, Pluto and Jupiter all show global warming.[577,578] Climate changes on other planets and their moons show that climate change elsewhere in the Solar System could not possibly be due to human activity on Earth. There must be a driving force outside the Earth. It is the Sun. If this is the case, we should see evidence of global warming in planets that orbit stars outside our Solar System. And we do. Jupiter-sized planets outside our Solar System show warming of the atmosphere related to orbital changes and changes in the energy emitted by the parent star.[579]

This is exactly what we see on Earth.

## Water, $CO_2$, temperature and the Sun

Over geological time, there is no observed relationship between global climate and atmospheric $CO_2$.[580] At times, $CO_2$ was up to 25 times higher than at present. At times, temperature was up to 10°C higher than at present. There were times when both temperature and $CO_2$ were high and there were times when the temperature was high and the atmospheric $CO_2$ was low. The time scale is not accurate enough over geological time to determine whether temperature drove the increase in $CO_2$, as is seen from ice cores covering the last 800,000 years.[581] Even in modern times, there is no relationship between temperature and $CO_2$ yet a close relationship between temperature and solar activity. The obvious question arises: If climate is unrelated to atmospheric $CO_2$ over the last 545 Ma, then how can today's climate be related to atmospheric $CO_2$?

---

[576] Fenton, L. K., Geissler, P. E. and Haberle, R. M. 2007: Global warming and climate forcing by recent albedo changes on Mars. *Nature* 446: 646-649.

[577] Marcus, P. S. 2004: Prediction of global climate change on Jupiter. *Nature* 428: 828-831.

[578] Hathaway, D. H. and Wilson, R. M. 2004: What the sunspot record tells us about space climate. *Solar Physics* 224: 5-19.

[579] Laughlin, G., Deming, D., Langton, J., Kasen, D., Vogt, S., Butler, P., Rivera, E. and Meschiari, S. 2009: Rapid heating of the atmosphere of an extrasolar planet. *Nature* 457: doi 10.1038/nature07649.

[580] Veizer, J., Godderis, Y. and Francois, L. M. 2000: Evidence for decoupling of atmospheric $CO_2$ and global climate during the Phaneroozoic eon. *Nature* 408: 698-701.

[581] Berner, R. A. 1990: Atmospheric carbon dioxide levels over Phanerozoic time. *Science* 249: 1382-1386.

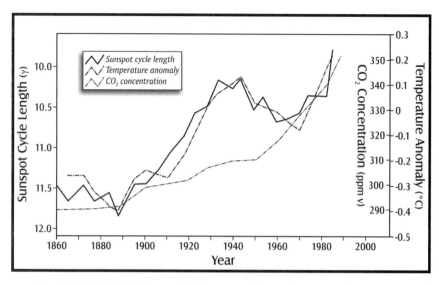

*Figure 15: Plot of the last 140 years of the increase and decrease of temperature (temperature anomaly) against the length of sunspots (a proxy for solar activity). This shows a correlation. There is no correlation between temperature and $CO_2$. The hypothesis that human emissions of $CO_2$ create global warming is shown to be false.*

The water cycle is the thermostat of climate change on Earth[582] with massive energy changes required for ice melting, water evaporation and precipitation. The carbon cycle does not drive climate, it piggybacks on the water cycle.[583,584,585]

Climate closely correlates with solar activity. Are climatic changes attributable to changes in solar activity much larger than can be expected from solar irradiance? When solar activity increases, the weak magnetic field that is carried by solar wind intensifies (providing more shielding of the Earth from low energy galactic cosmic rays), there is a reduction in cosmic ray-induced ion production in the lower atmosphere that results in fewer condensation nuclei there, and hence, less low-level cloud cover, which allows more solar radiation to impinge on the Earth, thereby increasing surface temperature.

[582] Shaviv, N. J. & Veizer, J. 2003: Celestial driver of Phanerozoic climate. *GSA Today* 13: 4-10.

[583] Lovett, R. A. 2002: Global warming: Rain might be the leading carbon sink factor. *Science* 296: 1787.

[584] Neff, U., Burns, S. J., Mangnini, A., Mudelsee, M., Fleitmann, D. and Matter, A. 2001: Strong coherence between solar variability and the monsoon in Oman between 9 and 6 kya ago. *Nature* 411: 290-293.

[585] Lee, D. and Veizer, J. 2003: Water and carbon cycles in the Mississippi river basin: potential implications for the northern hemisphere "residual terrestrial sink". *Global Biogeochemical Cycles* 17: doi:10.1029/2002GB001984.

Shaviv[586] identified six periods of the Earth's history (the entire Phanerozoic, Cretaceous, Eocene, the last glacial maximum, the 20th Century) and the 11-year solar cycle (as manifest over the last three centuries). He calculated on different time scales the changes in radiative forcing, temperature and cosmic ray flux. By using solar irradiance data,[587,588,589] Shaviv's result[590] validated the temperature increase (0.10°C) that was calculated by Idso (1998)[591] to be due solely to the 20th Century increase in the air's $CO_2$ concentration (75 ppmv; parts per million by volume).[592]  Shaviv's and Idso's independent analyses suggest that a maximum of only 15–20% (0.10°C of 0.57°C) of the observed warming of the 20th Century can be attributed to the concomitant rise in the air's $CO_2$ content.

Observations on historical, archaeological and geological time scales show that the principal driver of climate is celestial.[593]  Greenhouse gases act only as amplifiers. Solar activity accounts for some 80% of the global temperature trend over the last 150 years and the tiny carbon cycle has hitched a ride on the huge water cycles (including clouds). The natural climate trends can be amplified or modulated by $CO_2$, but $CO_2$ is not the principal driver of climate.

[586] Shaviv, N. J. 2005: On climate response to changes in the cosmic ray flux and radiative budget. *Journal of Geophysical Research* 110 (A8): A08105.1-A08105.15, doi:10.1029/2004JA010866.

[587] Hoyt, D. V. and Schatten, K. H., 1993: A discussion of plausible solar irradiance variations, 1700-1992. *Journal of Geophysical Research* 98, 18895-18906.

[588] Lean, J., Beer, J. and Bradley, R. 1995: Reconstruction of solar irradiance since 1610 – implications for climate change. *Geophysical Research Letters* 22: 3195-3198.

[589] Solanski, S. K., and Fligge, M. 1998: Solar irradiance since 1874 revisited. *Geophysical Research Letters* 25: 341-344.

[590] From these sets of data, Shaviv derived probability distribution functions of whole Earth temperature sensitivity to radiative forcing for each of the six time periods and combined them to obtain a mean planetary temperature sensitivity to radiative forcing of 0.28°C /W/m². Then, noting that the IPCC (2001) suggested that the increase in radiative forcing over the 20th Century was about 0.5 W/m², Shaviv calculated that the anthropogenic-induced warming of the globe over this period was approximately 0.14°C (0.5 W/m² x 0.28°C/W/m²). Based on information that indicated solar activity-induced increase in radiative forcing of 1.3 W/m² over the 20th Century (by way of cosmic ray flux reduction) plus the work of others, Shaviv calculated a globally averaged solar luminosity increase of approximately 0.4 W/m² for the same period. Shaviv calculated an overall and ultimately solar induced warming of 0.47°C (1.7W x 0.28 W/m²) and Shaviv calculated warming over the 20th Century as 0.61°C. This was noted by Shaviv to be very close to the 0.57°C temperature increase claimed by the IPCC to have been observed over the last century. Consequently, both Shaviv's and Idso's independent analyses suggest that a maximum of only 15-20% (0.10°C of 0.57°C) of the observed warming of the 20th Century can be attributed to the concomitant rise in the air's $CO_2$ content.

[591] Idso, S. B. 1998: $CO_2$-induced global warming: a skeptic's view of potential climate change. *Climate Research* 10: 69-82.

[592] Now abbreviated to ppmv for the rest of the book.

[593] Veizer, J. 2005: Celestial climate driver: A perspective from four billion years of the carbon cycle. *Geoscience Canada* 32: 13-28.

If solar activity is the principal driver of climate, we should see this in surface temperature measurements away from urbanisation and industry. We do. The 130-year measurement record of surface air temperatures in the Arctic shows variations in temperature that can be correlated with variations in the activity of the Sun.[594]

## Ancient signals of solar activity

Ice cores from Greenland show that the temperature was warmer at 1000 AD. The same cores show two very cold periods at 1550 and 1850 AD during the Little Ice Age. Temperature was 0.7 to 0.9°C colder than at present. After the Little Ice Age, temperature increased to 1930 and then decreased until 1995 (the year the study ended).[595] The Greenland continental climate is validated by sediment cores from a fjord in East Greenland that shows cooling after 1300 AD and very severe and variable climatic conditions from 1630 to 1900.[596] These temperature changes in Greenland may not seem great but in other areas temperature changes of up to 6°C over the last 8000 years have been recorded off the Alaskan coast by using floating single-celled organisms.[597] These provide a good indication to sea surface temperature and sea-ice cover, both of which changed in the Little Ice Age. Vegetation studies also show that the Alaskan land had temperature changes far greater than Greenland. Interglacial summer temperatures were 1 to 2°C higher than now and, in some locations, the summer temperature may have been as much as 5°C warmer.[598] In northern Quebec, soil deformations from the freezing and thawing of water show that there was severe cold from 1500 to 1900 AD.[599] In the Southern Hemisphere, evidence of solar changes controlling

---

[594] Soon, W. H. 2005: Variable solar irradiance as a plausible agent for multidecadal variations in the Arctic-wide surface air temperature record of the past 130 years. *Geophysical Research Letters* 32: L16712, doi:10.1029/2005GL023429.

[595] Dahl-Jensen, D., Mosegaard, K., Gundestrup, N., Clow, G. D., Johnsen, S. J., Hansen, A. W. and Balling, N. 1998: Past temperatures directly from the Greenland Ice Sheet. *Science* 282: 268-271.

[596] Jennings, A. E. and Weiner, N. J. 1996: Environmental change in eastern Greenland during the last 1,300 years: Evidence from foraminifera and lithofacies in Nansen Fjord, 68N. *The Holocene* 6: 171-191.

[597] Darby, D., Bischof, J., Cutter, G., de Vernal, A., Hillaire-Marcel, C., Dwyer, G., McManus, J., Osterman, L., Polyak, L. and Poore, R. 2001: New record shows pronounced changes in Arctic Ocean circulation and climate. *EOS* 82: 601-607.

[598] Muhs, D. R. , Ager, T. A. and Beget, J. E. 2001: Vegetation and palaeoclimate of the last interglacial period, Central Alaska. *Quaternary Science Reviews* 20: 41-61.

[599] Kasper, J. N. and Allard, M. 2001: Late Holocene climatic changes as detected by growth and decay of ice wedges on the southern shore of Hudson Strait, northern Quebec, Canada. *The Holocene* 11: 563-577.

the Little Ice Age are seen in the tropical Andes,[600] suggesting that the Little Ice Age was a global event driven by solar activity.

The advance and retreat of terrestrial glaciers in South America can directly be related to solar activity,[601] again showing that recent climate changes were global events driven by solar activity. Melting ice leaves debris, water is dammed by the debris and glacial lakes form. In winter, glacial lakes freeze and in summer the surface ice melts. This results in a winter-summer cycle of lake sediments and cycles of glacial and interglacial conditions. Glacial lakes in Sweden show a strong correlation with cosmic ray activity (and hence solar activity).[602]

The use of Northern Hemisphere temperature reconstructions,[603] three different proxies for solar energy output[604,605,606] and a method to model solar changes[607,608] show a good correlation between global temperature and solar-induced temperature curves.

In 641 AD a water level gauging structure was built on Rawdah (Rodda) Island in the Nile River at Cairo. It has provided the longest continuous record of river level measurements in the world. The analysis of 1080 years of measurements showed a 21-year cycle which correlates with sunspot activity.[609] An analysis of these records from 622–1470 AD showed 11-

[600] Polissar, P. J., Abbott, M. B., Wolfe, A. P., Bezada, M., Rull, V. and Bradley, R. S. 2006: Solar modulation of Little Ice Age climate in the tropical Andes. *Proceedings of the National Academy of Sciences* 103: 8937-8942.

[601] Douglass, D. C., Singer, B. S., Kaplan, M. R., Ackert, R. P., Mickelson, D. M. and Caffee, M. W. 2005: Evidence of early Holocene glacial advances in southern South America from the cosmogenic surface-exposure dating. *Geology* 33: 237-240.

[602] Snowball, I. and Sandgrena, P. 2002: Geomagnetic field variations in northern Sweden during the Holocene quantified from varved lake sediments and their implications for cosmogenic nuclide production rates. *The Holocene* 12: 517-530.

[603] Moberg, A., Sonechkin, D., Holmgren, K., Datsenko, N. and Karlen, W. 2005: Highly variable Northern Hemisphere temperatures reconstructed from low- and high-resolution proxy data. *Nature* 433: 613-617.

[604] Lean, J., Beer, J. and Bradley, R. 1995: Reconstruction of solar irradiance since 1610: implications for climate change. *Geophysical Research Letters* 22: 3195-3198.

[605] Lean, J. 2000: Evolution of the Sun's spectral irradiance since the Maunder Minimum. *Geophysical Research Letters* 27: 2425-2428.

[606] Wang, Y.-M., Lean, J. L. and Sheeley, Jr, N. R. 2005: Modelling the sun's magnetic field and irradiance since 1713. *The Astronomical Journal* 625: 522-538.

[607] Scarfetta, N. and West, B. J. 2005: Estimated solar contribution to the global surface warming using ACRIM TSI satellite composite. *Geophysical Research Letters* 32: doi:10.1029/2005GL025539.

[608] Scarfetta, N. and West, B. J. 2006: Phenomenological solar contribution to the 1900-2000 global surface warming. *Geophysical Research Letters* 33: doi:10.1029/2005GL025539.

[609] Hurst, H. E. 1951: Long term storage capacity of reservoirs. *Transactions of the American Society of Civil Engineers*, Paper 2447.

and 88-year cycles in accord with the Hale and Gleissberg solar cycles.[610] The 21-year Nile River cycles also correlated with other data from Africa including lake sediment data (over a period of 2000 years), tree rings (900 years), temperatures (175 years), rainfall (121 years) and wheat prices (1080 years).[611]

The Nile rises in the great equatorial lakes of East Africa (Lake Tana, Ethiopia; Lake Victoria, Tanzania, Uganda and Kenya). The past climate of Africa is interrelated to climate variability in the Indian and Atlantic Oceans and the history of African rainfall, rivers and lakes is directly related to solar activity.[612] There is a link between solar cycles and the water levels in Lake Victoria (the largest lake in Africa),[613] the Zambesi (Africa's third largest river)[614,615] and the Paraná River[616,617] in South America (the world's fifth largest river). The East African lakes have a large area and are shallow, hence they respond quickly to slight climate changes.

A later analysis of South Africa's largest river at the Vaal Dam showed a cyclicity of the Vaal River with sunspot activity.[618] These measurements were used successfully to predict changes from drought to flood in 1995[619] and again in 2006.[620] Major flood events were associated with the first half of the first sunspot cycle in a double sunspot cycle. This is when the rate of increase in sunspot activity is the greatest and is associated with global atmospheric and ocean turbulence. This generates the processes that produce heavy, widespread rainfall events that generate river flow. The second sunspot

[610] Ruzmaikin, A., Feynman, J. and Yung, Y. L. 2006: Is solar variability reflected in the Nile River? *Journal of Geophysical Research* 111: D21114, doi:10.1029/2006JD007462.

[611] Hurst, H. E. 1954: Measurement and utilisation of the water resources of the Nile Basin. *Proceedings of the Institution of Civil Engineers* 3: III, 1-26.

[612] Ruzmaikin, A., Feynman, J. and Yung, Y. L. 2006: Is solar variability reflected in the Nile River? *Journal of Geophysical Research* 111: D21114, doi:10.1029/2006JD007462.

[613] Sager, J. C., Ryves, D., Cumming, B., Meeker, L. D. and Beer, J. 2005: Solar variability and the levels of Lake Victoria, East Africa, during the last millennium. *Journal of Paleolimnology* 33: 243-251.

[614] Dyer, T. G. J. 1978: On the eleven-year solar cycle and river flow. *Water SA* 4: 157-160.

[615] Tyson, P. D., Cooper, G. R. J. and McCarthy, T. S. 2005: Millennial to multi-decadal variability in the climate of southern Africa. *International Journal of Climatology* 22: 1105-1117.

[616] Arpe, K., Cubasch, U. and Voss, R. 2000: Use of climate models for climate change investigations. *Proceedings of the 1st Solar and Space Weather Euroconference*, 25-29 September 2000, Tenerife, Spain. (ed. Wilson, A). Noordwijk, 233-241.

[617] Mauas, P. and Flamenco, E. 2003: Solar activity and the streamflow of the Paraná River. *Memoirs of the Astronomical Society of Italy* 76: 1002-1003.

[618] Alexander, W. J. R. 1978: Long range prediction of river flow – a preliminary assessment. *Technical Report TR80*, Department of Water Affairs, Pretoria.

[619] Alexander, W. J. R. 1995: Floods, drought and climate change. *South African Journal of Science* 91: 403-408.

[620] Alexander, W. J. R. 2005: Linkages between solar activity and climatic responses. *Energy and Environment* 16: 2.

cycle is a drought event. This cycle of drought and good seasons has been known for more than 100 years, whereby 10 good years are followed by 10 drought years.[621] The most extreme conditions occur at the beginning of the periods (floods) and the end of periods (droughts) with sudden reversals from drought to floods. This is not news – the ancient Egyptians were well aware of years of plenty followed by years of famine.

In northern Italy, more than 100 years of discharges from the Po River and rainfall records from the 75,000 square kilometre Po Basin show a correlation with solar activity. There are alternating wet and dry periods about every 20 years.[622]

There are claims that human-induced global warming will increase variability in hydrological processes, especially floods, droughts and water supplies. However, droughts commonly occur in solar-driven periods of global cooling.[623] Periods of drought are also windier and result in the deposition of wind-blown sand.[624] The claims of catastrophes resulting from warming have not been supported by a demonstration that global warming will change the alternating wet/dry sequences, the periodicity of such sequences and the drought/flood severities. Climate change models do not use the well documented, multi-year, alternating sequences of hydrometeorological processes which show a 21-year periodicity in the African data. A diversity of data sets has shown an unambiguous synchronous linkage between sunspot activity and climate. Why would this fundamental effect stop now and not continue to be a driver of climate change in the future?

Such predictions of solar variability and climate change are supported by studies of past climate.[625] The primary drivers of centennial- to decadal-scale changes in tropical rainfall and monsoon intensity 9000 to 6000 years ago in Oman were variations in solar intensity. Studies of changes in the Old World show that variations in solar activity have had a profound effect on human

---

[621] Hutchins, D. E., 1889: *Cycles of drought and good seasons in South Africa*. Wynberg.

[622] Zanchettin, D., Rubino, A., Traverso, P. and Tomasino, M. 2008: Impact of variations in solar activity on hydrological decadal patterns in northern Italy. *Journal of Geophysical Research* 113:, D12102, doi:10.1029/2007Jd009157.

[623] Booth, R. K., Jackson, S. T., Forman, S. L., Kutzbach, J. E., Bettis III, E. A., Kriegs, J. and Wright, D. K. 2005: A severe centennial-scale drought in midcontinental North America 4200 years ago and apparent global linkages. *The Holocene* 15: 321-328.

[624] Jackson, M. G., Oskarsson, N., Tronnes, R. G., McManus, J. F., Oppo, D. W., Gronvold, K., Hart, S. R. and Sachs, J. P. 2005: Holocene loess deposition in Iceland: Evidence for millennial-scale atmosphere-ocean coupling in the North Atlantic. *Geology* 33: 509-512.

[625] Neff, U., Burns, S. J., Mangini, A., Mudelsee, M., Fleitmann, D. and Matter, A. 2001: Strong coherence between solar variability and the monsoon in Oman between 9 and 6 kya ago. *Nature* 411: 290-293.

culture.[626] The use of climate proxies shows a correlation between rainfall in the Eastern Alps of Europe,[627] the monsoons of East Asia,[628] rainfall in the USA[629] and solar activity. There is a direct atmospheric link between the high-latitude climate of Greenland and that of the Aegean Sea. Aegean Sea surface temperature and the winter/spring intensity of the Siberian High follow solar cycles.[630]

Drilling of lake floor sediments in Cameroon's Lake Ossa has shown that floating micro-organisms responded to an oscillating climate. Oscillations are on a 1500-year cycle (Dansgaard-Oeschger Cycle) associated with the northern and southern movements of the Intertropical Convergence Zone.[631] The Intertropical Convergence Zone is a belt that circles equatorial regions. In this belt, solar energy and warm water raise air humidity. Drifting of the belt to the north and south affects rainfall-producing changes in the wet (flooding) and dry (drought) seasons in the tropics. Lake Ossa shows that the southward shift in the Intertropical Convergence Zone creates low rainfall in the northern tropics (e.g. Ghana, Nigeria) and high rainfall in the subequatorial zone (e.g. Zaire, Tanzania and Lake Malawi, Malawi[632]).

If the 1500-year cycle is global, then not only should lakes in Cameroon show cyclical features but lakes everywhere should show the cycles. Lake systems should reflect the 1500-year Dansgaard-Oeschger Cycle climate cycle with high lake levels during climate cooling (where there is high rainfall and little evaporation) and low lake levels during warming periods. Old beaches containing organic remains are one of the ways of documenting lake level changes. In the Great Lakes of the USA, the high water levels 1100 BC to 300 BC reflect cooling, the low water levels from 300 BC to 100 AD

---

[626] Feynman, J. 2007: Has solar variability caused climate change that affected human culture. *Advances in Space Research* 40: 1173-1180.

[627] Kofler, W., Krapf, V., Oberhuber, W. and Bortenschlager, S. 2005: Vegetation responses to the 8200 cal. BP cold event and to long-term climatic changes in the Eastern Alps: possible influence of solar activity and North Atlantic freshwater pulses. *The Holocene* 15: 779-788.

[628] Wang, Y., Cheng, H., Edwards, R. L., He, Y., Kong, X., An, Z., Wu, J., Kelly, M. J., Dykoski, C. A. and Li, X. 2005: The Holocene Asian monsoon: Links to solar changes and North Atlantic climate. *Science* 308: 854-857.

[629] Li, Y.-X., Lu, Z. and Kodama, K. P. 2007: Sensitive moisture response to millennial-scale climate variations in the Mid-Atlantic region, USA. *The Holocene* 17: 308.

[630] Rohling, E., Mayewski, P., Abu-Zied, R., Casford, J. and Hayes, A. 2002: Holocene atmosphere-ocean interactions: records from Greenland and the Aegean Sea. *Climate Dynamics* 18: 587-593.

[631] Nguestop, V. F., Servant-Vildary, S. and Servant, M. 2004: Late Holocene climatic changes in west Africa, a high resolution diatom record from equatorial Cameroon. *Quaternary Science Reviews* 23, 591-609.

[632] Filippi, M. L. and Talbot, M. R. 2004: The palaeolimnology of northern Lake Malawi over the last 25 ka based on the elemental and stable isotopic composition of sedimentary organic matter. *Quaternary Science Reviews* 24: 1303-1328.

reflect the Roman Warming, and the water level rise from 100 AD to 900 AD reflects the Dark Ages.[633] High water again between 1300 and 1600 AD reflects the Little Ice Age.[634]

Lake sediments from Lake Neufchatel in Switzerland show an abrupt temperature decrease of 1.5°C at the end of the Medieval Warming. These sediments also show that the mean annual temperatures in the Medieval Warming were higher than at present.[635]   In the Baltic Sea, the Medieval Warming allowed tropical and sub-tropical marine plankton to survive. Despite the Late 20th Century Warming, they have not returned because the Baltic Sea is still colder than it was in the Medieval Warming. At about 1200 AD, cold water micro-organisms replaced the warm water micro-organisms. This faunal replacement reflected the onset of the Little Ice Age.[636]

Lakes in Switzerland and France show 15 phases of high lake levels over the last 12,000 years. Correlations of European lake-level records[637,638] with the vegetation,[639,640] ice records,[641] the events of southward drift of North Atlantic ice[642] and $C^{14}$ records in tree rings and archaeological sites show that very slight changes in solar activity[643] produced large changes in climate in

[633] Thompson, T. A. and Baedke, S. J. 1999: Strandplain evidence for reconstructing later Holocene lake events in the Lake Michigan Basin. In: *Proceedings of the Great Lakes Palaeo-Levels Workshop: The last 4,000 years* (eds C Sellinger and F. Quinn). US Department of Commerce, Ann Arbor.

[634] Larsen, C. E. 1985: A stratigraphic study of beach features on the southwestern shore of Lake Michigan: New evidence of Holocene lake level fluctuations. *Illinois State Geological Survey Environmental Geology Notes* 112: 31.

[635] Filippi, F. L., Lambert P., Hunziker, J., Kübler, B. and Bernasconi, S. 1999: Climatic and anthropogenic influence on stable isotope record from bulk carbonates and Ostracodes in Lake Neufchatel, Switzerland, during the last two millennia. *Journal of Palaeolimnology* 21: 19-34.

[636] Andren, E., Andren, T. and Sohlenius, G., 2000: The Holocene history of the southwestern Baltic Sea as reflected in a sediment core from the Bornholm Basin. *Boreas* 29: 233-250.

[637] Harrison, S. P., Prentice, I. C. and Guiot, J. 1993: Climate controls on Holocene lake-level changes in Europe. *Climate Dynamics* 8: 189-200.

[638] Digerfeldt, G. 1988: Reconstruction and regional correlation of Holocene lake-level fluctuations in Lake Bysjön, South Sweden. *Boreas* 17: 165-182.

[639] Beaulieu, J.-L. de, Ruffaldi, R. H. and Clerc, J. 1994: History of vegetation, climate and human action in the French Alps and the Jura over the last 15,000 years. *Dissertationes Botanicae* 234: 253-276.

[640] Friedrich, M., Kromer, B., Spurk, M., Hofman, J. and Kaiser, K. F. 1999: Paleoenvironment and radiocarbon calibration as derived from Late Glacial/Early Holocene tree-ring chronologies. *Quaternary International* 61: 27-39.

[641] Finkel, R. C. and Nishiizumi, K. 1997: Beryllium 10 concentrations in the Greenland Ice Sheet Project 2 ice core for 3-40 ka. *Journal of Geophysical Research* 102: 26699-26706.

[642] Muscheler, R., Beer, G., Wagner, G. and Finkel, R. G. 2000: Changes in deep-water formation during the Younger Dryas event inferred from $^{10}Be$ and $^{14}C$ records. *Nature* 408: 567-570.

[643] Björck, S., Muscheler, R., Kromer, B., Andresen, C. S., Heineneier, J., Johnsen, S., Conley, D., Koç, N., Spurk, M. and Veski, S. 2001: High-resolution analysis of an early Holocene climate event may imply decreased solar forcing as an important climate trigger. *Geology* 29: 1107-1110.

the North Atlantic region over the last 12,000 years.[644]

Another way of looking at changes in solar activity document floating marine micro-organisms that are very sensitive to ocean surface temperature and sunlight. This correlation and a 1500-year cycle are seen in micro-organisms that lived in surface waters of the Sargasso Sea.[645] Surface winds and surface changes in the sub-polar North Atlantic Ocean have been influenced by solar input for the last 30,000 years. The production of $C^{14}$ and $Be^{10}$, derived from cosmic ray input, is also on 1500-year cycles. These cycles are seen in previous water temperatures measured from marine sediments, southward advances in drifting ice and changes in the North Atlantic Deep Water. A 3000-year temperature record has been reconstructed for the Norwegian Sea by using microfossils from sea floor sediment cores.[646] These cores show the major cool and warm periods over the last 3000 years and also show that it was common to have surface temperatures higher than the present temperature. Floating organisms preserved as a fossil record of the last 200,000 years in sea floor sediments in the Sulu Sea near the Philippines show a 1500-year cycle as well as the history of monsoon seasons.[647]

Marine sediment cores in the Gulf of California covering some 2000 years of deposition provide very high resolution, with each sediment layer representing 7 to 23 years of sediment deposition. Regularly spaced cycles of temperature-sensitive floating organisms for low sea surface temperature (*Octactis pulchra*) and warm tropical waters (*Azpeitia nodulifera*) show a solar influence on coastal upwelling (i.e. the rise of deep cold ocean waters to the surface).[648] Solar variability appears to be driving productivity cycles of these micro-organisms as increased intervals of $C^{14}$ production correlated with intervals of enhanced productivity. Increased winter cooling in the southwest USA during sunspot minima causes an intensification of the northwesterly winds that blow down the Gulf during late autumn to early spring. This leads to intensified overturn of surface waters and enhanced productivity.

The Earth's climate system is highly sensitive to extremely weak

[644] Magny, M. 2003: Holocene climate variability as reflected by mid-European lake-level fluctuations and its probable impact on prehistoric land settlements. *Quaternary International* 113: 65-79.

[645] Keigwin, L. 1996: The Little Ice Age and Medieval Warm Period in the Sargasso Sea. *Science* 274: 1503-1508.

[646] Andersson, C., Risebrobakken, B., Jansen, E., Dahl, S. O. and Meland, M. Y. 2003: Late Holocene surface ocean conditions of the Norwegian Sea (Vøring Plateau). *Palaeoceanography* 18, doi: 10.1029/2001PA000654.

[647] de Garidel-Thoron, T. and Beaufort, L. 2000: High frequency dynamics of the monsoon in the Sulu Sea during the last 200,000 years. *EGS General Assembly*, Nice, April 2000.

[648] Barron, J. A. and Bukry, D. 2006: Solar forcing of Gulf of California during the past 2000 yr suggested by diatoms and silicoflagellates. *Marine Micropalaeontology* 62: 115-139.

perturbations in the Sun's energy output on decadal and millennial scales. Changes of only 0.1% in solar activity over the 11-year sunspot cycle result in profound changes to climate.[649] The 1500-year cycle is a pervasive feature of the climate system linked to solar irradiance, predicted more than three decades ago.[650] This negates the need to suggest internally forced oscillations in the deep ocean circulation[651] or longer-term internally and orbitally forced modulations of atmospheric variability as primary forcing mechanisms.[652] For example, a statistical study of the El Niño-Southern Oscillation shows the 11-year solar cycle.[653] In the Atlantic, the weather in mid latitude areas is dominated by the variation in storm tracks (North Atlantic Oscillation).[654] Changes in solar activity can be the precursor of at least some of the North Atlantic Oscillation variability, part of a global pattern extending from the Earth's surface to the stratosphere.[655,656]

Records of rainfall preserved in stalagmites on the Arabian Peninsula also show solar cycles.[657] In areas closer to the equator where sea ice has not formed for more than 600 Ma, the oxygen chemistry of floating shells that, upon death, fell into deep ocean sediments and wind-blown dust give the same story of 1500-year cycles. Sea floor sediments are a goldmine of information. Organisms that float, plankton, are very sensitive to temperature changes. Surface water temperature changes of less than 1°C can induce the

---

[649] Bond, G., Kromer, B., Beer, J., Muscheler, R., Evans, M. N., Showers, W., Hoffmann, S., Lotti-Bond, R., Hajdas, I. and Bonani, G. 2001: Persistent solar influence on North Atlantic climate during the Holocene. *Science* 294: 2130-2136.

[650] Denton, G. H. and Karlen, W. 1973: Holocene climatic variations: their pattern and possible cause. *Quaternary Research* 3: 155.

[651] Broecker, W. S., Sutherland, S. and Peng, T.-H. 1999: Possible 20th-Century slowdown of Southern Ocean deep water formation. *Science* 286: 1132-1135.

[652] Cane, M. and Clement, A. 1999: A role for the tropical Pacific coupled ocean-atmosphere system in Milankovitch and millennial time scales. Part II. Global impacts. In: *Mechanisms of Global Climate Change at Millennial Time Scales* (eds Clark, P., Webb, R. and Keigwin, L. D.). Geophysical Monograph Series 112: 373-383.

[653] Kryjov, V. N. and Park, C.-Y. 2007: Solar modulation of the El Niño/Southern Oscillation impact on the Northern Hemisphere annular mode. *Geophysical Research Letters* 34: L10701, doi:10.1029/2006GL028015.

[654] Trigo, R. M., Osborn, T. J. and Corte-Real, J. M. 2002: The North Atlantic Oscillation influence on Europe: Climate impacts and associated physical mechanisms. *Climate Research* 20: 9-17.

[655] Appenzeller, C., Weiss, A. K. and Staehelin, J. 2000: North Atlantic Oscillation modulates total ozone winter trends. *Geophysical Research Letters* 27: 1131-1134.

[656] Marshall, J., Kushnir, Y., Battisti, D., Chang, P., Czaja, A., Dickson, R., Hurrell, J., McCartney, M., Saravan, R. and Visbeck, M. 2001: North Atlantic climate variability: Phenomena, impacts and mechanisms. *International Journal of Climatology* 21: 1863-1898.

[657] Neff, U., Burns, S. J., Mangini, A., Mudelsee, M., Fleitmann, D. and Matter, A. 2001: Strong coherence between solar availability and monsoon in Oman between 9 and 6 kyr ago. *Nature* 411: 290-293.

plankton to change to another species and the oxygen chemistry can be used as a highly accurate thermometer. In the Arabian Sea west of Karachi, two seabed cores give a 5000-year record and show the same 1500-year cycle recorded from Greenland ice cores.[658] These were interpreted as being of tidal rather than solar origin.

The global oxygen isotopic record[659] shows a 4000-year cyclic variation of the Antarctic climate and glacial cycles which result from the Earth's variable orbit around the Sun. Both temperature and $CO_2$ rise at the beginning of each cycle. This leaves little room for $CO_2$ produced by humans to affect climate.

Stalagmites in a cave in the Makapansgat Valley of South Africa show that the region was 1°C cooler than at present from 1300 to 1800 AD. The lowest temperatures recorded in South Africa were in the Maunder and Spörer Minima.[660] Similar cycles of climate and solar activity are seen from cave deposits in the USA.[661] Cave stalagmites from Sauerland in Germany record a 17,600-year history in which recent warmings (e.g. Medieval Warming, Roman Warming) and coolings (e.g. Dark Ages, Little Ice Age) are recorded as well as a 1500-year cycle.[662] The German cave record was very similar to that from Ireland,[663] indicating that climate changes were not local.

Vegetation is a very sensitive indicator of climate change. The measurement of $C^{14}$ in tree rings is a proxy for solar activity. The last 2000 years has shown 200-year cycles in $C^{14}$ in tree rings and wind intensity (calculated from the wind-blown component of lake sediments).[664] Changes in cyclonic activity and tropospheric winds have been reported a few days after solar flares,[665,666]

[658] Berger, W. H. and von Rad, U. 2002: Decadal to millennial scale cyclicity in varves and turbidites from the Arabian Sea: Hypothesis of tidal origin. *Global and Planetary Change* 34, 313-325.

[659] Raymo, M. E., Lisiecki, L. E. and Nisancioglu, K. H. 2006: Plio-Pleistocene ice volume, Antarctic climate, and the global $\partial O^{18}$ record. *Science* 313: 492-495.

[660] Tyson, D., Karlen, W., Holmgren, K. and Heiss, G. A. 2000: The Little Ice Age and Medieval Warming in South Africa. *South African Journal of Science* 96: 121-126.

[661] Asmerom, Y., Polyak, V., Burns, S. and Rasmussen, J. 2007: Solar forcing of Holocene climate: New insights from a speleotherm record, southwestern United States. *Geology* 35: 1-4.

[662] Niggemann, S., Mangini, A., Richter, D. K. and Wirth, G. 2003: A palaeoclimate record of the last 17,600 years in stalagmites from the B7 Cave, Sauerland, Germany. *Quaternary Science Reviews* 22: 555-567.

[663] McDermott, F. 1999: Centennial scale Holocene climate variability revealed by a high-resolution speleotherm O-18 record from SW Ireland. *Science* 294: 1328-1333.

[664] Anderson, R. Y. 1992: Possible connection between surface winds, solar activity and the Earth's magnetic field. *Nature* 358: 51-53.

[665] Damon, P. E., Cheng, S. and Linick, T. W. 1989: Possible connection between surface winds, solar activity and the Earth's magnetic field. *Radiocarbon* 31: 697-703.

[666] Stuiver, M., Braziunas, T. F., Grootes, P. M. and Zielinski, G. A. 1997: Is there evidence for solar forcing of climate in the GISP2 oxygen isotope record. *Quaternary Research* 48: 259-266.

again showing that the Sun rules the Earth.

Pollen fossils over the last 14,000 years in North America show that the vegetation has undergone nine separate events of re-organisation on a 1650 ± 500-year cycle. This is within the order of accuracy of the 1500 ± 500-year solar cycles, strongly suggesting a solar influence.[667] This again supports the view that there is climate stability during warm periods and wildly fluctuating climate in cold periods. The North American Pollen Database shows shifts in vegetation every 1650 years for the past 14,000 years. The most recent shift was 600 years ago, culminating in the Little Ice Age, with maximum cooling 300 years ago. The previous shift began about 1600 years ago and culminated in the maximum warming in the Medieval Warming 1000 years ago.[668] This again strongly suggests a cyclical event of about 1500 ± 500 years. Elsewhere in North America, the forests also responded to the extreme cold of the Little Ice Age.[669] The foxtail pine and western juniper of the southern Sierra Nevada Mountains indicate that it was warmer than current times from 1100 to 1375 AD and colder from 1450 to 1850 AD. Tree rings in the long-lived bristlecone pines on the California-Nevada border show that some trees are 5500 years old. From 800 AD to the present, the 100-year averages of bristlecone temperatures correlate statistically with the temperatures derived from central England.[670]

If solar activity drives climate, then these changes would be expected to be more widespread. And they are. Solar activity has influenced the North Atlantic and vegetation in the Eastern Alps of Europe.[671] It had a profound effect on the vegetation and population of China[672] and, in the higher latitudes of Europe, climate change was a matter of life and death for humans.[673]

---

[667] Viau, A., Gajewski, K., Fines, P., Aitkinson, D. E. and Sawada, M. C. 2002: Widespread evidence of a 1,500-yr climate variability in North America during the past 14,000 years. *Geology* 30: 455-458.

[668] Viau, A. E., Gajewski, K., Fines, P., Atkinson, D. E. and Sawada, M. C. 2002: Widespread evidence of 1,500-Yr climate variability in North America during the past 14,000 years. *Geology* 30: 455-458.

[669] Graumlich, I. D. 1993: A 1,000-year record of temperature and precipitation in the Sierra Nevada. *Quaternary Research* 39: 249-255.

[670] LaMarche, V. C. 1974: Palaeoclimatic interferences from long tree ring records. *Science* 183, 1043-1048.

[671] Kofler., W., Krapf, V., Oberhuber, W. and Bortenschlager, S. 2005: Vegetation responses to the 8200 cal. BP cold event and to long term changes in the Eastern Alps: possible influence of solar activity and North Atlantic freshwater pulses. *The Holocene* 15: 779-188.

[672] Nagovitsyn, Y. 2001: Solar activity during the last two millennia: Solar patrol in ancient and medieval China. *Geomagnetism and Aeronomy* 41: 680-688.

[673] Magny, M. 2004: Holocene climate variability as reflected by mid-European lake-level fluctuations and its probable impact on prehistoric human settlements. *Quaternary International* 113: 65-79.

Vegetation in South America also felt the effects of climate change, especially the Little Ice Age. Pollen from lake sediments in Peru provides a 4000-year record of climate. The Roman Warming was recorded, after which rainfall declined during the Dark Ages. In the Medieval Warming, increased pollen indicated warmer temperatures, more plants and greater plant diversity and there was a pollen decline in the Little Ice Age.[674] Elsewhere in South America, lake sediments from a high volcanic plateau showed that climate and rainfall changed quickly, with the Little Ice Age prominent.[675]

Bogs in the Netherlands,[676] Northern Hemisphere peats[677] and peat in Canada[678] all show climate cycles driven by solar activity. The stomata of fossil leaves can be used as a proxy measure of $CO_2$. The sampling of the stomata of sub-fossil leaves in 11,230- to 10,330-year-old bogs in the Faeroe Islands shows a $CO_2$ decrease at about 11,050 years ago, possibly due to expanding vegetation in the Northern Hemisphere. A consistent and steady decline in $CO_2$ occurred between 10,900 and 10,600 years ago and increased instability after 10,550 years ago was probably due to increased cooling of North Atlantic surface waters.[679] The reconstructed $CO_2$ changes show a distinct similarity to indicators of changing solar activity. This suggests that there was great sensitivity to changes in solar activity during this time and that atmospheric $CO_2$ concentrations fluctuated as a result of these rapid changes in climate.[680]

Peat bogs in northwest Europe show an abrupt climate cooling 2800 years ago. This cooling is recorded for many other places, such as on the other side of the world at Tierra del Fuego. The timing, nature and global

---

[674] Chepstow-Lusty, A. J. , Bennett, K. D., Fjeldsa, J., Kendall, A., Galiano, W. and Tupayachi Herrera, A. 1988: Tracing 4,000 years of environmental history in the Cuzco area, Peru, from the pollen record. *Mountain Research and Development* 18: 159-172.

[675] Valero-Garcés, V. L. , Delgado-Huertas, A., Ratto, N., Navas, A. and Edwards, L. 2000: Palaeohydrology of Andean saline lakes from sedimentological and isotopic records, northwestern Argentina. *Journal of Palaeolimnology* 24: 343-359.

[676] Blaauw, M., van Geel, B. and van der Plicht, J. 2004: Solar forcing of climate change during the mid-Holocene: indications from raised bogs in The Netherlands. *The Holocene* 14: 35-44.

[677] Mauquoy, D., van Geel, B., Blaauw, M., Speranza, A. and van der Plicht, J. 2004: Changes in solar activity and Holocene climate shifts derived from $^{14}C$ wiggle-match dated peat deposits. *The Holocene* 14: 45-52.

[678] Yu, Z., Campbell, I. D., Campbell, C., Vitt, D. H., Bond, G. C. and Apps, M. J. 2003: Carbon sequestration in western Canadian peat highly sensitive to Holocene wet-dry climate cycles at millennial time scales. *The Holocene* 13: 801-808.

[679] Jessen, C. A., Rundgren, M., Björck, S. and Muscheler, R. 2007: Climate forced atmospheric $CO_2$ variability in the early Holocene: A stomatal frequency reconstruction. *Global and Planetary Change* 57: 247-260.

[680] Jessen, C. A., Rundgren, M., Björck, S. and Muscheler, R. 2007: Climate forced atmospheric $CO_2$ variability in the early Holocene: A stomatal frequency reconstruction. *Global and Planetary Change* 57: 247-260.

characteristics of this cooling suggest solar forcing, amplified by oceanic circulation.[681] In fact, the numerous rapid high-magnitude climate changes in the Holocene, such as the Medieval Warming and the Little Ice Age, can really only be explained by changes in solar activity.[682]

The studies on plants, peat, pollen and spores show that $CO_2$ changed rapidly during the last 14,500 years of post-glacial climate fluctuations. Ice cores show little variability in atmospheric $CO_2$ over short time scales and may only give smoothed broad-scale trends.[683] This is because of the diffusion of air through ice during compaction.

## The Sun and climate

This data on solar forcing of climate shows that the climate system is far more sensitive to small variations in solar activity than previously thought. This has major implications for understanding short-term climate changes observed in deep time. Archaeological and historical records[684] and measurements of previous climate changes (e.g. Maunder Minimum[685,686]) all show that a very slight change in solar activity has a profound effect on the Earth's climate.

The 20th Century shows an overall warming of some 0.7°C, almost half of which was before 1945 when the Sun was more active and when there was far less emission of $CO_2$ into the atmosphere by humans. Although cosmic rays have been systematically measured only since 1937, the cosmic ray input resulting from past varying solar activity could be measured by the isotope fingerprints of $C^{14}$ and $Be^{10}$. The pronounced cooling from 1945 until the mid 1970s was at the time of a weakening of the Sun's magnetic activity and, after 1975, the upward trend in solar activity resumed. As a result, warming resumed. During the 20th Century, the Sun's magnetic shield more than doubled in strength. This larger magnetic shield reduced the bombardment

[681] Chambers, F. M., Mauquoy, D., Brain, S. A., Blaauw, M. and Daniell, J. R. G. 2007: Globally synchronous climate change 2800 years ago: Proxy data from peat in South America. *Earth and Planetary Science Letters* 253: 439-444.

[682] Bard, E. and Frank, M. 2006: Climate change and solar variability: What's new under the sun? *Earth and Planetary Science Letters* 248: 1-14.

[683] Monnin, E., Indermühle, A., Dällenbach, A., Flückiger, J., Stauffer, B., Stocker, T. F., Raynaud, D. and Barnola, J.-M. 2001: Atmospheric $CO_2$ concentrations over the last glacial termination. *Science* 291: 112-114.

[684] Perry, C. A. and Hsu, K. J. 1000: Geophysical, archaeological and historical evidence support a solar-output model for climate change. *Proceedings of the National Academy of Sciences* 97: 12433-12438.

[685] Soon, W. H. and Yaskell, S. H. 2004: *The Maunder Minimum and variable Sun-Earth connection.* World Scientific Publishing.

[686] Shindell, T., Schmidt, G. A., Mann, M. E., Rind, D. and Waple, A. 2001: Solar forcing of regional climate change during the Maunder Minimum. *Science* 294: 2149-2152.

of the Earth by cosmic rays, less cloud formed and the Earth was warmer. Warming of nearly 0.6°C has been calculated for the reduced cosmic ray input and cloudiness for the 20th Century.[687] If the atmospheric temperature in the 20th Century rose by 0.7°C, then most of this rise can be attributed to solar, cosmic ray and cloud influences. This trend may continue.[688] The cooling in the first decade of the 21st Century is almost as much as the whole warming of the 20th Century. Predictions of global warming driven by human emissions of $CO_2$ in the 21st Century look doomed. The most exaggerated predictions of the climate alarmists suggest an extraordinary temperature rise of up to 5°C.

However, all is not as simple as it seems. Anarchy exists in Antarctica. When the world as a whole seems to warm up, Antarctica cools. When Antarctica cools, the world warms. This paradox may be because in Antarctica, the clouds warm the highly reflective snow and ice surface whereas elsewhere, where the surface is land and water, clouds chill the surface. Antarctica normally warms when Greenland cools. Antarctica normally cools when Greenland warms.

During the Little Ice Age, Greenland was very cold and Antarctica was relatively warm. In the period from 1550 to 1700 AD, melting of Greenland ice was common with 8% of the years experiencing melting and elevated summer temperatures.[689] Furthermore, some 7000 years ago there was no ice melting and it was especially cold in Antarctica and it was unusually warm in Greenland at a period that had the greatest summer melting of ice over the last 10,000 years. It was suggested that there was a polar see-saw (i.e. Antarctic climate anomaly) with warmth flipping from one hemisphere to another.[690] What makes Antarctica even more isolated is that the Circum Antarctic Ocean Current isolates Antarctica from tropical currents. In the Antarctic stratosphere, the polar vortex winds are far more powerful than at the North Pole and again serve to isolate Antarctica.

The Antarctic climate anomaly is clearly not related to $CO_2$, which spreads almost uniformly in the atmosphere across the globe, including the poles. The numerous climate models based in increases in $CO_2$ predict warming of both hemispheres, including Antarctica and Greenland. Human-produced $CO_2$ or human influences on the ozone hole above Antarctica clearly did not

[687] Marsh, N. D. and Svensmark, H. 2000: Low cloud properties influenced by cosmic rays. *Physical Review Letters* 85: 5004-5007.

[688] Kin, Z.-S. and Xian, S. 2007: Multi-scale analysis of global temperature changes and trends of a drop in temperature in the next 20 years. *Meteorology and Atmospheric Physics* 95: 115-121.

[689] Joughun, I., Das, S. B., King, M. A., Smith, B. E., Howat, I. M. and Moon, T. 2008: Seasonal speedup along the western flank of the Greenland ice sheet. *Science* 320: 781-783.

[690] Shackleton, N. 2001: Climate change across the hemispheres. *Science* 291: 58-59.

influence the Antarctic climate anomaly 7000 years ago.

As we see in the next chapter, Milankovitch Cycle wobbles in the Earth's orbit are inadequate to explain the rapid flipping of Greenland and Antarctic temperatures but play a pivotal role in long-term climate change when combined with solar activity and other forcing factors.

If Popper's[691] falsification method is the process whereby science advances, then the Antarctic climate anomaly is sufficient to refute the hypothesis that humans change climate.

Clouds clearly have a large role to play in climate change. When cloud cover decreases, planet Earth warms and Antarctica cools because the snow cover of Antarctica is the whitest surface on Earth. It reflects more energy than Arctic snow, cloud tops, oceans and land masses. Cloud masses over Antarctica are such that warming takes place in both summer and winter and there is no predictive basis for climate change in Antarctica.[692] Studies in Greenland show that a decrease in cloud cover has a chilling effect although Greenland is considerably smaller than Antarctica and the Greenland snow is darker than Antarctic ice (i.e. absorbs more heat).

The NASA Earth Radiation Budget Experiment shows that if cloudiness increases by 4%, temperature should increase at the Equator by 1°C and decrease in Antarctica by 0.5°C. However, there are competing factors at work. Water vapour is the main greenhouse gas in the atmosphere. When times are warmer, water vapour evaporates more readily. This means that less heat escapes into space. With the slight decrease in cloudiness in the 20th Century, the planet would have warmed and with the consequent slight increase in atmospheric water vapour, Antarctica would have warmed slightly and this warming would have been greater than the cooling due to the loss of clouds.

Ice core measurements show that over the past 90,000 years there were seven warmings in Greenland. These were followed, 1500 to 3000 years later, by Antarctic warmings. It appears that as Antarctica was warming, Greenland was cooling and, when Antarctic warming ceased, rapid warming started in Greenland.[693] Because climate changes were over the scale of decades, it is highly unlikely that shifting ocean currents could have caused such variations for 90,000 years.

[691] Popper, K. R., 1979: *Objective knowledge*. Oxford University Press.

[692] Vaughan, D. G., Marshall, G. J., Connolley, W. M., Parklinson, C., Mulvaney, R., Hodgson, D. A., King, J. C., Pudsey, C. J. and Turner, J. 2003: Recent rapid regional climate warming on the Antarctic Peninsula. *Climate Change* 60: 243-270.

[693] Blunier, T. and Brook, E. J. 2001: Timing of millennial-scale climate change in Antarctica and Greenland during the last glacial period. *Science* 291: 109-112.

Are the changes in the energy emitted from the Sun unique? It appears not. Some Sun-like stars can lose 0.4% of their luminosity in a few years. Some 25 years of observation of Sun-like stars shows that Tau Ceti is now almost magnetically inert. Another Sun-like star, 54 Piscium, was a normal sun until 1980 and then magnetic activity suddenly dropped. If this happened to our Sun, we would enter another Dalton or Maunder Minimum.

It is strange that those who claim that human $CO_2$ emissions drive climate change are reluctant to consider the role of the Sun. Variations in solar activity correlating with climate have been known for hundreds of years since the time of Maunder and Herschel.

Occam's Razor[694] rules: That great ball of heat in the sky drives climate.

---

[694] *Entia non sunt multiplicanda*, roughly translated as "The simplest explanation of all the data is probably the best".

# Chapter 4

# THE EARTH

Question: Do volcanoes change climate?
Answer: Yes.

Question: Do wobbles in the Earth's orbit change climate?
Answer: Yes.

Question: Have past climate changes driven extinction?
Answer: Yes and no.

*Planet Earth is an evolving dynamic planet. Earth has less carbon and water than other planets, asteroids and comets. It is a warm wet volcanic greenhouse planet that has had ice for less than 20% of the time.*

*The carbon cycle has been operating for at least 4000 million years and has been controlled by chemical reactions between water, air and rocks. It still is. These reactions have stopped a runaway icehouse or a runaway greenhouse. Tipping points are a non-scientific myth.*

*Early life thrived in a $CO_2$-rich atmosphere. Bacteria rule the world and, together with water, are a key driver of the carbon cycle. The first $CO_2$ was added to the atmosphere from volcanoes. This process is still happening.*

*Extraordinary galactic and planetary processes are necessary to produce the environment for multicellular life. Since multicellular life appeared on Earth, there has been a constant draw-down of $CO_2$ from the atmosphere that once had more than 100 times the current $CO_2$ content. During the current ice age, the atmosphere has had the lowest $CO_2$ content ever.*

*There have been huge climate changes in the past, none of which are adequately explained. For example, in the Ordovician-Silurian glaciation (450–420 million years ago), the atmospheric $CO_2$ was more than 4000 ppmv, demonstrating that $CO_2$ does not drive warming. Desertification is a characteristic of global cooling and has occurred many times in the past. At present, the planet is becoming more vegetated and deserts are contracting.*

*There is a constant turnover of life by extinction, which creates environments for new species. Extinction is normal. Conservation of species is contrary to Nature. At times, there have been mass extinctions of life, mainly from gas. This gas is from supervolcanoes.*

*Some species extinctions result from cooling, whereas global warming and high $CO_2$ do not lead to extinction. Warming creates biodiversity, a thriving of life, species migration and adaptation.*

*Modern species turnover by competition has made some species extinct. An example is human migration driving the extinction of some macrofauna. Polar bears have survived numerous intense periods of past global warmings, and modern tropical species extinction is due to changing land use rather than climate change. Malaria and other tropical afflictions are diseases of poverty and show no relationship to warming.*

*Supervolcanoes shape the Earth. They induce extinction, change ocean currents, change climate and add monstrous amounts of particles, $CO_2$ and sulphur gases to the atmosphere. Unseen submarine supervolcanoes have yet to be understood. The loading and unloading of ice during past climate changes have triggered earthquakes and volcanoes. We live in a period when volcanoes are quiet.*

*Orbital wobbles place the Earth at a varying distance from the Sun. Past climate changes have been influenced by orbital wobbles but the trigger for climate change and the influence of the Earth's orbital changes are not yet understood.*

## Life on Earth

### Single-celled life

The dominant life form on Earth is bacteria. They, like us, are based on DNA. This means we are related to the bacteria in our gut, on our teeth, in our faeces and those that can kill us. Bacteria, mammals, reptiles, amphibians, plants and all life are children of DNA. We all have the same sort of cells and hence have a common origin.[695] Some 90% of the cells in a human body are

[695] Gu, X. 1997: The age of the common ancestor of eukaryotes and prokaryotes. Statistical inferences. *Molecular Biology and Evolution* 14: 861-866.

bacterial. Bacteria account for 15% of the weight of the human body. The greatest biomass on Earth is bacteria. They thrive in both hostile and benign environments. Bacteria can do many things we can't. For example, they can eat rock 3 km beneath our feet[696] and 5 km beneath the ocean floor.[697,698]

These single-celled organisms exist in clouds, ice, hot springs, salt, soil, water, cracks in rocks and every conceivable environment on Earth. At least half the world's biomass is below the Earth's surface,[699] mainly as extremophile bacteria that can tolerate high pressure, high or low temperatures and extremes of acidity, with or without oxygen.[700] Bacteria dominate hot, cold, dry, wet, oxygen-rich, oxygen-poor, dark, sunny, acid, alkaline, radioactive, high-pressure and low-pressure environments.[701,702] Liquid water allows the transfer of nutrients in bacterial cells and it is no coincidence that the first appearance of life on Earth was when there was liquid water on Earth. Bacteria can thrive in environments that normally are toxic to multicellular organisms. After hundreds of millions of years of being trapped in brine, bacteria can be resurrected.[703] Other bacteria can survive a 25 kV electron beam and resultant high temperatures in a vacuum, which stimulates speculation about extremophiles hitching a ride on meteorites.[704]

Bacteria are the ultimate survivors on planet Earth. The chances of finding modern or fossil life elsewhere in our Solar System are very high because of the resilience of bacteria[705] and the appropriate geological settings

---

[696] Lithautotrophs

[697] Stevens, T. O. and McKinley, J. P. 1995: Lithautotrophic microbial ecosystems in deep basalt aquifers. *Science* 270: 450-454.

[698] Santelli, C. M., Orcutt, B., Banning, E., Bach, W., Moyer, C. L., Sogin, M. L., Staudigel, H. and Edwards, K. J. 2008: Abundance and diversity of microbial life in ocean crust. *Nature* 453: 653-656.

[699] McCall, J. 2003: The deep biosphere. *Geoscientist* 13 (3): 11.

[700] Jørgensen, B. B. and Boetius, A. 2007: Feast and famine – microbial life in the deep seabed. *Nature Reviews Microbiology* 5: 770-781.

[701] Hei, D. J. and Clark, D. S. 1994: Pressure stabilization of proteins from extreme thermophiles. *Applied and Environmental Microbiology* 60: 932-939.

[702] Rothschild, L. J. and Mancinelli, R. L. 2001: Life in extreme environments. *Nature* 409: 1092-1100.

[703] Vreeland, R. H., Rosenzweig, W. D. and Powers, D. W. 2000: Isolation of a 250 million-year-old halotolerant bacterium from a primary salt crystal. *Nature* 409: 897-900.

[704] McKay, D. S., Gibson, E. K., Thomas-Keprta, K. L., Vali, H., Romanek, C. S., Clemett, S. J., Chillier, X. D. F., Maechling, C. M. and Zare, R. N. 1996: Search for past life on Mars: possible relic biogenic activity in Martian meteorite ALH84001. *Science* 273: 924-930.

[705] Baross, J. A. and Holden, J. F. 1996: Overview of hyperthermophiles and their heat-shock proteins. *Advances in Protein Chemistry* 48: 1-35.

for bacterial life.[706,707,708] Modern life may exist on Mars,[709] Europa[710] and Titan[711,712] and fossil life may be found on Mars.[713]

Was the rise of life inevitable on this planet? If we ran the story of the Earth again, would life itself again evolve? And if it did, would it evolve with DNA at its crux? Was ammonia the solvent necessary for life, only to be replaced by a better solvent (water)? We need a solvent for delivering chemicals in solution for cells to operate. Did the appearance of life happen just once and then spread across Earth, or did life appear many times, only to be wiped out many times by asteroid impacts? Did RNA evolve to DNA, or have mineral surfaces (iron sulphides, clays) been the site for the biological takeover of non-biological reactions? How quickly did life appear? If life appeared quickly, then there may well have been many occurrences of life elsewhere in the Solar System. Where did the appearance of life happen? Darwin suggested a Sun-warmed pond. Others suggest hot springs, especially in the deep oceans, whereas others suggest cracks deep in rocks. These early steps we do not see in the fossil record because the oldest rocks on Earth are 4200 Ma and ancient rocks have been deformed and heated many times. We see bacteria and archaea extremophiles as the oldest fossils. Were the individual stages in the origin of life (amino acids, nucleic acids, cells) dependent upon long-term changes in the Earth's environment? Did the origin of life change the environment such that life could never evolve again? At what stage did evolution take over to influence the development of life? Too many questions, not enough answers.

Colonies of bacteria are like a huge parallel processing machine, an

[706] Russell, M. J., Hall, A. J., Cairns-Smith, A. J. and Braterman, P. S. 1988: Submarine hot springs and the origin of life. *Nature* 336: 117.

[707] Romanek, C. S., Grady, M. M., Wright, I. P., Mittlefehldt, D. W., Socki, R. A., Pillinger, C. T. and Gibson, Jr, E. K. 1994: Record of fluid interactions on Mars from the meteorite ALH84001. *Nature* 372: 655-657.

[708] Russell, M. J., Ingham, J. J., Zedef, V., Maktav, D., Sunar, F., Hall, A. and Fallick, A. 1999: Search for signs of ancient life on Mars: expectations from hydromagnesite microbialites, Salda Lake, Turkey. *Journal of the Geological Society, London* 156: 869-888.

[709] Squyres, S. W. and Kasting, J. F. 1994: Early Mars – how warm and how wet? *Science* 265: 744.

[710] Jakosky, B. M. and Shock, E. L. 1998: The biological potential of Mars, the early Earth, and Europa. *Journal of Geophysical Research* 103: 19359.

[711] Lunine, J. I., Stephenson, D. J. and Yung, Y. L. 1983: Ethane ocean on Titan. *Science* 222: 1229-1230.

[712] Brown, R. H., Soderblom, L. A., Soderblom, J. M., Clark, R. N., Jaumann, R., Barnes, W., Sotin, C., Buratti, B., Baines, K. H. and Nicholson, P. D. 2008: The identification of liquid ethane in Titan's *Ontario Lacus*. *Nature* 454: 607-610.

[713] Kirschvink, J. L., Maine, A. T. and Vali, H. 1997: Palaeomagnetic evidence supports a low temperature origin of carbonate in the Martian meteorite ALH84001. *Science* 275: 1629-1633.

intelligent machine that does what no computer can do. Bacteria re-engineer their genome to survive in a new hostile environment, which is why they are the longest-running players in the game of evolution. The huge number of bacteria guarantees survival. Bacteria are the only organisms that remain on Earth from the beginning of life before 4000 Ma. Before that, there may have been RNA- or protein-based life.[714] Planet Earth formed on a Thursday[715] at 4567 Ma. By contrast, humans (*Homo sapiens*) have only been on Earth for just over 100,000 years.

Colonies of fossilised bacteria can still be found, despite the fact that 90% of the ancient rocks once deposited on Earth are now gone. By 3500 Ma we have very clear evidence of bacterial colonies (stromatolites) in Western Australia, Canada and South Africa. Although stromatolites are in the fossil record from 3500 Ma to the present, they reached their zenith at about 1000 Ma. Stromatolites extracted carbon dioxide ($CO_2$) from the atmosphere to secrete carbonate. Stromatolites still exist on Earth but in far more oxygenated conditions (e.g. Shark Bay, Western Australia). This clearly indicates that the first life on Earth (bacteria) is still with us and is the dominant life form on Earth. This again hints that if there is to be life elsewhere in the Universe, we could expect it to be bacterial.

The proportion of reduced to oxidised[716] carbon in life[717] and seawater[718] has remained the same since the earliest times on Earth. From this we conclude that the fundamentals of the carbon cycle on Earth were established as early as 4000 Ma.[719,720] This constant proportion is difficult to reconcile with a faint young Sun that was 30% less luminous at that time.[721,722] With such a weak Sun, the Earth should have been a snowball until a billion years ago.

---

[714] Lazcano, A. 1994: The RNA world, its predecessors, and its descendents. In: *Early life on Earth* (ed. Bengston, S.), 70-80, Columbia University Press.

[715] Obviously Thursday, named after Thor (the Norse god of thunder).

[716] Reduced carbon is biologically-bound carbon, methane and carbon monoxide whereas oxidised carbon is $CO_2$.

[717] Schidlowski, M., Eichmann, R. and Junge, C. E. 1975: Precambrian sedimentary carbonates: carbon and oxygen geochemistry and implications for the terrestrial oxygen budget. *Precambrian Research* 2: 1-69.

[718] Hayes, J. M., Kaplan, I. R. and Wedeking, K. W. 1983: Precambrian organic geochemistry. In: *The Earth's early biosphere: Its origin and evolution* (ed. Schopf, J. W.). Princeton University Press.

[719] Mojzsis, S., Arrhenius, G., McKeegan, K. D., Harrison, T. M., Nutman, A. P. and Friend, C. R. L. 1996: Evidence for life on Earth before 3,800 million years ago. *Nature* 385: 55-59.

[720] Veizer, J. 2005: Celestial climate driver: A perspective from four billion years of carbon cycle. *Geoscience Canada* 32: 13-28.

[721] Sagan, C. and Mullen, G. 1972: Earth and Mars: Evolution of atmospheres and surface temperatures. *Science* 177: 52-56.

[722] Crowley, T. J. and Baum, S. K. 1993: Effect of decreased solar luminosity on late Precambrian ice extent. *Journal of Geophysical Research* 98: 16723-16732.

Some argue that the Earth must have had a huge greenhouse effect at that time[723] with an atmospheric $CO_2$ content of 35% to prevent a snowball[724] yet this is not in accord with ancient limestone deposition and chemistry.[725] If there were decreased cloud cover,[726] the Sun's decreased luminosity would have kept Earth benign with neither ice sheets nor vaporised water.[727]

Although mass extinctions of multicellular life are the normal business of an evolving planet,[728] bacteria have survived all mass extinctions of complex life on Earth. Contrary to popular belief, there is no such thing as conservation of species. Life evolves, and part of evolution is a regular species turnover by extinction. Nothing can be conserved in a dynamic evolving environment. Extinction is normal. Species conservation is a romantic non-scientific view of planet Earth.[729]

Old 3800 Ma rocks in Greenland that were once gravels, sands and silts have a story to tell. Life appeared when the very heavy rain of large asteroids and comets stopped. These extraterrestrial impacts would have vaporised rocks and water. Calculations show that a 500 km diameter asteroid would vaporise oceans to a depth of 3000 metres. The planet's surface would have been sterilised. Compared to today, when life appeared there was still a higher frequency of asteroid impacting, the length of the day was shorter, Earth was rotating much faster, the Sun was much dimmer, the atmosphere was much denser, UV light entering Earth was more intense and the sky was orange to brick red. These large impacts reworked the surface of the Earth and were a driving force for the formation of the Earth's crust.[730]

The first rain on Earth was when the atmosphere was below 100°C. When this rain hit hot rocks, it would have vaporised. Rain was acid, had no oxygen and attacked the rocks to form thick soils (weathering). Because these soils were chemically reduced, they would have been green, compared to modern oxidised soils that are red-brown, or organic rich soils that are black. The first water on Earth was primarily from volcanic degassing although comets added a small amount. Once the surface of Earth cooled to below

[723] Kasting, J. F. 1993: Early earth's atmosphere. *Science* 259: 920-926.

[724] Kasting, J. F. and Ackerman, T. P. 1986: Climatic consequences of very high carbon dioxide levels in the Earth's early atmosphere. *Science* 234: 1383-1385.

[725] Shields, G. and Veizer, J. 2002: Precambrian marine carbonate isotope database. Version 1.1. *Geochemistry, Geophysics, Geosystems* 3: June 6, 2002.

[726] Rossow, W. B., Henderson-Sellers, A. and Weinreich, S. K. 1982: Cloud feed-back: A stabilising effect for the early earth? *Science* 217: 1245-1247.

[727] Ou, H.-W. 2001: Possible bounds on the Earth's surface temperature: From the perspective of conceptual global mean model. *Journal of Climate* 14: 2976-2988.

[728] Raup, D. 1979: A kill curve for Phanerozoic marine species. *Paleobiology* 17: 37-48.

[729] Raup, D. 1990: *Extinction: Bad genes or bad luck.* Norton.

[730] Glikson, A. Y. 1999: Ocean mega-impacts and crustal evolution. *Geology* 27: 387-390.

100°C, there was running water which would have removed soil (erosion) and deposited it as sediment (sedimentation).

Seas covered the Earth except for a few flat islands. The seas were acid and hot, and muddy with sediment. These sediments would have hardened to sedimentary rocks and the oldest sedimentary rocks are the most likely place to find clues about the first life on Earth. The planet went from being lifeless to having life. Although there was a faint Sun, the surface of planet Earth could not have been frozen because there is evidence of running water. The chemistry of black smudges in the old gravels of Greenland sediments shows that there was life[731] and oxidised uranium in the sediments shows that there was a minute trace of oxygen gas in the atmosphere.[732]

The conclusion is inescapable: we had a fully functioning biosphere at 3800 Ma. Bacteria were anaerobic (i.e. needed no oxygen) although there was a minute trace of oxygen in the atmosphere. The atmospheric gases derived from degassing of the planet by volcanoes.[733] The atmosphere at that time was dominated by nitrogen, $CO_2$, ammonia, $H_2O$ vapour, hydrogen sulphide, argon, helium and hydrogen.[734,735,736] Although carbon cycles have been continually refined,[737] the presence of a planetary thermostat is the only way that planet Earth has not suffered a runaway greenhouse effect as a result of the Sun emitting more energy over time.[738] Concepts of the Earth as a self-regulatory planet[739] fail because of the input of material from the Sun and the cosmos. For example, some 40,000 tonnes of extraterrestrial material falls each year and much of this contains organic molecules, including amino acids (the building blocks of DNA).[740,741] Amino acids can also be

---

[731] Rosing, M. T. 1999: C-13-depleted carbon microparticles in >3700-Ma sea-floor sedimentary rocks from west Greenland. *Science* 283: 674-676.

[732] Rosing, M. T. and Frei, R. 2004: U-rich Archaean sea-floor sediments from Greenland – indications of >3700 Ma oxygenic photosynthesis. *Earth and Planetary Science Letters* 217: 237-244.

[733] Hunten, D. M. 1993: Atmospheric evolution of the terrestrial planets. *Science* 259: 915-920.

[734] Holland, H. D. 1984: *The chemical evolution of the atmosphere and oceans*. Princeton University Press.

[735] Cloud, P. 1987: *Oasis in space*. Norton.

[736] Kasting, J. F. 1993: Earth's early atmosphere. *Science* 259: 920-926.

[737] Walker, J. C. G., Hays, P. B. and Kasting, J. F. 1981: A negative feedback mechanism for the long-term stabilization of Earth's surface temperature. *Journal of Geophysical Research* 86: 9776-9782.

[738] Sagan, C. and Mullen, G. 1972: Earth and Mars: Evolution of atmospheres and surface temperatures. *Science* 177: 52-56.

[739] Lovelock, J. E. 1979: *Gaia, a new look at life on Earth*. Oxford University Press.

[740] Chyba, C. F., Thomas, P. J., Brookshaw, L. and Sagan, C. 1990: Cometary delivery of organic molecules to the early earth. *Science* 249: 366-373.

[741] Chyba, C. F. and Sagan, C. 1992: Endogenous production, exogenous delivery, and impact-shock synthesis of organic molecules: An inventory for the origins of life. *Nature* 355: 125-131.

synthesised on Earth in hot springs,[742,743,744,745,746] by other simple processes[747] or at the site of an oceanic impact.[748] Hence an extraterrestrial origin of life is not necessary.

In order to balance the supply of $CO_2$ from a great diversity of natural processes, weathering may be the driving force.[749,750] Weathering is the process whereby rocks are converted to soil. The popular paradigm was that it is a chemical process involving addition of $H_2O$, oxygen and $CO_2$ to rocks. It is now thought of as a process where the decomposition of rock to soil and the addition of $H_2O$, $CO_2$ and oxygen to soils is driven by bacteria. On ancient Earth, oxygen was not necessary for weathering. A high $CO_2$ world has abundant energetic micro-organisms, abundant runoff and a resultant high rate of chemical weathering. However, land plants have been on Earth for only 10% of time. Before that, there was increased weathering, erosion and sedimentation because there were no land plants to hold soils to the underlying rocks. Chemical weathering (and hence the uptake of $H_2O$ and $CO_2$) in a glacial world is slow because of the lack of energetic micro-organisms, heat and acid rain. Atmospheric $CO_2$ then increases.

This weathering thermostat regulates planet Earth such that there is no permanent icehouse and no permanent runaway greenhouse. The thermostat is strong enough to create equilibrium although after occasional extreme events,[751] it took 100,000 years or so for $CO_2$ to reach equilibrium again.[752] Vostok ice cores in Antarctica show that, despite wild fluctuations

---

[742] Miller, S. L. and Bada, J. L. 1988: Submarine hot springs and the origin of life. *Nature* 334: 609-611.

[743] Marshall, W. L. 1994: Hydrothermal synthesis of amino acids. *Geochimica et Cosmochimica Acta* 58: 2099-2106.

[744] Fox, S. W. 1995: The synthesis of amino acids and the origin of life. *Geochimica et Cosmochimica Acta* 59: 1213-1214.

[745] Oro, J. 1994: Early chemical stages in the origin of life. In: *Early life on Earth* (ed. Bengtson, S.), 48-59. Columbia University Press.

[746] Nisbit, E. G. 1986: RNA, hydrothermal systems, zeolites and the origin of life. *Episodes* 9: 83-90.

[747] Miller, S. L. 1953: A production of amino acids under possible primitive Earth conditions. *Science* 117: 528-529.

[748] Furukawa, Y., Sekine, T., Oba, M., Kakegawa and Nakazawa, H. 2009: Biomolecule formation by oceanic impacts on early Earth. *Nature Geoscience* 2: 62-66.

[749] Gaillardet, J., Dupré, P., Louvat, P. and Allègre, C. J. 1999: Global silicate weathering and $CO_2$ consumption rates deduced from the chemistry of the large rivers. *Chemical Geology* 159: 3-30.

[750] Royer, D. L., Berner, R. A. and Park, J. 2007: Climate sensitivity constrained by $CO_2$ concentrations over the past 420 million years. *Nature* 446: 530-532.

[751] Such as the Palaeocene-Eocene Thermal Maximum at 55.8 Ma.

[752] Dickens, G. R., Castillo, M. M. and Walker, J. C. 1997: A blast of gas in the latest Paleocene: Simulating first-order effects of massive dissociation of oceanic methane hydrate. *Geology* 25: 259-262.

in temperature and atmospheric $CO_2$ over 650,000 years, the carbon fluxes must have been in balance to within 1–2%.[753] Measurements of modern atmospheric $CO_2$ show that there is a rapid uptake of $CO_2$ in the Northern Hemisphere growing season. The fact that the Earth's climate has been stable for far longer than the $CO_2$ equilibration time shows that since the first evidence of weathering, erosion and sedimentation at about 3800 Ma, $CO_2$ in the Earth's atmosphere has essentially been at equilibrium.

Over the history of time, volcanoes have been exhaling $CO_2$ into the atmosphere. For the first 4 billion years of Earth history, the $CO_2$ content of the atmosphere has been 3 to 100 times the current $CO_2$ content. Life has been extracting $CO_2$ for reefs and shells for billions of years. Cements holding grains together and sediments have also sequestered $CO_2$ over time. As a result, the $CO_2$ of the atmosphere has been decreasing and is near the lowest level it has been in the history of the planet. There was no "tipping" point or runaway greenhouse when $CO_2$ and temperature were high in the past. For billions of years, the Earth's thermostat has prevented runaway greenhouse or runaway icehouse. Nothing has changed.

In the Witwatersrand Basin of South Africa, gold has been mined from 2600 Ma ancient gravels since 1883. In the ancient gravels (conglomerate), rounded pieces of quartz, rock fragments and gold show that there was weathering, removal of weathered material and the deposition of this eroded material as sediment. In this process, there is an interaction of water, air, rocks and life. These conglomerates also contain rounded grains of an iron sulphide (pyrite, fools gold). Today, when the atmosphere has 21% oxygen gas, pyrite is easily oxidised in water to iron oxide and sulphuric acid. If pyrite has tumbled down river systems grinding it to rounded grains which were in contact with air and water, then the atmosphere at 2600 Ma had little (less than 1%) or no oxygen. Furthermore, these conglomerates contain rounded grains of a uranium oxide (uraninite). If there was oxygen in the atmosphere, the uranium would have oxidised and changed from the immobile reduced uranium to the highly mobile oxidised uranium. Because uraninite exists in the Witwatersrand conglomerates, we have a second line of evidence to suggest that there was little or no oxygen in the atmosphere at that time. Similar conglomerates in Brazil and Western Australia show that the global oxygen content of the atmosphere was low at 2600 Ma.

At about 2500–2700 Ma, there was a fundamental change to the Earth. Older crust, similar to that which is still present on the Moon, was replaced by a lighter more silica-rich crust. Earth now has a SIAL crust (silica-alumina)

---

[753] Zeebe, R. and Caldeira, K. 2008: Close mass balance of long-term carbon fluxes from ice-core $CO_2$ and ocean chemistry records. *Nature Geoscience* 1: 312-315.

which is continental and a SIMA crust (silica-magnesium) which is oceanic. This change in the Earth's crust is related to thickening of the crust, the lower heat flow and the expelling of very large and very small atoms from the Earth's mantle. The first validated evidence of plate tectonics is from rocks of this age. Continental evolution via plate tectonics had an influence on the evolution of life.

Before 2700 Ma, the continents occupied less than 5% of the surface of Earth. There was a rapid build-up of continents from 2700 to 2500 Ma (5% to 30%) due to an increased rate of plate tectonics. What is amazing is that around that time there was glaciation, a sudden diversification of life, a sudden increase in atmospheric oxygen, rusting of the oceans and the appearance of eukaryotic life.[754] The unanswered question is: Did this period of very rapid plate tectonics induce a diversification of life? The ocean volumes are clearly fundamental to life. If Earth's ocean volumes were greater, there would be no continents and Earth would be like Jupiter's moon Europa, with no land and 100 km deep oceans covered by ice. In this environment, only prokaryotic bacteria can exist. There would be no shallows for limestone formation and no continental weathering for the provision of nutrients for life in the oceans.

What if the continents covered 70% of the globe rather than the current 30%? If the continents occupied 70% of the globe, we would expect large and rapid temperature swings because large land areas create very high and very low seasonal temperatures. Large land areas reduce $CO_2$ draw-down because carbonate formation is in oceans and, in land-dominated systems, the opportunity for life to thrive is reduced. If plate tectonics were not operating, then we would have no continents. Where there is too much water, we get deep oceans and there is no natural brake on $CO_2$ build-up. This results in a runaway greenhouse. We remove $CO_2$ from the atmosphere into limestone, a rock that was very rare before 2500 Ma and a rock that has greatly increased in volume over time. This is why the atmospheric $CO_2$ has been decreasing over time and will continue to decrease. As limestone forms mainly in reefs that occur in water less than 6 metres deep, shallow water is essential to stop a runaway greenhouse. Continental weathering brings chemicals to the sea, changes seawater composition and adds nutrients into shallow water. Weathering is a process that consumes acid. The early oceans were acid and only after a long period of weathering could the oceans change from acid to alkaline. Although extremophiles can live in acid water, no shelled animal can live in acid water because its carbonate shell would dissolve. Again, it is

---

[754] Eukaryotic life is a cell with a nucleus within which genetic material is protected whereas prokaryotic life has no cell nucleus.

continents (formed from plate tectonics) and the resultant weathering that assisted the explosion of life.

Our Earth got it right. Without continents it would be too hot because main sequence stars like the Sun increase energy output over time and with too much continental landmass, the $CO_2$ draw-down during weathering would produce permanent glaciation.

Other fundamental and probably related processes also took place at this time. Isotopes of oxygen and carbon show that Earth was enjoying an equatorial glaciation[755] at around 2400 to 2100 Ma.[756] Life on Earth almost ceased.[757] These times of great dearth were interspersed with great bursts of high biological productivity. These bursts of productivity were assisted by high atmospheric $CO_2$, the release of nutrients into the oceans by glacial melt waters and warmer temperatures. Even today we see from satellites that the subtropical oceans contain less biota than the stormy mid-latitude and sub-polar seas which are better supplied with nutrients. Calm oceans and seas have a widespread scarcity of nutrients. There were a number of times in Earth history when the oxygen content of the atmosphere suddenly increased. These were immediately after glaciation. Melt waters bring huge amounts of nutrients to the oceans, stimulating a blossoming of photosynthetic micro-organisms, and these then add oxygen to the atmosphere.[758,759,760]

There was a global period from 2400–2200 Ma when the oceans rusted. All of the world's major iron ore deposits formed at this time, demonstrating that there was a sudden global increase in oxygen gas in the atmosphere. Furthermore, there was a diversification from single-celled prokaryotic to single-celled more complicated eukaryotic life. It was also a time of genocide by gas. Oxygen gas is poisonous for prokaryotic life and the cell nucleus protects eukaryotic life from oxidation. This period of increased global oxygenation would have led to the mass extinction of prokaryotic life, with

---

[755] Huronian glaciation

[756] Schopf, J. W. and Klein, C. (eds) 1992: *The Proterozoic biosphere: A multidisciplinary study.* Cambridge University Press.

[757] Kopp, R. E., Kirschvink, J. L., Hilburn, I. A. and Nash, C. Z. 2005: The Palaeoproterozoic snowball Earth: a climate disaster triggered by the evolution of oxygenic photosynthesis. *Proceedings of the National Academy of Sciences* 102: 1131-1136.

[758] Asmeron, Y., Jacobsen, S. B., Knoll, A. H., Butterfield, N. J. and Sweet, K. 1991: Strontium isotope variations of Neoproterozoic seawater: implications for crustal evolution. *Geochimica et Cosmochimica Acta* 55: 2883-2894.

[759] Jacobsen, S. B. and Kaufman, A. J. 1999: The Sr, C and O isotope evolution of Neoproterozoic seawater. *Chemical Geology* 161: 37-57.

[760] Higgins, J. A. and Schrag, D. P. 2003: The aftermath of snowball Earth. *Geochemistry, Geophysics, Geosystems* 4: doi: 10.1029/2002GC000403.

refugees hiding in oxygen-poor environments.[761] Such prokaryotic life still hides from oxygen in bogs, swamps, deep in rocks and in the folds of your stomach. Oxygen would have changed from less than 1% to more than 2% in the atmosphere. Why?

There are two theories. One invokes the breaking down of water into oxygen and hydrogen (which is lost to space) by UV radiation striking the oceans. This may give a slight increase in atmospheric oxygen but as soon as radiation changes oxygen to ozone, then the reaction would essentially stop. The other theory involves the sudden appearance of oxygen in the atmosphere by the invention of photosynthesis by bacteria. Photosynthesis uses $CO_2$ as plant food. This complex and efficient mechanism of survival had less risk than relying on energy from chemical reactions in rocks, hot springs and radioactive heat and allowed expansion into niches that could not be filled by extremophile life. The oxygenation of the atmosphere 2400–2200 Ma again shows that life, air, water and the rocks all interact. It happened then, and it still happens now. Most of the events that took place between 2200 Ma and now are a little uncertain.[762] The geological record hides many secrets that can help us understand the modern world. It may well be that a decrease in atmospheric methane led to an increase in atmospheric oxygen.[763] There were probably earlier failed attempts at atmospheric oxygenation or the collapse of atmospheric methane.

If the seawater temperature was above 100°C, the oceans would have boiled and life would not have existed. The earliest extreme thermophiles and methanogens can actually exist at temperatures greater than 100°C and were present at 3800 Ma. Cyanobacteria can exist at 70–73°C and first appeared at 3500 Ma. These prokaryotes liked it hot. Eukaryotes appeared at 2400–2200 Ma and can survive up to 60°C, multicellular soft-bodied animals appeared at 1500–1000 Ma and can exist up to 50°C, and land plants, which can survive up to 50°C, appeared at 470 Ma. The evolution of life on Earth is very much related to temperature. The planet had no animal life for the first 3500 Ma and was without animals large enough to leave fossils for the first 4000 Ma.

[761] Cranfield, D. E. 2005: The early history of atmospheric oxygen. *Annual Reviews of Earth and Planetary Sciences* 33: 1-36.

[762] Catling, D. C. and Claire, M. W. 2005: How the Earth's atmosphere evolved to an oxic state: a status report. *Earth and Planetary Science Letters* 237: 1-20.

[763] Zahnle, K., Claire, M. and Catling, D. 2006: The loss of mass-independent fractionation in sulfur due to a Palaeoproterozoic collapse of atmospheric methane. *Geobiology* 4: 271-283.

*What are the chances of multicellular life on Earth?*

What are the chances of finding multicellular life elsewhere in the Universe? Gravitational wobbles of stars, stellar eclipses and the bending of starlight have led to the discovery of Jupiter-sized planets outside the Solar System. Starlight passing through the atmospheres of these planets shows that there is no ozone, hence no oxygen. With no oxygen there would be neither photosynthesis nor complex life. This does not preclude bacterial life, which can exist without oxygen. A common view is that with the discovery now of more than 200 planets outside our Solar System, the presence of billions of stars in the Universe and the probability that planets are associated with these stars, then the chance of bacterial life elsewhere in the Universe is 100%.[764,765] However, this does not mean that the chance of multicellular life elsewhere in the Universe is 100%.[766]

I argue that the odds for multicellular life elsewhere in the Universe are very low and that planet Earth is a very rare environment. If indeed multicellular life on Earth is unique, then this opens up profound theological questions.

The Earth's position in our galaxy, the Milky Way, is a major factor.[767] In the star-packed interiors of galaxies, the frequency of supernovae and close stellar encounters may preclude stable stars. Outer regions of galaxies are metal-poor, hence the material for rocky planets is not present. This is important because if planets do not have a high uranium content, the planet cannot be internally heated for billions of years. Our Solar System moves through the plane of the Milky Way which reduces the chance of a star, planet, asteroid or comet impacting planet Earth. Although the Earth has been impacted in the past, the chances of impacting with a large body decrease with time and the Earth has had a very long time without catastrophic Earth-shattering impacts.[768]

The galaxy mass correlates with metal content.[769] Maybe the high metal content of the inner planets of our Solar System is an anomaly reflecting the

[764] Wetherill, G. W. 1991: Occurrence of Earth-like bodies in planetary systems. *Science* 253: 535-538.

[765] Williams, D. M., Kasting, J. F. and Wade, R. A. 1997: Habitable moons around extrasolar giant planets. *Nature* 385: 234-236.

[766] McKay, C. 1996: Time for intelligence on other planets. In: *Circumstellar habitable zones* (ed. L. Doyle), 405-419. Travis House.

[767] Chang, S. 1994: The planetary setting of prebiotic evolution. In: *Early life on Earth* (ed. Bengston, S.), 10-33. Columbia University Press.

[768] Maher, K. A. J. and Stevenson, D. J. 1988: Impact frustration on the origin of life. *Nature* 331: 612-614.

[769] Hart, M. H. 1979: Habitable zones around main sequence stars. *Icarus* 33: 23-39.

unusual nature of our galaxy. The most distant galaxies are too young to have enough metals for the formation of rocky planets. They also have frequent quasar-like activity and supernovae which emit life-destroying radiation. Globular clusters are also metal-poor despite having up to a million stars. Solar mass stars have evolved into giants that are too hot for life on the inner planets and catastrophic stellar encounters are far too common. Elliptical galaxies are also too metal-poor and too hot, and small galaxies are also too metal-poor. In galaxy centres, there is far too much radiation for life to exist. The galactic address is vital for single-celled life and especially multicellular life.

Planet Earth orbits a star which has a relatively stable output of energy.[770] Although bacterial life may exist in more extreme environments on other planets and moons in the Solar System, multicellular plant and animal life on Earth needs the benign conditions that existed in almost a steady-state environment for billions of years. It took more than 2 billion years of Earth history before the atmosphere became oxygenated. However, $CO_2$ is a common gas on all planets in our Solar System, in comets and other extraterrestrial bodies. Just because there is $CO_2$ in the Earth's atmosphere today does not mean that it derived from plant life or human activities. Even if multicellular life did evolve on other worlds, it would not be expected to survive because of the varying stellar energy flux. We therefore would not expect complex life to be associated with double and triple stars.[771] We may never know how many planets poised for evolution from bacterial to multicellular life have been wiped out by cosmic impact or by the energy released from a proximal star. We clearly have the right address in our Solar System.[772] We are the right distance from the Sun to stop the permanent freezing or boiling off of liquid water. We are far enough away from the Sun to avoid tidal lock, the Sun's mass gives it a long life of emitting energy and little UV energy, and Earth has a stable planetary orbit around the Sun.[773] Many recently discovered giant planets do not have a stable orbit.

Some Solar System planets are too close (Mercury) or too far (Jupiter) to allow liquid water to exist at the surface. Liquid water can only exist if the Sun emits a stable amount of energy for a long time and if the Earth's orbit is not

[770] Ksanfomaliti, L. V. 1998: Planetary systems around stars of late spectral types: A limitation for habitable zones. *Astronomicheskii Vestnik* 32: 413.

[771] Whitmore, D. P., Matese, J. J., Criswell, L. and Mikkola, S. 1998: Habitable planet formation in binary star systems. *Icarus* 132: 196-203.

[772] Wetherill, G. W. 1994: Provenance of terrestrial planets. *Geochimica et Cosmochimica Acta* 58: 4513-4520.

[773] Kasting, J. F., Whitmore, D. P. and Reynolds, R. T. 1993: Habitable zones around main sequence stars. *Icarus* 101: 108-128.

eccentric.[774] Furthermore, a massive impact would vaporise surface water on Earth and, because such massive impacts have not taken place over the last 4000 Ma, Earth has kept its precious liquid water.[775] We are fortunate to have Jupiter, with two thirds of the mass of all the planets in the Solar System, acting as a giant gravitational vacuum cleaner sucking up the material that was left in the Solar System after initial planetary formation. This is especially the case as the Asteroid Belt lies between Mars and Jupiter. Without Jupiter, the impact rate on Earth would have been higher and multicellular life may never have had the chance to evolve.

We live in a good neighbourhood. Earth is the only planet with a close moon of an appreciable size.[776] Not only does the Moon gravitationally attract impacts, it also stabilises the orbit of the Earth. Without such stability, the Earth would wobble far more than it does and there would be far greater seasonal changes.[777] Size counts. The Earth's size allows the atmosphere and oceans to be gravitationally glued to Earth. The size allows enough heat for plate tectonics and enough heat to keep the outer core molten for billions of years.[778,779] The molten outer core has given Earth a magnetic field. This magnetic field protects Earth because of a magnetic shield high in the atmosphere.[780] Without a magnetic field, multicellular life on Earth would fry from incoming solar and cosmic radiation and the Earth's oceans and atmosphere would be blasted away by solar wind. Furthermore, chemical reactions in the upper atmosphere create ozone which also behaves as a protective shield.

Our nearest planetary neighbour, Mars, is far smaller than Earth. Its core froze. This led to the loss of Mars' magnetic field which had protected Mars from cosmic radiation and solar particle bombardment. As a result, solar wind swept away the atmosphere and oceans.[781] Mars possibly had oceans before Earth did. It had volcanoes, and the new volcanic rocks were cooled

[774] Kasting, J. F. 1988: Runaway and moist greenhouse atmospheres and the evolution of Earth and Venus. *Icarus* 74: 472-494.

[775] Sleep, N. H., Zahnie, K. J., Kasting, J. F. and Morowitz, H. J. 1989: Annihilation of ecosystems by large asteroidal impacts on the early Earth. *Nature* 342: 139.

[776] Pluto has misbehaved and has now been stripped of its planetary status. It also has a large proximal moon.

[777] Laskar, J., Joutel, F. and Robutel, P. 1993: Stabilization of the Earth's obliquity by the Moon. *Nature* 361: 615-617.

[778] Condie, K. C. 1984: *Plate tectonics and crustal evolution*. Pergamon.

[779] Solomatov, V. and Moresi, L. 1997: Three regimes of mantle convection with non-Newtonian viscosity and stagnant lid convection on the terrestrial planets. *Geophysical Research Letters* 24: 1907-1910.

[780] McElhinny, M. W. 1973: *Paleomagnetism and plate tectonics*. Cambridge University Press.

[781] Carr, M. H. 1998: Mars: Aquifers, oceans and the prospects for life. *Astronomichskii Vestnik* 32: 453.

by water. This water would have been emitted from Mars as geysers, hot springs and boiling mud pools. This is the perfect environment for bacterial life. Fossil bacterial life should be present on Mars and the last Martian refugees may still live in cracks in rocks deep in Mars.

Multicellular life on Earth is maintained by a slowly evolving Sun providing us with constant energy input and shielding us from galactic attack derived from high energy particles emanating from supernovae. In order to understand multicellular life on Earth we need to understand the solar magnetic system, its variability, its large solar eruptions and the interaction between the heliosphere and Earth's magnetosphere and atmosphere.

Radioactivity is vital for multicellular life on Earth.[782,783] The radioactive decay of uranium, thorium and potassium deep in the Earth creates heat. Heat from the core and radioactive heat is convected from deep in the Earth to near the surface. Large plumes of heat break into two parts beneath more rigid rocks and pull apart the Earth.[784] Elsewhere, descending plumes pull the Earth downwards. These conveyor belts of ascending hot plumes and descending cooler plumes drive the pulling apart and stitching back together of parts of the Earth.[785] This process, plate tectonics, has been operating for thousands of millions of years. Earth is the only planet in the Solar System that has plate tectonics.

Without plate tectonics, Earth might have only had deep oceans and no land. Continental landmasses and shallow water settings are vital for multicellular life on Earth. Plate tectonics provides a diversity of habitats for life on Earth. Without plate tectonics, there is no mechanism for weathering, erosion, sedimentation, storage of carbon and the recycling of water, $CO_2$ and other earth materials. Plate tectonics provides the thermostat that prevents a runaway greenhouse. Weathering also releases nutrients into the oceans and water becomes trapped in weathered materials. Weathering occurs in both continental and submarine environments. Water in weathered rocks is pushed down into the Earth, it fluxes partial melting to produce light continental melts (SIAL). The continental SIAL "floats" on a basaltic crust (SIMA). Continents can drift, fragment and be inhabited by fools, but they do not disappear. Continents reflect energy, can be glaciated, can change ocean water circulation patterns and provide nutrients to the sea.

---

[782] Broecker, W. 1985: *How to build a habitable planet.* Eldigio Press.

[783] Ward, P. D. and Brownless, D. 2000: *Rare Earth: Why complex life is uncommon in the Universe.* Copernicus.

[784] Hill, R. I., Campbell, I. H., Davies, G. F. and Griffiths, R. W. 1992: Mantle plumes and continental tectonics. *Science* 256: 186-193.

[785] Larsen, R. L. 1991: Geological consequences of superplumes. *Geology* 19: 963-966.

If plate tectonics stopped, the recycling of $H_2O$ and $CO_2$ would cease and our planet would freeze, mountains would cease to form and the nutrient supply to the oceans would stop. Radioactive heat drives plate tectonics. Radioactivity is necessary to create the environment for life on Earth. Moving heat from deep in the Earth produces earthquakes and volcanoes. Radioactivity, magnetism, earthquakes and volcanoes tell us that the planet is healthy, evolving and changing.

Water is vital for life but water is really a very unusual substance. For a molecule that has an atomic weight of 18, it should have a very low boiling point. For example, ammonia has a similar atomic weight (17) yet has a boiling point of -33°C. Water should boil at a similar temperature. Water ice should be denser than liquid water. It is not. Water is stable at high temperatures, other materials of similar atomic weight are not. To change states from ice to water or water to steam, a large amount of energy is needed. The surface of liquid water behaves as if it was covered by an elastic sheet. Water does not behave like other liquids because it has an extensive network of hydrogen bonds, cohesive forces that hold the water molecules together. If water did not have a V-shaped molecule and hydrogen bonds, nutrients could not feed cells, heat would not be held in the oceans, the surface of the planet would be at -18°C and waterways would freeze from the bottom to the surface. There would be no multicellular life on Earth. There is life on Earth because water is really weird.

Because life is carbon based, we tend to think that Earth is a carbon-rich planet. Compared to other rocky planets and meteorites, Earth has a low water and a low carbon content. It appears that a small amount of water and carbon is necessary for life. Too much water would leave no terrestrial and shallow water habitats and too much carbon as $CO_2$ would create a runaway greenhouse as there would be little $H_2O$ to act as a thermostat and stop the runaway. Earth has just the right amount of water and carbon and recycling via plate tectonics keeps Earth in a steady state. Earth contains only 0.5% water, but this is enough to sustain sizeable oceans. Although small amounts of water are constantly added to Earth from upper atmosphere vaporisation of comets, the planet's water has derived from deep in Earth. Some of this water is from multiple events of recycling and some of it is original water that was trapped in the planet at the time of Earth's formation.

When we look at the number of extraordinary events needed to have multicellular life on Earth, then multicellular life in the Universe is either very rare or unique.

Dead multicellular life leaves cholesterol-type compounds in sediments. There are chemical hints that there were a number of attempts to start multicellular life at around 1500 to 1000 Ma. No fossils are preserved and, although the attempts were successful, they were not sustainable. It took the world's biggest climate change to kick-start multicellular life.

*The biggest climate change of all time*

In many parts of the world, there are two distinct sequences of glacial rocks between 750 and 580 Ma.[786] Chemical techniques show that there were three glaciations.[787] This is called the Neoproterozoic glaciation and some have suggested that at this time the Earth was a snowball[788] or maybe a slush ball.[789] Glacial debris deposits are 1500 m thick.[790] Later glaciations occurred at 450–420 Ma (minor), 300–260 Ma (major), 151–132 Ma (minor) and the current major glaciation which saw Antarctic ice at 34 Ma and Greenland ice at 2.67 Ma.

The proof that $CO_2$ does not drive climate is shown by previous glaciations. The Ordovician-Silurian (450–420 Ma) and Jurassic-Cretaceous (151–132 Ma) glaciations occurred when the atmospheric $CO_2$ content was more than 4000 ppmv and about 2000 parts per million by volume (ppmv) respectively.[791] The Carboniferous-Permian glaciation (360–260 Ma) had a $CO_2$ content of about 400 ppmv, at least 15 ppmv higher than the present figure. If the popular catastrophist view is accepted, then there should have been a runaway greenhouse when $CO_2$ was more than 4000 ppmv. Instead, there was glaciation. Clearly a high atmospheric $CO_2$ does not drive global warming and there is no correlation between global temperature and atmospheric $CO_2$. This has never been explained by those who argue that human additions of $CO_2$ to the atmosphere will produce global warming.

The clues to a glaciation are written in the rocks. Retreating ice leaves gravel at the edge and end of ice sheets.[792] This can be reworked by melt water. Terrestrial glacial rocks such as tillite are common, as are lithified

---

[786] Dunn, P. R., Thomson, B. P. and Rankama, K. 1971: Late Pre-Cambrian glaciation in Australia as a stratigraphic boundary. *Nature* 231: 498-502.

[787] Halverson, G. P., Hoffman, P. F., Schrag, D. P., Maloof, A. C. and Rice, A. H. N. 2005: Toward a Neoproterozoic composite carbon-isotope record. *Geological Society of America Bulletin* 117: 1181-1208.

[788] Hoffman, P. F. and Schrag, D. P. 2002: The snowball Earth hypothesis: Testing the limits of global change. *Terra Nova* 14: 129-155.

[789] Meert, J. G. and van der Voo, R. 1994: The Neoproterozoic (700-540 Ma) glacial intervals: no more snowball Earth. *Earth and Planetary Science Letters* 123: 1-13.

[790] Young, G. M. and Gostin, V. A. 1989: An exceptionally thick late Proterozoic (Sturtian) glacial succession in the Mount Painter area, South Australia. *Geological Society of America Bulletin* 101: 834-845.

[791] Berner, R. A. and Kothavala, Z. 2001: Geocarb III: A revised model of atmospheric $CO_2$ over Phanerozoic time. *American Journal of Science* 301: 182-204.

[792] Till.

permafrost soils.[793,794] Glacial lakes were covered by ice in winter and only a few grains of clay sank to form thin layers whereas melt waters in summer produced thicker sandy layers. These annually layered rocks (varve shales) are common. The fact that deep marine and terrestrial rocks occur together suggest that during the Neoproterozoic glaciation there was so much water locked up in ice that there was no continental shelf. This means that sea level rises and falls were at least 400 and maybe 600 metres. Marine sequences contain drop boulder beds. These derive from melting icebergs that drop terrestrial boulders far out to sea into deep marine muds.[795] In some places, there are iron oxide deposits. These suggest pack ice. When ice covers the sea surface, no oxygen can penetrate, the sea becomes reduced and iron oxides in sea floor sediments dissolve. When the ice starts to break up and oxygen can dissolve in seawater, the dissolved iron oxidises and precipitates as insoluble iron oxides. Some of these iron oxide deposits contain drop boulders.[796] Many Neoproterozoic tillites and drop boulder beds are capped by carbonate sediments formed during an interglacial.[797,798] The carbonate chemistry shows that seawater was at least 40°C, and there was a thriving of cyanobacteria which removed $CO_2$ from the atmosphere by building massive algal reefs. Sea level must have very rapidly changed by hundreds of metres.[799] Another suggestion is that there was a climate-driven explosion of methane hydrate, the atmosphere warmed from an excess of this greenhouse gas and upon oxidation the $CO_2$ was locked into algal reefs by cyanobacteria.[800] Some suggest that this draw-down of $CO_2$ may have stimulated the next glaciation.

[793] Williams, G. E. 1986: Precambrian permafrost horizons as indicators of palaeoclimate. *Precambrian Research* 32: 233-242.

[794] Maloof, A. C., Kellog, J. B. and Anders, A. M. 2002: Neoproterozoic sand wedges: crack formation in frozen soils under diurnal forcing during snowball Earth. *Earth and Planetary Science Letters* 204: 1-15.

[795] Christie-Blick, N., Dyson, I. A. and van der Borch, C. C. 1995: Sequence stratigraphy and the interpretation of Neoproterozoic earth history. *Precambrian Research* 73: 3-26.

[796] Lottermoser, B. G. and Ashley, P. M. 2000: Geochemistry, petrology and origin of Neoproterozoic ironstones in the eastern part of the Adelaide Geosyncline. *Precambrian Research* 106: 21-63.

[797] Kennedy, M. J. 1996: Stratigraphy, sedimentology, and isotope geochemistry of Australian Neoproterozoic post-glacial cap dolostones: deglaciation, $^{13}C$ excursions, and carbonate precipitation. *Journal of Sedimentary Research* 66: 1050-1064.

[798] Kennedy, M. J., Runnegar, B., Pave, A. R., Hoffman, K. H. and Arthur, M. 1998: Two or four Neoproterozoic glaciations? *Geology* 26: 1059-1063.

[799] James, N. P., Narbonne, G. M. and Kyser, T. K. 2001: Late Neoproterozoic cap carbonates: Mackenzie Mountains, northwestern Canada: Precipitation and global glaciation. *Canadian Journal of Earth Sciences* 38: 1229-1262.

[800] Hoffman, P. F., Halverson, G. P. and Gotzinger, J. P. 2002: Are Proterozoic cap carbonates and isotope excursions the record of gas hydrate destabilization following Earth's coldest intervals? Comment. *Geology* 30: 286-287.

The Neoproterozoic glaciation is even more intriguing. The magnetic field of the Neoproterozoic rocks shows that at that time, the continents were clustered around the equator. Not only was the glaciation at the equator, it was also at sea level.[801] What needs to be explained is how the planet lurched into a long biting glaciation, came into a hot wet interglacial, lurched back into a long glaciation again and then came into warm wet interglacial conditions that persisted for hundreds of millions of years before a minor glaciation at 450–420 Ma.

The Sturtian glaciation (~730 Ma) is contemporaneous with the break-up of a giant supercontinent, Rodinia. The continental break-up produced large amounts of basalt in supervolcano provinces between 825 and 755 Ma. Basalt eruptions exhale monstrous amounts of $CO_2$.[802] Fresh basalt, especially at tropical latitudes, is extremely chemically reactive and is quickly weathered, accelerating the draw-down of $CO_2$ from the atmosphere, inducing a long-term climatic cooling.[803] The filling of the atmosphere with dust (from volcanicity, impacting and comets) results in the reflection of sunlight and the cooling of the planet. This has happened in the historical past (e.g. the 1815 Tambora eruption in Indonesia). However, there is no record in the rocks of volcanicity or impacting that might have initiated the Neoproterozoic glaciation. With equatorial volcanoes and impacts, both hemispheres are affected. By the same methods, it has been shown that there was a long cold period in the 17th and 18th Centuries (Little Ice Age), which was coincidental with a lack of sunspot activity. Therefore, changes in the Sun's radiation may induce a rapid climate change. However, in Neoproterozoic times, our Sun, a main sequence star which heats up with time, was dim and emitted far less radiation than now. Maybe the fact that the Earth did not totally freeze was because the Neoproterozoic atmosphere had 100 to 1000 times the current atmospheric $CO_2$ content? Maybe there was less cloud?

During the Neoproterozoic glaciation, the Solar System was in the Sagittarius-Carina Arm. During the next but minor glaciation, there was an encounter with the Perseus Arm. The large Permo-Carboniferous glaciation occurred when there was an encounter with the Norma Arm and, in the Jurassic to early Cretaceous glaciation, the planet was in the Period Scutum-

---

[801] Sohl, L. E., Christie-Blick, N. and Kent, D. V. 1999: Paleomagnetic polarity reversals in Marinoan (ca 600 Ma) glacial deposits of Australia: implications for low-latitude glaciation in Neoproterozoic time. *Geological Society of America Bulletin* 111: 1120-1139.

[802] Marty, B. and Tolstikhin, I. N. 1998: $CO_2$ fluxes from mid-ocean ridges, arcs and plumes. *Chemical Geology* 145: 233-248.

[803] Goddéris, Y., Donnadieu, Y., Nédélec, A., Dupré, B., Dessert, C., Grard, A., Ramstein, G. and François, L. M. 2003: The Sturtian 'snowball' glaciation: fire and ice. *Earth and Planetary Science Letters* 211: 1-12.

Crux Arm. Later cooling in the Miocene (Sagittarius-Carina Arm) was quickly followed by the Pleistocene glaciation (Orion Spur).[804] There is a suggestion that the onset of the most intense period of the Pleistocene glaciation at 2.75 Ma was from cosmic rays emitted from a nearby supernova.[805] There is some evidence that wobbles in the Earth's orbit (Milankovitch Cycles) have influenced climate for hundreds of millions of years[806,807] and there is a good history of changing lunar tides from the Neoproterozoic to the present.[808,809] Is an extraterrestrial origin of glaciation viable or do we need a combination of factors to produce glaciation?

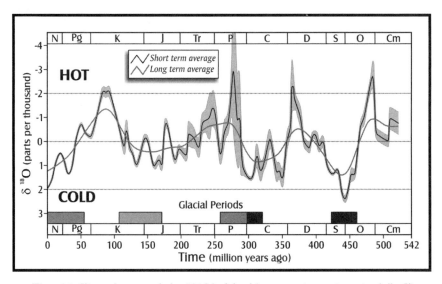

*Figure 16: Climate change over the last 530 Ma deduced from oxygen isotopes in marine shells. Climate has changed over time and Earth is currently in a period of glaciation. For most of the last 530 Ma, planet Earth has been warmer than at present.*

[804] Shaviv, N. J. and Veizer, J. 2003: Celestial driver of Phanerozoic climate. *GSA Today* 13: 4-10.

[805] Knie, K., Korschinek, G., Faestermann, T., Dorfi, E. A., Rugel, G. and Wallner, A. 2004: $^{60}Fe$ anomaly in deep-sea manganese crust and implications for a nearby supernova source. *Physical Review Letters* 93: 171103-171107.

[806] Pannella, G. 1972: Paleontological evidence of the Earth's rotational history since the Precambrian. *Astrophysics and Space Science* 16: 121-137.

[807] Park, J. and Oglesby, R. J. 1991: Milankovitch rhythms in the Cretaceous: A GCM modelling study. *Palaeogeography, Palaeoclimatology, Palaeoecology* 4: 329-356.

[808] Goldreich, P. 1966: History of the lunar orbit. *Reviews in Geophysics* 4: 411-439.

[809] Williams, G. E. 1998: Precambrian tidal and glacial eustatic clastic deposits: implications for Precambrian Earth-Moon dynamics and palaeoclimate. *Sedimentary Geology* 120: 55-974.

If global volcanicity was reduced, then the amount of $CO_2$ entering the atmosphere would have been less and the planet's atmosphere would have cooled. There is some evidence that the Neoproterozoic glaciations were in a period of reduced volcanicity. However, if the planet were a snowball, the ice would have reflected light and heat and, unless the atmosphere had a dramatic increase in $CO_2$, it would have stayed a snowball. Cyclical volcanicity would be needed to explain the two events of Neoproterozoic glaciation. So, if reduced global volcanicity was the cause of the Neoproterozoic glaciation, then it would have had to be singing from the same hymn sheet as the cyanobacteria. Clearly life did influence climates, but by how much?

Moving continents can produce glaciation if they happen to wander over a pole[810] but the evidence suggests that the continents in the Neoproterozoic were equatorial and not polar so we can dismiss this idea to explain the Neoproterozoic glaciation. Glaciation might occur during periods of increased mountain building. Mountain building increases weathering and erosion rates and $CO_2$ is removed during these processes.[811] This is a highly controversial mechanism, which has been suggested for the Pleistocene glaciation[812] and fails in both the Pleistocene and the Neoproterozoic because of the timing of the necessary sequence of events. Increased mountain building leaves its mark, such as rocks that have been heated up at high pressures, increased rates of erosion and sedimentation and increased volcanic and sub-volcanic activity. There is no such evidence in the Neoproterozoic.

Maybe if early Earth was like Saturn, then it might have had ice rings that reflected sunlight. The problem is, how did Earth acquire ice rings, lose them, acquire them and then lose them permanently? This is the only scenario that can explain cycles of glaciation.

The Earth's orbit can influence climate. At present its tilt is 23.45°, which gives us seasons. If the Earth's tilt was 70° and there was a weak Sun, then calculations show that there could be equatorial ice and warmer poles.[813] The Earth wobbles on its axis rather like a spinning top with wobbles at 100,000, 41,000 and 21,000 years (Milankovitch Cycles). These can give cyclical climates during a large event of glaciation (like the current event) but cannot explain how a massive cycle of glaciation is initiated.[814]

---

[810] Such as the Permo-Carboniferous glaciation from 300 to 260 Ma.

[811] West, A. J., Galy, A. and Bickle, M. 2005: Tectonic and climatic controls on silicate weathering. *Earth and Planetary Science Letters* 235: 211-228.

[812] Raymo, M. E. and Ruddiman, W. F. 1992: Tectonic forcing of late Cenozoic climate. *Nature* 359: 117-122.

[813] Williams, G. E. 2000: Geological constraints on the Precambrian history of Earth's rotation and the Moon's orbit. *Reviews of Geophysics* 38: 37-59.

[814] Berger, A. 1988: Milankovitch theory and climate. *Reviews of Geophysics* 26: 624-657.

There are four mutually exclusive theories for the origin of the Neoproterozoic glaciation. The zipper-rift Earth theory involves glaciation at the same time as volcanic activity at the rift margin of a continent. The high-tilt Earth theory has the Earth with an axis of 70° and preferential glaciation at low altitude. Snowball Earth theory has extreme runaway glaciation related to ice reflecting solar energy. The slush ball Earth theory has extreme glaciation coexisting with unfrozen oceans and glaciers at sea level in the tropics.[815]

The Neoproterozoic glaciation had a profound effect on planet Earth. There were at least two major glacial events (Sturtian and Marinoan) separated by interglacial events over more than a hundred million years. There may even have been four glacial events.[816] More detailed analysis shows that climate was changing very quickly, sea level went up and down rapidly and during the maximum interglacial, bacterial-precipitated carbonate rocks formed in water at more than 40°C by extracting huge amounts of $CO_2$ from the atmosphere. During glaciation, bacteria hung on to life and the atmospheric $CO_2$ rose.

If we cannot understand the biggest climate changes of all times, then we have to be very circumspect about claiming that we understand modern climate.

## The first multicellular life

Before multicellular life, the sea floor was covered with an algal mat with an elephant-skin texture. Sediments deposited after multicellular animals arose have been continually turned over by burrowing and crawling animals. This has destroyed delicate layering and allowed more sediment to be shifted by currents, waves and tides. Sediments deposited before the first multicellular life are extremely delicately layered, and layers only a millimetre thick can be traced for tens of kilometres. Almost immediately after the first evidence of churning over of sea floor sediments, all sorts of alien-looking soft-bodied animals appeared. Although small soft-bodied multicellular life might have appeared at 1500-1000 Ma, it was in the Ediacaran Period (583–542 Ma) that we saw larger multicellular life with hard parts. These hard parts were not shells, teeth or skeletons but the same material as our fingernails. In Ediacaran times there were no trees, shrubs, grass or stem plants. This lack of rooted vegetation led to high erosion rates with sediments being dumped in the shallow seas and waterways. This added nutrients for the first multicellular

[815] Fairchild, I. J. and Kennedy, M. J. 2007: Neoproterozoic glaciation in the Earth System. *Journal of the Geological Society, London* 164: 895-921.
[816] Kennedy, M. J., Runnegar, B., Pave, A. R., Hoffman, K. H. and Arthur, M. 1998: Two or four Neoproterozoic glaciations? *Geology* 26: 1059-1063.

animals. Although stromatolites appeared at 3500 Ma and are still with us, their domination was in decline in the Ediacaran Period, suggesting that they may have been meals for grazing multicellular animals.

There is great biological confusion regarding the Ediacarans. The Ediacaran fauna are now found on all continents, showing that their distribution was global. More than 80 species have been identified. They look like bizarre jellyfish, mutated worms and quilted air mattresses. Some are up to a metre long. Some look like sponges, some look like molluscs,[817] some are disc-like, others show an internal anatomy, and grazing and locomotion tracks are rare.[818] Trails of faecal pellets show that they had a one-way gut. In the type locality[819] they occur in silty to sandy rocks. It appears that they grazed on algal mats on the sea floor and in stormy times were wrapped in the algal mat which operated like a death mask. This preserved the Ediacarans from bacterial decay, the death mask was rapidly covered by thick layers of sediment, and the biological material became fossilised. The Ediacarans may be ancestors to modern animals or they may have been a failed experiment with multicellular life that is now extinct. The Ediacaran fauna coexisted with other animal fauna in the earliest Cambrian, hence may not be a precursor to modern life and may indeed be an extinct side branch. They probably had the seas to themselves as there is no evidence of predation until 542 Ma.

The Ediacarans grazed their way through the sea floor algal mats. They had no competition. As soon as there was competition, they became extinct. The Ediacarans were overwhelmed by Cambrian explosion of life and the resultant competition. The disappearance at 542 Ma of the Ediacaran fauna was the first mass extinction of multicellular life. The first Cambrian organisms filled the niches left by Ediacarans and, because the algal mats had disappeared, devised new ways of eating.

## The explosion of life

The explosion of life took place over a very short interval of time when all animal phyla[820] appeared. This rate of evolution has never been equalled. This leads to questions. Is the Cambrian explosion of life the cause or effect of multicellular life? What would Earth have been like with a small pop rather than an explosion of life? Was multicellular life inevitable? If we ran the tape of the history of life again, would we have the same events in the

---

[817] *Kimberella.*

[818] Vickers-Rich, P and Komarower, P. (eds) 2007: *The rise and fall of the Ediacaran biota.* Geological Society of London Special Publication 286.

[819] Ediacaran Hills, near Lake Torrens, South Australia.

[820] Organism with unique body plans.

same order?

Clearly features such as temperature, the change from an acid to alkaline ocean, the appearance of continents that underwent weathering and produced shallow water settings and a biotic boom after more than 100 million years of a very intense ice age are key aspects of the Cambrian explosion of life (540–520 Ma). The Cambrian explosion of life laid down the fundamental architecture of body types of multicellular animals.

Forming an animal is more difficult than forming a bacterium. It involves time. Over the history of time, there have been some great leaps forward (RNA- to DNA-dominated life, appearance of eukaryotic cells, appearance of multicellular life) but there are some fundamental questions. Why didn't the Cambrian explosion take place much earlier? It was the genuine appearance of 100 phyla that exploited the environment to make hard parts. These hard parts were mainly shells made out of calcium carbonate. This produced a huge draw-down of atmospheric $CO_2$. The Cambrian explosion went on for 20–30 million years. Since then no new phyla have appeared on Earth. Even after mass extinction events, no new phyla appeared. Since the Cambrian explosion of life, there has been a huge diversification of life as shown by the number of species, genera and families of life.

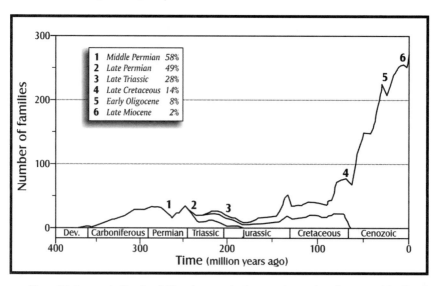

*Figure 17: Increase in diversity of life and mass extinctions over time as shown by preserved families of tetrapods.*[821] *Mass extinctions are numbered 1 to 6 with percentages of tetrapods rendered extinct. The three lines represent the three dominant tetrapod faunas.*

[821] Benton, M. J. 1985: Mass extinction among non-marine tetrapods. *Nature* 316: 811-814.

What, if anything, triggered the Cambrian explosion? These are some of the suggestions. There may have been an environmental stimulus with high phosphorus, calcium, iron and $CO_2$ dissolved in ocean water after intense post-glacial weathering and erosion. The high iron would have stimulated algal blooms, which are photosynthetic organisms, and the atmospheric oxygen content would have increased. Oxygen and phosphorus are necessary for muscles and the high atmospheric $CO_2$ and calcium are necessary for shells. Certainly the break-up of the giant supercontinent Rodinia would have changed ocean circulation, changed the shape of the ocean floor and brought phosphorus to the ocean surface. The process could only have been helped by oceans that had been warm for a long time after the rapidly changing freezing and heating of the oceans during the Neoproterozoic glaciation. Plate tectonics may have played a part as there was rapid continental drifting in the Cambrian. The position of the continents would have had an effect on ocean currents and helped the upwelling of deep nutrient-rich ocean waters.

Shells offer protection from predation, desiccation and UV energy. Muscles allow attachment to the floor of the sea, locomotion and maintenance of body form. Oxygen was critical. Large skeletons restricted the access of seawater to the soft body parts and respiration took place by direct adsorption of dissolved oxygen in seawater across a body wall. Shells prevent a large area of the body from having access to oxygen. Therefore, a high dissolved oxygen content in seawater was required. Hard skeletons are not an addition to body plans, they modify the plans. With the ability to produce hard parts, new animal groups could use these hard parts for jaws, legs or body support and thus enabled them to exploit entirely new ways of life and locomotion and new environments. Up to the Cambrian explosion, most multicellular animals were normally less than 1 mm in size and soft-bodied. A large size was required for improved circulatory, respiratory and excretory systems and each had to evolve before the larger body size was attainable. This evolution may have taken place in the few rare large Ediacaran animals.

Predation appeared in the Cambrian explosion and forced new ways of living. Survival was enhanced by animals that evolved the ability to defend themselves by producing shells, burrowing or rapidly moving from danger. These organisms found themselves able to exploit under-utilised food resources such as filter feeding in shells or algal meals from burrowing.

The number and diversity of organisms rapidly increased. This was matched by a sharp increase in the churning over of sediments by burrowers. As a result, buried organic matter that would normally be locked up in sediments was recycled through new life. The proportion of black organic-rich sediments deposited decreased and the deposition of limey rocks increased. This modified the carbon cycle, as did the huge draw-down of $CO_2$ for the formation of shells. This draw-down continued for the next 542 Ma. It is still taking place.

After the Cambrian explosion of life, there was a period of great biodiversity from about 470 Ma. Biodiversity normally rapidly increases after a mass extinction event, so what triggered the great increase in biodiversification? The explosion in the number of species is almost at the same time as an increased frequency of meteorite impacts. This constant bombardment by meteorites increased the global rate of landsliding.[822] Maybe this increased rate of impacting created many regional mass extinctions of life, and accelerated biodiversity.[823]

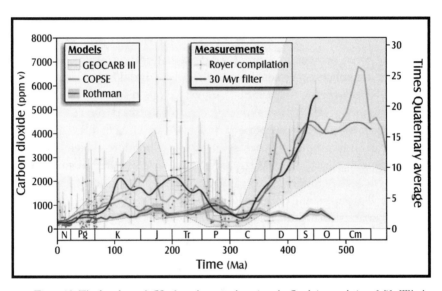

*Figure 18: The drawdown of $CO_2$ from the atmosphere since the Cambrian explosion of life. Whether models or measurements are used, the atmsopheric $CO_2$ content has been up to 25 times higher than at present and it has been extracted from the atmosphere into carbonate rocks (limestone, dolomite, carbonate cement, carbonate fossils) such that the modern $CO_2$ content is the lowest in geological history. Most of the planet's $CO_2$ is held in rocks and the smallest amount of planetary $CO_2$ is currently held in the atmosphere.*

---

[822] Parnell, J. 2009: Global mass wasting at continental margins during Ordovician high meteorite flux. *Nature Geoscience* 2: 57-61.

[823] Schmitz, B., Harper, D. A. T., Peucker-Ehrenbrink, B., Stoude, S., Alwmark, C., Cronholm, A., Bergström, S., Tassinari, M and Xiaofeng, W. 2008: Asteroid breakup linked to the Great Ordovician Biodiversification Event. *Nature Geoscience* 1: 49-53.

For some time, the theory that massive impacting created mass extinctions of life was the popular paradigm.[824,825] This theory has now been modified because it has been shown by later scientific work that few mass extinctions of multicellular life result from extraterrestrial impacts.

On any topic, the science is never settled.

## Extinction

*Minor and major mass extinctions of life*

Planet Earth has enjoyed five major mass extinctions of complex life and numerous minor mass extinctions.[826] A sobering thought is that 99.99% of all species that have ever existed in planet Earth are now extinct. There is a view that extinction is with us all the time as part of species turnover and, at times, there is accelerated sudden extinction that we view as a mass extinction. Life always carries on in a time of extinction, and conservation of species has never occurred in the history of the planet. There is no reason why it should occur just because humans live on the planet. Change is normal and the process of speciation requires extinction, which vacates ecosystems for new species. A major mass extinction wipes out at least 50% of species.

A major mass extinction at 450–440 Ma was actually two events, and the major mass extinction at 375–360 Ma destroyed 70% of species in a series of extinctions that may have lasted for 20 million years.[827] The biggest extinction of all, at 251.4 Ma, killed 53% of marine families, 84% of marine genera and 96% of all marine species.[828] There may have actually been two extinction events.[829] Another extinction at 205 Ma[830] was followed by the best-known of all mass extinctions at 65 Ma. There were numerous minor mass extinctions. The loss of the Ediacaran fauna at 542 Ma was followed

[824] Alvarez, W. L. 1986: Towards a theory of impact crises. *EOS* 67: 649-658.

[825] Hut, P., Alvarez, W., Elder, W. P., Hansen, T., Kauffman, E. G., Keller, GH., Shoemaker, E. M. and Weissman, P. R. 1987: Comet showers as a cause of mass extinction. *Nature* 329: 118-126.

[826] Hallam, A. and Wignall, P. B. 1997: *Mass extinctions and their aftermath*. Oxford University Press.

[827] McGhee, G. R. 2006: Extinction: Late Devonian mass extinction. *Encyclopedia of Life Sciences,* John Wiley.

[828] Elewa, A. M. T. 2008: *Mass extinction*. Springer-Verlag.

[829] Stanley, S. M. and Yang, X. 1994: A double mass extinction at the end of the Paleozoic era. *Science* 266: 1340-1344.

[830] Hallam, A. 1990: The end-Triassic mass extinction event. *Geological Society of America Special Paper* 247: 577-583.

by more than a dozen minor mass extinctions.[831,832,833,834,835] There is a view
that we live in the sixth major mass extinction.[836] The causes of the five
previous major mass extinctions and the dozens of minor mass extinctions
are unknown.[837]

Some 100 new groups of animals (phyla) appeared in the Cambrian
explosion of life. Despite many mass extinctions since the loss of the Ediacaran
fauna at 542 Ma, no new phyla have appeared. Phyla have disappeared. After
the Cambrian explosion of life, the seas were teeming with bacterial and
multicellular life. Then, another remarkable irreversible event occurred. The
continental masses became populated with terrestrial plants (470 Ma) and
simultaneously footprints appeared on tidal flats. It appears that reptilian
animals went walkabout to exploit the new food source. The continents have
been vegetated for only 10% of time and this is shown in the geological
record by decreased rates of weathering and erosion.

The fossil record is an incomplete record and the chances of an
organism becoming fossilised are slim. For example, if you drop dead in the
desert, what is your chance of becoming fossilised? You undergo bacterial
decomposition, carnivores eat your flesh, your bones are eaten, disarticulated,
spread and bleached and it is a case of atoms-to-atoms. The best chance of
being fossilised is if you die in an oxygen-free swamp. Some individual cloud
mountain Central American rainforest trees have hundreds of endemic
insects. If this one tree dies, what are the chances of these insects becoming
fossilised and escaping bacterial decomposition in the rainforest leaf litter?
Furthermore, how do we measure a mass extinction? Do we count the change
in phyla, families, groups, genera or species? In addition, palaeontology
continues to find new species, so we really do not know what awaits us in the
next fossil-bearing rock.

[831] Bralower, T. J., Arthur, M. A., Lecke, R. M., Sliter, W. V., Allard, D. J. and Schlanger, S.
O. 1994: Timing and paleoceanography of oceanic dysoxia/anoxia in the Late Barremian to
Early Aptian (Early Cretaceous). *Palaios* 9: 335-369.

[832] Aberhan, M. and Fürsich, F. T. 1996: Diversity analysis of Lower Jurassic bivalves of the
Andean Basin and the Pliensbachian-Toarcian mass extinction. *Lethaia* 9: 181-195.

[833] Abramovich, S., Almogi-Labin, A. and Benjamini, C. 1998: Decline of the Maastrichtian
pelagic ecosystem based on planktonic foraminifera assemblage change: implication for the
terminal Cretaceous faunal crisis. *Geology* 26: 63-66.

[834] Copper, P. 1998: Evaluating the Frasnian-Famennian mass extinction: comparing
brachiopod faunas. *Acta Palaeontologica Polonia* 43: 137-154.

[835] Hallam, A. 1999: Discussion on oceanic plateau formation: a cause of mass extinction
and black shale deposition around the Cenomanian-Turonian boundary. *Journal of the Geological
Society, London* 156: 208.

[836] Leakey, R. and Lewin, R. 1996: *The sixth extinction: Patterns of life and the future of humankind.*
Anchor.

[837] Vines, G. 1999: Mass extinctions. *New Scientist*, 11th September 1999, Supplement 126: 1-4.

Despite this, palaeontologists have undertaken species counts and assemblages of species. If the host rocks contain volcanic ash which can be dated precisely by radioactive dating methods to determine the timing of events, then we can ascertain the exact life span of species. The exact life span of many species is not known. Furthermore, in ascertaining mass extinctions, the palaeoenvironment of species is documented. Events that affect terrestrial organisms may not affect shallow marine or deep marine animals. Some events affect terrestrial plants and not terrestrial animals. Whatever happens, bacteria are the ultimate survivors of mass extinctions although it is hard to determine whether there has been bacterial species extinction. There are some suggestions that after the Acraman impact in South Australia at 580 Ma, there was a minor mass extinction of bacteria.

Because it is an exciting concept, the relationship between sudden catastrophic events and mass extinctions on Earth has attracted a great deal of scientific research, speculation and opinions from those with little scientific knowledge. Better age measurements require a refining of hypotheses because the dates of high-velocity impact craters are now more accurate. The new dates show that there is no one-for-one correlation for preserved high velocity impact craters, global mass extinctions and large igneous provinces.[838]

Mass extinctions may occur for a number of reasons. The best way to kill life is by gassing. The best example of extinction by gassing is the rusting of the oceans at 2400–2200 Ma. This took place because, in an explosion of bacterial life, bacteria consumed $CO_2$ and exhaled oxygen which killed most of the prokaryotic organisms. The main gases released from volcanoes are $H_2O$ vapour and $CO_2$. Neither gas could cause a mass extinction. However, some volcanoes derive from a sulphur-rich part of the mantle or incorporate seawater sulphate and exhale huge volumes of hydrogen sulphide (rotten egg gas). Above 10 ppmv, hydrogen sulphide is deadly poisonous. Hydrogen sulphide oxidises in the atmosphere to sulphuric acid, rain becomes very acid, and runoff waters change from alkaline to acid, as do the oceans. In a terrestrial environment, land plants are killed by acid rain and the food chain collapses. Sulphur dioxide, another common volcanic gas, behaves in a very similar way.

In the oceans, gassing by hydrogen sulphide removes oxygen from the water, there is a brief period of acidity, the shells of animals are dissolved, and life cannot accommodate the sudden change from alkaline to acid conditions

[838] Kelley, S., 2007: The geochronology of large igneous provinces, terrestrial impact craters, and their relationship to mass extinctions on Earth. *Journal of the Geological Society, London* 164: 923-936.

or from oxygen-rich to oxygen-poor conditions. In deep ocean basins today, there are catastrophic fish kills which may be from gassing. A good clue to gassing is the presence of black carbon- and sulphide-rich mud and there are suggestions that at least three of the mass extinctions are associated with events that formed black muds.[839,840] Clouds of sulphuric acid in the atmosphere reflect light and heat and hence these types of extinctions may be associated with cooling.

Gassing may be local or global, hence the distinction between a local extinction, a minor mass extinction and a major mass extinction becomes blurred. The most common type of volcano for this process is a basalt volcano such as in the mid ocean ridges, Hawaii and Iceland. Five mass extinctions occurred at the same time as massive basalt eruptions from the eleven basalt supervolcano provinces. Six of the eleven basalt supervolcano provinces coincide with events of global warming and loss of oxygen from marine waters.[841] This suggests that volcanic gases have had a profound effect on both life and climate. We have one example of local gassing that many readers may remember. The gas was $CO_2$. In Cameroon on 21 August 1986, Lake Nyos burped out a cloud of $CO_2$ derived from molten rocks at depth. Lake Nyos is a crater lake formed above what was thought to be an extinct gas volcano. It was only a small exhalation of gas. Because $CO_2$ is heavier than air, it displaced air in a valley and some 1700 people and their livestock were asphyxiated.[842]

One mass extinction is well known. That is the K-T extinction, which we are told led to the extinction of dinosaurs at 65 Ma. The popular paradigm was that an asteroid struck Chicxulub (Mexico) at 65 Ma, vaporised rock, created choking clouds of sulphurous gases, and the dust and gas fallout destroyed plant life. Once plants were destroyed, there was a collapse in the ecosystem, loss of animal life and a major mass extinction.[843] However, far more accurate dating of rocks has shown that the smoking gun was not in Mexico but in India. The K-T major mass extinction at 65.5 Ma occurred

[839] Wignall, P. B. and Twitchett, R. J. 1996: Oceanic anoxia and the end Permian mass extinction. *Science* 272: 1155-1158.

[840] Kump, L. R., Pavlov, A. and Arthur, M. A. 2005: Massive release of hydrogen sulfide to the surface ocean and atmosphere using intervals of oceanic anoxia. *Geology* 33: 397-400.

[841] Wignall, P. B. 2001: Large igneous provinces and mass extinctions. *Earth Science Reviews* 53: 1-33.

[842] Kling, G. W., Clark, M. A., Wagner, G. N., Compton, H. R., Humphrey, A. M., Devine, J. D., Evans, W. C., Lockwood, J. P., Tuttle, M. L. and Koenigsberg, E. J. 1987: The 1986 Lake Nyos disaster in Cameroon, West Africa. *Science* 236: 169-175.

[843] Morgan, J., Lana, C., Kersley, A., Coles, B., Belcher, C., Montanari, S., Diaz-Martinez, E., Barbossi, A. and Neumann, V. 2006: Analysis of shocked quartz at the global K-T boundary indicate an origin from a single, high-angle, oblique impact at Chicxulub. *Earth and Planetary Science Letters* 251: 264-279.

during the 800,000-year eruption period of basalts that formed the Deccan Traps (India).[844,845] The Chicxulub impact occurred 300,000 years before the mass extinction and dinosaurs lived before, during and after the impact.[846] It is probable that the K-T extinction, as with other extinctions, was the result of the combination of a number of catastrophic events. The Deccan eruptions emitted a huge volume of sulphur dioxide ($SO_2$) into the atmosphere, which combined with water producing widespread acid rain which killed vegetation and possibly even made the oceans acid for a short time.

Some mass extinctions of life are coincidental with many continental flood basalt events[847] such as the Siberian Traps, the Brazilian Highlands and Karoo-Ferrar basalt provinces (South Africa-Antarctica).[848] The biggest mass extinction of all time occurred when basalt supervolcanoes were at their peak.[849] Up to 96% of all complex life on Earth was wiped out at 251.4 Ma when the Siberian Traps basalts were erupted in the Talnakh Basin.[850] At the Triassic-Jurassic boundary (205 Ma), there was a mass extinction of life. The following 2 million years had increasing atmospheric $CO_2$ levels, increased burial of organic carbon and a long-term shift in the carbon chemistry of marine limestones and organic matter. This shift was probably due to increasing amounts of $CO_2$ released into the atmosphere derived from volcanism associated with the opening of the Atlantic Ocean by sea floor spreading to form the Central Atlantic Magmatic Province.[851] This

[844] Duncan, R. A. and Pyle, D. G. 1988: Rapid eruption of the Deccan flood basalts at the Cretaceous/Tertiary boundary. *Nature* 333: 841-843.

[845] Keller, G., Adatte, T., Gardin, S., Bartolini, A. and Bajpai, S. 2008: Main Deccan volcanism phase ends near the K-T boundary: Evidence from the Krishna-Godavari Basin, SE India. *Earth and Planetary Science Letters* 268: 293-311.

[846] Keller, G., Adatte, T., Stinnesbeck, W., Rebolledo-Vieyra, M., Fucugauchi, J. U., Kramar, U. and Stüben, D. 2004: Chicxulub impact predates the K-T boundary mass extinction. *Proceedings of the National Academy of Sciences* 101: 3753-3758.

[847] Wignall, P. B. and Hallam, A. 1997: *Mass extinctions and their aftermath*. Oxford University Press.

[848] Marsh, J. S., Hooper, P. R. Rahacek, J., Duncan, R. A. and Duncan, A. R. 1997: Stratigraphy and age of Karoo basalts of Lesotho and implications for correlations within the Karoo igneous province. In: (eds J. J. Mahoney and M. F. Coffin) *Large igneous provinces: continental, oceanic and planetary flood volcanism.* Geophysical Monograph 100, American Geophysical Union, 247-272.

[849] Knoll, A. H., Bambach, R. K., Canfield, D. E. and Grotzinger, J. P. 1996: Comparative Earth history and Late Permian mass extinction. *Science* 273: 452-456.

[850] Reichow, M. K., Saunders, A. D., White, R. V., Pringle, M. S., Al'Muhkhamedov, A., Medvevdev, A. I., Kirda, N. P. 2002: $^{40}Ar/^{39}Ar$ dates from the West Siberian Basin: Siberian flood basalt province doubled. *Science* 296: 1846-1849.

[851] van de Schootbrugge, B., Payne, J. L., Tomasovych, A., Pross, J., Fiebig, J., Benbrahim, M., Föllmi, K. B. and Quan, T. M. 2008: Carbon cycle perturbation and stabilization in the wake of the Triassic-Jurassic boundary mass-extinction event. *Geochemistry, Geophysics and Geosystems* 9: Q04028, doi: 10.1029/2007GC001914.

example again serves to show that one single volcanic event can have a long-term effect on atmospheric $CO_2$.

Continental flood basalt provinces have formed by numerous eruptions over a short period of geologic time, characteristically a few million years. Within this period, a short-lived eruption phase lasts about 1 million years. It typically erupts a large proportion of the lava volume which is in the order of 1000–10,000 cubic kilometres of lava. Each eruption lasts in the order of a decade or so and builds up immense lava flow fields by eruptive activity along fissures tens to hundreds of kilometres long. Fire fountains emanating from vents along the fissures eject lava many kilometres into the air. At times, the fire fountains sustain eruption columns that loft gas and ash tens of kilometres into the atmosphere at the same time as lava flowing over huge areas.

Based on recent eruptions and determination of volatile contents of ancient flood basalt volcanism, it is calculated that individual eruptions were capable of releasing 10,000 billion tonnes of sulphur dioxide, resulting in atmospheric loadings of at least 100 billion tonnes per annum during a sustained decade-long eruptive event. Atmospheric changes associated with sulphur dioxide emissions from just one of these long-lasting eruptions were likely to have been severe. By contrast, the amounts of $CO_2$ released would have been smaller with only tens of billions of tonnes exhaled per annum. This is far more $CO_2$ than human activity currently emits into the atmosphere. At present, each year 186 billion tonnes of $CO_2$ enter the atmosphere from all sources, of which 3.3% comes from human activities. More than 100 billion tonnes (57%) is given off by the oceans and 71 billion tonnes is exhaled by animals (including humans). Individual eruptions were followed by periods of no eruptions of hundreds of thousands of years during which the gas contributions to the atmosphere would be recycled. Global warming certainly did not cause the mass extinction at 65 Ma.[852] It was probably the continual addition of sulphur gases to the atmosphere during eruptions of the supervolcanoes in India.

There have been few unusual events of gassing over the last 200 Ma when change was extreme, rapid, unpredictable and had short-lasting consequences.[853] There were only two major events of note, the Late

[852] Self, S., Widdowson, M., Thordarson, T. and Jay, A. J. 2007: Volatile fluxes during basalt eruptions and potential effects on the global environment. A Deccan perspective. *Earth and Planetary Science Letters* 248: 518-532.

[853] Cohen, A. S., Coe, A. L. and Kemp, D. B. 2007: The Late Palaeocene-Early Eocene and Toarcian (Early Jurassic) carbon isotope excursions: a comparison of their time scales, associated environmental changes, causes and consequences. *Journal of the Geological Society, London* 164: 1093-1108.

Palaeocene-Early Eocene (55.8 Ma) and Early Jurassic, (183 Ma) events of extinction, intense rapid global warming, changes to the hydrologic cycle, widespread loss of oxygen from the oceans and major changes in carbon chemistry. However, the continued rifting of the North Atlantic Ocean along a 3000 km fissure looks as if it produced a period of cooling before the short warming event at 55.8 Ma.[854] At present in deep cool high-pressure alkaline seawater, carbonate shells dissolve at about 3.8 km depth. In the 55.8 Ma event, the depth at which shells dissolved was 1.8 to 2 km because the oceans were slightly more acidic. There was a 100,000-year period when the mean annual temperature in tropical areas was 26–27°C.[855] There were also changes in flora and fauna at high latitudes[856,857] and in both hemispheres.[858,859,860,861] This warming had a profound effect on marine and terrestrial life and there was a minor mass extinction of tropical life at this time. Land and sea surface temperatures increased by 5 to 10°C[862] with associated extinctions of life[863] over a period of 10,000 to 20,000 years.[864] The Arctic sea surface

[854] Jolley, D. W. and Widdowson, M. 2005: Did Paleogene North Atlantic rift-related eruptions drive early Eocene cooling? *Lithos* 79: 355-366.

[855] Harrington, G. J. and Jaramillo, C. A. 2007: Paratropical floral extinction in the Late Palaeocene-Early Eocene. *Journal of the Geological Society, London* 164: 323-332.

[856] Clyde, W. C. and Gingerich, P. D. 1998: Mammalian community response to the latest Palaeocene thermal maximum: an isotaphonomic study in the northern Bighorn Basin, Wyoming. *Geology* 26: 1011-1014.

[857] Wing, S. L., Harrington, G. J., Smith, F. A., Bloch, J. I., Boyer, D. M. and Freeman, K. H. 2005: Transient floral change and rapid global warming at the Palaeocene-Eocene boundary. *Science* 310: 993-996.

[858] Moran, K., Backman, J., Brinkuis, H., Clemens, S. C., Cronin, T., Dickens, G. R., Eynaud, F., Gattacceca, J., Jakobsson, M., Jordan, R. W., Kaminski, M., King, J., Koc, N., Krylov, A., Martinez, N., Matthiessen, J., McInroy, D., Moore, T. C., Onodera, J., O'Reagan, M., Pälike, H., Rea, B., Rio, D., Sakamoto, T., Smith, D. C., Stein, R., St John, K., Suto, I., Suzuki, N., Takahashi, K., Watanabe, M., Yamamoto, M., Farrell, J., Frank, M., Kubik, P., Jokat, W. and Kristoffersen, Y. 2006: The Cenozoic palaeoenvironment of the Arctic Ocean. *Nature* 441: 601-605.

[859] Storey, M., Duncan, R. A. and Swisher, C. C. 2007: Palaeocene-Eocene Thermal Maximum and the opening of the Northeast Atlantic. *Science* 316: 587-589.

[860] Poore, P. Z. and Matthews, R. K. 1984: Late Eocene-Oligocene oxygen and carbon isotope record from the South Atlantic DSDP site 522. *Initial Reports of the DSDP 73*: 725-735. Government Printing Office, Washington.

[861] Wilf, P., Cúeno, N. R., Johnson, K. R., Hicks, J. F., Wing, S. L. and Obradovich, J. D. 2003: High plant diversity in Eocene South America: Evidence from Patagonia. *Science* 300: 122-125.

[862] Pearson, P. N., Ditchfield, P. W., Singano, J., Harcourt-Brown, K. G., Nicholas, C. J., Olsson, R. K., Shakleton, N. J. and Hall, M. A. 2001: Warm tropical sea surface temperatures in the Late Cretaceous and Eocene epochs. *Nature* 413: doi: 10.1038/35097000.

[863] Kennett, J. P. and Stott, L. D. 1991: Abrupt deep-sea warming, palaeoceanographic changes and benthic extinctions at the end of the Palaeocene. *Nature* 353: 225-229.

[864] Fricke, H. C., Clyde, W. C., O'Neil, J. R. and Gingerich, P. D. 1999: Evidence for rapid climate change in North America during the latest Palaeocene thermal maximum: oxygen isotope compositions of biogenic phosphate from the Bighorn Basin (Wyoming). *Earth and Planetary Science Letters* 160: 193-208

temperature soared to 24°C[865] and warm water species dominated the Arctic fossil record from that time. Fresh melt waters were flushed into the Arctic Ocean.[866] Temperature then decreased to warm conditions for the next 100,000 years.[867] Some 1200 to 5000 billion tonnes of carbon as methane was suddenly released into the atmosphere and oceans at this time.[868]

Methane hydrates form under pressure from methane combining with water to make a complex molecule.[869] There are vast amounts of methane hydrate at certain depths in the ocean, created by the decomposition of the rain of planktonic bodies and other organic material from areas where life flourishes. Methane was suddenly released to the atmosphere. Ignition of methane to $CO_2$ and $H_2O$ was probably triggered by volcanic activity[870] but could have also been triggered by an impact, lightning or spontaneous combustion. For about 10,000 years the oceans were more acidic while the $CO_2$ in the oceans was rapidly and permanently sequestrated by weathering of rocks.[871] Oceans then went back to being alkaline and buffered by processes of weathering on the land and in the seas.

These gassing processes at 183 and 55.8 Ma probably derive from the explosive loss of methane from shallow water ocean sediments. Methane is a potent greenhouse gas that quickly oxidises in air to $H_2O$ and $CO_2$. How sudden methane degassing occurs is not really known. Suggested mechanisms are the heating of oceans by volcanicity, ocean current change or climate change to release methane hydrate from sediments, the lowering of sea level

[865] Brinkhuis, H. Schouten, S., Collinson, M. E., Sluijs, A., Sinninghe-Damste, J. S., Dickens, G. R., Huber, M., Cronin, T. M., Onodera, J., Takahashi, K., Bujak, J. P., Stein, R., van der Burgh, J., Eldrett, J. S., Harding, I. C., Lotter, A. F., Sangiorgi, F., van Konijnenburg-van Cittert, H., de Leeuw, J., Matthiessen, J., Backman, J., Moran K. and the Expedition 302 Scientists, 2006: Episodic fresh surface waters in the Eocene Arctic Ocean. *Nature* 441: 606-609.

[866] Brinkhuis, H., Schouten, S., Collinson, M. E., Sluijs, A., Sinninghe Damsté, J. S., Dickens, G. R., Huber, M., Cronin, T. M., Onodera, J., Takahashi, K., Bujak, J. P., Stein, R., van der Burgh, J., Eldrett, J. S., Harding, I. C., Lotter, A. F., Sangiori, F., van Kronijnenburg-von Cittert, H., de Leeuw, J. W., Mattiessen, J., Backman, J., Moran, K. and the Expedition 302 Scientists 2005: Episodic fresh surface waters in the Eocene Arctic Ocean. *Nature* 441: 606-609.

[867] Röhl, U., Bralower, T. J., Norris, R. D. and Wefer, G. 2000: New chronology for the late Palaeocene thermal maximum and its environmental implications. *Geology* 28: 927-930.

[868] Zachos, J. C., Pagani, M., Sloan, L. C., Thomas, E. and Billups, K. 2001: Trends, rhythms, and aberrations in global climate 65 Ma to present. *Science* 99: 686-693.

[869] Lu, H., Seo, Y-T., Lee, J.-W., Moudrakovski, I., Pipmeester, J. A., Chapman, N. R., Coffin, R. B., Gardner, G. and Pohlman, J. 2007: Complex gas hydrate from the Cascadia margin. *Nature* 445: 303-306.

[870] Svensen, H., Planke, S., Malthe-Sorenssen, A., Jamtveit, B., Myklebust, R., Eidem, T. R. and Rey, S. S. 2004: Release of methane from a volcanic basin as a mechanism for initial Eocene global warming. *Nature* 427: 542-545.

[871] Zachos, J. C., Röhl, U. Schellenberg, S. A., Sluijs, A., Hodell, D. A., Kelly, D. C., Thomas, E., Nicolo, M., Raffi, I., Lourens, L. J., McCarren, H. and Kroon, D. 2005: Rapid acidification of the ocean during the Palaeocene-Eocene Thermal Maximum. *Science* 308: 1611-1615.

by global cooling to depressurise sediment-bound methane hydrate or the triggering of methane hydrate release by earthquake, volcanic, cometary or meteorite activity. There are suggestions that the gassing may not have derived material from methane hydrate but from global forest fires or from burning coal seams, but little evidence exists and the total vegetation biomass on Earth is too low to account for so much carbon in the atmosphere. Another suggestion is the impact of a carbon-rich meteorite, but the impact craters and fallout have not been found. Maybe there was an overturning of the oceans which brought oxygen-poor $CO_2$-rich waters to the surface, resulting in an increase in atmospheric $CO_2$ and killing of oxygen-breathing marine life.

Both the 183 and 55.8 Ma events are probably best explained by the sudden large-scale breakdown of methane hydrate that followed more gradual changes linked to the intrusion of the Karoo Igneous Province (183 Ma)[872] and the North Atlantic Igneous Province.[873,874] Although yet to be quantified, the environmental affects following the emplacement of a large igneous province may well differ depending upon whether emplacement was continental or submarine.

The Late 20th Century Warming is insignificant when compared with these events. In order to understand modern climate, we need an understanding of past climate. This need for much improved knowledge of the durations and ages of climatic and geological events is constrained by the accuracy of geological and astronomical dating. There needs to be a systematic and co-ordinated detailed intercalibration of radioisotope clocks, the rock standards and the tie points with astronomical events. For example, the details of the 55.8 Ma Palaeocene-Eocene thermal event are not well known. We know little about the exact dating and timing of fluxes in and out of the marine carbon reservoir. Unless such information is known, it is very hard to assess competing hypotheses for extinction or climate change.[875]

Terrestrial supervolcanoes are explosive and occur along the Mediterranean-trans Asiatic Belt and the circum-Pacific "Ring of Fire". Many of these volcanoes are equatorial and belch dust and sulphurous gases

---

[872] Cox, K. G. 1989: The role of mantle plumes in the development of continental drainage patterns. *Nature* 342: 873-877.

[873] Saunders, A. G., Fitton, J. G., Kerr, A. C., Norry, M. J. and Kent, R. W. 1997: The North Atlantic Igneous Province. In: *Large Igneous Provinces: Continental, Oceanic and Planetary Flood Volcanism* (eds Mahoney, J. J. and Coffin, M. F.), 45-93, American Geophysical Union Monograph 100.

[874] Thomas, D. J. and Bralower, T. J. 2005: Sedimentary trace element constraints on the role of the North Atlantic Igneous Province volcanism in the late Palaeocene-early Eocene environmental change. *Marine Geology* 217: 233-254.

[875] Pälike, H. and Hilgren, F. 2008: Rock clock synchronization. *Nature Geoscience* 1: 282.

into both hemispheres. This material reflects heat and light, the planet cools, photosynthesis cannot take place, plants die and the food chain collapses. In 1980 Mt St Helens ejected a tiny 0.5 to 1 cubic kilometre of dust, Krakatoa in 1883 ejected 30 cubic kilometres, Toba 74,000 years ago ejected 2800 cubic kilometres of dust and, in the not so distant past, volcanoes in the Yellowstone area of USA and New Zealand ejected 10,000–100,000 cubic kilometres of dust. Humans almost became extinct after the Toba supervolcano eruption.

There is something evocative about an asteroid or comet impacting Earth and getting rid of those scary dinosaurs.[876] The problem is that this scenario probably did not happen. Furthermore, most dinosaurs were about the size of chickens and the rare species are the huge, scary museum attractions for children. Tsunamis, choking dust, sonic booms and falling heated impact target material gives the story of impacting a Hollywood touch. There have been many large impacts on Earth and it is hard to correlate these with all mass extinctions. If the target zone is limestone or gypsum-rich rocks, then huge amounts of $CO_2$ and sulphur gases will be released into the atmosphere and we get the same effects as gassing by volcanicity. Mass extinctions from asteroids may be cyclical and may occur when the Sun's motion is at right angles to the galactic plane.[877] Rattling the Earth with an impact can induce volcanism, continental drift and a mass extinction. However, don't lose sleep. There are many extraterrestrial bodies out there with your name written on them and premature death most commonly results from human stupidity.

Climate changes certainly cause biological stress. Contrary to popular view, it is global cooling that leads to biological stress and, in the case of a prolonged period of glaciation, species diversity drops and there are minor mass extinctions. Furthermore, sea level drops and there is a restriction or loss of shallow water habitats. This was especially the case in the Neoproterozoic (750–635 Ma) and Permian (300–260 Ma) glaciations. However, during the Permian glaciation, although life on the supercontinent Gondwana may have been stressed while it went for a trip over the South Pole, the rest of the continents were equatorial and it is in these environments that life thrives. Hence, although climate change may be global it does not affect all species on planet Earth. Moving continents also open and close seaways, change the shape of the sea floor, change ocean currents and change winds, all of which have an effect on climate.

---

[876] Alvarez, L. W., Alvarez, L., Asaro, F. and Michel, H. V. 1980: Extraterrestrial cause for the Cretaceous-Tertiary extinction: experimental results and theoretical interpretation. *Science* 208: 1095-1108.

[877] Rampino, M. R. and Stothers, R. H. 1984: Terrestrial mass extinctions, cometary impacts and the Sun's motion perpendicular to the galactic plane. *Nature* 308: 709-712.

Global warming actually produces a thriving of life, not extinction. Superplumes of molten rock beneath the sea floor created extreme warmth in the middle of the Cretaceous period.[878] Not only did these superplumes release huge amounts of heat but they also released large quantities of $H_2O$ vapour, $CO_2$ and methane into the atmosphere. Global temperature was 6 to 14°C higher than today,[879] the temperature difference between the poles and the equator was far less than today,[880] there were no ice sheets in mountains or at the poles[881], cold-blooded reptiles occurred at high latitudes[882] and coral reefs moved polewards.[883] The Arctic and Antarctica enjoyed a temperate to tropical climate.[884] Coral reefs were drowned and killed a number of times by oxygen-poor waters and migrated to higher latitudes (e.g. 36°N)[885] in fits and starts.[886,887,888] Antarctica contained broad-leafed evergreen conifer forests with cycads, ferns and flowering plants at 88–112 Ma.[889] These grew on flood plains. Winter temperatures were 0° to -4°C and summer temperatures 20–24°C.[890] This was only possible because the planet was in its normal state – warm, wet and volcanic.

During the Mesozoic, and especially the Cretaceous, global temperatures

[878] Larsen, R. L. 1991: Geological consequences of superplumes. *Geology* 19: 963-966.

[879] Barron, E. J. 1983: A warm, equable Cretaceous: the nature of the problem. *Earth Science Reviews* 29: 305-338.

[880] Bice, K. L., Huber, B. T. and Norris, R. D. 2003: Extreme polar warmth during the Cretaceous greenhouse? Paradox of the late Turonian $\partial^{18}O$ record at Deep Sea Drilling Project Site 511. *Palaeoceanography* 18: 1-11.

[881] Frakes, L. A. 1979: *Climate changes throughout geologic time.* Elsevier.

[882] Tarduno, J. A., Brinkman, D. B., Renne, P. R., Cottrell, R. D., Scher, H. and Castillo, P. 1998: Evidence of extreme climatic warmth from Late Cretaceous arctic vertebrates. *Science* 282: 2241-2244.

[883] Johnson, C. C., Barron, E. J., Kaufman, E. G., Arthur, M. A., Fawcett, P. J. and Yasuda, M. K. 1966: Middle Cretaceous reef collapse linked to ocean heat transport. *Geology* 24: 376-380.

[884] Huber, B. T. 1998: Tropical paradise at the Cretaceous poles? *Nature* 282: 2199-2200.

[885] Takashima, R., Sano, S-I., Iba, Y. and Nishi, H., 2007: The first Pacific record of the Late Aptian warming event. *Journal of the Geological Society, London* 164: 333-339.

[886] Föllmi, K. B., Weissert, H., Bisping, M. and Funk, H. 1994: Phosphogenesis, carbon-isotope stratigraphy and carbonate platform evolution along the lower Cretaceous northern Tethyan margin. *Geological Society of America Bulletin* 106: 729-746.

[887] Weissert, H., Lini, A., Föllmi, K. B and Kuhn, O. 1998: Correlation of Early Cretaceous carbon isotope signature and platform drowning events: a possible link? *Palaeogeography, Palaeoclimatology, Palaeoecology* 137: 189-203.

[888] Wilson, P. A., Jenkyns, H. C., Elderfield, H. and Larson, R. L. 1998: The paradox of drowned carbonate platforms and origin of Cretaceous Pacific guyots. *Nature* 392: 88-94.

[889] Falcon-Lang, H. J., Cantrill, D. J. and Nichols, G. J. 2007: Biodiversity and terrestrial ecology of a mid-Cretaceous, high-latitude floodplain, Alexander Island, Antarctica. *Journal of the Geological Society, London* 158: 709-724.

[890] Smith, A. G., Smith, D. G. and Funnel, B. M. 1994: *Atlas of Mesozoic and Cenozoic coastlines.* Cambridge University Press.

were considerably warmer than now. This was not a time of extinction. There was a great flourishing of life. The global warmth of the Cretaceous has been attributed to elevated levels of $CO_2$ in the atmosphere. The $CO_2$ derived from continental fragmentation, rifting and associated sea floor basalt volcanism. The rifting eventually formed the Atlantic Ocean. However, there are some suggestions that the Cretaceous climate was decoupled from the $CO_2$ content of the atmosphere.[891,892]

Supernoval explosions or gamma ray bursts are the great unknown for ancient extinctions. Additional cosmic radiation and solar radiation can destroy life by radiation damage. This is where species' palaeoenvironments are important because burrowing terrestrial and marine organisms and deep-sea organisms can survive. Furthermore, there are chemicals produced that have a short life, hence we can have a smoking gun for such processes. Once we are looking at sequences older than about 10 Ma, these chemicals have decayed and cannot be measured. It has been suggested that a supernova or gamma ray burst was responsible for the extinction at 450–440 Ma.

We primates thrived in warm wet times at 5 Ma. Sea level was at least 10 metres above its present level. Numerous species of hominids coexisted at the same time in Africa. This warm world suddenly became cold. What happened to the Earth's thermostat? There was a profound cooling of the Earth starting at around 2.67 Ma. Many species, including many hominids, became extinct. The Earth was already cooling and each little perturbation in orbit, solar activity, galactic travel and global geography affected climate. There were large changes to the Earth's environment, rainfall redistribution took place and tropical ocean currents were redistributed. This led to us, *Homo*, the first appearance of man-made tools, language, fire and human genes.

Drilling ocean floor sediments in the North Atlantic by the JOIDES *Resolution* has shown a sudden change in ocean floor sediments, heralding the onset of very cold conditions at 2.67 Ma.[893] Changes in oxygen chemistry in floating marine organisms, the shrinking of forests and the increase in the abundance of wind blown sands indicated the onset of a cold climate and aridity.[894] Forests changed to grasslands, hominids started to walk as there

---

[891] Veizer, J., Goddéris, Y. and François, L. M. 2000: Evidence for decoupling of atmospheric $CO_2$ and global climate during the Phanerozoic era. *Nature* 408: 698-701.

[892] Donnadieu, Y., Pierrehumbert, R., Jacob, R. and Fluteau, F. 2006: Cretaceous climate decoupled from $CO_2$ evolution. *Earth and Planetary Science Letters* 248: 426-437.

[893] Prueher, L. M. and Rea, D. K. 1998: Rapid onset of glacial conditions in the subarctic North Pacific region at 2.67 Ma: Clues to causality. *Geology* 26: 1027-1030.

[894] Baumann, K.-H. and Huber, R. 1996: Sea-surface gradients between the North Atlantic and the Norwegian Sea during the last 3.1 m.y.: comparison of sites 982 and 985. In: *Proceedings of the Ocean Drilling Program, Scientific Results*, Vol. 162 (eds Raymo, M.E., Jansen, E., Blum, P. and Herbert, T.D.), College Station, TX, 179–190.

were fewer trees, big game adapted to bushy grassland thrived on the extended grasslands and hominids had to change from eating mainly fruit to becoming omnivores.[895] Tools with razor sharp edges were created for hunting and cutting meat, cooking was able to make meat more digestible and hominid vegetarians headed for the eternal bliss of a meat-free extinction.

HMS *Challenger* dredged manganese-rich nodules from the ocean floor in the 19th Century. A century later, this discovery led to evaluating sea floor mining, the dredging of manganese nodules from the ocean floor and the discovery of submarine hydrothermal systems on the mid ocean ridges. These manganese nodules have been chemically analysed for everything under the Sun and this was exactly what was found. The isotope of interest is a heavy isotope of iron ($Fe^{60}$).[896] Because the nodules are only millions of years old and grow very slowly (0.23 cm every million years[897]), they still contain the short-lived isotopes created in supernoval explosions. Furthermore, if the layers in the nodule can be dated, then the timing of $Fe^{60}$ formation can be determined. One particular nodule, lovingly called 237kd, gave the vital clue.[898] There was a massive increase in the amount of $Fe^{60}$ around 2.8 Ma. Almost coincidentally, the NASA satellite RRHESS (Reuven Ramaty High Energy Solar Spectroscopic Imager) spotted decaying $Fe^{60}$ from the embers of a supernova nuclear disintegration.[899] A cosmic ray winter from the nearby supernova could have triggered a climate change that led to accelerated human evolution and the extinction of many other hominid species. Cosmic rays would have been emitted for hundreds of thousands of years after the supernoval explosion.[900]

The location of the supernova is unknown. The best candidates are in Orion in an area called the Orion OB1 Association. The NASA Compton satellite (1991–2000) detected the afterglow of a supernova, Orion OB1. Orion OB1 stars are hotter and brighter and 10 to 50 times bigger than the

---

[895] deMenocal, P. B. 1995: Plio-Pleistocene African climate. *Science* 270: 53-59.

[896] Knie, K., Korschinek, G., Faestermann, T., Wallner, C., Scholten, J. and Hillebrandt, W. 1999: Indication for supernova produced $^{60}Fe$ activity on Earth. *Physics Reviews Letters* 83: 18-21.

[897] Kinoshita, N., Sato, Y., Yamagata, T., Nagai, H., Yokoyama, A. and Nakanishi, T. 2007: Incorporation rate measurements of $^{10}Be$, $^{230}Th$, $^{231}Pa$ and $^{239,240}Pu$ radionuclides in manganese crust in the Pacific Ocean: A search for extraterrestrial material. *Journal of Oceanography* 63: 813-820.

[898] Smith, D. M. 2003: The Reuven Ramaty High Energy Solar Spectroscopic Imager observation of the 1809 keV line from galactic 26Al. *The Astrophysical Journal Letters* 589: L55-L58.

[899] Gallagher, P. T., Dennis, B. R., Krucker, S., Schwartz, R. A. and Tolbert, A. K. 2002: Rhessi and trace observations of the 21 April 2002 x1.5 flare. *Solar Physics* 210: 341-356.

[900] Knie, K., Korschinek, G., Faestermann, T., Wallner, C., Scholten, J. and Hillebrandt, W. 1999: Indication for supernova produced $^{60}Fe$ activity on Earth. *Physics Review Letters* 83: 18-21.

Sun. These blue stars have a short life (30 to 100 million years). This is the most likely place for a supernova in Orion. The cosmic radiation from a supernova in OB1 would take four centuries to reach Earth. The understanding of the effects of exploding stars on life on Earth is in its infancy and is one of the great future interdisciplinary areas of scientific research. Gould's Belt is the first place to investigate. The adventures of the Solar System in its passage through the Milky Way and the influence of neighbouring stars in our galaxy are still only sketchy.

Biological competition has always been with us and probably constantly creates extinction. Pandemics of trans-species viruses and bacteria may create turnover of species or even a mass extinction. These are extremely hard to ascertain because fossil bacteria and viruses are extremely rare, trans-species diseases are not related to environment (e.g. humans and armadillos both suffer from leprosy) and there are few clues in the fossil record to help us with this scenario. However, algal blooms in terrestrial waterways and red algal blooms in the seas suggest that this is a possible mechanism.

The selfish gene is constantly striving in this world of competing evolutionary pressures to achieve a longer life, a better habitat and to survive. Considering that terrestrial mammals have a species life of less than 2 Ma, we have a constant turnover of species that may be driven by nothing extraordinary. Watch this spot while you speciate.

*Recent macrofauna extinction*

During the last ice age, much of the planet's water was trapped in continental ice sheets, sea level was 130 metres lower than at present, forests contracted to grasslands and cold winds swept across grasslands. Because of the low sea level, humans were able to walk to areas that had previously not been hunted. For example, the entry of humans into North America was from Asia via the Bering Strait during an extremely cold post-glacial period. What a wonderful sight it must have been. In the grip of bitterly cold times, there was a promised land with huge numbers of wild birds and mammals. Within a very short time, more than 40 edible species were rendered extinct.[901]

Hunting is the method by which humans were traditionally able to efficiently obtain protein. Cooked meat protein gives a better chance for survival than a vegetarian diet. It was the bitter cold of the Younger Dryas elsewhere that probably led to the rise of agriculture and animal husbandry. The more fertile lands were used for grazing and grain cropping and the

---

[901] Diamond, Jared, 1997: *Guns, germs and steel.* W. W. Norton, New York.

infertile lands were left to Nature. Hunting became less necessary.

Travel, immigration and trade transport alien species. This has made survival amongst species more global and, in many places endemic species (especially on islands) become extinct. A rapid rate of extinction, especially of macrofauna, occurred in the Late Pleistocene (Australasia), terminal Pleistocene (Americas), Early to Middle Holocene (West Indies and Mediterranean islands) and Late Holocene (Madagascar, New Zealand and Pacific Islands). These extinction events coincided with human colonisation.[902] The different timing and places of extinction show that global climate changes could not have forced extinction.[903] In Australia, the continent-wide extinction did not coincide with extreme climatic events, thereby suggesting hunting overkill as the cause.

Except for the extinction in the Americas during the Younger Dryas, none of these extinction events are associated with climate change.[904] In the Americas, there was an extinction of a species every 40,000 years until 12,000 years ago.[905] Around 12,000 years ago there was a 2000-year long extinction event in which 57 species were lost, including three genera of elephants, mammoths, mastodons, the giant ground sloth, horses and camels. Extinction occurred at one species per 30 years. It appears that the slow-moving megafauna of the grasslands did not adapt to their new and aggressive predators, who over-killed.[906,907] Ecological stresses driven by humans such as fire, introduction of competing species and hunting on a

[902] Steadman, D. W., Martin, P. S., MacPhee, R. D. E., Jull, A. J. T., McDonald, H. G., Woods, C. A., Iturralde-Vinent, M. and Hodgins, G. W. L. 2005: Asynchronous extinction of late Quaternary sloths on continents and islands. *Proceedings of the National Academy of Sciences* 102: 11763-11768.

[903] Brook, B., Bowman, D. M. J. S., Burney, D. A., Flannery, T. F., Gagan, M. K., Gillespie, R., Johnson, C. N., Kershaw, P., Magee, J. W., Martin, P. S., Miller, G. H., Peiser, B. and Roberts, R. G. 2007: Would the Australian megafauna have become extinct if humans had never colonised the continent? Comment on "A review of the evidence for the human role in the extinction of Australian megafauna and an alternative explanation" by S. Wroe and J. Field. *Quaternary Science Reviews*, 26, 3-4.

[904] Burney, D. A. and Flannery, T. F. 2005: Fifty millennia of catastrophic extinctions after human contact. *Trends in Ecology and Evolution* 20: 395-401.

[905] Bulte, E., Horan, R. D. and Shogren, J. 2006. Megafauna extinction: A palaeoeconomic theory of human overkill in the Pleistocene. *Journal of Economic Behaviour and Organisation*, 59, 297-323.

[906] Brook, B. W. and Bowman, D. M. J. S. 2002: Explaining the Pleistocene megafaunal extinctions: Models, chronologies, and assumptions. *Proceedings of the National Academy of Sciences* 99: 14624-14627.

[907] Stuart, A. J., Kosintsev, P. A., Higham, T. F. and Lister, A. M. 2004: Pleistocene to Holocene extinction dynamics in giant deer and woolly mammoth. *Nature* 431: 684-689.

local[908] or regional scale[909] assisted extinction. As humans were migrating south from North America to South America, the Amazon Basin was changing from grasslands with copses of trees to rainforest. This habitat change would also have assisted extinction of macrofauna.

There was a heavy extinction of megafauna through America, Australia, Madagascar, the Mediterranean islands, New Zealand, the Pacific islands and the West Indies. In each case heavy extinction followed human invasion and colonisation.[910] For example, in the Americas, large animals from Alaska to Argentina vanish at about the same time around 12,000 to 11,000 years ago. The exception is the Greater Antilles where all sloths and other endemics disappeared some 6000 years ago when humans first colonised the islands.[911] The Younger Dryas was not a time of extinction in Australia, Madagascar, New Zealand and other Pacific islands.

There are few well-dated large mammal fossils but in Alaska and the Yukon Territory fossils show that many species survived the human invasion 13,500 to 11,500 years ago and began to increase in population before and during human colonisation. Examples are *Bison priscus*, which evolved into *Bison bison*, wapiti and moose, and these increases were before the extinction of the horse and mammoth. This suggests that something other than blitzkrieg human extinction, such as a subtler human impact, displacement or the cold of the Younger Dryas may be more likely.[912] Mammoths and horses, unlike elk and bison, were equipped physically to digest large quantities of low-nutrient grass when the Alaska-Yukon region was cold, treeless and arid. When the thaw came and the rains began, the grass became greener and richer, attracting elk and bison. Evergreen forests began to replace pasture, trees leached nutrients from soil and armed themselves with resins and other chemicals that made

[908] Prideaux, G. J., Roberts, R. G., Megirian, D., Westaway, K. E., Hellstrom, J. C. and Olley, J. M. 2004: Mammalian responses to Pleistocene climate change in southeastern Australia. *Geological Society of America* 35: 33-36.

[909] Brook, B. W., Bowman, D. M. J. S., Burney, D. A., Flannery, T. F., Gagan, M. K., Gillespie, R., Johnson, C. N., Kershaw, P., Magee, J. W., Martin, P. S., Miller, G. H., Peiser, B. and Roberts, R. G. 2005: Would the Australian megafauna have become extinct if humans had never colonised the continent? Comments on "A review of the evidence for a human role in the extinction of Australian megafauna and an alternative explanation" by S, Wroe and J. Field. *Quaternary Science Reviews* 26: 560-564.

[910] Martin, P. M., 2005: *Twilight of the mammoths: Ice Age extinctions and the rewilding of America.* University of California Press.

[911] Steadman, D. W., Pregill, G. K. and Olson, S. L. 1984: Fossil vertebrates from Antinqua, less Antilles: Evidence for Late Holocene human-caused extinctions in the West Indies. *Proceedings of the National Academy of Sciences* 81: 4448-4451.

[912] Guthrie, R. D. 2007: New carbon dates link climatic change with human colonization and Pleistocene extinctions. *Nature* 441: 207-209.

them unpalatable. Horses died off first, mammoths lasted a thousand years longer, elk and bison declined dramatically but survived, while moose, the only bark eaters among the animals, survived unaffected.

While temperature changes during the Younger Dryas were global, fauna adapted. However, the precipitation changes were rapid. In Siberia, dry steppes changed into marshes and swamps in the Preboreal, following the Younger Dryas, causing mammoths to die out around 11,200 years ago. The American megafauna perished at the beginning of the Younger Dryas around 12,500 years ago with an extreme dry period.[913]

There is a popular view that global warming will lead to extinction of species. Although climate changes affect geographical ranges and population persistence, little is known about the genetic response to climate change. Ancient DNA gives an equivocal answer for two widespread mammal species during the late-Holocene climate change. Despite some population sizes decreasing at times of climate change, some species will show declining gene diversity, as expected, whereas others will not.[914]

*Modern climate change and extinction*

There is a huge amount of emotion associated with extinction.[915,916,917] Some speculations suggest that a mere 0.8°C temperature rise over 50 years will result in extinction of 20% of the world's species. If this were the case, we should have seen a mass extinction of life in the Minoan Warming, the Roman Warming and the Medieval Warming. We did not. We may actually be living in a period of low extinctions, with relatively few species becoming extinct over the last 2.5 million years.[918] Current projections of extinctions may be an overestimation as we focus on terrestrial vertebrates and not the spectrum of life on Earth.

---

[913] Flessa, K. W. and Jablonski, D. 1983: Extinction is here to stay. *Paleobiology* 9: 315-321.

[914] Hadly, E. A., Ramakrishnan, U,, Chan, Y. L., van Tuinen, M., O'Keefe, K., Spaeth, P. A. and Conroy, C. J. 2004: Genetic response to climate change: Insights from ancient DNA and phytochronology. *Public Library of Science Biology* 2, 1600-1609.

[915] Thomas, C. D., Cameron, A., Green, R. E., Bakkenes, M., Beaumont, L. J., Collingham, Y. C., Erasmus, B. F. N., de Siqueira, M. F., Grainger, A., Hannah, L., Hughes, L., Huntley, B., van Jaarsveld, A. S., Midgley, G. F., Milkes, L., Ortega-Huerta, M. A., Peterson, A. T., Phillips, O. L. and Williams, S. E. 2004: Extinction risk from climate change. *Nature* 427: 145-148.

[916] Root, T., Price, J. T., Hall, K. R., Schneider, S. H., Rosenzweig, C. and Pounds, J. A. 2003: Fingerprints of global warming on wild animals and plants. *Nature* 421: 57-60

[917] Parmersan, C. and Yohe, G. 2003: A globally-coherent fingerprint of climate change impacts across natural systems. *Nature* 421: 37-42.

[918] Botkin, D. B., Saxe, H., Araújo, Betts, R., Bradshaw, R. H. W., Cedhagen, T., Chesson, P., Dawson, T., Etterson, J. R., Faith, D. P., Ferrier, S., Guisan, A., Skjoldborg Hansen, A., Hilbert, D. W., Loehle, C., Margules, C., New, M., Sobel, M. J. and Stockwell, D. R. B. 2007: Forecasting the effects of global warming on biodiversity. *BioScience* 57: 227-236.

The IPCC has promoted the view that global warming creates extinction. This was based on one study. Suggestions that human-induced global warming results in extinction is, at best, scientifically flawed. It was suggested that rising sea surface temperature in the equatorial Pacific Ocean led to the disappearance of 22 of the 50 known species of frogs and toads in the Montverde cloud forest of Costa Rica.[919] However, the authors also suggested that lowland deforestation may have a major influence on preservation of the cloud forests. This reservation was ignored as human-induced global warming was given as the reason for extinction. Although 21 of these species are known from elsewhere, one species (the golden toad) lost its only habitat and became extinct. It was the loss of this species that led to the conclusion that human-induced global warming could create an extinction of 20% of species over 50 years with a 0.8°C temperature rise. However, the trade winds that bring moist air from the Caribbean spend 5 to 10 hours over lowlands before they reached the golden toad's habitat. By 1992, 18% of the lowland vegetation remained after land clearing, resulting in an increase in the altitude of the cloud base thereby depriving the cloud forests of their moisture.[920] Land clearing created the extinction of the golden toad, not global warming.

Estimates of the impact of climate change on wildlife using the modelling method endorsed by the IPCC are a good example. When the models are run, they are not in accord with the modern distribution of wildlife.[921] This agrees with other studies on the effect of climate change on wildlife.[922,923,924] They are worthless because, despite the advances in mathematical simulation, the assumptions made are simplistic and lack critical variables. The scope for error is huge and cannot be reliably estimated. The same models have warned us that there will be massive extinction if the temperature rises yet

[919] Pounds, J. A. and Schneider, S. H. 1999: *Present and future consequences of global warming for highland tropical forests exosystems: The case of Costa Rica.* U.S. Global Change Research Program Seminar, Washington D.C., 29.9.99.

[920] Lawton, R. O., Nair, U. S., Pielke, R. A. Snr and Welch, R. M. 2001: Climate impact of tropical lowland deforestation on nearby mountain cloud forests. *Science* 294: 584-587.

[921] Beale, C. M., Lennon, J. J. and Gimona, A. 2008: Opening the climate envelope reveals no macroscale associations with climate in European birds. *Proceedings of the National Academy of Sciences* 105: 14908-14912.

[922] Pearson, R. G. & Dawson, T. E. 2003: Predicting the impacts of climate change on the distribution of species: are bioclimate envelope models useful? *Global Ecology and Biogeography* 12: 361-371.

[923] Thuiller, W. 2006: Patterns of uncertainties of species' range shifts under climate change. *Global Change and Biology* 12: 2020-2027.

[924] Hampe, A. 2004: Bioclimate envelope models: what they detect and what they hide. *Global Ecology and Biogeography* 13: 469-476.

these predictions of extinction are the opposite of what is seen with past warmings. A consultation with the Oracle at Delphi would be a more useful method of prediction.

Other emotive speculations on extinction are misleading, such as those regarding foxes in the Arctic, as touted by the IPCC. The key publication on foxes does not deal with extinction but migration of red foxes into the range of Arctic foxes in North America and Eurasia. The Arctic foxes survived the last interglacial warming and the Roman and Medieval Warmings and hence it is hardly likely that the milder Late 20th Century Warming would lead to their extinction. There are a great diversity of reasons for species migration which seem to be related to hunting or competition rather than human-induced global warming.[925]

The emotion about human-induced global warming is underpinned by the assumption that a future climate change will be so rapid that plants and animals would not be able to adapt to with the rate of temperature change. This view ignores the past, where there have been large climate changes on the scale of decades which have not led to plant or animal extinction. For example, the Fremont Glacier in Wyoming records a substantial warming from 1840 to 1850.[926] A substantial warming in less than a decade is far faster than the most speculative catastrophist models for human-induced global warming, yet there is no evidence that there was an extinction in North America at that time. Multicellular plants and animals have been on Earth at least 500 million years, so they have enjoyed at least 20 major climate changes. If we took the emotive argument to its logical conclusion, then there would be no multicellular life on Earth, as previous global warmings would have rendered life extinct and planet Earth would be a moonscape. Why was there not an extinction with the Eemian interglacial, the Younger Dryas, the Roman Warming, the Dark Ages, the Medieval Warming or the Little Ice Age? Why is it that only the Late 20th Century Warming will produce extinction whereas previous times when it was far warmer did not produce an extinction?

A key argument is that plants are immobile, hence a rapid global warming will push them into extinction. The scenario is that when plants become extinct, then animals that feed off plants will also become extinct. Because

[925] Hertsteinsson, P. and Macdonald, D. W. 1992: Interspecific competition and the geographic distribution of red and arctic foxes, *Vulpes vulpes* and *Alopex lagopus. Oikos* 64: 505-515.

[926] Schuster, P. F., White, D. E., Naftz, D. L. and Cecil, L. DeW. 2000: Chronological refinement of an ice core record at Upper Fremont Glacier in south central North America. *Journal of Geophysical Research* 105: 4657-4666.

this was not seen in previous warmings, the alarm bells should have been ringing for those speculating about extinction due to warming. What is observed is that plants in the Arctic have adapted to the frigid conditions but their distribution is rarely limited by warm conditions.[927,928,929,930] Many Arctic and alpine plants are extremely tolerant to high temperatures, adult trees harvest the light and it is only when an adult tree dies that a plant is replaced. This process takes time thereby giving lichen, fungi and animals time to move with the migrating plant ecosystem.

Geology shows that in times of global warming, there is an explosion of life, diversity increases and speciation is rapid. The Cambrian explosion of life (542–520 Ma) took place in the post-glacial warm times when atmospheric $CO_2$ was at least 25 times greater than today. Other great diversifications have taken place in the past. Species-rich forests existed during the warm Tertiary times in the western USA where many mountain species grew amongst mixed conifers and broad leaf sclerophylls.[931,932] It is only by completely ignoring the history of the planet can it be claimed that global warming can produce extinction.

In modern settings, it is also suggested that increased temperature will bring more species diversity[933] by extending the ranges of plants and animals. Replacement of high altitude forests by mixing with low altitude forests to create greater species diversity has happened in previous times of warming and would be expected in another warming event. Furthermore, if a future warmer climate had a higher atmospheric $CO_2$ content, plant life would be far more vigorous because increased $CO_2$ enables plants to grow better in nearly all temperatures, especially at higher temperatures. Both animals and plants are limited by the latitude and altitude cold-boundaries of their range and

---

[927] Loehle, C. 1998: Height growth rate tradeoffs determine northern and southern range limits for trees. *Journal of Biogeography* 25: 735-742.

[928] Gauslaa, Y. 1984: Heat resistance and energy budget in different Scandinavian plants. *Holarctic Ecology* 7: 1-78.

[929] Levitt, J. 1980: *Responses of plants to environmental stresses. Vol 1: Chilling, freezing and high temperature stresses.* Academic Press, New York.

[930] Kappen, L. 1981: Ecological significance of response to high temperature. In: *Physiological plant ecology. I. Response to the physical environment.* (eds O. L. Lange *et al.*). Springer-Verlag, New York.

[931] Axelrod, D. I. 1956: Mio-Pliocene floras from west-central Nevada. *University of California Publications in the Geological Sciences* 33: 1-316.

[932] Axelrod, D. I. 1987: The Late Oligiocene Creede Flora, Colorado. *University of California Publications in the Geological Sciences* 130: 1-235.

[933] Idso, S., Idso, C. and Idso, K. 2003: *The specter of species extinction.* The Marshall Institute, Washington, D.C., 1-39.

are not limited by the heat-limited boundaries of their range.[934,935,936,937,938] If atmospheric $CO_2$ is doubled, plant growth is unaffected at 10°C and growth is doubled at 38°C.

On a global scale, satellite measurements of vegetation between 1982 and 1999 showed that plant growth increased by 6% in response to slightly increased rainfall and slightly increased temperature but the major change was due to slightly increased $CO_2$. If the $CO_2$ content is doubled, the net productivity rise of herbaceous plants is 30 to 50%, while of woody plants it is 50 to 80%. In the European Alps, there are plant species counts from 1895 to the present. Mountaintop temperatures have increased by 2°C since 1920 with 1.2°C of that rise over the last 30 years. Of the 30 mountaintops, nine showed no change in the species count, 11 gained 59% more species and one had a 143% increase in species. The 30 mountaintops showed a mean species loss of 0.68 out of an average of 15.57 species.[939] The loss of a species from a particular mountain does not mean extinction but shows local mobility of plants. There are numerous other studies of lichen,[940] plants,[941,942,943] butterflies,[944] birds,[945]

---

[934] Idso, K. E. and Idso, S. B. 1994: Plant responses to atmospheric $CO_2$ enrichment in the face of environmental constraints: A review of the past 10 years' research. *Agriculture and Forest Meteorology* 69: 153-203.

[935] Cannell, M. G. R. and Thorley, H. H. M. 1998: Temperature and $CO_2$ responses of leaf and canopy photosynthesis. A clarification using the non-rectangular hyperbola model of photosynthesis. *Annals of Botany* 82: 883-892.

[936] Nemani, R. R., Keeling, C. D., Hashimoto, H., Jolly, W. M., Piper, S. C., Tucker, C. J., Myneni, R. B. and Running, S. W. 2003: Climate-driven increases in global terrestrial net primary production from 1982 to 1999. *Science* 300: 1560-1563.

[937] Kimball, B. A. 1983: Carbon dioxide and agricultural yield: An assemblage and analysis of 430 prior observations. *Agronomy Journal* 75: 779-788.

[938] Saxe, H. E., Ellsworth, D. S. and Heath, J. 1998: Tree and forest fluctuating in an enriched $CO_2$ atmosphere. *New Phytologist* 139: 395-436.

[939] Pauli, H., Gottfried, M. and Grabherr, G. 1996: Effects of climate change on mountain ecosystems – upward shifting of mountain plants. *World Resources Review* 8: 382-390.

[940] Van Herk, C. M., Aptroot, A. and Dobben, H. F. 2002: Long term monitoring in The Netherlands suggests that lichens respond to global warming. *Lichenologist* 34: 141-154.

[941] Sobrino, E., González-Moreno, A., Sanz-Elorza, M., Dana, D., Sánchez-Mata, D. and Gavilán, R. 2001: The expansion of thermophilic plants in the Iberian Peninsula as a sign of climate change. In: *"Fingerprints" of climate change: Adapted behaviour and shifted species ranges* (eds Walther, G. R., Burga, C. A. and Edwards, P. J.) Plenum, New York, 163-184.

[942] Sturm, M., Racine, C. and Tape, K. 2001: Increasing shrub abundance in the Arctic. *Nature* 411: 546-547.

[943] Smith, R. I. L. 1994: Vascular plants as bioindicators of regional warming in Antarctica. *Oecologia* 99: 322-328.

[944] Pollard, E., Moss, D. and Yates, T. J. 1995: Population trends of common British butterflies at monitored sites. *Journal of Applied Ecology* 32: 9-16.

[945] Jackson, N. K. 1994: Pioneering and natural expansion of breeding distributions of western North American birds. *Studies in Avian Biology* 15: 27-44.

plankton,[946] marine systems[947,948] and fish[949] to show that a slight temperature rise induces species diversity, species migration and adaptation.

These observations show that planet Earth is dynamic. Life is constantly adapting to change. Life will adapt to change when temperature and atmospheric $CO_2$ rise slightly. Whether the changes are natural or human-induced is irrelevant. Far greater changes occurred in pre-industrial times without extinction of life. Detailed studies in a specific area may record a local extinction. However, this is misleading, as an extinction is the total loss of a species whereas local extinction could mean that a species has migrated to another area. Some species that were thought to be extinct in one area have been found decades later in another area.

To argue that increasing temperature and atmospheric $CO_2$ will result in extinction of plants is to argue that $CO_2$ is not plant food. Even if the planet warms due to increased atmospheric $CO_2$, it is clear that plants will not feel the need to migrate to cooler parts of our planet. In fact it is the very opposite. Young plants adsorb more $CO_2$ than old plants. If we wanted an effective carbon plant sequestration scheme, then we would cut down all our old growth forests and plant saplings or even leave what was forest as grass, both of which would adsorb far more $CO_2$ than mature trees.

What will happen to crop yields if there is global warming? As with most matters with science, there are conflicting data and no simple answers.[950] Several decades of research on the effects of elevated $CO_2$ concentration on crop growth and yield have produced a wealth of information.[951,952,953] A 300 ppmv increase in atmospheric $CO_2$ enhances growth of $C_3$ cereals

[946] Southward, A. J. 1995: Seventy years' observation of changes in distribution and abundance of zooplankton and intertidal organisms in the Western English Channel in relation to rising sea temperatures. *Journal of Thermal Biology* 20: 127-155.

[947] Smith, R. C., Ainley, D., Baker, K., Domack, E., Emslie, S., Fraser, B., Kennett, J., Leventer, A., Mosley-Thompson, E., Stammerjohn, S. and Vernet, M. 1999: Marine ecosystem sensitivity to climate change. *BioScience* 49: 393-404.

[948] Sagarin, R. D., Barry, J. P., Gilman, S. E. and Baxter, C. H. 1999: Climate-related change in an intertidal community over short and long time scales. *Ecological Monographs* 69: 465-490.

[949] Collins, Simon, 2004: Antarctic fish set to survive warmer seas. *New Zealand Herald*, 16th April 2004.

[950] Tubiello, F. N., Amthor, J. S., Boote, K. J., Donatti, M., Easterling, W., Fischer, G., Gifford, R. M., Howden, M., Reilly, J. and Rosenweig, C. 2007: Crop response to elevated $CO_2$ and world food supply. A comment on "Food for thought..." by Long et al., Science 312:1918-1921, 2006. *European Journal of Agronomy* 26: 215-223.

[951] Eamus, D. 1996: Responses of field grown trees to $CO_2$ enrichment. *Commonwealth Forestry Review* 75: 39-47.

[952] Eklundh, L. and Olsson, L. 2003: Vegetation index trends for the African sahel 1982-1999. *Geophysical Research Letters* 30: 1430-1433.

[953] Saxe, H., Ellsworth, D. S. and Heath, J. 1998: Tree and forest functioning in an enriched $CO_2$ atmosphere. *New Phytologist* 139: 395-436.

(49%), $C_4$ cereals (20%), fruits and melons (24%), legumes (44%), roots and tubers (48%) and vegetables (37%). It must be remembered that $CO_2$ is a plant food, not a pollutant, and is a great stimulus to plant life on Earth. An enrichment in atmospheric $CO_2$ is not even a little bit bad for life on Earth. It is wholly beneficial.

Although glasshouses with higher $CO_2$ contents stimulate growth, free air carbon dioxide enrichment outside such controlled environmental chambers presents a different story. Despite the decades of research, there is still uncertainty concerning the effects of elevated $CO_2$ especially in the light of air pollution, changes in moisture availability and mineral nutrition and altered incidence of pests, diseases and weeds. There are global warming models that suggest increased temperature and decreased soil moisture will act to reduce crop yields by 2050 AD, but such decreases are more than offset by the direct fertilisation effect of elevated $CO_2$.[954] However, free air concentration enrichment experiments in large fields shows that elevated $CO_2$ enhances yield but the enhancement is some 50% lower than similar experiments in glasshouses.

Crops vary over time with changing climate and weather. If the climate becomes warmer, the loss of arable land will be replaced by the gain in arable land at higher latitudes. Climate conditions best suited for growing wheat will shift away from the tropics during global warming. Not only will there be adaptation, but new crops genetically engineered for future conditions will be the norm.

By contrast, the Stern Review suggests from one source that carbon fertilisation is "weak" and "smaller than previously thought".[955] The footnotes of the Stern Review show that it does not assume weak fertilisation by $CO_2$ but "no fertilisation effect".[956] The basis for this opinion is just one study[957] which looked at $CO_2$ fertilisation under field conditions and suggested that $CO_2$ fertilisation might be a third to half what is suggested from growth chambers. The Review's assumption of $CO_2$ fertilisation led to the headline,

---

[954] Long, S. P., Ainsworth, E. A., Leakey, A. D. B., Nösberger, J. and Ort, D. R. 2006: Food for thought: Lower than expected crop yield stimulation with rising $CO_2$ concentrations. *Science* 312: 1918-1921.

[955] Stern Review, p. 67-68, Box 3.4 (p. 70) and Fig. 3.6 (p. 73). The numbers used in the A2 scenario, used by the Stern Review as the base case, are also far higher than in any other scenario.

[956] Stern Review, Page 72, Footnote 43.

[957] Long, S. P., Ainsworth, E. A., Leakey, A. D. B., Nösberger, J. and Ort, D. R. 2006: Food for thought: Lower than expected crop yield stimulation with rising $CO_2$ concentrations. *Science* 312: 1918-1921.

"250–550 million additional people may be at risk".[958]

Polar bears are cute, people care about them and they sell. Just ask the marketing executives of Coca-Cola and Bundaberg Rum. Well, they are not really that cute. Try to hug a polar bear and suffer the consequences. Claims that the polar bears are facing extinction due to human activity packs an emotional punch, as we can imagine a cuddly furry polar bear painfully drifting into extinction, whereas we don't suffer too much emotional damage if scorpions, snakes and spiders become extinct. The polar bear has become an icon of the global warmers. However, the polar bear has survived the Medieval Warming, the Roman Warming, the Minoan Warming and numerous previous warmings when the temperature was higher than today. If the ice distribution changes, polar bears move. They are not glued to the one place on a static planet.

It is claimed that the polar bear habitat is shrinking hence polar bears will become extinct. Al Gore's movie *An Inconvenient Truth* claimed that, as there is now less sea ice, polar bears are dying while trying to find ice. In fact, the sea ice has expanded and high winds during an Arctic storm killed four polar bears in an area where sea ice was growing.[959] This was used as "evidence" by Gore that global warming was killing polar bears. The coastal temperature measuring stations on Greenland show cooling and the average summer temperatures at the summit of the Greenland Ice Sheet have decreased 2°C per decade since measurements began in 1987. Furthermore, Russian coastal stations show the extent and thickness of sea ice has varied greatly over 60- to 80-year periods during the last 125 years.

Ice core from Baffin Island and sea core sediments from the Chukchi Sea also show that even if there is warming, it has occurred many times before the Industrial Revolution, which is heralded as the cause of the current warming. In Alaska, the onset of a climate shift in 1976–1977 ended a multidecadal cold trend in the mid 20th Century. This warming only returned temperatures to those of the early 20th Century. The Canadian Fisheries and Oceans Department stated: "the possible impact of global warming appears to play a minor role in changes to Arctic sea ice".

The US Fisheries and Wildlife Service[960] showed that there are some 22,000 polar bears in about 20 distinct populations worldwide. Only two bear populations, accounting for 16.4% of the total, are decreasing, and they are in areas where air temperatures have actually fallen, such as the

---

[958] Stern Review, Page 72.

[959] Amstrup, S. C., McDonald, T. L. and Durner, G. M. 2004. Using satellite radio-telemetry data to delineate and manage wildlife populations. *Wildlife Society Bulletin.* 32: 661-679.

[960] US Fisheries and Wildlife Service, Report 6th April 2006.

Baffin Bay region. By contrast, another two populations, about 13.6% of the total number, are growing and they live in areas where air temperatures have risen, near the Bering Strait and the Chukchi Sea. As for the rest, 10 populations, comprising about 45.4% of the total, are stable and the status of the remaining six is unknown. The US National Biological Service found that polar bear populations in western Canada and Alaska were thriving to the point that some were at optimum sustainable levels.[961]

Based on the evidence, there is little reason to suggest that polar bears are on the path to extinction. Polar bears have survived for thousands of years, during both colder and warmer periods, and their populations are by and large in good shape. Polar bears suffer many threats, but global warming is not one of them. The main threats are ecotourists, bureaucrats and hunters.

Similar arguments are given for humans, with claims that higher temperature will increase mortality. This is contrary to history and modern observations. For example, the daily mortality rate for the last 30 years in southwest Germany shows that cold spells lead to increased mortality as do heat waves.[962] Those that suffered heat stress were old or sick, whereas others adapted to temperature changes far more rapidly.

## Global warming and infectious diseases

There are predictions that current global warming will result in the emergence of malaria and other diseases from the tropics to Europe and North America.[963] Projections are based on scientific studies that do not take account of changes in technology and increases in adaptive capacities as developing nations become richer.[964] Once the *per capita* income reaches $3100, malaria is functionally eliminated.[965] Furthermore, malaria is common in cold climates as well as warm climates. Techniques to eradicate malaria have been available for decades. It is a disease of poverty, not a disease of climate change. For more than a decade, malaria has held a prominent

---

[961] Amstrup, S. C., McDonald, T. L. and Durner, G. M. 2004. Using satellite radio-telemetry data to delineate and manage wildlife populations. *Wildlife Society Bulletin*. 32: 661-679.

[962] Laschewski, G. and Jendritzky, G. 2002: Effects of the thermal environment on human health: an investigation of 30 years of daily mortality from SW Germany. *Climate Research* 21: 91-103.

[963] Hay, S. I., Cox, J., Rogers, D. J., Randolph, S. E., Stern, D., Shanks, D. G., Myers, M. F. and Snows, R. W. 2002: Climate change and the resurgence of malaria in the East African Highlands. *Nature* 415: 905-909.

[964] Goklany, I. M. 2003: Relative contributions of global warming to various climate sensitive risks, and their implications for adaptation and migration. *Energy and Environment* 14: 797-822.

[965] Tol, R. S. J. and Dowlatabadi, H. 2001: Vector borne diseases, development and climate change. *Environmental Science and Policy* 8: 572-578.

place in exaggerated alarming speculations regarding climate change. Highly publicised mathematical models predict increases in the geographical distribution of malaria[966,967] despite efforts to put the issue into perspective and data to the contrary.[968]

Malaria, also known as ague, tertian or quartan was well known in Europe over the last 1000 years. During the Medieval Warm Period, the European literature from Christian Russia to Muslim Spain recorded malaria-like illnesses. In *The Inferno*, Dante (1265–1321) describes the symptoms of malaria, as did Geoffrey Chaucer (1342–1400) in *The Nun's Priest's Tale*. William Shakespeare (1564–1616) mentions ague in eight of his plays. In the 16th Century, many English marshlands were notorious for their ague-stricken populations and remained this way until well into the 19th Century.[969] Malaria in England during the Little Ice Age and its demographic, epidemiologic and social impacts[970] are related to the association of brackish water where the mosquito *Anopheles* breeds (especially *A. atroparvus*).[971] Along the Thames River from downstream Upchurch, Iwade and Medway, it was well known in the 17th Century that the estuarine and marsh conditions led to an increase in illness, where further upstream at Burnham, there was less disease. It was then thought that ague derived from noxious vapours emitted by marshes. Demographic data is similar to those of malaria-endemic populations in the tropics today.

Although malaria was endemic in 17th Century England, not all summers were cool. Samuel Pepys (1633–1703), an ague sufferer, described the dry hot summers of 1661, 1665 and 1666. This may well have enhanced the transmission of malaria. Drought malaria arises when rivers and ponds are reduced to smaller ponds and puddles, where anopheline mosquitos prefer to breed. There is a correlation between the warm dry summers in 1660 and 1810 and seasonal burial rates in the Bradwell-juxta-Mare, a marsh parish of

---

[966] Rogers, D. J. and Randolph, S.E. 2000: The global spread of malaria in a future, warmer world. *Science* 289: 1763-1766.

[967] Tanser, F. C., Sharp, B. and le Sueur, D. 2003: Potential effect of climate change on malaria transmission in Africa. *The Lancet* 362: 1792-1798.

[968] Reiter, P., Thomas, C. J., Atkinson, P. M., Hay, S. I., Randolph, S. E., Rogers, D. J., Shanks, G. D., Snow, R. W. and Spielman, A. J. 2004: Global warming and malaria: a call for accuracy. *The Lancet Infectious Diseases* 4: 323-324.

[969] Reiter, P. 2000: From Shakespeare to Defoe: Malaria in England in the Little Ice Age. *Emerging Infectious Diseases* 6: 1-11.

[970] Dobson, M. J. 1997: *Contours of death and disease in early modern England*. Cambridge University Press.

[971] Dobson, M. J. 1994: Malaria in England: a geographical and historical perspective. *Parasitologia* 36: 35-60.

Essex[972] and, even in the mid 19th Century towards the end of the Little Ice Age, there is a correlation between high summer temperatures and ague in Kent.[973] Daniel Defoe (1660–1731) travelled extensively in southern England from 1685 to 1690 and describes ague in the Dengie marshes of Essex (70 km east of London) that remained notorious for ague until the end of the 18th Century.[974]

Malaria in the Little Ice Age was not restricted to England. Even as far north as Inverness (57°20'N),[975] malaria had been recorded and the northern limits in southern Sweden, Finland, Scotland, the Gulf of Bothnia near the Arctic Circle,[976] through most of the United States[977] and some parts of Canada[978] can be related to the 15°C July isotherm.[979] After the 1880s, transmission dropped precipitously and malaria became relatively rare except following a short period after World War I. The decline occurred during the warming at the end of the Little Ice Age. Mosquito habitat was reduced by land reclamation and intensive drainage, winter root crops such as turnip that allowed greater numbers of animals to be kept thereby diverting *A. atroparvus* from feeding on humans, and the increased use of mechanisation in farming thereby reducing the number of humans as hosts for both mosquitos and parasites. Houses were more mosquito proof, better medicine and a drop in the price of quinine also reduced the incidence of malaria. A similar decline occurred in the prosperous countries of Europe such as Norway, Sweden, Denmark, Germany, Holland, Belgium and Italy. By contrast, malaria was still prevalent in the poorer countries of Eastern Europe, the Black Sea coast and the eastern Mediterranean.

Malaria is unstable in regions that normally have abundant rainfall, and epidemics can occur during drought. The 1934–1935 epidemic in what is now Sri Lanka killed 100,000 people. Worst hit was the southwest of the island

---

[972] Dobson, M. J. 1980: "Marsh fever" – the geography of malaria in England. *Journal of Historical Geography* 6: 357-389.

[973] Macdonald, A. 1920: On the relation of temperature to malaria in England. *Journal of the Royal Army Medical Corps 35*: 99-119.

[974] Defoe, D. 1986: *A tour through the whole island of Great Britain.* Penguin, London.

[975] Russell, P. F. 1956: World-wide malaria distribution, prevalence and control. *American Journal of Tropical Medicine and Hygiene* 5: 937-965.

[976] Ekblom, T. 1938: Suédoises de l'Anopheles manulipennis et leur rôle épidémiologique. *Bulletin de Société Pathologie Exotique* 31: 647-655.

[977] Faust, E. C. 1941. The distribution of malaria in North America, Mexico, Central America and the West Indies. In: *A symposium on human malaria with special reference to North America and the Caribbean region.* Washington. American Association for the Advancement of Science, 8-18.

[978] Fish, G. H. 1931: Malaria and the Anopheles mosquito in Canada. *Canadian Medical Association Journal* Dec. 1931; 679-683.

[979] Patz, J. A., Epstein, P. R., Burke, T.A. and Balbus, J. M. 1996: Global climate change and emerging infectious diseases. *Journal of the American Medical Association* 275: 217-223.

where rainfall is greater than 250 cm per annum and malaria is endemic but relatively infrequent. *A. culicifacies* breeds along river banks and is scarce in normal years. There was greater than average annual rainfall between 1928–1933 and river flow was high. After the failure of two successive monsoons, the drying rivers produced a population explosion of *A. culicifacies* resulting in a malaria epidemic. In the drier parts of the island where *A. culicifacies* was dominant but transmission more stable, immunity protected the population from the worst ravages of the epidemic.[980]

After World War II, attempts were made to eliminate malaria from all of Europe[981] and from the United States. In 1975, the WHO declared that Europe was free of malaria and by 1977, 83% of the world's population was living in areas from which malaria had been eradicated or control activities were in progress. This trend has been reversed, principally from the lack of use of DDT, population increase, forest clearance, irrigation, changes in ecology, population movement, urbanisation, deterioration of public health systems, resistance to insecticides, disruptions from war and natural disasters. The suggestion that global warming may lead to malaria ascending to new altitudes[982] is contradicted by records of its distribution from 1880 to 1945.[983,984,985] The increase in incidence of malaria and other diseases is probably also related to the exponential increase in international air travel.[986] Malaria has been on the increase in Bulgaria, Romania, Moldova, Italy, Corsica, Kazakhstan, Kyrgyzstan, Turkmenistan and Uzbekistan.[987] Maybe concerns should be with the realities of malaria and other disease transmission rather than global warming.

There are extraordinary claims made that increased atmospheric $CO_2$, even to 425 ppmv, will cause human health problems[988] such as acidosis,

[980] Dunn, C. 1937: *Malaria in Ceylon: an enquiry into its causes*. Bailliere, Tindall and Cox, London.

[981] Russell, P. F. 1955: *Man's mastery of malaria*. Oxford University Press.

[982] McMichael, A. J., Patz, J. and Kovats, R. S., 1998: Impacts of global environmental change on future health and health care in tropical countries. *British Medical Bulletin* 54: 475-488.

[983] Hackett, L. W. 1945: The malaria of the Andean region of South America. *Revisto del Instituto de Salubridad y Enfermedades Tropicales* 6: 239-252.

[984] Reiter, P. 1998: Global warming and vector-borne disease in temperature regions and at high altitude. *Lancet* 351: 839-840.

[985] Mouchet, J., Manguin, S., Sircoulon, J., Laventure, S., Faye, O., Onapa, A. W., Carnevale, P., Julvez, J. and Fontinelle, D. 1998: Evolution of malaria in Africa for the past 40 years: impact of climatic and human factors. *Journal of the American Mosquito Control Association* 14: 121-130.

[986] Zuletta, J. 1973: Malaria eradication in Europe. *Journal of Tropical Medicine and Hygiene* 76: 279-282.

[987] Sabatinelli, G. 1998: Malaria situation and implementation of the global malaria control strategy in the WHO European region. *WHO Expert Committee on Malaria* MASL/EC20/98.9.

[988] Robertson, D. S. W. 2001: The rise in atmospheric concentration of carbon dioxide and the effects on human health. *Medical Hypotheses* 56: 513-519.

restlessness and mild hypertension. This is based on a US Navy study in nuclear submarines where $CO_2$ in the air was 15 times the present atmospheric level.[989] It was claimed that the poor performance of some athletes at the 2003 World Games and fatalities in France in 2003 were due to increased $CO_2$ in the atmosphere.[990] Suggestions are that in 50 years time, the increased $CO_2$ in the atmosphere will lead to widespread health problems. What was not stated was that the $CO_2$ measurements of the mid 19th Century show that $CO_2$ was above 425 ppmv[991] and that there were no widespread health problems at that time from high atmospheric $CO_2$.

## Desertification

The geological record shows long periods of desertification. These are characterised by solidified red dune sands and salt beds. For example, the huge salt deposits of northern England, Germany, Poland and Siberia all formed during a time[992] when the Southern Hemisphere (and India) were covered with ice. This has been known for a century. Alfred Wegener suggested continental drift in 1912.[993] However he needed more data and so he mapped the global location of Permian (280–251.4 Ma) organisms that could not swim or fly (e.g. *Glossopteris* flora). He chose the Permian because he was a glaciologist and the literature showed that there were Permian glacial rocks in South Africa, India, Australia and South America. They were later discovered in Antarctica after Wegener's time. This was perplexing to Wegener because these places were clearly not glaciated landmasses like Greenland or Antarctica. By plotting the striations in rocks over which ice travelled, Wegener showed a radial pattern if he put the continents back together. He correctly concluded that there must have been a large polar landmass that enjoyed glaciation. He then plotted the position of red sandstones (i.e. desert dune sands) and evaporites (i.e. salt formed from evaporation) and found that these deposits, typical of mid latitude areas, were common in the Permian rocks of Siberia, Poland, north Germany, France and north England. None of these areas are now mid latitude areas.

Wegener then plotted the distribution of Permian coral reefs which could only have formed in tropical areas and showed that these no longer

---

[989] Lambert, R. J. W. 1972: The nuclear submarine environment. *Proceedings of the Royal Society of Medicine* 65: 795-800.

[990] Robertson, D. S. 2006: Health effects of increase in concentration of carbon dioxide in the atmosphere. *Current Science* 90: 1607-1609.

[991] Beck, E. G. 2007: 180 years of atmospheric $CO_2$ gas analysis by chemical methods. *Energy and Environment* 18: 259-282.

[992] Permian Period, 280 to 251.4 Ma.

[993] Wegener, A. 1912: Die Entstehung der Kontinente. *Geologische Rundschau* 3: 276-292.

are in tropical areas. He concluded that the continents moved. He suggested that there was a giant supercontinent Pangea[994] which split into two other supercontinents, Laurasia and Gondwana. Wegener noted that this was in the Late Triassic (217–204 Ma) when there was a sudden change; plants like *Glossopteris* disappeared and there was a rapid rate of plant evolution. We now know that at that time there was a mass extinction, massive impacting in North America and Russia and the break-up of Laurasia to form the Atlantic Ocean. Wegener suggested that the continents ploughed across the basalt ocean crust like a boat or slid across the interface between basalt ocean crust and the continental crust. These ideas were fairly quickly demolished and Wegener's theory and data were almost lost in history. Some 50 years later, a greater understanding of Earth magnetism, earthquakes, volcanoes and distribution of fossils was used to construct the theory of plate tectonics. It built on Wegener's work and showed that the continents actually drift. Wegener died on the Greenland ice sheet in 1930 and never saw his theory of continental drift validated, albeit by another mechanism.

To blithely state that global warming will lead to increased desertification ignores previous validated science that shows that desertification occurs during glaciation, and ignores the history of the development of scientific ideas. Modern climate change can lead to widespread and irreversible desertification, or so we are told.[995,996,997] This is contrary to all we know about the history of the planet. Evidence is now coming to light that shows that Africa is not undergoing severe desertification.[998,999,1000] In fact, recent evidence shows the opposite.[1001] These studies used vegetation greenness and rainfall in the African Sahel derived from satellite measurements and concluded that rainfall appears to be the dominant causative factor in the dynamics of vegetation greenness and long-term trends suggest that there might be another weaker causative factor. This other weaker causative factor may be slightly increased

[994] From the Greek, meaning all lands.

[995] Dregne, H. E. 1983: *Desertification of arid lands*. Harwood Academic Publishers, New York.

[996] Lamprey, H. F. 1988: Report on desert encroachment reconnaissance in Northern Sudan: 21 October to 10 November 1975. *Desertification Control Bulletin* 17: 1-7.

[997] Middleton, N., Thomas, D. and United Nations Environment Program, 1997: *World Atlas of Desertification*. Arnold, London.

[998] Eklundh, L. and Olsson, L. 2003: Vegetation index trends for the African Sahel 1982-1999. *Geophysical Research Letters* 30: 10.1029/2002GL016772.

[999] Anyamba, A. and Tucker, C. J. 2005: Analysis of vegetation dynamics using NOAA-AVHRR NDVI data from 1981-2003. *Journal of Arid Environments* 63: 596-614.

[1000] Olsson, L., Eklundh, L. and Ardo, J. 2005: The recent greening of the Sahel – trends, patterns and potential causes. *Journal of Arid Environments* 63: 556-566.

[1001] Herrmann, S. M., Anyamba, A. and Tucker, C. J. 2005: Recent trends in vegetation dynamics in the African Sahel and their relationship to climate. *Global Environmental Change* 15: 394-404.

atmospheric $CO_2$. Maybe increased warmth and $CO_2$ are actually good for life on Earth and do not constitute the greatest threat facing the planet today. News reports suggest that deserts are decreasing in China by 7500 square kilometres per year.[1002] In the 2001 *Summary for Policymakers*, the IPCC stated: "Increased summer continental drying and associated risk of drought is likely over most mid-latitude continental interiors".

This *Summary* is for the public. The science tells a different story. There are a number of papers that suggest increased drought in central USA with global warming and others that show that, in the 20th Century in central USA:[1003] "droughts have, for the most part, become shorter, less frequent, less severe, and cover a smaller proportion of the country".

Two different data sets derived from radiometers on the US National Oceanic and Atmospheric Administration satellites showed that between 1982 and 1999, there was a worldwide increase in photosynthesis.[1004] This is despite increased deforestation in Third World countries[1005,1006] and urbanisation. Plants and meat derived from photosynthesis are ultimately the food source of most of the biosphere. This increased photosynthesis may be due to slight global warming during this period, increased $CO_2$ or to associated increased rainfall.[1007,1008] In fact, there are studies[1009] that suggest a "theoretical expectation" that global warming will result in significant increases in global precipitation.

The Global Precipitation Climate Project has merged satellite and ground measurements of rainfall to produce a record beginning in 1979.[1010,1011] An

[1002] http://upi.com/NewsTrack/view.php?StoryID=200060602-103610-9168r

[1003] Andreadis, K. and Lettenmaier, D., 2006: Trends in 20th Century drought over continental United States. *Geophysical Research Letters* 33: L10403, doi:10.1029/2006GL025711.

[1004] Young, S. S. and Harris, R. 2005: Changing patterns of global-scale vegetation photosynthesis, 1982-1999. *International Journal of Remote Sensing* 26: 4537-4563.

[1005] Skole, D. and Tucker, C. J. 1993: Tropical deforestation and habitat fragmentation in the Amazon: satellite data from 1978 to 1988. *Science* 260: 1905-1909.

[1006] Steininger, M. K., Tucker, C. J., Ersts, P., Killen, T. J., Villegas, Z. and Hecht, S. B. 2001: Clearance and fragmentation of tropical deciduous forest in the Tierras Bajas, Santa Cruz, Bolivia. *Conservation Biology* 15: 856-866.

[1007] Myneni, R. C., Keeling, C. D., Tucker, C. J., Asrar, G. and Nemani, R. R. 1997: Increased plant growth in the northern high latitudes from 1981 to 1991. *Nature* 386: 698-702.

[1008] Ichii, K., Kawabata, A. & Yamaguchi, Y. 2002: Global correlation analysis for NDVI and climate variables & NDVI trends: 1982-1990. *International Journal of Remote Sensing* 23: 3873-3878.

[1009] Huntington, T. G. 2006: Evidence for intensification of the global water cycle: Review and synthesis. *Journal of Hydrology* 319: 83-95.

[1010] Huffman, G. J., Adler, R. F., Chang, A., Ferraro, R., Gruber, A., McNab, A., Rudolf, B., and Schneider, U. 1997: The Global Precipitation Climatology Project (GPCP) combined data set. *Bulletin of the American Meteorological Society* 78: 5-20.

[1011] Adler, R. F., Susskind, J., Huffman, G. J., Bolvin, D., Nelkin, E., Chang, A., Ferraro, R., Gruber, A., Xie, P.-P., Janowiak, J., Rudolf, B., Schneider, U., Curtis, S. and Arkin, P. 2003: The version-2 global precipitation climatology project (GPCP) monthly precipitation analysis (1979-present). *Journal of Hydrometeorology* 4: 1147-1167.

analysis of global precipitation data shows that precipitation variations are associated with El Niño and have no trend.[1012] Increased tropical precipitation over the Pacific and Indian Oceans is associated with local warming of the sea and this is balanced by decreased precipitation in other regions, hence the global average is near zero.

One of the great scare campaigns is that global warming is so fast that plants and animals may not have enough time to migrate, hence there will be untold extinction. A 0.7°C temperature increase in the last 100 years is very slow – the change from the Younger Dryas to warmer times was 100 times faster. It is distinctly possible that the globe will not warm as fast as predicted, that the increased $CO_2$ will allow plants to flourish in warmer temperatures and that climate change may rapidly impose natural selection thereby allowing life to better adapt.[1013] Such natural selection can occur over a few decades.[1014,1015] During the Early Eocene Climatic Optimum (52–50 Ma), the atmospheric $CO_2$ content was probably 1125 ppmv,[1016] some three times the current figure. Erosion and chemical weathering increased due to the enhanced reaction between atmospheric $CO_2$ and rocks.[1017] There was a rapid expansion in vegetation, there were huge forests from pole to pole and new families of plants evolved to capitalise on the excess $CO_2$ in the atmosphere. By these processes, additional $CO_2$ is rapidly scrubbed from air and ultimately sequestered in rocks.

A study of the modern world is not the best way to understand the modern climate. The modern climate is just one frame in the 4567-million-year movie of changing Earth climates.

[1012] Smith, T. M., Yin, X. and Gruber, A. 2006: Variations in annual global precipitation (1979-2004), based on the Global Precipitation Climate Project 2.5 analysis. *Geophysical Research Letters* 33:10.1029/2005GL025393.

[1013] Franks, S. J., Sim, S. and Weis, A. E. 2007: Rapid evolution of flowering time by an annual plant in response to a climate fluctuation. *Proceedings of the National Academy of Sciences USA* 104: 1278-1282

[1014] Kinnison, M. T. and Hendry, A. P. 2001: The pace of modern life II: from rates of contemporary microevolution to pattern and process. *Genetica* 112: 145-164.

[1015] Reznick, D. N. and Ghalambor, C. K. 2001: The population ecology of contemporary adaptations: what empirical studies reveal about the conditions that promote adaptive evolution. *Genetica* 112: 183-198.

[1016] Lowenstein, T. K. and Demicco, R. V. 2007: Elevated Eocene atmospheric $CO_2$ and its subsequent decline. *Science* 313: 1928.

[1017] Smith, M. E., Carroll, A. R. and Mueller, E. R. 2008: Elevated weathering rates in the Rocky Mountains during the Early Eocene Climatic Optimum. *Nature Geoscience* 1: 370-374.

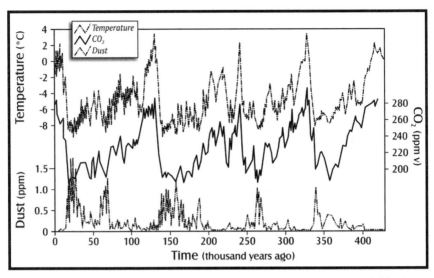

*Figure 19: Ice core record of the last 440,000 years showing cyclical temperature and $CO_2$ changes. During times of glaciation, the dust content of polar ice increases showing that desertification takes place during cool times, not warm times.*

## One volcano can ruin your day

There are two main types of volcanoes. Those in mid ocean rifts are unseen and by far the most abundant volcanoes. Some 85% of the world's volcanoes are unseen, unmeasured, quietly erupting deep in the ocean and ignored in climate models. Most eruptions take place on the deep ocean floor along the global 64,000 km of oceanic ridge systems and pose little in the way of volcanic hazards. These mid-oceanic active volcanic ridges quietly play their own game while the rest of the world goes by without noticing. They are characterised by basalt lava flows and emission of very large volumes of the main volcanic gases ($H_2O$ vapour, $CO_2$, methane, hydrogen sulphide, sulphur dioxide, hydrogen and nitrogen). The second most abundant gas in the modern atmosphere, oxygen, is not emitted from volcanoes. Oxygen in the atmosphere derives from life by converting $CO_2$ into oxygen. Major submarine basalt provinces formed and are still forming in the ocean basins. These basalt volcanoes are normally non-explosive.

Basalt volcanoes also occur in continental areas. These can vary from small cinder cones and immense lava flows from fractures to supervolcanoes. Basalt supervolcanoes have released enormous volumes of lava and gases over a short period. Examples are in the Talnakh Basin (Russia) at 251 Ma and the Deccan Traps (India) at 65 Ma. Basalt volcanism occurs where the crust of the Earth is being pulled apart or where a superplume of hot material has risen beneath the crust.

Submarine volcanoes are poorly understood because of the lack of continuous observation and measurement. They emit huge amounts of very hot gases. Exchange of heat and mass between ocean waters and submarine volcanic rocks is concentrated at mid-ocean ridges and ridge flanks where seawater circulation releases heat and fluids into the ocean.[1018] This affects global heat and geochemical budgets of the oceans. Seamounts away from mid ocean ridges also act as pathways for the exchange of heat.[1019] The $CO_2$ from tens of thousands of submarine hot springs associated with these submarine basalt volcanoes quietly dissolves in the cold high-pressure deep ocean water and does not bubble to the surface. Water at the bottom of the oceans is undersaturated in dissolved $CO_2$, hence very large volumes of $CO_2$ can dissolve. One hot spring can release far more $CO_2$ than a 1000 mW coal-fired power station yet they are neither seen nor measured. Submarine volcanic gas does not even figure in calculations of the sources and sinks of atmospheric $CO_2$ in the IPCC climate models.

Changes in the rate of sea floor spreading affect the rate of submarine and terrestrial volcanic activity. When the floor of the ocean is pulled apart, the depressurisation deep in the Earth results in partial melting, the melts ascend to the mid ocean ridge and may become solid beneath the sea floor or spew out as submarine lava. Gases such as $H_2O$ vapour and $CO_2$ flux the process of melting. The dissolved gases make the melts buoyant and they rise along fractures.

Circulating seawater cools these new melts. This circulating seawater becomes heated. This hot seawater leaches metals from the new volcanic rocks, adds heat and gases to the oceans and precipitates metal sulphide ore deposits from hot springs in the medial rift of mid ocean ridges.[1020] These hot springs can be from 80 to 420°C and are at water depths of 2 to 4 km.[1021] Since the first discovery of these new ore deposits on the sea floor in 1979, more than 200 have been found.

[1018] Mottl, M. and Wheat, G. G. 1994: Hydrothermal circulation through mid-ocean ridge flanks: Fluxes of heat and magmatism. *Geochemica et Cosmochimica Acta* 58: 2225-2237.

[1019] Villinger, H. 2007: Heat flow at mid-ocean ridges and ridge flanks: methods and challenges. *Geophysical Research Abstracts* 9: 07710.

[1020] Elderfield, H. and Schultz, A. 1996: Mid-ocean ridge hydrothermal fluxes and the chemical composition of the ocean. *Annual Review of Earth and Planetary Sciences* 24: 191-224.

[1021] Edmond, J. M., von Damm, K. L., McDuff, R. E. and Measures, C. I. 1982: Chemistry of hot springs on the East Pacific Rise and their effluent dispersal. *Nature* 297: 187-191.

As the oceans contain 22 times more heat than the atmosphere, ocean heat contributes greatly to driving climate and the unseen submarine volcanism can have a profound effect on the surface heat of the Earth.[1022] Because of the lack of measurement, there is no average spreading rate of volcanic heat transfer that can be used to predict modern climate. Although the average geothermal heat distributed over the globe is far less than incoming solar heat, geothermal heat flow is essentially concentrated at points along the mid ocean ridges and volcanic island arcs. This focus of heat, especially in mid ocean ridge areas, can affect the surface heat balance on Earth because the planet's surface heat is in the oceans and not the atmosphere.[1023] If the pulling apart of the sea floor increases slightly, then the rate of submarine volcanicity increases. A slow pull-apart rate, such as in the Gakkal Ridge in the Arctic Ocean, does not necessarily mean that there are no volcanoes or geothermal activity.[1024] There is no constant rate of volcanicity and there are long periods of quiescence and periods of frequent large eruptions. In the South Pacific Ocean, there is a point[1025] that clearly shows heat transfer from submarine volcanoes into the oceans which affects the surface heat balance of the Earth.[1026] Even areas remote from the mid ocean ridges exhale large volumes of very hot fluids.[1027] Because we know so little about the oceans, any attempt to calculate the amount of heat added to oceans from submarine hydrothermal activity is speculative.

We need to get basalt volcanicity into perspective. Some basalt eruptions have an enormous rate of discharge, up to 10,000 cubic metres per second.[1028] Fragmentation of the molten rock occurs by volatile degassing. As the molten

[1022] Corliss, J. B., Dymond, J., Gordon, L. I., Edmond, J. M., von Herzen, R. P., Ballard, R. D., Green, K., Williams, D., Bainbridge, A., Crane, K. and van Andel, T. H. 1979: Submarine thermal springs on the Galápagos Rift. *Science* 203: 1073-1083.

[1023] Covey, C. and Thompson, S. L. 1989: Testing the effects of ocean heat transport on climate. *Global and Planetary Climate Change* 75: 331-341.

[1024] Sohn, R. A., Willis, C., Humphris, S., Shank, T. M., Singh, H., Edmonds, H. N., Kunz, C., Hedman, U., Helmke, E., Jakuba, M., Liljebladh, B., Linder, J., Murphy, C., Nakamora, K., Saato, T., Schlindwein, V., Stranne, C., Taisenfreund, M., Upschurch, L., Winsor, P., Jakobsson, M. and Soule, A. 2008: Explosive volcanism on the ultraslow-spreading Gakkal ridge, Arctic Ocean. *Nature* 453: 1236-1238.

[1025] The South Pacific Isotopic and Thermal Anomaly, above a triple point.

[1026] Staudigel, H., Davies, G. R., Hart, S. R., Marchant, K. M. and Smith, B. M. 1995: Large scale Sr, Nd and O isotopic anomaly of altered oceanic crust: DSDP/ODP sites 417/418. *Earth and Planetary Science Letters* 130: 169-185.

[1027] Melchert, B., Devey, C. W., German, C. R., Lackschewitz, K. S., Seifert, R., Walter, M., Mertens, C., Yoerger, D. R., Baker, E. T., Paulick, H and Nakamura, K. 2008: First evidence for high-temperature off-axis venting of deep crustal/mantle heat: The Nibelungen hydrothermal field, south Mid-Atlantic Ridge. *Earth and Planetary Science Letters* 275: 61-69.

[1028] Swanson, D., Wright, T. and Helz, R. 1975: Linear vent systems and estimated rates of eruption for the Yakima basalt on the Columbia Plateau. *American Journal of Science* 275: 877-905.

rock rises to the surface, gas comes out of solution and coexists with melt, the proportion of gas increases as bubbles grow larger. The expanding gas squirts lava out at the surface as fountains. The quicker the gas is released and the faster the molten rock rises, the higher the lava fountain. Lava fountains are in the order of 1 km high. As the main gas is $H_2O$, a degassing of only 1–5% of $H_2O$ is sufficient to produce lava exit velocities of 100 to 500 metres per second.[1029]

In contrast to the submarine pulling apart of the oceans, a different type of volcano forms where fragments of the Earth's crust are being stitched together. These are a terrestrial or subaerial setting such as in the "Pacific Ring of Fire" or the Mediterranean-trans-Asiatic Belt. These are rhyolitic to andesitic in composition. They are explosive, release very little lava, and belch out massive quantities of gas. Boulders, rock fragments, pieces of pumice, crystals, molten rock and glass are all blasted out from explosive volcanoes. Again, dissolved gas makes the molten rock lighter, the buoyant lava rises and in contrast to basalt, is explosively blasted out at the surface as dissolved gases instantaneously leave the molten rock. Explosive volcanic eruptions create global ash clouds that leave ash in polar ice. All volcanic eruptions release large volumes of a diversity of gases, especially $H_2O$ vapour, $CO_2$ and sulphur gases and some of this material falls as sulphuric acid in rain and snow. If a large volcanic eruption is equatorial (e.g. Central America, Indonesia, Papua New Guinea), then both hemispheres will be blanketed in a very fine-grained volcanic ash, whereas at medium to high latitude eruptions may only affect the hemisphere in which they erupt.

Over 1500 subaerial volcanoes have been active over the last 10,000 years and more than one third of these have erupted one or many times in recorded history.[1030] These constitute 1% of the world's surface area but only 15% of the world's volcanism and 80% of the documented historic eruptions.[1031] Some 57% of the world's visible 600 active volcanoes are islands or are in coastal settings and 38% are within 250 km of continental landmasses. Some 500 million people live within proximity of active or potentially active volcanoes.[1032] This was the entire global population in the 17th Century. Most

[1029] Carey, S. N. 2005: Understanding the physical behaviour of volcanoes. In: *Volcanoes and the Environment* (eds J. Marti and G. Ernst), 1-54. Cambridge University Press.

[1030] Simkin, T. and Siebert, L. 1994: *Volcanoes of the world: A regional directory, gazetteer, and chronology of volcanism during the last 10,000 years.* Geoscience Press, Tuscon, AZ.

[1031] Tilling, R. I. 1996: Hazards and climatic impact of subduction-zone volcanism: a global and historical perspective. In *Subduction: top to bottom* (eds G. E. Bebout, D. W. Scholl, S. H. Kirby et al.). Geophysical Monograph 96, American Geophysical Union, 331-115.

[1032] Tilling, R. I. 2005: Volcano hazards. In: *Volcanoes and the Environment* (eds J. Marti and G. Ernst), 55-89. Cambridge University Press.

of these people live in the Pacific region. If explosive volcanism occurs, it can produce tsunamis and choking ash clouds which have a profound effect on islands and adjacent continental landmasses. We currently live in a time of volcanic quiescence.

If you really want a bad hair day, then a supervolcano can do it for you.

## Supervolcanoes

Supervolcanoes occur in large igneous provinces in continental and submarine areas.[1033,1034,1035,1036,1037] Most are basalt volcanoes, although areas such as Taupo (New Zealand), Toba (Indonesia) and Yellowstone (USA) are explosive supervolcano provinces. Widespread volcanic events or a single supervolcano can change climate.[1038] Large volcanic eruptions emit the same amount of energy as an asteroidal impact and are more frequent. Deep sea sediments from the North Pacific Ocean show that sediment changed from non-glacial to glacial at about 2.67 Ma over a period of 2000 years. The rate of change was too fast to be in direct response to either mountain building extracting $CO_2$ from the atmosphere or Milankovitch orbital forcing. The number and thickness of volcanic ash layers in this deep sea sediment increased ten-fold at this same time, suggesting a widespread period of explosive volcanism, possibly from many volcanoes. The rapid intensification of glaciation was likely to be associated with this widespread volcanic episode which began at 2.67 Ma.[1039] At this time, volcanoes closed the Central American seaway between the Pacific Ocean and the Caribbean, there was an exchange of vertebrates between the Americas and the changed circulation of ocean water accelerated Northern Hemisphere cooling.[1040] At

---

[1033] Anderson, D. L. 2005: Large igneous provinces, delamination and fertile mantle. *Elements* 1: 271-275.

[1034] Campbell, I. H. 2005: Large igneous provinces and the plume hypothesis. *Elements* 1: 265-269.

[1035] Coffin, M. and Eldholm, O. 1994: Large igneous processes: crustal structure, dimensions, and external consequences. *Reviews in Geophysics* 32: 1-36.

[1036] Saunders, A. D. 2005: Large igneous provinces: origins and environmental consequences. *Elements* 1: 293-297.

[1037] Wignall, P. 2005: The link between large igneous provinces eruptions and mass extinctions. *Elements* 1: 293-297.

[1038] Mason, B. G., Pyle, D. M. and Oppenheimer, C. 2004: The size and frequency of the largest volcanic explosions on Earth. *Bulletin of Volcanology* 66: 735-748.

[1039] Prueheer, L. M. and Rea, D. K. 1998: Rapid onset of glacial conditions in the subarctic North Pacific region at 2.67 ma; clues to causality. *Geology* 26: 1027-1030.

[1040] Molnar, P. 2008: Closing of the Central American Seaway and the Ice Age: A critical review. *Paleoceanography* 23: PA2201 doi:10.1029/2007PA001574.

2.67 Ma, there was also a starburst that flooded Earth with cosmic rays,[1041] hence a combination of events probably drove climate change.

Since 16 Ma, North America has been moving over a hot spot, which is now centred under Yellowstone. The last major eruptions were at 2.12 Ma (2450 cubic kilometres of ash) and 0.64 Ma (1000 cubic kilometres of ash).[1042] If such eruptions occurred in an equatorial region, they would have created global cooling because such a huge volume of ash high in the atmosphere would have spread globally. Solar energy would be reflected. The 2.12 and 0.64 Ma eruptions affected only the Northern Hemisphere. At that time humans in Africa were struggling to survive in the alternating glacial-interglacial conditions. Gas explosions occur about every 20,000 years and an explosion 13,000 years ago left a 5 km wide crater at Mary Bay on the edge of Yellowstone Lake.[1043] Concurrently, beneath Yellowstone is a chamber of molten rock which contains gases dissolved at very high pressures.[1044] If this explosively erupted as a supervolcano, it could destroy the world's biggest economy.

Eruptions like Tambora (1815) and Krakatoa (1883) were modest in size[1045] whereas those of Mt St Helens (1980), El Chichón (1982) and Mt Pinatubo (1991) were comparatively very small eruptions. What would be the effect on climate of a supervolcano? We have one relatively recent supervolcano, Toba (Sumatra, Indonesia). Its eruption 74,000 years ago released 2800 cubic kilometres of dust, compared to the 0.5 to 1 cubic kilometre of dust from the eruption of Mt St Helens. Toba released at least 1000 million tonnes of sulphuric acid aerosols, as shown in a marked acidity peak in the Greenland ice cores. Aerosols remained in the atmosphere for at least six years and the eruption took place during a natural cooling event when sea level was dropping.[1046] The eruption may have triggered a millennium of cooling before

[1041] Knie, K., Korschinek, G., Faestermann, T., Wallner, C., Scholten, J. and Hillebrandt, W. 1999: Indication for supernova produced 60Fe activity on Earth. *Physics Reviews Letters* 83: 18-21.

[1042] Marsh, B. D. 1989: Magma chambers. *Annual Review of Earth and Planetary Sciences* 17: 439-472.

[1043] Morgan, L. A., Shanks III W. C., Lovalvo, D. A., Johnson, S. Y., Stephenson, W. J., Pierce, K. L., Harlan, S. S., Finn, C. A., Lee, G., Webring, M., Schulze, B., Dühn, J., Sweeney, R. and Balistrieri, L. 2003: Exploration and discovery in Yellowstone Lake: results from high-resolution sonar imaging, seismic reflection profiling and submarine studies. *Journal of Volcanology and Geothermal Research* 122: 221-242.

[1044] Eaton, G. P., Christiansen, R. L., Iyer, H. M., Pitt, A. D., Mabey, D. M., Blank, H. R., Zietz, I, and Gettings, M. E. 1975: Magma beneath Yellowstone National Park. *Science* 188: 787-796.

[1045] Self, S., Gertisser, R., Thordarson, T., Rampino. M. R. and Wolff, J. A. 2004: Magma volume, volatile emissions and stratospheric aerosols from the 1815 eruption of Tambora. *Geophysical Research Letters* 31: L20608, doi: 10.1029/2004GL020925.

[1046] Zielinski, G. A. 1996: Potential atmospheric impact of the Toba mega-eruption. *Geophysical Research Letters* 23: 837-840.

a natural cool event,[1047] a six-year volcanic winter[1048] and an acceleration of glaciation.[1049] The Northern Hemisphere temperature decreased by about 10°C.[1050]

Besides a huge volume of sulphuric acid aerosols, some 15 cm of volcanic ash fell over India and Asia.[1051,1052] As little as 1 cm of ash can destroy agriculture and Toba certainly destroyed forests and grasslands over much of the low latitude lands. Most hunter-gatherer humans lived in low latitude areas, so all life would have been greatly stressed. This global ecological crisis resulted in a massive depopulation of humans. The total human population on planet Earth was reduced to between 4000 and 10,000 individuals.[1053] We humans very nearly became extinct. This genetic bottleneck is contested on the basis of abundant widespread stone tool artefacts in India.[1054] Others argue that there was indeed a human genetic bottleneck and that the devastation resulting from the Toba eruption encouraged human migration.[1055] The Toba eruption was followed by a rapid population increase, technological innovations and the spread of humans over Europe, India, Asia and Australia.

Supervolcanoes are a threat to civilisation on Earth[1056,1057] as they occur at twice the frequency of impacting asteroids and comets larger than 1 km diameter. Asteroid and comet impacts can have similar effects on climate as a supervolcano. Current climate models do not even consider the possibility

---

[1047] Dansgaard-Oeschger Event 19.

[1048] Oppenheimer, C. 2002: Limited global change due to the largest known Quaternary eruption. *Quaternary Science Reviews* 21: 1593-1609

[1049] Rampino, M. R. and Self, S. 1992: Volcanic winter and accelerated glaciation following the Toba super eruption. *Nature* 359: 50-52.

[1050] Jones, G. S. and Stott, P. A. 2002: Simulation of climate response to a super-eruption. *American Geophysical Union Chapman Conference on Volcanism and the Earth's Atmosphere*, Santorini, Greece, 17-21 July 2002, Abstracts, p.45.

[1051] Rose, W. I. and Chesner, C. A., 1990: Worldwide dispersal of ash and gases from Earth's largest known eruption. *Paleogeography, Paleoclimatology, Paleoecology* 89: 269-275.

[1052] Song, S.-R., Chen, C.-H., Lee, M.-Y., Yang, T. F., Iizuka, Y. and Wei, K.-Y. 2000: Newly discovered eastern dispersal of the youngest Toba Tuff. *Marine Geology* 167: 303-312.

[1053] Rampino, M. R. and Ambrose, S. H. 1999: Volcanic winter in the Garden of Eden: the Toba supereruption and the late Pleistocene human population crash. *Geological Society of America Special Paper* 345: 1-12.

[1054] Petraglia, M., Korisettar, R., Boivin, N., Clarkson, C., Ditchfield, P., Jones, S., Koshy, J., Lahr, M. M., Oppenheimer, C., Pyle, D., Roberts, R., Schwenninger, J.-L., Arnold, L. and White, K. 2007: Middle Palaeolithic assemblages from the Indian subcontinent before and after Toba super-eruption. *Science* 317: 114-116.

[1055] Gathorne-Hardy, F. J. and Harcourt-Smith, W. E. H. 2003: The super-eruption of Toba, did it cause a human bottleneck? *Journal of Human Evolution* 45: 227-230.

[1056] McGuire, W. J., Griffiths, D. R., Hancock, P. L. and Stewart, I. S. 2000: The archaeology of geological catastrophes. *The Geological Society of London Special Publication* 171.

[1057] Rampino, M. P. 2001: Supereruptions as a threat to civilizations on Earth-like planets. *Icarus* 156: 562-569.

of another supervolcano eruption. They sleep restlessly in Indonesia, Papua New Guinea and the Pacific region.[1058] And these are the small supervolcanoes that we can see. The ones we don't see are a greater threat.

Continental basalt supervolcanoes flood large areas of continents with lava as in the Ethiopian Highlands, the Columbia River basalts (USA), Emeishan Large Igneous Province (western China) and the Parana-Etendeka basalts (Brazil-Namibia).[1059] Mass extinctions of life have coincided with many of these events[1060] such as in the Siberian Traps, the Brazilian Highlands and Karoo-Ferrar basalt provinces (South Africa-Antarctica).[1061] The greatest mass extinction of all time occurred when basalt supervolcanoes were at their peak.[1062]

Basalt supervolcanoes on the land can create a mass extinction of life whereas submarine supervolcanoes change climate. There are huge supervolcano provinces lurking beneath the oceans. These submarine oceanic plateaux such as the Caribbean-Columbian Plateau (Caribbean Sea), Kerguelen Plateau (Indian Ocean), Ontong Java Plateau (SW Pacific Ocean), Manihiki Plateau (SW Pacific) and Hikurangi Plateau (SW Pacific). The pulling apart of continents produces oceans. During this process, large volumes of basalt lava and volcanic gases are erupted. The lavas form the floor of all the major oceans of the Earth. The Atlantic Ocean formed by the rifting and pulling apart of an ancient continent. Land masses at the edge of the rifted continent also contain basalt. For example, the North Atlantic Igneous Province includes basalts in Greenland, Iceland, Ireland, Scotland and the Faeroes. The Central Atlantic Magmatic Province of eastern USA, northern South America and northwest Africa is also part of the same event, the opening up of the Atlantic Ocean. These large basalt eruptions of the Central Atlantic Magmatic Province occurred at the same time as a mass extinction of complex life at the end of the Triassic Period.[1063]

Throughout time, there have been sudden events of basalt eruptions

[1058] Simkin, T. and Siebert, L. 1994: *Volcanoes of the world.* Geoscience Press.

[1059] Peate, D. W. 1997: The Parana-Etendeka province. In (eds J. J. Mahoney and M. F. Coffin) *Large igneous provinces: continental, oceanic and planetary flood volcanism.* Geophysical Monograph 100, American Geophysical Union, 272-281.

[1060] Wignall, P. B. and Hallam, A. 1997: *Mass extinctions and their aftermath.* Oxford University Press.

[1061] Marsh, J. S., Hooper, P. R. Rahacek, J., Duncan, R. A. and Duncan, A. R. 1997: Stratigraphy and age of Karoo basalts of Lesotho and implications for correlations within the Karoo igneous province. In: (eds J. J. Mahoney and M. F. Coffin) *Large igneous provinces: continental, oceanic and planetary flood volcanism.* Geophysical Monograph 100, American Geophysical Union, 247-272.

[1062] Knoll, A. H., Bambach, R. K., Canfield, D. E. and Grotzinger, J. P. 1996: Comparative Earth history and Late Permian mass extinction. *Science* 273: 452-456.

[1063] Bambach, R. K., Knoll, A. H. and Wang, S. C. 2004: Origination, extinction, and mass depletions of marine diversity. *Paleobiology* 30: 522-542.

associated with the stretching and breaking of the Earth's crust. In ancient rocks, basalt dyke swarms are also common (e.g. MacKenzie Dyke Swarm, Coppermine basalts, Canada)[1064] showing that the fragmentation and stitching back together of continents (i.e. plate tectonics) has taken place for a long time.

Despite the abundance of basalt volcanoes, plate tectonics has not taken place on Mars. Therefore, any climate change on Mars cannot be related to plate tectonics, $CO_2$, water, life, the oceans or the atmosphere and can only be related to the Sun. Mars enjoys global warming yet has almost no atmosphere. To my knowledge, there is no industry or human emissions of $CO_2$ on Mars. This is strong evidence that the Sun drives climate on Mars. The Sun is probably also the main driving force for climate on Earth.

Basalt supervolcano provinces occupy a few million square kilometres and contain in the order of a million cubic kilometres of basalt which spewed out as lava in less than one million years. Basalt lava erupts at 1100°C and commonly at a higher temperature. Major eruptions derive from a gas-charged plume of heat and molten rock that buoyantly rises in the Earth's mantle. The rise of the plume may cause continental break-up and can leave identical basalts on each side of an ocean (e.g. Parana, South America, and Etendeka, Africa). Observation of the rate of lava outpourings from Icelandic volcanoes, especially the 1783 Laki eruption, show that it would take only 10 years for a basaltic supervolcano to erupt 1500 cubic kilometres of lava. This can be validated by looking at older basalt volcanoes.

The 14.7 Ma Roza Flow of the Columbia River flood basalt province is 1000 cubic kilometres in volume. This eruption also released at least 10,000 million tonnes of sulphur dioxide together with large amounts of hydrofluoric and hydrochloric acids.[1065] At maximum eruption rates, similar to those measured for Laki, the Roza eruption produced lava fountains that spurted 1.5 km into the sky and a column of gas and heat rising 15 km above the vents. If the Roza eruption had lasted 10 years, the aerosol burden of the atmosphere would have been massive. The conversion of such large quantities of sulphur dioxide into sulphuric acid would have created dry air. At present, the atmosphere contains some 1000 million tonnes of water vapour and the atmosphere would have been greatly depleted in water vapour. The effects of

---

[1064] Barager, W. R. A., Ernst, R. E., Hulbert, L. and Peterson, T. 1996: Longitudinal petrochemical variation in the Mackenzie dyke swarm, northwestern Canadian Shield. *Journal of Petrology* 37: 317-359..

[1065] Thordarson, T. and Self, S. 1996: Sulfur, chlorine and fluorine degassing and atmospheric loading by the Roza eruption, Columbia River Basalt group, Washington, USA. *Journal of Volcanological and Geothermal Research* 74: 49-73.

a Roza-type eruption cannot be calculated with the available models. Flood basalt release of aerosols can cool the atmosphere by 5 to 10°C but they can also potentially warm the atmosphere by adding $CO_2$. There are suggestions that the Deccan flood basalts raised atmospheric $CO_2$ by some 200 ppmv. There was a warming of some 2°C.[1066] However, the atmosphere may have been warmed by the Deccan flood basalts as circulating air was the only way the flood basalts could cool.

A supervolcano beneath the sea heats the ocean, adds $CO_2$ to ocean water (which is later released to the atmosphere), temporarily turns ocean water from alkaline to acid, deoxygenates ocean water and creates a local, minor or mass extinction. Most supervolcanoes are submarine, we do not see them, they are not factored into climate models, and they emit monstrous volumes of $CO_2$.

## Volcanic gases

White Island, in the offshore extension of the Taupo Volcanic Zone of New Zealand, has many active and fossil geothermal systems.[1067] Every day it pumps out 4800 to 18,000 tonnes of $H_2O$ vapour, 1150 to 4120 tonnes of $CO_2$, 320 to 1200 tonnes of sulphur dioxide, 96 to 360 tonnes of hydrochloric acid, 2.6 to 10 tonnes of hydrofluoric acid and 0.16 to 0.36 tonnes of ammonia.[1068] Some hot springs even deposit liquid mercury on the adjacent sea floor such as in the offshore extension of the Taupo Volcanic Zone.[1069] This is nothing unusual. The Kudovskaya volcano on Iturup Island in the Kuriles each day pumps out gas at 150 to 940°C containing 1800 tonnes of $H_2O$, 50 tonnes of $CO_2$, 62 tonnes of sulphur dioxide, 7 tonnes of hydrogen sulphide, 8 tonnes of hydrochloric acid, 0.1 tonnes of hydrogen fluoride, 0.4 tonnes of nitrogen and 30 grams of helium.[1070] The gases precipitate a

[1066] Caldeira, K. and Rampino, M. R. 1990: Carbon dioxide emission from Deccan volcanism and a K/T boundary greenhouse effect. *Geophysical Research Letters* 17: 1299-1302.

[1067] Hedenquist, J. W. and Browne, P. R. L. 1989: The evolution of Wairotapu geothermal system, New Zealand, based on the chemical and isotopic composition of its fluids, minerals and rocks. *Geochimica et Cosmochimica Acta* 53: 2235-2257.

[1068] Giggenbach, W. 1995: Variations in the chemical and isotopic composition of fluids discharged from the Taupo Volcanic Zone, New Zealand. *Journal of Volcanology and Geothermal Research* 68: 89-116.

[1069] Stoffers, P., Hannington, M., Wright, I., Herzig, P., de Ronde, C., Arpe, T., Battershill, C., Botz, R., Britten, K., Browne, P., Cheminee, J. L., Fricke, H. W., Garbe-Schoenberg, D., Hekinian, R., Hissman, K., Huber, R., Robertson, J., Schauer, J., Schmitt, M., Scholten, J., Schwarz-Schampera, U. and Smith, I. 1999: Elemental mercury at submarine hydrothermal vents in the Bay of Plenty, Taupo volcanic zone, New Zealand. *Geology* 27: 931-934.

[1070] Korzhinskii, M. A., Tkachenko, S. I., Bulgakov, R. F. and Shmulovich, K. I. Condensate compositions and native metals in sublimates of high temperature gas streams of Kudryavyi Volcano, Iturup Island, Kuril Islands. *Geokhimiya* 12: 1175-1182.

sublimate which is mined to produce annually 12 tonnes of rhenium,[1071] 25 tonnes of indium, 6 tonnes of germanium and 1.4 tonnes of gold.[1072] The Mammoth Hot Spring at Yellowstone, USA, pumps out 160 to 190 tonnes per day of $CO_2$.[1073] On 15 June 1991, Mt Pinatubo in the Philippines released 20 million tonnes of sulphur dioxide (which reacted with water vapour in the atmosphere to form a 30 million-tonne sulphuric acid climate-modifying aerosol[1074]) and very large quantities of chlorofluorocarbons, the gases that destroy the ozone layer.[1075]

Although the gas types and abundances vary from volcano to volcano depending upon the chemistry of the molten rock, the typical gas composition by volume would be 70–80% $H_2O$, 8–12% $CO_2$, 3–5% nitrogen, 5–8% sulphur dioxide and minor proportions of hydrogen, carbon monoxide, sulphur, chlorine and argon. Volcanic gases are free of oxygen gas, yet the atmosphere contains 21% oxygen and only 0.03% $CO_2$. The atmosphere has contained abundant oxygen for half of time. This derives from the removal of $CO_2$ from the atmosphere by photosynthesis.[1076]

Sulphur gases emitted from volcanoes have a profound effect on atmospheric composition. Other gases, such as chlorine and chlorofluorocarbons, may temporarily alter the atmospheric composition but are scavenged by ash fallout processes.[1077,1078] Sulphur is released into the atmosphere mainly as sulphur dioxide ($SO_2$) but also as hydrogen sulphide ($H_2S$) from oxygen-poor melts. $H_2S$ rapidly oxidises to $SO_2$ by photochemical processes within several days. Sulphuric acid forms droplets or nucleates onto

[1071] Tessalina, S. G., Yudovskaya, M. A., Chaplygin, I. V., Brick, J.-L. and Capmas, F. 2007: Sources of unique rhenium enrichment in fumaroles and sulphides at Kudryavy volcano. *Geochimica et Cosmochimica Acta* 72: 889-909.

[1072] Kremenetsky, A. A. 2006: High-temperature Re, In, Ge, Bi-including volcanic gases are actually parental to modern ore genesis. *IAGOD Symposium, Understanding Ore Genesis*, Moscow 2006.

[1073] Lowenstern, J. B., Smith, R. B. and Hill, D. P. 2006: Monitoring super-volcanoes: geophysical and geochemical signals at Yellowstone and other large caldera systems. *Philosophical Transactions of the Royal Society* A364: 2055-2072.

[1074] McCormick, M. P., Thomason, L. W. and Trepte, C. R. 1995: Atmospheric effects of the Mount Pinatubo eruption. *Nature* 373: 399-404.

[1075] Brasseur, G. and Granier, C. 1992: Mount Pinatubo aerosols, chlorofluorocarbons, and ozone depletion. *Science* 257: 1239-1242.

[1076] Cloud, P. 1988: *Oasis in space: Earth history from the beginning.* Norton, New York.

[1077] Tabazadeh, A. and Turco, R. P. 1993: Stratospheric chlorine injection by volcanic eruptions: hydrogen chloride scavenging and implications for ozone. *Science* 20:1082-1086.

[1078] Rose, W. I., Jr., Stoiber, R. E. and Malinconico, L. L. 1982: Eruptive gas compositions and fluxes of explosive volcanoes: budget of S and Cl emitted from Fuego volcano, Guatemala. In *Andesites* (ed. R. S. Thorpe) John Wiley, 669-676.

tiny ash or ice particles for at least a month after an eruption.[1079,1080] Aerosol particles are icy to liquid spheres from 0.1 to 1.0 micrometres (μm) with an average size of 0.5 μm. This is about the middle of the wavelength size of incoming solar radiation. Volcanic and other aerosols scatter incoming short-wave solar radiation, resulting in cooler surface and troposphere temperatures and stratosphere warming. Aerosols and short-lived ash and ice clouds derived from eruption columns provide nuclei for upper troposphere and stratosphere cirrus clouds. These reflect solar radiation, resulting in cooling. Increases in atmospheric water vapour in tropical regions associated with an eruption have a similar effect.[1081] Coolings of 0.3 to 1.0°C for one to three years have been measured for modern and historical explosive eruptions[1082,1083,1084] which, at times, may be balanced by the warming effects of El Niño-Southern Oscillation warming.[1085,1086] These temperature measurements are for everyday volcanic eruptions, not supervolcanoes.[1087]

Light and heat passing through the atmosphere are affected by the transparency of the atmosphere. Most of the variation in transparency was correlated with volcanic eruptions.[1088,1089,1090,1091] It had been generally assumed that up until the 1980 eruption of Mt St Helens, the decrease in

[1079] Thomason, L. W. 1991: A diagnostic aerosol size distribution inferred from SAGE II measurements. *Journal of Geophysical Research* 96: 501-522.

[1080] Zhao, J., Turco, R. P. and Toon, O. B., 1996: A model simulation of Pinatubo volcanic aerosols in the stratosphere. *Journal of Geophysical Research* 100: 7315-7328.

[1081] Sodon, B. J., Wetherald, R. T., Stenchikov, G. L. and Robock, A. 2002: Global cooling following the eruption of Mt Pinatubo: a test of climate feedback by water vapor. *Science* 296: 727-730.

[1082] Self, S., Rampino, M. R. and Barbera, J. J. 1981: The possible effects of large 19th and 20th century volcanic eruptions on zonal and hemispheric surface temperatures. *Journal of Volcanological and Geothermal Research* 11: 41-60.

[1083] Bradley, R. S. 1988: The explosive eruption signal in Northern Hemisphere continental temperature records. *Climate Change* 12: 221-243.

[1084] Mass, C. F. and Portman, D. A. 1989: Major volcanic eruptions and climate: a critical evaluation. *Journal of Climate* 2: 566-593.

[1085] Angell, J. K. 1988: Impact of El Niño on the delineation of tropospheric cooling due to volcanic eruptions. *Journal of Geophysical Research* 93: 3697-3704.

[1086] Angell, J. K. 1997: Stratospheric warming due to Agung, El Chichón and Pinatubo taking into account the quasi-biennial oscillation. *Journal of Geophysical Research* 102: 947-948.

[1087] Angell, J. K. and Korshover, J. 1985: Surface temperature changes following the six major volcanic episodes between 1780-1980. *Journal of Climate and Applied Meteorology* 24: 937-951.

[1088] Bryson, R. A. and Goodman, B. M. 1980: Volcanic activity and climate changes. *Science* 27: 1041-1044.

[1089] Bryson, R. A. and Goodman, B. M. 1980: The climatic effect of explosive volcanic activity: Analysis of the historical data. *Atmospheric Effects and Potential Climatic Impact of the 1980 Eruptions of Mt St Helens Symposium*, November 18-19, 1980. NASA Conference Publication 2240.

[1090] Bryson, R. A.,1982: Volcans et climat. *La Recherche* 13 (135): 844-853.

[1091] Bryson, R. U. and Bryson, R. A. 1998: Application of a global volcanicity time-series on high-resolution palaeoclimatic modeling of the Eastern Mediterranean. In *Water, Environment and Society in times of climatic change* (eds Issar, A. S. and Brown, N.) Kluwer Academic Publishers, 1-19.

the transparency of the atmosphere was due to a dust of glass and fine rock particles. Although Mt St Helens was a very minor eruption, it was intensely studied. One of the findings was that droplets of sulphuric acid blasted from the eruption had a long residence time in the atmosphere and reflected energy, resulting in atmospheric cooling. An almost eruption-free period from 1912 to 1963 coincided with an average global warming of 0.5°C. It is quite possible that the atmosphere warmed due to the lack of a normal quota of volcanic aerosols.[1092,1093]

Stratosphere residence times are from weeks to years whilst troposphere residence times are relatively unknown but are expected to be lower. After spreading around the Earth, aerosol clouds gradually diminish in concentration by fallout of the particles. Some falls with polar snow and hence ice cores can be used to sample some of the more sulphur-rich volcanic events.[1094] Other atmospheric perturbations such as depletion of stratospheric ozone in temperate-polar latitudes are also associated with periods of enhanced volcanic aerosols.[1095]

In addition to the reduced sunspot activity, there were periods of decades to centuries during the Little Ice Age when eruptions from $SO_2$-emitting volcanoes were more frequent.[1096,1097] In Europe, 1783 was referred to as *Annus Mirabilis* (Year of Awe) because of the coincidence of a number of large-scale disasters. A massive haze of dry sulphuric acid hung over Europe and it was Benjamin Franklin, in his capacity as US Ambassador in Paris, who suggested that this was due to a volcanic eruption in Iceland.[1098] Between June 1783 and February 1784 the Lagagigar (Laki) volcano in Iceland erupted basalt lava fountains and at least 150 million tonnes of sulphur dioxide aerosols into the atmosphere. This eruption almost completely depopulated Iceland. Historical records report an acrid odour, difficulty in breathing, dry

[1092] Crowley, T. J. 2000: Causes of climate change over the past 1000 years. *Science* 289: 270-277.

[1093] Robock, A. 1991: The volcanic contribution to climate change of the past 100 years. In: *Greenhouse-gas-induced climate change: A critical appraisal of simulations and observations.* (ed. M. E. Schlesinger). Elsevier, 429-444.

[1094] Zielinski, G. A., Mayewski, P. A., Meeker, L. D., Whitlow, S. and Twickler, M. S. 1996: A 110 000-yr record of explosive volcanism from the GISP2 (Greenland) ice core. *Quaternary Research* 45: 109-118.

[1095] Vogelmann, A. M., Ackerman, T. P. and Turco, R. P. 1992: Enhancements in biologically effective ultra-violet radiation following volcanic eruptions. *Nature* 359: 47-49.

[1096] Porter, S. C. 1986: Pattern and forcing of northern hemisphere glacier variations during the last millennium. *Quaternary Research* 26: 27-48.

[1097] Crowley, T. J. and Kim, K.-Y. 1999: Modelling the temperature response to forced climate change over the last six centuries. *Geophysical Research Letters* 26: 1901-1904.

[1098] Franklin, B. 1784: Meteorological imaginations and conjectures. *Manchester Literary and Philosophical Society Memoirs and Proceedings* 2: 373-377.

deposition of sulphate and vegetation damage. All this indicates a high $SO_2$ content in the air. This led to higher than normal human mortality[1099,1100] and gives us a window into the effects of a supervolcano. In July 1783, temperatures in Western Europe were 3°C warmer and the greenhouse effect of the $SO_2$-rich volcanic haze could have led to significant regional short-term warming. Europe and North America were covered in a haze, the Sun's rays were dimmed and when the $SO_2$ formed sulphuric acid, both summer and winter became bitterly cold and there were crop failures the next year. If this is what happens with a modest eruption such as Laki, imagine what would happen with a supervolcano. A very large emission of $SO_2$ into the atmosphere, such as from the Toba eruption or a supervolcano, will dehydrate the atmosphere. Although sulphuric acid droplets will fall to Earth, there would be so much excess $SO_2$ in the atmosphere and so little $H_2O$, that the process will be prolonged.

Large explosive volcanic eruptions and large-scale aerosol blankets have a great effect on climate and weather but even small events cause short-term disruption to climate and weather. Sulphuric acid aerosol clouds have the most significant effect on the radiation budget, surface temperatures and atmospheric circulation patterns. The effects can be regional, hemispherical or global depending upon the location of the volcano and the atmospheric circulation patterns at the time of the eruption.[1101,1102,1103] However, our understanding of aerosols in the atmosphere is really only based on the last 100 years of measurement, a period of relative quiescence. Nevertheless, volcanic eruptions that release large quantities of sulphur gases correlate with global cooling at the Earth's surface,[1104] There appears to be no relationship between the size of an eruption and the amount of sulphur belched into the atmosphere.[1105] A number of very minor volcanic eruptions may add more

[1099] Courtillot, V. 2005: New evidence for massive pollution and mortality in Europe in 1783-1784 may have bearing on global change and mass extinctions. *Comptes Rendes Geosciences* 337: 635-637.

[1100] Grattan, J., Rabartin, R., Self, S. and Thordarson, T. 2005: Volcanic air pollution and mortality in France 1783-1784. *Comptes Rendes Geosciences* 337: 641-651.

[1101] Kelly, P. M., Jones, P. D. and Pengqun, J. 1996: The spatial response of the climate system to explosive volcanic eruptions. *International Journal of Climatology* 16: 537-550.

[1102] Halmer, M. M., Schmincke, H-U. and Graf, H.-U. 2002: The annual volcanic gas input into the atmosphere, in particular into the stratosphere: a global data set for the past 100 years. *Journal of Volcanology and Geothermal Research* 115: 511-528.

[1103] Robock, A. 2000: Volcanic eruptions and climate. *Reviews of Geophysics* 38: 191-219.

[1104] Palais, J. M. and Sigurdsson, H. 1989: Petrologic evidence of volatile emissions from major historic and pre-historic eruptions. In: *Understanding climate change* (eds A. Berger, R. E. Dickinson and J. W. Kidson). American Geophysical Union, 31-53.

[1105] Self, S. 2005: Effects of volcanic eruptions on the atmosphere and climate. In: *Volcanoes and the Environment* (eds J. Marti and G. Ernst), 152-174. Cambridge University Press.

sulphur gases to the atmosphere than a massive eruption.

The cyclical regularity in volcanic events varies significantly during Heinrich and Dansgaard-Oeschger events. It does not seem logical that variations in volcanicity are related to oceanic and atmospheric variations. The inverse may be the case, such that atmospheric transparency resulting from volcanic dust drives Heinrich and Dansgaard-Oeschger events.[1106] Oceanic and atmospheric responses to the modulation of incoming radiation might drive these cycles.[1107] The recognition of 1430-year periodicity in the volcanic and climate records, now called the Dansgaard-Oeschger Cycle, is not new. It was first recognised in 1914.[1108]

The August 1883 eruption of Krakatoa in Indonesia was the first large natural disaster to be reported worldwide and almost instantaneously due to the telegraph. This allowed the correlation of sunsets and other optical phenomena with the eruption. It was thought that there was a "dust veil" around the globe, derived from the eruption.[1109] It was later ascertained that the "dust veil" was aerosol droplets and not fine ash. Data from this eruption prompted measurement of decreasing incoming solar radiation following the 1902 Santa Maria (Guatemala) and 1912 Katmai (Alaska) eruptions. There was then a lull of more than 50 years with no significant stratospheric volcanic aerosols until Gunung Agung (Bali, Indonesia) erupted in 1963.[1110] Although this was a small eruption producing some 7 million tonnes of $SO_2$, the seasonal timing dispersed most of the aerosol cloud over the Southern Hemisphere. The troposphere cooled by 0.5°C and the stratosphere warmed by a few degrees. Another small eruption, the 1982 El Chichón (Mexico) eruption only released 0.5 cubic km of lava but some 11–13 million tonnes of $SO_2$ were released in a short period. This coincided with an El Niño event so it was difficult to calculate the amount of cooling due to the aerosol emissions.

Many eruptions far bigger than those recorded in historic times are inferred from tree ring data which reflects the slower growth during cool times and from dust and acid in Greenland and Antarctic ice cores. Although

---

[1106] Bryson, R. U. 1988: Late Quaternary volcanic modulation of Milankovitch climate forcing. *Theoretical and Applied Climatology* 39: 115-139.

[1107] Bond, G. C. and Lotti, R. 1995: Iceberg discharges into the North Atlantic in millennial time scales during the last glaciation. *Science* 267: 1005-1010.

[1108] Pettersson, O. 1914: Climate variations in historic and prehistoric time. *Svensk Hydrografisk-Biologiska Komm.,* Skrifter 4: 1-25.

[1109] Lamb, H. H. 1970: Volcanic dust in the atmosphere with its chronology and assessment of its meteorological significance. *Philosophical Transactions of the Royal Society London, Series A* 266: 425-533.

[1110] Self, S. and King, A. J. 1996: Petrology and sulfur and chlorine emissions of the 1963 eruption of Gunung Agung, Bali, Indonesia. *Bulletin of Volcanology* 58: 263-286.

historically recorded eruptions such as the 1815 Tambora (Indonesia) eruption relate to peaks in non-sea-spray sulphate in polar ice, some volcanic aerosol concentrations at 1809 and 1258 AD have no known source volcano.[1111] These mystery eruptions may have been local eruptions like El Chichón that produced a small volume of lava with little damage and few casualties. However, they can produce a huge volume of $SO_2$. Perhaps the coincidence of a massive volcanic eruption with a decrease in sunspot activity started the Little Ice Age in the late 13th Century?

The 1991 Mt Pinatubo eruption on Luzon (Philippines) erupted 5 cubic kilometres of molten rock in 3.5 hours on 12 June 1991. Only Katmai (Alaska) has produced more molten rock in the 20th Century (11 km³). These were small eruptions. Mt Pinatubo emitted the largest stratospheric $SO_2$ cloud ever observed by modern instruments.[1112] The eruption columns of ash and dust reached 35 km in height and 28 million tonnes of sulphate aerosols formed in the atmosphere. This was about the same amount as that derived from Krakatoa in 1883. The aerosol clouds spread around the Earth in three weeks and, because Mt Pinatubo is equatorial, the whole globe was covered within seven months of eruption. These persisted for more than 18 months, causing spectacular sunsets and sunrises worldwide, a hazy Sun (Bishop's Ring) and global cooling of 0.5 to 0.7°C.[1113] The lower stratosphere warmed 2 to 3°C due to adsorption of incoming radiation by the aerosols. There were floods along the Mississippi River in 1993 and drought in the Sahel area of Africa. In 1992, the USA had its third coldest and third wettest summer in 77 years. It took four years for the atmosphere to return to its normal aerosol level. Cooling manifested itself globally as changing regional climate and weather patterns until mid 1995.[1114] Mt Hudson in southern Chile erupted in August 1991 at the time when the Mt Pinatubo aerosol cloud reached Antarctica. There was a dramatic decrease in ozone above Antarctica and the Southern Hemisphere "ozone hole" increased to an unprecedented 27 million square kilometres in area. In late 1992, warm tropical ozone-rich air entered the Antarctic atmosphere and halted the ozone depletion.

---

[1111] Strothers, R. B. 2000: Climatic and demographic consequences of the massive volcanic eruption of 1258. *Climate Change* 45: 361-374.

[1112] Bluth, G. J. S., Doiron, S. D., Schnetzler, C. C., Kreuger, A. J. and Walter, L. S. 1992: Global tracking of the $SO_2$ clouds from the June 1991 Mount Pinatubo eruptions. *Geophysical Research Letters* 19: 151-154.

[1113] Hansen, J., Lacis, A., Ruedy R. and Sato, M. 1992: Potential climate impact of the Mount Pinatubo eruption. *Geophysical Research Letters* 19: 215-218.

[1114] Hansen, J., Sato, M. K. I., Ruedy, R. and Lacis, A. 1996: A Pinatubo climate modelling investigation. In: *The Mount Pinatubo eruption: Effects on the atmosphere and climate.* NATO ASI series 142, Springer-Verlag, 233-272.

Mt Pinatubo was a very small eruption. However, it was the first time that modern instrumentation was used to evaluate volcanic $SO_2$ aerosols and this natural experiment provides a window into larger eruptions.

The 1815 eruption of Tambora (Indonesia) is one of the best-known examples of a volcanically induced cooling event.[1115] At least 30 cubic kilometres of molten rock and 100 million tonnes of sulphuric acid aerosols were erupted in a 20-hour period on 11 April 1815. Some 90,000 people perished on and around Sumbawa Island in the Java Sea. The eruption was so powerful that Tambora was broken in half and the upper part of the volcano was blown away, decreasing its height from 4300 to 2850 metres. The area was dark for days and hot and cold air pockets drifted around. Dimming of stars was observed in the Northern Hemisphere between 6 and 20 September 1815. Strikingly spectacular banded red-yellow-white sunsets resulted from aerosols and dust blasted into the troposphere and stratosphere. Even though Benjamin Franklin had noted the connection between volcanism and cool dark years in 1784, no connection was made between volcanism and the unusual weather of 1816. Volcanic aerosols lowered air pressure at mid latitudes above the North Atlantic Ocean and mid latitude cyclones were pushed southwards. The low-pressure zone that currently sits above Iceland was pushed southwards over England.

Tambora was one of the larger historical eruptions but, on a scale of volcanic eruptions, it was modest.[1116] Ice cores show a period of volcanic aerosol fallout in 1809–1810 from a mystery volcano and a slight cooling.[1117] These predated the cooling due to Tambora. Both coolings took place during the Dalton Minimum (1795–1823 AD) when sunspot activity was low.[1118] Furthermore, in both the Dalton and Maunder Minima, the Sun shifted position in the Solar System. This is a cyclical process and happens every 180 to 200 years. During this gravitational sleight of hand, the Sun gyrates around the Solar System's centre of mass. The effect of this wobbling[1119] on the Earth's climate is not known. The gravitational wobble of planets outside the Solar System has also been measured. These wobbles occurred

---

[1115] Harington, C. R. 1992: *The year without a summer? World climate in 1816*. Canadian Museum of Nature, Ottawa.

[1116] Self, S., Gertisser, R., Thondorson, T., Rampino, M. R. and Wolff, J. A. 2004: Magma volume, volatile emissions and stratospheric aerosols from the 1815 eruption of aerosols. *Geophysical Research Letters* 31: L20608, doi.1029/2004GL020925.

[1117] Legrand, M. and Delmas, R. J. 1987: A 200-year continuous record of volcanic $H_2SO_4$ in the Antarctic Ice Sheet. *Nature* 327: 671-676.

[1118] Lean, J. and Rind, D. 1999: Evaluating Sun-climate relationships since the Little Ice Age. *Journal of Atmospheric and Solar-Terrestrial Physics* 61: 25-36.

[1119] Inertial solar motion, caused by the gravitational pull of the large outer planets, mainly Jupiter and Saturn.

in 1632, 1811 and 1990 AD. During the Maunder Minimum, sunspots persisted for a number of solar rotations indicating slow decay, reduced solar convection and solar luminosity. During the Dalton Minimum, 1810 was the last year without sunspots. Solar cycles then lasted about 14 years, compared with the modern average of 11 years. This is consistent with reduced solar luminosity and a cooler climate on Earth. After the eruption of Tambora, astronomers all over the world could see sunspots with the naked eye. These large sunspots derived from the same magnetic activity that produces bright solar regions.[1120]

The coincidental combination of at least two of these events (Tambora eruption, Dalton Minimum) and possibly a third event (intertidal solar motion) would have exacerbated the cooling in the last stages of the Little Ice Age. The peak of the Sun's 11-year sunspot cycle was in 1816. However, there were only 35 sunspots per annum as opposed to about 100 for a normal solar maximum peak. Volcanic aerosol sulphuric acid in both Greenland and Antarctica ice cores show that the aerosol veil covered both hemispheres. Global temperature dropped by at least 1°C. The following year was called the "year without a summer".[1121] In Europe it was cold, hazy and wet, crops failed and there was famine.[1122] Around the Hudson Bay, temperatures dropped 5–6°C, and the bay froze so early that summer shipping was disrupted.[1123] The New England region of the USA was frozen and had heavy snow.[1124] The snow reflected heat and it was a close call as to whether a permanent snow cover would develop that could have led to a longer colder period. In the Northern Hemisphere, the year of 1816 was miserable. Contemporary records, diary entries[1125] and newspaper accounts describe famine, drought, destructive snows, sleet, howling winds, lightning storms and rains after an abnormally cold spring and summer. Summer snows did not melt. People even noted the sky's abnormal colour and the large sizes of sunspots. Most of the Northern Hemisphere agriculture was subsistence farming so the consequent crop failure meant hardship, famine and death.[1126]

---

[1120] Plages or faculae.

[1121] Stommel, H. and Stommel, E. 1983: *Volcano weather.* Seven Seas Press.

[1122] Post, J. A. 1985: *The last great subsistence crisis of the western world.* John Hopkins University Press.

[1123] Catchpole, A. J. W. and Faurer, M. A. 1983: Summer sea ice severity in Hudson Straight 1751-1870. *Climate Change* 5: 115-139.

[1124] Oppenheimer, C. 2003: Climatic, environmental and human consequences of the largest known historic eruption: Tambora volcano (Indonesia) 1815. *Progress in Physical Geography* 27: 230-259.

[1125] Thomas Jefferson kept records at his home in Monticello, Virginia. Mary Wollstonecraft Shelley recorded (16th June 1816) the foul weather, confinement indoors at Geneva led to the telling of ghost stories and, as a result, in 1818 Mary Shelley published the Gothic chiller *Frankenstein: Or, the Modern Prometheus.*

[1126] Many interpreted the harrowing year of 1816 as punishment from an angry God.

It is quite possible that the dust and sulphuric acid emitted from the 17th–16th Century BC Bronze Age eruption of Santorini[1127,1128] caused the first of the eight biblical plagues.[1129] Humans, cattle, game and harvests would have been reduced, drinking water was contaminated with ash and acid, many humans suffered burns, and eye, respiratory and skin problems were widespread. Medical methods were developed at the time to try to treat these ailments.[1130] This volcano led to the collapse of the Minoan Empire and severe socio-political disruption in Egypt.[1131,1132]

We live in a period of volcanic quiescence. If climate models use only 15% of the world's volcanoes with low eruption rates over the last 100 years, then the models are invalid.

## Ice, volcanoes and earthquakes

Over the past 650,000 years, the polar ice caps have grown to be far bigger than at present on seven different occasions. This locks up huge volumes of water in frozen oceans and continental ice caps and glaciers. Each time there is a thaw, the ice retreats and sea level rises. These changes in distribution of the Earth's water have resulted in repeated rapid changes in sea level by up to 130 metres below and 7 metres above today's level. This unloading and loading has an effect on volcanoes and faults.[1133,1134] During the last 2.67 Ma of glaciations, explosive volcanicity increased. Was this a coincidence or is there a connection? As the ice sheets melted, the Earth's crust rebounded. Faults were reactivated, seismic activity increased and the load on volcanoes decreased. In the current interglacial, this rebound is occurring in Scandinavia,

[1127] Pyle, D. M. 2007: Ice-core acidity peaks, retarded tree growth and putative eruptions. *Archeometry* 31: 88-91.

[1128] Manning, S. W., Ramsey, C. B., Kutschera, W., Higham, T., Kromer, B., Steier, P. and Wild, E. M. 2006: Chronology for the Aegean late bronze age 1700-1400 B.C. *Science* 312: 565-569.

[1129] Trevisanto, S. I. 2006: Six medical papyri describe the effects of Santorini's volcanic ash, and provide Egyptian parallels to the so-called biblical plagues. *Medical Hypotheses* 66:193-196.

[1130] Trevisanto, S. I. 2005: Ancient Egyptian doctors and the nature of the biblical plagues. *Medical Hypotheses* 65: 811-813.

[1131] Marinatos, S. 1939: The volcanic destruction of Minoan Crete. *Antiquity* 13: 425-439.

[1132] Sigurdsson, H., Carey, S. and Devine, J. D. 1990: Assessment of mass, dynamics, and environmental effects of the Minoan eruption of Santorini Volcano. In: *Thera and the ancient world – Proceedings of the Third International Congress* (eds Hardy, D. A. et al.), 100-112, Thera Foundation.

[1133] McGuire, W. J., Howarth, R. J., Firth, C. R., Solow, A. R., Pullen, A. D., Saunders, S. J., Stewart, I. S. and Vita-Finzi, C. 1997: Correlation between the rate of sea-level change and frequency of explosive volcanism in the Mediterranean. *Nature* 399: 473-476.

[1134] McGuire, W. J. 2008: Changing sea levels and erupting volcanoes: cause and effect? *Geology Today* 8: 141-144.

Scotland and Canada after ice sheets up to 5 kilometres thick melted over the last 14,000 years. The warming that heralded the start of the current interglacial period around 10,000 years ago brought forth a burst of volcanic activity in Iceland.[1135] The decrease of load on the chamber of molten rock below as the ice caps melted stimulated volcanic activity. On a world scale, the incidence of high latitude volcanism increased between 12,000 and 7000 years ago as a result of the melting of ice sheets. This produced an increase in atmospheric $CO_2$ of 40 ppmv so elevated volcanic activity during interglacials helps to maintain a high $CO_2$ content.[1136] In California, volcanic activity increased in interglacial periods during the last 800,000 years. A similar pattern has been measured in the ice-covered volcanoes of the Cascades Ranges (USA) and the Andes.[1137]

A similar trend is seen in the Mediterranean where sea level changes over the past 80,000 years are linked to bursts of volcanic activity[1138] with the most pronounced activity being in the last 15,000 years. Climate change as a trigger for large volcanic eruptions was proposed as early as 1979.[1139,1140] The association of melting of ice sheets and increased volcanic activity is seen in the ice cores from the Greenland ice sheet, thus showing it is a global feature.[1141]

Unstable volcanoes can also collapse. Most collapses result from volcanoes erupting material to form a widespread blanket so that the loss of molten material beneath the volcano induces a collapse.[1142,1143,1144] Many experiments

[1135] Sigvaldason, G. E., Annertz, K. and Nilsson, M. 1992: Effect of glacier loading/deloading on volcanism: post-glacial volcanic production rate of the Dyngjfjöll area, central Iceland. *Bulletin of Volcanology* 54: 385-392.

[1136] Huybers, P and Langmuir, C. 2007: Feedback between deglaciation and volcanic emissions of $CO_2$. http://environment.harvard.edu/docs/faculty_pubs/huybers-feedback.pdf

[1137] Glazner, A. F., Manley, C. R., Marron, J. S. and Rojstaczer, S. 1999: Fire or ice: Anticorrelation of volcanism and glaciation in California over the past 800,000 years. *Geophysical Research Letters* 26: 1759-1762.

[1138] McGuire, W. J., Howarth, R. J., Firth, C. R., Solow, A. R., Pullen, A. D., Saunders, S. J., Stewart, I. S. and Vita-Finzi, C. 1997: Correlation between the rate of sea-level change and frequency of explosive volcanism in the Mediterranean. *Nature* 399: 473-476.

[1139] Rampino, M. R., Self, S. and Fairbridge, R. W. 1979: Can rapid climate change cause volcanic eruptions? *Science* 206: 826-829.

[1140] McGuire, W. J. 2008: Changing sea levels and erupting volcanoes: cause and effect? *Geology Today* 8: 141-144.

[1141] Zielinski, G. A. 2000: Use of paleo-records in determining variability within volcanism-climate system. *Quaternary Science Reviews* 19: 417-438.

[1142] McGuire, W. J., Howarth, R. J., Firth, C. R., Solow, A. R., Pullen, A. D., Saunders, S. J., Stewart, I. S. and Vita-Finzi, C. 1997: Correlation between the rate of sea-level change and frequency of explosive volcanism in the Mediterranean. *Nature* 399: 473-476.

[1143] Van Wyk de Vries, B., Self, S., Francis, P., W. and Keszthelyl, L. 2001: A gravitational spreading origin for the Socompa debris avalanche. *Journal of Volcanology and Geothermal Research* 105: 225-247.

[1144] Clavero, J. E., Sparks, S. J., Polanco, E. and Pringle, M. S. 2002: Evolution of Parinacota volcano, Central Andes, Northern Chile. *Revista Geológica de Chile* 31: 317-347.

have been performed to show that after eruption, faults deform volcanoes and this leads to the eventual collapse.[1145,1146,1147,1148] Such collapses are on a local scale, however major collapses in glacial regions are due to decreasing ice load, melt waters entering the chamber of molten rock and heavy rains.[1149] It appears that since 5 Ma, most volcanic collapses in ice-capped volcanoes occurred after the main glacial peaks and were induced by global warming. Periods of heavy rains, such as those associated with Hurricane Mitch in 1998, devastated Central America. There were massive landslides over large areas and the Casita volcano in Nicaragua collapsed resulting in the deaths of 2500 people.[1150]

Some volcanic activity is even influenced by the weather. Most of the eruptions from the Pavlof volcano in the Bering Straits occur in the autumn and winter months.[1151] Eruptions are probably triggered by the low-pressure weather systems. A fall in pressure raises the water level around Pavlof by some 30 centimetres and storm winds may raise the water level even more. Increased loading of the volcano probably squeezes the molten rock upwards like toothpaste out of a tube. In a more comprehensive study of some 3000 eruptions between 1700 and 1990, there appears to be a seasonal trend of volcanic eruptions across the planet with more eruptions in the November–March period.[1152] This is related to the annual movements of water from oceans to continents. Even typhoons can cause earthquakes. An unusual type of earthquake occurs in eastern Taiwan when there are typhoons.[1153]

Land masses loaded with ice sink. The sinking is associated with increased earthquake activity. As the ice melts, the land rebounds and earthquakes again

---

[1145] Van Wyk de Vries, B. and Francis, P. W. 1997: Catastrophic collapse at stratovolcanoes induced by gradual volcano spreading. *Nature* 387: 387-390.

[1146] Lagmay, A. M. F., van Wyk de Vries, B., Kerle, N. and Pyle, D. M. 2000: Volcano instability induced by strike-slip faulting. *Bulletin of Volcanology* 62: 331-346.

[1147] Acocella, V. and Tibaldi, A. 2005: Dike propagation driven by volcano collapse: A general model tested at Stromboli, Italy. *EOS* 86 (52).

[1148] Norini, G. and Lagmay, A. M. F. 2005: Deformed symmetrical volcanoes. *Geology* 33: 605-608.

[1149] Capra, L. 2006: Abrupt climate changes as triggering mechanisms of massive volcanic collapses. *Journal of Volcanology and Geothermal Research* 155: 329-333.

[1150] Scott, K. M., Vallance, J. W., Kerle, N., Macias, J. L., Strauch, W. and Devoli, G. 2004: Catastrophic precipitation-triggered lahar at Casita volcano, Nicaragua: occurrence, bulking and transformation. *Earth Surface Processes and Landforms* 30: 59-79.

[1151] McNutt, S. R. and Beavan, R. J. 1981: Volcanic earthquakes at Pavlof Volcano correlated with the solid earth tide. *Nature* 294: 615-618.

[1152] Mason, B. G., Pyle, D. M., Dade, W. B. and Jupp, T. 2004: Seasonality of volcanic eruptions. *Journal of Geophysical Research* 109: B04206, doi: 10.1029/2002JB002293.

[1153] Wang, W.-N., Wu, H.-L., Nakamura, H., Wu, S.-C., Ouyang, S. and Yu, M.-F 2003: Mass movements caused by recent tectonic activity: The 1999 Chi-chi earthquake in central Taiwan. *Island Arc* 12: 325-334.

occur. Northern Europe and northern North America were covered by ice sheets up to 5 km thick. These ice sheets started to melt 14,700 years ago, and Scandinavia, Scotland and North America are currently rising and producing earthquakes.[1154,1155] The melting of ice caps triggers earthquakes. Unloading by rapid glacial melting in southwest Alaska may have produced the 1979 earthquake of Richter magnitude 7.2. Areas such as the Alps, Himalayas, Rocky Mountains, Andes and Southern Alps of New Zealand have a high earthquake risk. One reason is that melting glaciers and active faults coincide[1156] and the other is that compression causes uplift of mountains. Strong earthquakes can dislodge huge piles of sediment accumulated around the edge of land masses, such as Greenland. The resulting underwater landslides could generate large tsunamis, such as those that followed the Storegga slide 8000 years ago off the west coast of Norway. The tsunami from Storegga was 20 metres high when it hit the Shetland Islands and 6 metres high along the east coast of Scotland.[1157]

Since the last glaciation ended 14,000 years ago, sea level has risen some 130 metres. The loading of continental margins with melt water has reactivated faults and triggered earthquakes around the rims of all ocean basins. Many of these earthquakes have triggered submarine landslides which cause tsunamis. The West Antarctic Ice Sheet is not on a continental land mass and is pinned to islands under the ice. Some two thirds of the West Antarctic Ice Sheet has already melted over the last 14,000 years since the last Ice Age and the remaining ice will melt, detach and cause a 7-metre sea level rise. Not only will this cause inundation of low-lying coastal areas but also there will be increased earthquake activity and an increased number of tsunamis. This process is not a result of the Late 20th Century Warming, it is the end result of the 130-metre increase in sea level since the last Ice Age.[1158]

With changes in weather and climate, trillions of tonnes of water are redistributed around the oceans and between the hemispheres. November is a time of increased cometary activity. In the Northern Hemisphere, November is a particularly dangerous month with autumnal weather likely

---

[1154] Mörner, N. A. 1991: Intense earthquakes and seismotectonics as a function of glacial isostasy. *Tectonophysics* 188: 407-410.

[1155] Wu, P., Johnston, P. and Lambeck, K. 2002: Glacial rebound and fault instability in Fennoscandia. *Geophysical Journal International* 139: 657-670.

[1156] Syvitski, J. P. M., Stoker, M. S. and Cooper, A. K. 1997: Seismic facies of glacial deposits from marine and lacustrine environments. *Marine Geology* 143: 1-4.

[1157] Dawson, A. G., Long, D. and Smith, D. E. 1988: The Storegga slides: evidence from eastern Scotland for a possible tsunami. *Marine Geology* 82: 271-276.

[1158] Sauber, J. M. and Molnia, B. F. 2004: Glacier ice mass fluctuations and fault instability in tectonically active Southern Alaska. *Global and Planetary Change* 42: 279-293.

to increase both earthquake frequency and volcanic activity. So, if you are paranoid, seeking alarmism and frightened of your environment, just don't do November.

## Milankovitch wobble theory wobbles

In the 1920s, Milutin Milankovitch refined ideas on how ice ages form. He argued that sunshine falling on different parts of the Earth changes over time. Changes were due to rotational forces. Antarctica has enjoyed a partial ice cover for tens of millions of years. However, ice sheets in the Northern Hemisphere have waxed and waned. Milankovitch suggested that the waxing and waning depend on whether the Sun in summer was able to melt the ice and snow that had accumulated in winter. At times, the Sun is close and high in the sky and melting is quick. A more distant lower Sun can leave snow all summer, and snow and ice accumulate year after year.[1159]

The Earth's eccentricity is the shape of the Earth's orbit around the Sun. It is not circular. The orbital path varies from slightly elliptical (0.005) to slightly more elliptical (0.058) over a cycle of about 100,000 years. The ellipticity is currently 0.017. This alters the distance between the Sun and the Earth, which changes the amount of energy received at the Earth's surface in different seasons. Differences of about 3% occur between the farthest point (aphelion) and the closest point (perihelion). This 3% difference means that the Earth enjoys a 6% change in solar activity between January and July. When the Earth's orbit is the most elliptical, the solar energy at the perihelion is some 20–30% more than at the aphelion. These cyclical changes in the amount of solar energy received by the Earth influence climate. At present, the orbital eccentricity is nearly at the minimum of the cycle. In 2008, at perihelion Earth was 147,100,000 km from the Sun and, at aphelion, 152,100,000 km from the Sun. This change in distance of 5 million km is thirteen times the distance between the Earth and the Moon. A change of 5 million kilometres has a huge effect on the amount of solar energy that the Earth receives, and greatly influences climate. In about 9000 years time, perihelion will occur in the Northern Hemisphere and aphelion will occur in the Southern Hemisphere, the reverse of today.

The Earth's axis is tilted. The tilt is the inclination of the Earth's axis in relation to its plane of orbit around the Sun. Today this tilt is at 23.5° but it varies from 24.5 to 21.5°. Oscillations in the Earth's axial tilt occur on a periodicity of 41,000 years thereby changing the differences between the equator and poles in the amount solar energy received.

---

[1159] Berger, A. 1988: Milankovitch theory and climate. *Reviews in Geophysics* 26: 624-657.

The Earth wobbles as it spins on its axis. This is called precession and is like the wobble of a spinning top. The Earth wobbles from pointing at the North Star (Polaris) to the star Vega, hence, when the axis is pointing towards Vega, this star can be considered the North Star. This slow wobble of the Earth's axis in relation to background stars is due to the gravitational pull of nearby planets. Precession has a periodicity of 21,000 years, advances by 25 minutes per year and exaggerates seasonal contrasts. The Moon's gravity prevents a huge variation in the Earth's wobbles. If it were not for the Moon, the Earth would have a far greater seasonal range.

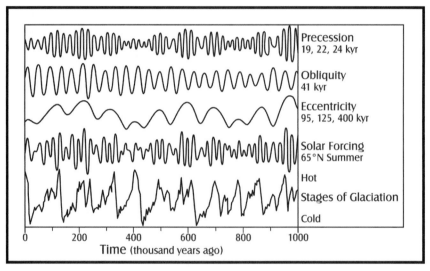

*Figure 20: Milankovitch Cycles showing relationship between orbital changes, solar forcing and climate cycles over the last million years. The combination of Milankovitch Cycles, solar activity and other natural features has driven cyclical climates for at least the last million years.*

The combination of these three Milankovitch Cycles influences climate because the closer we are to the Sun, the hotter it is. When the Sun is above or near the tropics, the radiative energy is 1412.9 watts per square metre (perihelion) and 1321.5 watts per square metre (aphelion). The mean figure of 1367 watts per square metre is used in climate models, thereby omitting the effects of orbit on the change in solar input. This is important because of the greater proportion of ocean water in the Southern Hemisphere. It is this water that adsorbs solar energy, stores solar heat and moves it around in currents. At times when the Northern Hemisphere summers are coolest (i.e. farthest from the Sun due to precession and the greatest orbital eccentricity) and winters warmest (i.e. minimum tilt), snow can accumulate over large areas of the Northern Hemisphere. At present, only precession is in the glacial mode whereas tilt and eccentricity are not favourable for glaciation. However, in the Holocene when it was warmer than now, the tilt was at 24° (23.5° today) and there was a 7% (6% today) difference in solar activity between January and July.

Since 2.67 Ma, the Earth has undergone major environmental changes in climate, seasons, humidity and ice caps. Ice volume and mean temperature variations are well displayed from the oxygen chemistry of foraminifer shells,[1160] variations in ice volume are related to alternating glacial and interglacial periods,[1161,1162] changes in sedimentation[1163] and there are many suggestions in the scientific literature that such changes are only driven by the Earth's orbit (roughly 100,000-year eccentricity, the 41,000-year obliquity and the 21,000-year precession[1164]). The duration of the climate cycles may be controlled by obliquity or precession. From 2.67 to 1 Ma, the climate system responded to obliquity, resulting in 41,000-year climate cycles.[1165,1166,1167]

However, from about 900,000 years ago to the present, climate has responded to a combination of eccentricity and precession[1168] and there are some authors who suggested that this change took place 620,000 years ago.[1169]

[1160] Lisiecki, L. E. and Raymo, M. E. 2005: A Pliocene-Pleistocene stack of 57 globally distributed benthic $\partial O^{18}$ records. *Paleoceanography* 20: 1-17.

[1161] Ruddiman, W. F. 2003: Orbital forcing ice volume and greenhouse gases. *Quaternary Science Reviews* 22: 1597-1629.

[1162] Berger A. and Loutre, M. F. 2004: Théorie astronomique des paléoclimats. *Comptes Rendus Géoscience* 336: 710-709.

[1163] Tian, J., Zhao, Q., Wang, P., Li, Q. and Cheng, Q. 2008: Astronomically modified Neogene sediment records from the South China Sea. *Paleoceanography* 23: PA3210, doi: 10.1029/2007PA001552.

[1164] Maslin M. A. and Ridgewell, A. J. 2005: Mid-Pleistocene revolution and the 'eccentricity myth'. In: Early-Mid Pleistocene Transitions: the Land-Ocean Evidence (eds Head, M. J and Gibbard, P. L.), *Geological Society London Special Publication* 247: 19-34.

[1165] Pisias, N. G. and Moore, T. C. 1981: The evolution of Pleistocene climate: A time series approach. *Earth and Planetary Science Letters* 52: 450-458.

[1166] Kroon, D., Alexander, I., Little, M., Lourens, L. J., Matthewson, A. H. F., Robertson, A. H. F. and Sakamoto, T. 1998: Oxygen isotope and sapropel stratigraphy in the Eastern Mediterranean during the last 3.2 million years. In: *Proceedings of the Ocean Drilling Program, Scientific Results 160* (eds Robertson, A. H. F., Emeis, K.-C., Richter, C. *et al*). Ocean Drilling Program, College Station, TX: 181-189.

[1167] Ruddiman, W. F. 2003: Orbital forcing ice volume and greenhouse gases. *Quaternary Science Reviews* 22: 1597-1629.

[1168] Von Grafenstein, R., Zahn, R., Tiedemann, R. and Murat, A. 1999: Planktonic $\partial O^{18}$ records at sites 976 and 977, Alboran Sea: stratigraphy, forcing and paleoceanographic implications. In: *Proceedings of the Ocean Drilling Program, Scientific Results 161* (eds Zahn, R., Comas, M. C. and Klaus, A.). Ocean Drilling program, College Station, TX, 469-479.

[1169] Mudelsee, M. and Stattegger, K. 1997: Exploring the structure of the mid-Pleistocene revolution with advance methods of time-series analysis. *Geologische Rundschau* 86: 499-511.

This reason for this change in frequency of climate cycles is debatable.[1170] Pollen records, marine faunal associations, sedimentation and isotopes in shells and marine sediments tracked these changes and showed the icehouse-greenhouse cycles,[1171] whatever their origin.

Milankovitch Cycles of 100,000 years have been detected in 25 Ma lignite in Ireland.[1172] Milankovitch Cycles have been detected from the oxygen chemistry of shells in deep-sea sediments, even those hundreds of millions of years old. However, were Milankovitch Cycles the dominant 100,000-year cycles of climate change switches in the current ice age? After all, these cycles are somewhat weak and it is more likely that if a Milankovitch Cycle was superimposed on another change, then there could be brief periods of quick warming or cooling. Effects of cosmic rays would be much more pronounced during cooler Milankovitch Cycles than during warmer cycles. During glaciation, sea level drops. This exposes large areas of continental shelf that change the pattern of ocean currents. Land rises and the Earth's orbit changes. Straits such as the English Channel, Bass Strait, Torres Strait and Bering Strait would be closed and ocean currents would be diverted, resulting in the redistribution of heat by the oceans. Furthermore, although the increased ice and snow would reflect more energy back into space the increased amount of land exposed would result in less reflection of energy back into space from the enlarged land masses.

In the 1970s, we all knew that Milankovitch Cycles were the makers and pace makers of climate change. That was the consensus. However, as with all science, new data has destroyed this popular paradigm. It does not appear that Milankovitch Cycles alone are enough to initiate glaciation.[1173] Solar activity and the reflection of heat from ice and snow can trigger glaciation. Increased winter snowfall and decreased summer melting result in a greater reflection of energy from the Sun back into space. By contrast, soils, outcrop and vegetation reflect far less energy back into space.

[1170] Maslin M. A. and Ridgewell, A. J. 2005: Mid-Pleistocene revolution and the 'eccentricity myth'. In: Early-Mid Pleistocene Transitions: the Land-Ocean Evidence (eds Head, M. J and Gibbard, P. L.) *Geological Society London Special Publication* 247: 19-34.

[1171] Joannin, S., Cornée, J.-J., Moissette, P., Suc, J.-P., Koskeridou, E., Lécuyer, C., Buisine, C., Kouli, K. and Ferry, S. 2007: Changes in vegetation and marine environments in the eastern Mediterranean (Rhodes, Greece) during the Early and Middle Pleistocene. *Journal of the Geological Society of London* 164: 1119-1131.

[1172] Large, D. J. 2007: A 1.16 Ma record of carbon accumulation in western European peatland during the Oligocene from Ballymoney lignite, Northern Ireland. *Journal of the Geological Society of London* 164: 1233-1240.

[1173] Roe, G. 2006: In defense of Milankovitch. *Geophysical Research Letters* 33: L24703, doi: 10.1029/GL027817.

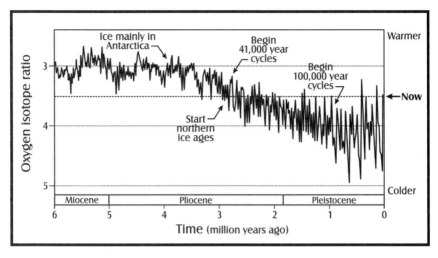

*Figure 21: The last 6 million years of climate showing cooling over the last 3 million years and changes from 41,000 to 100,000-year Milankovitch Cycles. Since the start of the last glaciation at 2.67 Ma, there were very large variations in temperature (as shown by the oxygen isotope ratios) and the unexplained shift from the 41,000-year Milankovitch to the weaker 100,000-year Milankovitch Cycle at about 1 Ma. Climate at present is cooler than in the early Pliocene and Miocene when the planet was its normal warm wet greenhouse self (despite having Antarctic ice sheets).*

The problem with the Milankovitch Cycle theory may be that orbital wobbles are not the main driver of climate. The annual amount of solar energy between the hemispheres is on a 21,000-year cycle. Yet, this is not the major glacial cycle. The 41,000-year cycles affect the amount of solar radiation entering the tropics and the poles. This is a glacial cycle and the glacial cycles switched from 41,000 to 100,000-year cycles at about 1 Ma. The 41,000-year cycle is no longer a glacial cycle. Why? Despite the almost uniform amount of solar energy striking the Earth in the 100,000-year cycle, it is the 100,000-year cycle that is the major glacial cycle. Why? The 100,000-year cycle is the weakest of the three Milankovitch Cycles and hence may be hardly enough to drive climate change. The Earth's orbital eccentricity also shows a 400,000-year as well as a 100,000-year cycle. The two cycles are of comparable strength yet the 400,000-year cycle is not recorded in climate records.[1174] A warming climate predates by about 10,000 years the change

---

[1174] If you think this is a confusing explanation, try the IPCC's tortured explanation at IPCC, 1990: *Climate change: The IPCC scientific assessment* (eds Houghton, J. T. et al.). Cambridge University Press. It is clear that the IPCC and many others, including the author, do not understand the processes of Milankovitch Cycles and glaciation.

in incoming solar radiation that supposedly had been its cause. What is remarkable and cannot be explained well by Milankovitch Cycle theory is that the transition from an interglacial to a glaciation occurs at peak temperature and when the melting of ice is at a maximum. There are great variations in temperature during the transition. At present, the Earth is close to the level at which past transitions occurred.

The Milankovitch Cycle theory suggests that changes in incoming solar radiation drive the Earth's ice ages. The mechanism whereby very slight changes in incoming radiation drive major climate change is unclear. Furthermore, detailed palaeoclimate measurements show that the end of a glaciation 135,000 years ago was not at the same time as orbital changes.[1175]

Climate is related to Milankovitch Cycle wobbles – we just don't know how.

| Known Cycles | |
|---|---|
| variable | tectonic |
| 143 million year | galactic |
| 100,000 years | orbital |
| 41,000 years | orbital |
| 21,000 years | orbital |
| 1,500 years | solar |
| 210 years | solar |
| 87 years | solar |
| 22 years | solar |
| 18.6 years | lunar |
| 11 years | solar |

*Figure 22: The known galactic, Milankovitch Cycle, solar and tidal cycles that drive the Earth's climate. Climates always change and are driven by a diversity of natural cyclical and random processes. What we do not see with past climate changes is climate change driven by changes in $CO_2$.*

---

[1175] Karner, D. B. and Muller, R. A. 2000: A causality problem for Milankovitch. *Science* 288: 2143-2144.

A new theory[1176] has described how resonant diffusion waves in the Sun can explain the palaeotemperature record over the last 5.3 Ma such as climate cycle periodicities, the relative strength of each cycle and the emergence of the 100,000-year periodicity. This new theory has its own unresolved difficulties.

A combination of variations in rotation rate combined with the rate of continent formation, the behaviour of the Sun and the amount of atmospheric $CO_2$ may have an effect on climate.[1177] The changes in the Earth's rotation have an effect on life and may also influence climate.[1178] The energy from lunar tides is dissipated in the oceans. To compensate, the Earth's rotation slows down, amounting to about 28 seconds per century. If the Earth is slowing down, in order to preserve angular momentum, the Moon moves away from the Earth. The length of the day is theoretically 86,400 seconds and there have been measurements of the length of the day since about 700 BC. There is an increase in the length of the day (2.3 milliseconds each century) because of the combined gravitational forces of the Sun and the Moon, and there is a steady decrease by 0.6 milliseconds per century as a result of the post-glacial land rise.[1179] The length of the day is increasing by 1.7 milliseconds per century. Every now and then, such as 2008, the clocks have to be adjusted by one second because of the change in the shape in the Earth and its atmosphere.

Did you notice that you had an additional one second of life in 2008? I hope you used it well.

[1176] Ehrlich, R. 2007: Solar resonant diffusion waves as a driver of terrestrial climate change. *Journal of Atmospheric and Solar-Terrestrial Physics* 69: 759-766.

[1177] Kuhn, W. R., Walker, J. C. G. and Marshall, H. G. 1989: The effect on the Earth's surface temperature from variations in rotation rate, continent formation, solar luminosity, and carbon dioxide. *Journal of Geophysical Research* 94: 11129-11136.

[1178] Scrutton, C. T. 1978: Periodic growth features in fossil organisms and the length of the day and month. In: *Tidal friction and the Earth's rotation* (eds Brosche, P. and Sdermann, J.), 154-196, Springer-Verlag.

[1179] Stephenson, F. R. 2003: Historical eclipses and Earth's rotation. *Astronomy and Geophysics* 44: 2.22-2.27.

# Chapter 5

## ICE

Question: Is warming melting the polar ice caps and alpine
valley glaciers?
Answer: Yes but no.

*Planet Earth has been a warm wet greenhouse volcanic planet for 80% of time. There have been numerous ice ages in which there are fluctuating greenhouse and icehouse conditions. The Earth is currently in an ice age. During the biggest ever ice age, atmospheric $CO_2$ was hundreds of times higher than at present.*

*Since the end of the last glaciation 14,000 years ago, ice sheets have largely melted and created a 130-metre sea level rise. During previous glaciations and interglacials, ice sheets expanded and contracted due to changes in temperature, humidity, slope, snowfall, landslides, solar radiation and ice evaporation. Many changes to ice sheets are due to local features. Temperature was warmer and sea level was higher during past interglacials.*

*At present, some ice sheets are advancing, others are retreating, and others are static. Flow is due to ice crystal recrystallisation and not by lubrication by water at the base of the ice sheet.*

*Ice surging, a temporary increase in flow rate, passes down a glacier and results from thicker ice that formed long ago. Calving of ice sheets to form icebergs is the end result of a process that started thousands of years ago. Armadas of icebergs from large-scale surging of the ice sheets are a permanent feature of every glaciation and do not reflect global warming.*

*In Antarctica, Canada and Iceland there are sub-glacial volcanoes which continually provide heat, hot water and hot gases. At times they erupt. Furthermore, Antarctica is rising. Under the Arctic Ocean is the Gakkal Ridge where there are large explosive eruptions, lava, hot springs and hot gas vents. The role of polar volcanoes is unknown but they are never considered in climate models.*

*There is neither a significant loss nor a gain to polar ice, alpine valley glaciers and sea ice. Drill holes in Antarctic and Greenland ice are a window into the past. During previous interglacials when it was far warmer than now, the ice sheets did not completely melt. Nor was there a runaway greenhouse. Dust from meteors, comets, volcanoes, droughts and interstellar activity is preserved in ice, as is dust from Greek and Roman industrial activity. This dust can give a record of changes in solar and supernoval activity. Ice sheets provide a guide to past temperature and a less reliable guide to past air composition.*

*In Antarctica, ice sheets and sea ice are expanding. There is a millennial-scale polar see-saw. When Antarctica is cooling, the Arctic is warming and vice versa. The Arctic was warmer than now between 1920 and 1940.*

## Ice

Water is weird. So too is ice. Ice is a rock and behaves like many other rocks. If the temperature of ice rises to 0°C, then ice will not necessarily melt. Heat needs to be added to the ice such that it will convert to water at 0°C.[1180] It takes one calorie to raise the temperature of water by 1°C, but 80 calories are needed to melt the equivalent amount of ice. Ice needs a large amount of energy to break loose a water molecule and water needs an even larger amount of energy to break loose a molecule to form steam. It takes a lot of heat to make water hot and, once heated, it takes a long time to cool. Normally when a solid forms, the molecules are more tightly packed together and when that solid is melted, the molecules move apart and a liquid is formed. No so with water. Water ice is less dense than liquid water.

Glaciers are rivers of ice that flow due to gravity. Glaciers advance and contract, sometimes they are static and at other times they surge.[1181] They are not thermometers that expand or contract depending upon air temperature. If glaciers are to be melted by an increase in temperature, a monstrous amount of heat needs to be added just to convert ice to water. Ice sheets do not melt from the surface down; they melt at the edges and underneath and become smaller when ice converts directly to water vapour (sublimation). Glacial melting also takes time, which creates a lag between global warming and ice melting. Ice is a good insulator and the transfer of heat in ice is a slow process. Ice also reflects a large amount of radiation back into the sky.

---

[1180] Latent heat of fusion of ice 333.55 J/g (79.72 cal/g). Latent heat of vapourisation of water (at 0°C) 2500 J/g (598 cal/g) and (at 100°C) 2260 J/cal (539 cal/g). Specific heats (cal/g°C) are water 1.00, ice 0.50 and steam 0.47. By contrast wood is 0.12 and gold is 0.03.

[1181] Dowdeswell, J. A., Hamilton, G. S. and Hagen, J. O. 1991: The duration of the active phase on surge-type glaciers: contrasts between Svalbard and other regions. *Journal of Glaciology* 37: 388-400.

Increased solar radiation does not necessarily mean that ice will melt. If ice is at -30°C and the temperature warms by 10°C, the ice will still be ice although solar radiation causes ice to sublimate. Just because ice melts, it does not necessarily mean that sea level will rise, as there are a large number of other processes that influence sea level change.

Ice is less dense than water. If ice sank, lakes, seas and oceans would freeze from the bottom up. This would prevent ice melting and would eventually produce permanent ice on Earth. This ice would reflect radiation, planet Earth would not be able to escape from being an ice ball and, except for some extremophile micro-organisms, all life on Earth would freeze. It is indeed fortunate that water is weird.

Ice forms hexagonal crystals. These crystals have a preferred plane of weakness. Ice in lakes has the plane of weakness parallel to the surface and can be sheared with far less stress than ice in glaciers, which has randomly distributed crystals. Ice in glaciers moves by creep. This involves the constant movement of atoms from one ice crystal to another. Crystals grow in size from minute crystals in snow to large crystals in ice, especially at the snout of the glacier where the ice crystals are hundreds of times larger than ice crystals in snow. The rate of creep is proportional to temperature and the weight of the overlying ice. Creep can only operate when there is a considerable weight of ice. Ice sheets in Greenland and Antarctica have very cold ice so flow is limited in most of the ice. However, at the base of a glacier the ice is warmed by the Earth's heat and most flow is close to the base of the glacier. The thicker the ice, the greater the rate of flow.

Are the polar ice sheets expanding or contracting? The only answer to this question is yes. At times they expand, at times they contract. During the last glaciation, which started in the Northern Hemisphere at 2.67 Ma, the ice sheets have waxed and waned hundreds of times. Over the last 2.67 Ma, the ice sheets have not completely disappeared during long periods when the temperature was far higher than at present.[1182] For example, in the Holocene maximum 6000 years ago when sea level was 2 metres higher than now and temperature was at least 6°C warmer, the Greenland ice sheet did not disappear,[1183] polar bears did not become extinct and the Antarctic ice sheet

---

[1182] Dowdeswell, J. A., Hagen, J. O., Björnsson, H., Glazovsky, A. F., Harrison, W. D., Holmlund, P., Jania, J., Koerner, R. M., Lefauconnier, B., Ommaney, C. S. L., and Thomas, R. H. 1997: The mass balance of circum-Arctic glaciers and recent climate change. *Quaternary Research* 48: 1-14.

[1183] Weidick, A., Oerter, H., Reeh, N., Rhomsen, H. H. and Thorning, L. 1990: The recession of the inland ice margin during the Holocene climatic optimum in the Jakobhavn Isfjord area of West Greenland. *Palaeogeography, Palaeoclimatology, Palaeoecology* 82: 289-299.

grew.[1184]  The Antarctic ice sheet is much older and is largely isolated from the rest of the world, including other parts of the Southern Hemisphere.

Correlation of Antarctic marine sediments with Greenland ice cores over the last 4000 years shows that there are cycles of warm and cold events every 200 years. The DeVries-Suess solar cycle occurs every 200 years.

## Ice ages

One of the oldest great scientific puzzles is what caused the ice sheets to come and go. In 1840, Louis Agassiz proposed that there were once great ice sheets across Europe and North America and in 1842 Joseph Adhémar proposed that glaciation occurs when the winters are very long (i.e. the Earth's orbit is farthest from the Sun). In the Northern Hemisphere when the Earth is close to the Sun, the Earth orbits quicker and summer is about a week shorter.

This idea was modified by James Croll who suggested that long winters that could stimulate glaciation derive from weaker solar radiation. In 1934, Milutin Milankovitch turned Croll's ideas around and suggested that glaciation in the Northern Hemisphere occurs when the amount of solar radiation in summer time is low. This was driven by orbital cycles. Snow and ice persist throughout the summer and gradually accumulate into an ice sheet. Milankovitch's ideas were confirmed with data showing that over the past 800,000 years, the ice sheets took about 90,000 years to grow and about 10,000 years to shrink.[1185] Glaciologists still do not understand how subtle shifts in solar radiation at the top of the atmosphere are converted to massive changes in the ice volume on the ground.[1186]  The 100,000-year glacial cycles recorded in the late Pliocene to early Pleistocene (at 1–3 Ma) are far more regular than those of the late Pleistocene, typically lasting about 41,000 years.[1187,1188]  The precession orbital cycles (23,000 to 19,000-year intervals) are observed in ice volume and sea level records over the last 700,000 years.[1189]

[1184] Domack, E. W., Ishman, S. E., Stein, A. B., McClennan, C. E. and Jull, A. J. T. 1995: Late Holocene advance of the Müller Ice Shelf, Antarctic Peninsula: sedimentologic, geochemical, and paleontological evidence. *Antarctic Science* 7: 159-170.

[1185] Hays, J. D., Imbrie, J. and Shackleton, N. J. 1976: Variations in the Earth's orbit: pacemaker of the ice ages. *Science* 194: 1121-1131.

[1186] Raymo, M. E. and Huybers, P. 2008: Unlocking the mysteries of the ice ages. *Nature* 451: 284-285.

[1187] Pisias, N. G. and Moore, T. C. 1981: The evolution of Pleistocene climate: a time series approach. *Earth and Planetary Science Letters* 52: 450-458.

[1188] Huybers, P. J. and Tziperman, E. 2007: Integrated summer insolation and forcing and 40,000 year glacial cycles: the perspective from ice sheet/energy balance model. *Paleoceanography* 23:doi:10.1029/2007PA001463.

[1189] Huybers, P. J. 2006: Early Pleistocene glacial cycles and the integrated summer insolation forcing. *Science* 313: 508-511.

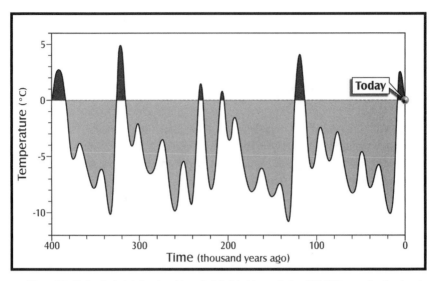

*Figure 23: Cycles of glacials (grey) and interglacials (black) over the last 400,000 years showing that the current interglacial is not as warm as previous interglacials, that the current interglacial should be followed by glaciation and that there is nothing extraordinary about the modern climate. Polar bears lived in all previous interglacials, some of which were warmer and longer than the present interglacial.*

It is widely accepted that the Earth's orbit affects glaciation but a better and more detailed understanding of this process is needed. How can the 41,000-year glacial cycles of the early Pleistocene be explained, let alone the 100,000-year cycles of the late Pleistocene? Why was there a change? How do subtle changes in radiation relate to the glacial cycles? What are climate proxy records actually measuring? These questions are vital because the Earth's modern climate is tied to the last remaining vestige of northern ice sheets that 14,000 years ago covered the Northern Hemisphere down to 38°S and the remaining Southern Hemisphere ice sheet on Antarctica. If we can't understand how previous large cyclical climates occurred, we have little hope of creating models to predict future climate changes.

Planet Earth is a warm wet volcanic planet and, over the history of time, there have been ice sheets for only 20% of time. During a glaciation, the ice sheets come and go, sea level rises and falls, life retreats and advances and climate changes are abrupt and quick. Short hot interglacials occur when there are ice sheets and short sharp cold periods occur during interglacials.

There was an ice age at 2400 to 2100 Ma.[1190] Glaciation was at sea level

---

[1190] Huronian Glaciation.

and in the tropics. Just after this event, there was a diversification of bacterial life and the atmosphere became oxygenated due to photosynthesis. Another tropical sea level glaciation occurred again between 750 and 635 Ma.[1191] After this glaciation, the atmosphere again increased in oxygen and multicellular life appeared. Ice sheets were waxing and waning.[1192]

There have been numerous mechanisms promoted for explaining the origin of the biggest climate change of all time including reduced solar luminosity,[1193] increased rotation rate of the Earth[14], a steep tilt to the Earth's axis,[1194,1195,1196] different atmospheric concentrations,[1197] different geography and topography governed by plate tectonic conditions,[1198,1199] varying reflection of solar radiation by ice, incorporation of ocean circulation and heat transport and sea ice dynamics. The origin of the biggest glaciation ever experienced on planet Earth is obscure. Ice sheets were at the Equator at the end of the Proterozoic eon, 750 to 635 Ma.[1200] There were alternating glaciations and long interglacials when the sea surface temperature was at least 40°C. No sea or ocean on Earth today is as warm as this. Sea level rose and fell by 600 metres. No such sea level changes have ever been found since. Sea chemistry changed. Life on Earth then was only bacterial. Life fluctuated in the oceans from being highly productive in interglacials to just surviving during glaciation. This led to suggestions that the Earth was a snowball.

[1191] Cryogenian or Neoproterozoic Glaciation.

[1192] Allen, P. A. and Etienne, J. L. 2008: Sedimentary challenge to Snowball Earth. *Nature Geoscience* 1: 817-825.

[1193] Jenkins, G. S. and Frakes, L. A. 1998: GCM sensitivity test using increased rotation rate, reduced solar forcing and orography to examine low latitude glaciation in the Neoproterozoic. *Geophysical Research Letters* 25: 3525-3528.

[1194] Williams, G. E., Kasting, J. F. and Frakes, L. A. 1998: Low-latitude glaciation and rapid changes in the Earth's obliquity explained by obliquity-oblateness feedback. *Nature* 396: 433-455.

[1195] Jenkins, G. S. 2000: Global climate model high-obliquity solutions to the ancient climate puzzles of the Faint Young Sun Paradox and low-latitude Proterozoic glaciation. *Journal of Geophysical Research* 105: 7357-7370.

[1196] Donnadieu, Y., Ramstein, G., Fluteau, F., Besse, J. and Meert, J. 2002: Is high obliquity a plausible cause for Neoproterozoic glaciations. *Geophysical Research Letters* 29: doi:10.1029/2002GL015209.

[1197] Crowley, T. J., Hyde, W. H. and Peltier, W. R. 2001: $CO_2$ levels required for a deglaciation of a 'Near-Snowball' Earth. *Geophysical Research Letters* 28: 283-286.

[1198] Donnadieu, Y., Goddéris, Y., Ramstein, G., Nedelec, A. and Meert, J. A. 2004: A 'snowball Earth' climate triggered by continental break-up through changes in run-off. *Nature* 428: 303-306.

[1199] Goddéris, Y., Donnadieu, Y., Dessert, C., Dupre, B., Fluteau, F., François, L. M., Meeert, J., Nédélec, A. and Ramstein, G. 2007: Coupled modelling of global carbon cycle and climate in the Neoproterozoic: Links between Rodinia break-up and major glaciations. *Compes Rendes Géosciences* 339: 212-222.

[1200] Fairchild, I. J. and Kennedy, M. J. 2007: Neoproterozoic glaciation in the Earth System. *Journal of the Geological Society* 164: 895-921.

If we do not understand the cause or causes for the biggest climate change of all time, what hope have we of understanding modern climate change?

The mild Ordovician-Silurian high-latitude glaciation (450–420 Ma) was followed by a very intense Permo-Carboniferous (300–260 Ma) high-latitude glaciation. The mild Jurassic-Cretaceous (151–132 Ma) glaciation was at high latitude as is our current glaciation which started at 2.67 Ma (Pleistocene glaciation). Previous glaciations occurred at times when atmospheric $CO_2$ was far higher than at present, which raises doubt about $CO_2$ creating global warming. We are currently in a period of fluctuating icehouse and greenhouse conditions. We came out of glaciation 14,000 years ago and it is no surprise that temperature and sea level have risen.

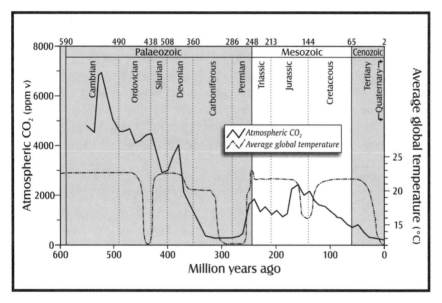

*Figure 24: Plot of atmospheric $CO_2$, time and temperature showing that the Ordovician, Permo-Carboniferous and Jurassic ice ages occurred when atmospheric $CO_2$ was higher than at present. The hypothesis that a high atmospheric $CO_2$ produces global warming is therefore invalid.*

Furthermore, during the biggest glaciation of all time the atmospheric $CO_2$ was far higher than today. The super continent Rodinia started to break up at 830 Ma. This would have resulted in large amounts of $CO_2$ being released to the atmosphere from basalt volcanoes.[1201] The increased rainwater runoff on numerous landmasses would have consumed $CO_2$ by continental weathering, and increased weathering increased $CO_2$ consumption. There

---

[1201] Goddéris, Y., Donnadieu, Y., Nédélec, A., Dupré, C., Dessert, C., Grard, A., Ramstein, G. and François, L. M. 2003: The Sturtian 'snowball' glaciation: fire and ice. *Earth and Planetary Science Letters* 211: 1-12.

would have been increased runoff, especially as there was no vegetation at that time. Atmospheric $CO_2$ concentration was decreased by 1320 ppmv[1202] thereby allowing factors other than $CO_2$ to drive climate. Nevertheless, this $CO_2$ content was more than four times the current atmospheric $CO_2$ content yet there was glaciation, not global warming. After the glaciation, nutrients added to the warm oceans during interglacials stimulated the expansion of single-celled life which increased the atmospheric oxygen content. This then led to the appearance of multicellular life[1203] at 583 Ma, the draw-down of $CO_2$ from the atmosphere, the locking of $CO_2$ into rocks and fossils by lime-secreting plants and the moderating of climate.

Glaciation causes massive removal of soil and fractured rocks beneath the soils. In some areas, glaciers erode more material than the big rivers traversing the Himalayas.[1204] This is seen by the volume of material that accumulates in the fjords of Alaska.[1205] What is even stranger is that the advance of some ice sheets can change the processes of faulting, folding and vertical movement of mountain chains and the sediments that shed from mountain chains.[1206] Rivers can do the same.

The latest ice age did not just happen. There has been profound global cooling for the last 50 million years with various short- and long-lived warming events within the cooling trend. Once the planet has cooled, then combinations of orbital perturbations, variations in the Sun and geological factors trigger glaciation.

## Ice advance and retreat

Snow falls on higher ground, it becomes more and more compact with time, air is extruded and it turns to solid ice. Some air bubbles are trapped and can be used to determine the air composition at the time of snowfall. However, compressed ice squeezes air bubbles and moves them around. More precipitation of snow forms layer upon layer, which goes through the same process. The youngest layer is on top, the oldest is at the base, and such

---

[1202] Donnadieu, Y., Goddéris, Y., Ramstein, G., Nédélec, A. and Meert, J. 2004: A 'snowball' Earth climate triggered by continental break-up through changes in runoff. *Nature* 428: 303-306.

[1203] Ediacaran fauna.

[1204] Hallet, B., Hunter, I. and Bogen, J. 1996: Rates of erosion and sediment evacuation by glaciers: A review of field data and their implications. *Journal of Global Planetary Change* 12: 213-235.

[1205] Molnar, P. and England, P. 1990: Late Cenozoic uplift of mountain ranges and global climate change: chicken or egg? *Nature* 346: 29-34.

[1206] Berger, A. L., Gulick, S. P. S., Spotila, J. A., Upton, P., Jaeger, J. M., Chapman, J. B., Worthington, L. A., Pavlis, T. L., Ridgway, K. D., Willems, B. A. and McAleer, R. J. 2008: Quaternary tectonic response to intensified glacial erosion in an orogenic wedge. *Nature Geoscience* 1: 793-799.

thick layers of ice are a basic source of broad-scale temperature and $CO_2$ data over the last few hundred thousand years. Ice sheets are not just great layers of ice, they often have basal layers of sediment with ice above and below these layers.[1207] The deepest ice core is more than 700,000 years old. The preservation of this stratified ice shows conclusively that the ice caps have not melted during any of the warm phases over this long time span.

There is not a simple relationship between ice and temperature. If the regional climate becomes too dry, there will be little precipitation and the glacier will retreat. This has been happening at Mt Kilimanjaro since the late 19th Century. This could occur if the region became cold enough to reduce evaporation from the ocean. If temperatures rises, evaporation is increased. As a result, there is increased snowfall. Paradoxically, a rise in temperature may lead to increased growth of glaciers and ice sheets. This has happened many times in Greenland and Antarctica. However, evaporation requires energy, which comes from the land surface and the air. It is this water vapour in the air and the cycle of evaporation and precipitation that limits air temperature. Water vapour in the air provides summers in the humid tropics with a lower temperature than summers in the deserts at higher latitudes. Rather than $CO_2$, it is water that drives climate.

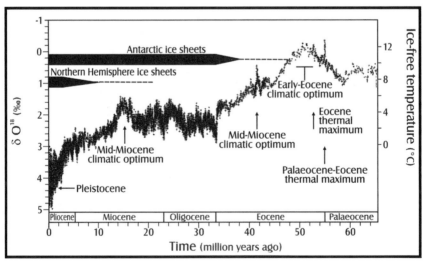

*Figure 25: Climate reconstruction over the last 65 million years from measurement of oxygen isotopes in fossilised floating animal shells. The last 65 Ma has been one of cooling with warm spikes, warm periods and temperature variations within an overall cooling trend. The Earth is currently cool, for 80% of time it has been warmer than now.*

[1207] Murton, J. B., Whiteman, C. A., Waller, R. I., Pollard, W. H., Clark, I. D. and Dallimore, S. R. 2004: Basal ice facies and supraglacial melt-out till of the Laurentide Ice Sheet, Tuktoyaktuk Coastlands, western Arctic Canada. *Quaternary Science Reviews* 24: 681-708.

Glaciers advance and retreat, for many reasons such as creep (all glaciers), temperature (Arctic, Antarctic), humidity (Kilimanjaro), water flow beneath ice (Greenland and Antarctica), change in land slope (Antarctica) and sub-glacial volcanoes (Iceland, Antarctica, Canada). There are balance sheets kept showing the losses and gains of ice.[1208] They show that glacial retreat and advance is not global. However, we really do not know how glaciers and ice sheets melt. There may be a transfer of surface melt water to the bed of the glacier resulting in lubrication or loss of buttressing ice shelves at the end of the glacier.[1209,1210,1211] Measurements and observations of ice sheet and glacier melting are not in agreement with calculations and models.[1212] When ice sheets melt, they can do so quickly.[1213]

The physics of ice flow has been known for some 70 years. The idea that glaciers slide on a lubricated base can be tested. A line of sticks across a glacier makes a convex downstream curve showing that ice does not simply slide on a lubricated base. If the ice slid on a lubricated base, the sticks would still be in a straight line. Ice grain size increases towards the terminus as does the preferred orientation of ice crystals. This shows that ice movement is from plastic flow or creep[1214] by continual recrystallisation of ice.[1215,1216,1217] This is how metals, rocks and other substances that appear solid can bend and flow and this process has been known for decades.[1218,1219,1220] There is a

[1208] Braithwaite, R. L. 2002: Glacier mass balance: The first 50 years of international monitoring. *Progress in Physical Geography* 26: 76-95.

[1209] Alley, R. B., Clark, P. U., Huybrechts, P. and Joughin, I. 2005: Ice-sheet and sea-level changes. *Science* 310: 456-460.

[1210] Stearns, L. A., Smith, B. E. and Hamilton, G. S. 2008: Increased flow speed on a large East Antarctic outlet glacier caused by sub-glacial floods. *Nature Geoscience* 1: 827-831.

[1211] Bell, R. E. 2008: The role of sub-glacial water in ice-sheet mass balance. *Nature Geoscience* 1: 297-304.

[1212] Siddall, M. and Kaplan, M. R. 2008: A tale of two ice sheets. *Nature Geoscience* 1: 570-571.

[1213] Carlson, A. E., Legrande, A. N., Oppo, D. W., Came, R. E., Schmidt, G. A., Anslow, F. S., Licciardi, J. M. and Obbink, E. A. 2008: Rapid early Holocene deglaciation of the Laurentide ice sheet. *Nature Geoscience* 1: 620-624.

[1214] Determined from X-ray crystallography.

[1215] Perutz, M. F. 1940: Mechanism of glacier flow. *Proceedings of the Physical Society* 52: 132-135.

[1216] Nye, J. F. 1953: The flow law of ice from measurements in glacier tunnels, laboratory experiments and the Jungfraufirn borehole experiment. *Proceedings of the Royal Society of London, Series A* 219: 477-489.

[1217] Nye, J. F. 1955: The creep of polycrystalline ice. *Proceedings of the Royal Society of London, Series A* 207: 554-572.

[1218] Benioff, H. 1951: Earthquakes and rock creep. *Bulletin of the Seismological Society of America* 1951: 31-62.

[1219] Mitra, S. K. and McLean, D. 1966: Work hardening and recovery in creep. *Proceedings of the Royal Society of London, Series A* 295: 288-299.

[1220] Boukharov, G. N., Chanda, M. W. and Boukharov, N. G. 1995: The three processes of brittle crystalline rock creep. *International Journal of Rock Mechanics and Mining Sciences* 32: 325-335.

view that ice creep can be very slow, over thousands of years. Other evidence suggests that major changes in some parts of ice sheets take place over only a few years to decades.[1221] This shows that there is a great range in the rate of glacier flow. Some scientific papers suggesting that melt waters reaching the base of an ice sheet speed the movement of ice's seaward flow. Yet these papers do not even mention creep as the tried and proven alternative explanation of ice movement.[1222,1223,1224]

When the ice is thick enough, it starts to flow like a plastic substance under the force of gravity.[1225] Wherever liquid water is present, as in melt water, because of the huge difference between the latent and specific heat of water, the glacier is at freezing point. The rate of ice flow is proportional to stress and to temperature. If the glacier is all at freezing point, temperature is unimportant and stress controls flow. In ice sheets, stress causes the ice to flow uphill and out of deep basins. Even alpine valley glaciers can flow uphill as the original alluvial basin is deepened to form a rock-carved basin.[1226] The fact that ice flows uphill shows that another mechanism besides the lubrication by melt waters at the base of the ice sheet controls ice movement. Most ice sheet flow is near the base, where geothermal heat reduces the ice temperature. Some deep ice cores run out of stratified ice at depth. This is where the ice is beginning to flow under stress though still well below freezing point.

Glaciers flow downstream and ice sheets flow outwards from the snow depositional centre which is towards the edge of the ice sheet.[1227] These depositional centres are the uplands of Greenland and Antarctica. The Greenland, West Antarctic and East Antarctic ice sheets occupy basins kilometres deep, hence the ice flow is actually uphill. Flow rates can be

---

[1221] Bamber, J. L., Alley, R. B. and Joughin, I. 2007: Rapid response of modern day ice sheets to external forcing. *Earth and Planetary Science Letters* 257: 1-13.

[1222] Stearns, L. A., Smith, B. E. and Hamilton, G. S. 2008: Increased flow speed on a large East Antarctic outlet glacier caused by sub-glacial floods. *Nature Geoscience* 1: 827-831.

[1223] Joughin, I., Das, S. B., King, M. A., Smith, B. E., Howat, I. M. and Moon, T. 2008: Seasonal speedup along the western flank of the Greenland ice sheet. *Science* 320: doi: 10.1126/science.1153288.

[1224] Das, S. B., Joughin, I., Behn, M. D., Howat, I. M., King, M. A., Lizarralde, D. and Bhatia, M. P. 2008: Fracture propagation to the base of the Greenland ice sheet during supraglacial lake drainage. *Science* 320: doi: 10.1126/science.1153360.

[1225] Hagen, J. O., Melvold, K., Pinglot, F. and Dodeswell, J. 2003: On the net mass balance of the glaciers and ice caps in Svalbard, Norwegian Arctic. *Arctic, Antarctic and Alpine Research* 35: 264-270.

[1226] Corrie or cirque.

[1227] Dowdeswell, J. A., Hodges, R., Nutall, A.-M., Hagen, J. O. and Hamilton, G. S. 1995: Mass balance change as a control on the frequency and occurrence of glacier surges in Svalbard, Norwegian High Arctic. *Geophysical Research Letters* 22: 2909-2912.

variable and up to 40 metres per day (e.g. Upernavik Glacier, Greenland[1228]), which is as much as many alpine glaciers flow in a year. When the ice reaches a lower altitude where temperature is warmer, it starts to melt and evaporate. If growth and evaporation are balanced, the glacier appears stationary. If precipitation and growth exceed evaporation, the glacier grows. If evaporation exceeds precipitation, the glacier recedes. In the Northern Hemisphere, the movement of glaciers is responsible for the short-term variations in glacier velocity.[1229,1230,1231,1232] Ice sheets move by creep at the base of the ice whereas in most alpine valley glaciers, the movement is above the ice base because of drag.

In alpine valley glaciers, there is frictional drag at the base of the glacier and no flow at the top because of the lack of overlying ice. The upper ice of a glacier is brittle and the lowest part is plastic. Maximum flow of valley glaciers is in the middle. The surface ice does not flow by creep and it cracks. Glacial flow rates are variable and, at times, there are surges of ice. For example, the sudden doubling of the rate of glacier melting in 2004 in Greenland was food for the catastrophists. Two of the largest glaciers have now slowed and, in 2006, one of them[1233] stopped thinning and thickened in the main trunk, and the mass loss from Greenland glaciers is now near previous rates. These glaciers were responding to the loss of some restraining ice at their lower ends, much the same as a river's flow would change due to the destruction of a dam.[1234] Surging, retreat and lack of flow have also been measured in Antarctic glaciers.[1235] The thicker the ice at the source and the greater the slope, the faster the flow of ice. The increasing pace passes downstream. Surges do not relate to air temperature, melt water or sea temperature, they relate to changes that took place long ago.[1236]

---

[1228] Rignot, E. and Kanagaratnam, P. 2006: Changes in velocity structure of the Greenland ice sheet. *Science* 311: 986-990.

[1229] Ringot, E. and Kanagaratnam, P. 2006: Changes in the velocity structure of the Greenland Ice Sheet. *Science* 311: 986-990.

[1230] Iken, A. and Bindschadler, R. A. 1986: Combined measurements of sub-glacial water pressure and surface velocity of Findelengletscher, Switzerland: Conclusions about drainage system and sliding mechanism. *Journal of Glaciology* 32: 101-119.

[1231] MacGregor, K. R., Riihimaki, C. A. and Anderson, R. S. 2005: Spatial and temporal evolution of rapid basal sliding on Bench Glacier, Alaska, USA. *Journal of Glaciology* 51: 49-63.

[1232] O'Neel, S., Pfeffer, W. J., Krimmel, R. and Meier, M. 2005: Evolving force balance at Columbia Glacier, Alaska, during its rapid retreat. *Journal of Geophysical Research* 110: F03012.

[1233] Kangerdlugssuaq Glacier.

[1234] Howat, I., Joughin, I. R. and Scambos, T. A. 2007: Rapid changes in ice discharge from Greenland outlet glaciers. *Science* 315: 1559-1561.

[1235] Budd, W. F. and McInnes, B. J. 1979: Periodic surging of the Antarctic ice sheet – an assessment by modelling. *Hydrological Studies Bulletin* 24-1: 95-103.

[1236] Fookes, P. G. 2008: Some aspects of the geology of Svalbard. *Geology Today* 24: 146-152.

In the Little Ice Age (1280–1850 AD), there were at least six phases of glacial advance and retreat.[1237,1238,1239,1240] All glaciers advanced quickly (1560–1610) to a maximum (1640–1650 in Switzerland). There was a glacial maximum in Austria (1670–1705) and a glacial maximum in Norway (1720–1750). Between 1816 and 1825 there were minor advances of all glaciers and the glacial maximum was at 1850–1890. In the 1930s and 1940s, there was a rapid shrinkage of glaciers, then an expansion from the 1950s to the 1980s. In the past, we also see that glacial advance and retreat may not be related to temperature. For example, glacial expansions in the mountains of Patagonia in the Younger Dryas (12,900–11,500 years ago) are probably unrelated to cooling and are a response to increased amounts of easterly-sourced precipitation.[1241] To argue that any modern glacial retreat is a result of humans warming the planet is to ignore history.

Measurement of the mass loss of glaciers and the flow rate of glacial ice as a guide to climate change is misleading because the flow rate may rapidly decrease or increase in fits and starts for many reasons. An increase in the flow rate is most commonly unrelated to temperature and is a result of increased precipitation thousands of years ago. Glaciers build up to a surge with the higher parts thickening as a result of snow accumulation. The slope of the ice increases to a critical point and surge is initiated by the shearing of ice crystals and the accumulation of large amounts of pressurised water on the glacier bed. Most of the most vivid symbols, albeit erroneous, used to show global warming are the torrents of glacial melt water that drain from the lakes that form each summer on the Greenland ice sheet. This water flows along natural drainpipes.[1242] During the surge, ice masses move down slope with a wave-like motion creating great variations in the rate of ice flow. When the surge has ceased, the upper masses of the ice sheet start to accumulate snow again and the surge repeats itself. These surges take place every 10 to 100 years.[1243]

Ice grinding across the land surface plucks off soil and rocks. Rock

[1237] Nesje, A. and Kvamme, M. 1991: Holocene glacier and climate variations in western Norway: Evidence for early Holocene glacier demise and multiple Neoglacial events. *Geology* 19: 610-612.

[1238] Luckman, B. H., Holdsworth, G. and Osborn, G. D. 1993: Neoglacial glacier fluctuations in the Canadian Rockies. *Quaternary Research* 39: 144-153.

[1239] Svendsen, J. I. and Mangerud, J. 1997: Holocene glacial and climate variations on Spitsbergen, Svalbard. *The Holocene* 7: 45-57.

[1240] Nesje, A. and Dahl, S. O. 2000: *Glaciers and environmental change.* Arnold.

[1241] Ackert, R. P., Becker, R. A., Singer, B. S., Kurz, M. D., Caffee, M. W. and Mickelson, D. M. 2008: Patagonian glacier response during the late glacial-Holocene transition. *Science* 321: 392-395.

[1242] Moulins.

[1243] Fookes, P. G. 2008: Some aspects of the geology of Svalbard. *Geology Today* 24: 146-152.

fragments are ground to a powder, some are polished and scratched and some can be carried by ice hundreds of kilometres from where they were plucked from outcrops. A retreating glacier can leave behind large boulders the size of a building. In antiquity many of these were interpreted as material left by giants and devils.

When glaciers melt, this material is dumped as a moraine at the end and edges of the glacier. If the glacier is advancing to the sea, icebergs calved from the glacier carry soil and rocks from the land. Glaciers are not static piles of ice. They are constantly flowing rivers of ice. It is normal for glaciers to calve off large blocks of ice when they reach the sea regardless of how warm or cold it is. The ice calving process is related to many factors, one of which is air temperature.

Contrary to popular belief, ice is shed from glaciers during periods of global cooling. At these times, the ice sheets are very thick, and alpine valley glaciers flow down slope picking up surface soil, silt, sand and boulders and break off to form icebergs. Ice sheets may flow down slope or upslope. Once the iceberg melts, it drops material scoured from the land surface by glaciers. Because of the lack of strong currents and the distance from shorelines, sediments in the ocean deeps are very fine grained. Sea floor sediments also contain ash from distant volcanoes, wind blown dust, extraterrestrial dust and the shells from dead organisms (especially floating micro-organisms) as well as the material dropped from melting icebergs. Every time icebergs melt, there is a cascade of coarse-grained sand, pebbles and boulders onto the fine-grained sea floor muds. Ice rafting and melting icebergs occur all the time but, during periods of cold climate, ice rafting of armadas of icebergs define that cold period.[1244] Drop boulder beds have long been recognised in sediments that formed in every ice age. Once icebergs drift to lower latitudes, they melt. It was one of these southerly-drifting icebergs that led to the sinking of the *Titanic*. During cold periods, shipping becomes far more dangerous, even at low latitudes, because the ice front is at a lower latitude.

There is evidence that icebergs drifted much further south from the Arctic to coastal North Africa during the last cold period (118,000–14,000 years ago). Icebergs contain messages from the land. These layers of coarse-grained material dropped from melting icebergs, discovered in the 1980s by Hartmut Heinrich, are called Heinrich Events. Heinrich Events have now been measured as far back as 60,000 years ago, showing the last glacial period

---

[1244] Hulbe, C. L., MacAyeal, D. R., Denton, G. H., Kleman, J. and Lowell, T. V. 2004: Catastrophic ice shelf breakup as the source of Heinrich event icebergs. *Paleoceanography* 19: doi:10.1029/2003PA000890.

was far from one where there was just ice sheet cover.[1245] Material from the last five Heinrich Events in the North Atlantic Ocean derived from ancient continental sources surrounding the Labrador Sea[1246,1247] dumped into the sea by the surging Laurentide ice sheet.[1248] Other parts of the North Atlantic Ocean show that the European ice sheets predated[1249] and provided material that was later included in the Laurentide ice sheet.[1250] Heinrich layers take 50 to 1250 years of iceberg delivery to produce a 10 cm Heinrich layer in the North Atlantic Ocean.[1251]

Each time there was an ice armada and the deposition of Heinrich Events, there was a sudden drop in global temperature of a few degrees. Not only did these cold periods affect the oceans, they also affected inland lakes. For example, the level of the Dead Sea fell greatly every time there was a Heinrich Event, showing that there was decreased rainfall. This shows that Heinrich Events are a reflection of periods of global cooling. The 1500-year Bond Cycles have been recorded in the deepest lake on Earth, Lake Baikal, showing that cooling periods such as Heinrich Events are not restricted to the oceans.[1252] The calving of an iceberg is a rapid event. However, it is not related to temperature at the time of calving but results from precipitation, ice accumulation and ice flow that took place long ago.

---

[1245] Broecker, W. S. and Hemming, S. 2001: Climate swings come into focus. *Science* 294: 2308-2309.

[1246] Hemming, S. R., Broecker, W. S., Sharp, W. D., Bond, G. C., Gwiazda, R. H., McManus, J. F., Klas, M. and Hajdas, I. 1998: Provenance of Heinrich layers in core V28-82, northeastern Atlantic: $^{40}Ar/^{39}Ar$ ages of ice-rafted hornblende, Pb isotopes in feldspar grains, and Nd-Sr-Pb isotopes in the fine sediment fraction. *Earth and Planetary Science Letters* 164: 317-333.

[1247] Huon, S., Grousset, F. E., Burdloff, D., Bardoux, G. and Mariotti, A. 2002: Sources of fine-sized organic matter in North Atlantic Heinrich layers: $\partial^{13}C$ and $\partial^{15}N$ tracers. *Geochimica et Cosmochimica Acta* 66: 223-239.

[1248] Broecker, W., Bond, D., Klas, M., Clark, E. and McManus, J. 1992: Origin of the North Atlantic's Heinrich events. *Climate Dynamics* 6: 265-273.

[1249] Grousset, F. E., Pujol, C., Labeyrie, L., Auffret, G. and Boelaert, A. 2000: Were the North Atlantic Heinrich events triggered by the behaviour of the European ice sheets. *Geology* 28: 123-126.

[1250] Scourse, J. D., Hall, I. R., McCave, I. N., Young, J. R. and Sugdon, C. 2000: The origin of Heinrich layers: evidence from H2 for European precursor events. *Earth and Planetary Science Letters* 182: 187-195.

[1251] Dowdeswell, J. A., Maslin, M. A., Andrews, J. T. and McCave, I. N. 1995: Iceberg production, debris rafting, and the extent and thickness of Heinrich layers (H-1, H-2) in North Atlantic sediments. *Geology* 23: 310-304.

[1252] Prokopenko, A. A., Williams, D. F., Karabanov, E. B. and Khursevich, G. K. 2001: Continental response to Heinrich events and Bond cycles in sedimentary record of Lake Baikal, Siberia. *Global and Planetary Change* 28: 217-226.

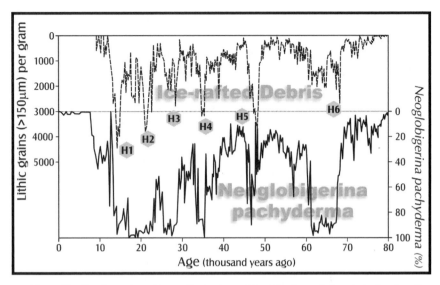

*Figure 26: Abundance of a cold water floating foram (Neoglobigerina pachyderma) showing the coldest periods and temperature fluctuations during the last glaciation. A prolonged cold period (70,000 to 63,000 years ago) was after the Toba eruption (74,000 years ago). There were cold spells (48,000, 36,000 and 27,000 years ago) and extreme cold from 25,000 to 15,000 years ago. The last 6000 years of the interglacial has not had the wild fluctuations of temperature typical of glaciation. Superimposed are periods of ice-rafted debris (Heinrich layers) showing that armadas of icebergs calved from ice sheets are a characteristic of cold times, not warm times.*

Over the last 100,000 years, Dansgaard-Oeschger cycles[1253] and Heinrich Events have been the dominant signal of climate variability over Greenland and the North Atlantic. The succession of cold and warm times associated with these cycles has been documented over the entire Northern Hemisphere, South America, New Zealand, Antarctica, the South Atlantic and the Southern Ocean. Dansgaard-Oeschger cycles affect both hemispheres whereas Heinrich Events are a response to the sporadic surging of the Laurentide and Scandinavian ice sheets.[1254]

To try to use calving of glaciers to show that there is global warming is naïve and misleading. Ice sheets just do not simply grow or melt in response to global temperature. The Northern Hemisphere has had ice sheets for 2.5 million years, whereas Antarctica has had ice sheets for 37 million years. The simple melting or growing of ice sheets in response to temperature cannot

---

[1253] High frequency climate oscillations with 1,000-, 1,450- and 3,000-year cyclicities in Greenland ice cores.

[1254] Leuschner, D. C. and Sirocko, F. 2000: The low-latitude monsoon climate during Dansgaard-Oeschger cycles and Heinrich events. *Quaternary Science Reviews* 19: 243-254.

explain this. Glaciers grow, flow and melt continuously. There is a budget of gains and losses. For example, in 1997 the Jakobshavn Isbræ glacier that feeds a deep ocean fjord on Greenland's west coast changed from slow thickening to rapid thinning, associated with a doubling of the glacier growth speed. One model erroneously suggests that there might have been more summer melt water lubricating the bottom of the glaciers and another model suggests that the floating ice in the fjord at the end of the glacier started to break up allowing more ice to flow into the fjord. In this model, there is no mention of creep as an alternative mechanism of ice flow. A recent work has suggested that the arrival of warm water from the Irminger Sea off Iceland triggered the increase in flow of the Jakobshavn Isbræ glacier.[1255] Again, there was no mention of creep or the fact that ice flow starts at the source of the glacier. The melting of glaciers is not the bellwether of global warming because ocean and atmosphere circulation is not well understood. However, a calving iceberg makes good television for a catastrophic message.

Ice sheets in Greenland and Antarctica can be up to 3 km thick. They are a continent-wide pile of snow that has been compressed into ice. As snow falls on the ice sheet, the ice sheet spreads and thins under its own weight. As the ice flows, small amounts can melt at the edge of the ice flow or glacier and at the leading edge at sea level, the ice breaks off to form icebergs that drift away. However, the water that forms the snow comes from evaporation of ocean water. Hence a temperate rise that gives an increase in evaporation results in an increase in snowfalls, which can result in a fall in sea level. Sea level can also fall if there is a decrease in ice melting or a decrease in the flow rate of ice to the sea. If the ice-melting rate is faster than the accumulation of snow, sea level will rise. Sea level can also rise with expansion of ocean water. Sea level can rise or fall if the shape of the land or the ocean floor is changed. Ice sheets cover about 10% of the world's land surface. If they were to completely melt, global sea level would increase by 70 metres. It's happened before, it will happen again.

There are many stories hidden in the ice sheets. Ice is a sedimentary rock. Ice cores are like layered sedimentary rocks and present us with a record of the past. The latest snowfall is preserved, then is covered by another snowfall which is in turn covered by another snowfall. The uppermost layers tell us about recent events and the deeper layers give us information about events long ago. Ice is a frozen horizontal archive of past climates, volcanic eruptions, dust storms, strong winds, supernoval eruptions, meteor shows,

---

[1255] Holland, D. M., Thomas, R. H., de Young, B., Ribergaard, M. H. and Lyberth, B. 2008: Acceleration of Jakobshavn Isbræ triggered by warm subsurface ocean waters. *Nature Geoscience* 1: 659-664.

cometary dust and solar activity.

Evaporation of water preferentially favours water with lighter oxygen ($O^{16}$). This water falls as snow and accumulates as ice sheets. During glaciation, light oxygen has been removed from the oceans and the water in the oceans becomes relatively concentrated in heavy oxygen ($O^{18}$). So too does life.[1256] Floating marine organisms in glacial times are enriched in $O^{18}$ whereas in warmer times, they are enriched in $O^{16}$. Furthermore, by measuring oxygen in ice, the past temperature at the site of snowfall can be calculated. Trace amounts of other isotopes in ice such as $C^{14}$ and $Be^{10}$ give us a history of the activity of the Sun and the varying input of cosmic radiation.

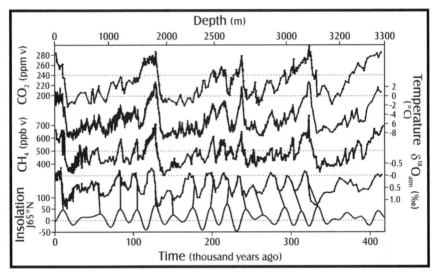

*Figure 27: A story from the last 400,000 years of ice core showing past $CO_2$, calculated temperature, methane ($CH_4$), oxygen isotope composition and solar energy input. These climate cycles show that the present times are no different from numerous past interglacial times.*

Occasional volcanic ash falls can be used as clocks in the ice layers because they result from events that are almost instantaneous. These volcanic ashes can be dated very accurately by a great diversity of isotopic dating techniques, and the chemistry of the volcanic ash can also be used to determine which volcano erupted. Some volcanoes belch out huge amounts of hydrogen sulphide (rotten egg gas) and sulphur dioxide, which oxidise and combine with water to form sulphuric acid, and this is preserved as a sulphate-rich layer in the ice.

---

[1256] Shackleton, N. 1987: Oxygen isotopes, ice volumes and sea level. *Quaternary Science Reviews* 6: 183-190.

Cold windy times add dust and sea spray to snow and these are also preserved in ice. The increased dust during periods of aridity are also preserved in ice. Some 40,000 tonnes of extraterrestrial dust fall to Earth each year and dust from space, comets and meteorites can be used to determine changes in extraterrestrial activity. Each year there is more cometary dust in November, so dust can be used to measure the time. Even supernoval eruptions are recorded in ice because the great flux of radiation creates nitrogen-bearing acids in the upper atmosphere. These have a unique chemical signature. These acids fall to Earth and are preserved in ice. Lakes also preserve dust from comets, meteors, droughts and volcanoes, and a comparison of dust events can be used to show if they were global or local.

Trapped dust and air bubbles in ice can be used to measure the past $CO_2$, methane, ozone and nitrogen compounds.[1257] However, air bubbles move upwards in ice, ice recrystallises as it flows and releases air inclusions. Information on ancient air using air inclusions derives from a greatly smoothed broad-brush modified sample. The Greek silver-lead mining at Lavrion (Attika) was used to finance the Punic Wars. Greek mining in Attika and Roman base metal mining in Spain left its mark on the Northern Hemisphere. Tourists see the aqueducts, roads, walls, arches, theatres and memorials but the greatest Roman legacy is mining and smelting. Romans needed lead as a corrosion-resistant metal for plumbing, shipbuilding and storage of foods and wines. The smelting of lead ores quadrupled the lead content of the atmosphere and contaminated all of Europe. Unusually high quantities of lead occur in the Greenland ice sheet that formed between 600 BC and 300 AD.[1258] This lead has the chemical fingerprint of the mines in the Iberian Pyrite Belt that were exploited by the Celts, King Solomon, Carthaginians and Romans.

Trapped air in Antarctic ice from the Law Dome shows that over the last 2000 years there has been an increase in $CO_2$ (29%), methane (150%) and nitrogen oxide ($N_2O$, 21%), that $CO_2$ stabilised at 310–312 ppmv from 1940–1955 concurrent with a decrease in the growth rate of methane and nitrogen oxides. There was not a decrease in $CO_2$ emissions from 1940–1955 and the oceans probably sucked up $CO_2$ rather than plants.[1259] Although the $CO_2$ stabilisation occurred during a shift from El Niño to La Niña conditions,

---

[1257] Etheridge, D. M., Pearman, G. I. and de Silva, F. 1988: Atmospheric trace-gas variations as revealed by air trapped in an ice core from Law Dome, Antarctica. *Annals of Glaciology* 10: 28-33.

[1258] Hong, S., Candelone, J. P., Patterson, C. C. and Boutron, C. F. 1994: Greenland ice evidence of hemispheric lead pollution two millennia ago by Greek and Roman civilisations. *Science* 265: 1841-1843.

[1259] MacFarling Meure, C., Etheridge, D., Trudinger, C., Steele, P., Langenfelds, R., van Ommen, T., Smith, A. and Elkins, J. 2006: Law Dome $CO_2$, $CH_4$ and $N_2O$ ice core records extended to 2000 years BP. *Geophysical Research Letters* 33: doi:10.1029/2006GL026152.

other shifts from El Niño to La Niña did not result in $CO_2$ stabilisation. A sudden change in nitrogen oxide between 610 and 870 AD is unexplained. The Law Dome ice records lower $CO_2$, methane and nitrogen oxides in the Little Ice Age, probably as a result of the cooler terrestrial biosphere. Many studies show that the gas composition responds to changes in climate and does not initiate climate change.[1260,1261]

Sometimes, the end of glaciers exhibit numerous retreats and advances.[1262] At other times, the ice front remains static.[1263] For example, the retreat and advance at the margins of the ice sheet that covered Scotland can be measured by dating marine microfauna.[1264] Numerous advances and retreats occurred before Scotland was finally free of ice. The load of ice pushed Scotland down, and when the ice melted Scotland started to rise. It is still rising, with numerous earth tremors as the tell-tale sign that rocks are bending and breaking.[1265] Similar results from Ireland show that the British-Irish ice sheet that formed during the last glaciation was advancing and retreating as a response to a large regional climate change.[1266]

The melting of the North American Laurentide ice sheet may have been a rapid event. The ice sheet reached as far south as New York and Ohio and was up to 3 km thick. Some 20,000 years ago it was one of the largest ice sheets on Earth. The processes of rapid melting of ice sheets is poorly understood and observations of modern ice sheets are indirect and not in agreement.[1267] A novel approach using $Be^{10}$ in debris left behind by ice and $C^{14}$ ages of plant material showed that the Laurentide ice sheet retreat was rapid and contributed to a sea level rise of 1.3 metres per century (9000 to 8500 years ago) and 0.7 metres per century (7600 to 6800 years ago).[1268] Such

---

[1260] Ruddiman, W. F. 2003: The anthropogenic greenhouse era began thousands of years ago. *Climate Change* 61: 261-293.

[1261] Bauer, E., Claussen, M., Brovkin, V., and Huenerbein, A. 2003: Assessing climate forcings of the Earth system for the past millennium. *Geophysical Research Letters* 30: doi: 10.1029/2002GL016639.

[1262] Browne, M. A. E., 1980: Late-Devensian marine limits and pattern of deglaciation of the Strathern area, Tayside. *Scottish Journal of Geology* 16: 221-230.

[1263] Everest, J. and Kubik, P., 2006: The deglaciation of eastern Scotland: cosmogenic [10]Be evidence for a late glacial stillstand. *Journal of Quaternary Science* 21: 95-104.

[1264] McCabe, A. M., Clark, P. U., Smith, D. E. and Dunlop, P., 2007: A revised model for the last deglaciation of eastern Scotland. *Journal of the Geological Society, London* 164: 313-316.

[1265] Sissons, J. B., Smith, D. E. and Cullingforth, R. A. 1966: Late-glacial and post-glacial shorelines in south-east Scotland. *Transactions of the British Institute of Geographers* 39: 9-18.

[1266] Sejrup, H. P., Haflidason, H., Aarseth, I., King, E. and Forsberg, C. F. 1994: Late Weichselian glaciation history of the northern North Sea. *Boreas* 35: 231-243.

[1267] Siddall, M., Kaplan, M. R. 2008: A tale of two ice sheets. *Nature Geoscience* 1: 570-571.

[1268] Carlson, A. E., Legrande, A. N., Oppo, D. W., Came, R. E., Schmidt, G. A., Anslow, F. S., Licciardi, J. M. and Obbink, E. A. 2008: Rapid early Holocene deglaciation of the Laurentide ice sheet. *Nature Geoscience* 1: 620-624.

sea level changes are commensurate with the 1 metre per century in post-glacial times (14,000 years to the present).

The East Antarctic Ice Sheet responded to a well-documented period of global warming at 3 Ma.[1269] One group of scientists argue that there was a dry polar climate at this time (e.g. Dry Valleys, East Antarctica)[1270] whereas another group suggest that the East Antarctic Ice Sheet retreated.[1271] Fossil 3 Ma soils in East Antarctica are tundra soils, and river sediments contain fossil plants (grass, shrubs, lichen moss), insects and invertebrates.[1272] These show that East Antarctica then was like modern Siberia. This suggests that Antarctica was considerably warmer than now only at 3 Ma with mean winter temperatures of -12°C and short summer seasons with temperatures up to 5°C.[1273]

The West Antarctic Ice Sheet occupies a basin that extends well below the modern sea level. The ice at the edge of the sheet is susceptible to collapse. Studies of biological chemicals in the Ross Sea sediments show that a large amount of melt water cascaded into the Ross Sea 18,000 years ago, 10,500 years ago, 5500 years ago, 2500 years ago and 1500 years ago.[1274] This has affected Antarctic water circulation with southern waters moving north.[1275] If the West Antarctic Ice Sheet collapses again and melts, then we should not be surprised. It happened many times before industrialisation, and it will happen again. The numerous events of collapse of the West Antarctic Ice Sheet are shown by ages of the raised beaches,[1276] surface exposure age of

[1269] Barrett, P. J., Adams, C. J., McIntosh, W. C., Swisher, C. C. and Wilson, G. S., 1992: Geochronological evidence supports Antarctic deglaciation three million years ago. *Nature* 359: 816-818.

[1270] Denton, G. H., Sugden, D. E., Marchant, D. R., Hall, B. L. and Wilch, T. I. 1993: East Antarctic ice sheet sensitivity to Pliocene climate change from a Dry Valleys perspective. *Geografiska Annaler Stockholm* 75a: 155-204.

[1271] Webb, P. N., Harwood, D. M., McKelvey, B. C., Mercer, J. H. and Stott, L. D. 1984: Cenozoic marine sedimentation and ice volume variation on the East Antarctic craton. *Geology* 12: 287-291.

[1272] Francis, C. E., Haywood, A. M., Ashworth, A. C. and Valdes, P. J. 2007: Tundra environments in the Neogene Sirius Group, Antarctica: evidence from the geological record and coupled atmosphere-vegetation models. *Journal of the Geological Society, London* 164: 317-322.

[1273] Dowsett, H. J., Barron, J. A., Poore, R. Z., Thompson, R. S., Cronin, T. M., Ishman, S. E. and Willard, D. A. 1999: Middle Pliocene palaeoenvironmental reconstruction: PRISM2. *US Geological Survey Open-File Report* 99-535.

[1274] Ohkouchi, N., Toyoda, M., Yokoyama, Y., Miura, H., Chikaraishi, Y., Tokuyama, H. and Kitazato, H. 2006: Massive melting of West Antarctic ice sheet during the latest Pleistocene and Holocene: hydrogen isotopic records of sedimentary biomarkers in Ross Sea. *Geochemica et Cosmochimica Acta* 70: 453.

[1275] Pahnke, K., Goldstein, S. L. and Hemming, S. R. 2008: Abrupt changes in Antarctic Intermediate Water circulation over the past 25,000 years. *Nature Geoscience* 1: 870-874.

[1276] Conway, H., Hall, B. L., Denton, G. H., Gades, A. M. and Waddington, E. D. 1999: Past and future grounding-line retreat of the West Antarctic ice sheet. *Science* 286: 280-283.

glacial deposits[1277] and sea floor sediments.[1278,1279]

## The Arctic

To understand the Arctic, we need to understand the Arctic Ocean. The history of the Arctic Ocean is largely unknown and is deduced from indirect evidence. Evidence from deep drill holes on the Lomonosov Ridge in the Arctic Ocean shows a transition from warm wet times to cooler times at about 55 Ma.[1280] Other work shows that this transition took place at about 34 Ma.[1281] There was a warm period at about 14 Ma, and since then the planet has had large temperature fluctuations during a long-term cooling trend. Ice rafted debris shows that there were icebergs at 45 Ma, some 35 million years earlier than thought, and the Greenland ice sheet expanded at 3.2 Ma. If this can be validated with further work, then the timing of the Arctic cooling coincides with the Antarctic cooling. There is a body of evidence that shows that Antarctic cooling commences at 37 Ma[1282] and hence the Arctic icebergs at 45 Ma are unexplained. This is contrary to other information which shows that Arctic warming takes place when there is Antarctic cooling, and vice versa.[1283,1284] Matters of science are rarely settled.

Arctic climate has always been complex[1285] yet we are constantly bombarded with glib explanations of Arctic climate variability. We often

[1277] Stone, J. O., Balco, G. A., Sugden, D. E., Caffee, M. W., Sass, L. C. III, Cowdery, S. G. and Siddoway, C. 2003: Holocene deglaciation of Marie Byrd Land, West Antarctica. *Science* 299: 99-102.

[1278] Pahnke, K., Zahn, R., Elderfield, H. and Schulz, M. 2003: 340,000-year centennial-scale marine record of Southern Hemisphere climatic oscillations. *Science* 301: 948-952.

[1279] Ohkouchi, N., Toyoda, M., Yokoyama, Y., Miura, H., Chikaraishi, Y., Tokuyama, H. and Kitazato, H. 2006: Massive melting of West Antarctic ice sheet during the latest Pleistocene and Holocene: hydrogen isotopic records of sedimentary biomarkers in Ross Sea. *Geochemica et Cosmochimica Acta* 70: 453.

[1280] Moran, K., Backman, J., Brinkhuis, H., Clemens, S. C., Cronin, T., Dickens, G. R., Eynaud, F., Gattacceca, J., Jakobsson, M., Jordan, R. W., Kaminski, M., King, J., Koc, N., Krylov, A., Martinez, N., Mattiessen, J., McInroy, D., Moore, T. C., Onodera, J., O'Reagan, M., Pälike, H., Rea, B., Rio, D., Sakamoto, T., Smith, D. C., Stein, R., St John, K., Suto, I., Suzuki, N., Takahashi, K., Watanabe, M., Yamamoto, M., Farrell, J., Frank, M., Kubik, P., Jokat, W. and Kristtoffersen, Y. 2006: The Cenozoic palaeoenvironment of the Arctic Ocean. *Nature* 441: 601-605.

[1281] Pälike, H. and Hilgen, F. 2008: Rock clock synchronization. *Nature Geoscience* 1: 282.

[1282] Billups, K. 2008: A tale of two climates. *Nature Geoscience* 1: 294-295.

[1283] Shackleton, N. 2001: Climate change across the hemispheres. *Science* 291: 58-59.

[1284] Blunier, T., Chappelaz, J., Schwander, J., Dällenbach, A., Stauffer, B., Stocker, T. F., Raynaud, D., Jouzel, J., Clausen, H. B., Hammer, C. U. and Johnsen, S. J. 1998: Asynchony of Antarctic and Greenland climate change during the last glacial period. *Nature* 394: 739-743.

[1285] O'Brien, S. R., Mayewski, P. A., Meeker, L. D., Meese, D. A., Twickler, M. S. and Whitlow, S. I. 1995: Complexity of Holocene climate as reconstructed from Greenland ice core. *Science* 270: 1962-1964.

hear in the media about unprecedented warming of the Arctic. A good way to test this claim is to go to Baffin Island, Canada, one of the coldest parts of the world, and measure pollen, fossils and oxygen isotopes. Pollen from some of the Baffin Island lakes shows that it was some 5°C warmer 10,000 and 8500 years ago than now.[1286] This is in accord with previous pollen measurement[1287] and diatom fossils[1288] in Baffin Island lakes. It is also in accord with work from elsewhere in the Canadian Arctic[1289] and it agrees with oxygen isotope measurements of summer temperature in Greenland lakes,[1290] Greenland ice sheet borehole calculations[1291] and fossils[1292] from northern Baffin Bay marine cores. The evidence from Baffin Island and the adjacent Greenland is clear. We are not in an era of unprecedented warming of the Arctic and this part of the Arctic was far warmer a few thousand years ago. The question then remains: was this warming of Baffin Island and Greenland a local feature some 10,000 and 8500 years ago or was it more widespread?

The tree line history of Siberia shows that boreal forest development commenced about 10,000 years ago, advanced until about 7000 years ago, and then retreated to its current position some 4000 to 3000 years ago.[1293] At peak forest development, the summer temperature of northern Siberia would have been 2.5 to 7.0°C warmer than at present. A review of all the research undertaken on physical, chemical and biological proxies in the

[1286] Briner, J. P., Michelutti, N., Francis, D., Miller, G. H., Axford, Y., Wooller, M. J. and Wolfe, A. P., 2006: A multi-proxy lacustrine record of Holocene climate change on northeastern Baffin Island, Arctic Canada. *Quaternary Research* 65: 431-442.

[1287] Kerwin, M. W., Overpeck, J. T., Webb, R. S. and Anderson, K. H. 2004: Pollen-based summer temperature reconstructions for the eastern Canadian boreal forest, subarctic, and Arctic. *Quaternary Science Reviews* 23: 1901-1924.

[1288] Joynt, E. H. III and Wolfe, A. P. 2001: Paleoenvironmental inference models from sediment diatom assemblages in Baffin Island lakes (Nunavut, Canada). *Canadian Journal of Fisheries and Aquatic Sciences* 58: 1222-1243.

[1289] Bradley, R. S. 1990: Holocene palaeoclimatology of the Queen Elizabeth Islands, Canadian high Arctic. *Quaternary Science Reviews* 9: 365-384.

[1290] Murton, J. B., Whiteman, C. A., Waller, R. I., Pollard, W. H., Clark, I. D. and Dallimore, S. R. 2005: Basal ice facies and supraglacial melt-out till of the Laurentide Ice Sheet, Tuktoyaktuk Coastlands, western Arctic Canada. *Quaternary Science Reviews* 24: 681-708.

[1291] Dahl-Jensen, D., Mosegaard, K., Gundestrup, N., Clow, G. D., Johnsen, S. J., Hansen, A. W. and Balling, N. 1998: Past temperatures directly from the Greenland ice sheet. *Science* 282: 268-271.

[1292] Levac, E., de Vernal, A. and Blake, W. 2001: Holocene palaeoceanography of the North Water Polynya. *Journal of Quaternary Science* 16: 353-363.

[1293] MacDonald, G. M., Velichko, A. A., Kremenetski, C. V., Borisova, O. K., Goleva, A. A., Andreev, A. A., Cwynar, L. C., Riding, R. T., Forman, S. L., Edwards, T. W. D., Aravena, R., Hammarlund, D., Szeicz, J. M. and Gattaulin, V. N. 2000: Holocene treeline history and climate change across northern Eurasia. *Quaternary Research* 53: 302-311.

Arctic[1294] shows that 10,000 years ago it was warmer than present conditions for 120 of the 140 measuring sites.

The Arctic was considerably warmer than now from 1920 to 1940. Much of the evidence for this derives from old literature from the former Soviet Union.[1295] In 1905–1933, glaciers retreated, the permafrost boundary shifted 40 km, sea ice decreased, ice drift accelerated, cyclone paths changed, air temperature increased, passage through northern waterways was easier and there was an increase in the temperature and heat content of the Atlantic waters entering the Arctic basin. On the 1934 cruise of the *Persey*, Zubov noted that the glaciers of Jan-Mayen and Spitsbergen were reduced compared with British observations of 1911. Glaciers at Novaya Zemlya were retreating and the ice bridges between some of the Franz-Joseph Islands had melted. The Spitsbergen glaciers were in retreat, as were the Icelandic glaciers in the period 1935–1938. Vasilievsky Island in the Laptev Sea and the Lyakhovsky Islands were composed of ice.

In the Greenland Sea, the area of sea ice between April and August in 1921–1939 was 15 to 20% less than in 1898–1920 and, over the same time periods, the area of the sea ice in the Barents Sea decreased by 12%. Last century, ice often came close to Iceland. During 1915–1940, no ice was observed except for minor ice in 1929. Before the 1920–1940 Arctic warming, the Strait of Jugorsky Shar froze in winter two months later than in other times. Near Disco and Franz-Joseph Land, tides increased by 20 to 30% because of the decreasing amount of ice. In Varde (northeast Norway), the annual air temperature in 1918 was higher that it had been for the previous century. In 1926, it was slightly lower. From 1930, the annual air temperature in the area from Greenland to Cape Tchelsukin was higher than in previous times. In 1934–1935, the Cape Tchelsukin air temperature was 4 to 10°C higher and, at Spitsbergen, 10°C higher than previous times. Annual average temperature measured on the *Fram* cruise (November 1893–August 1895) was 4.1°C lower than during the *Sedov* cruise (November 1937–August 1939) despite being in almost the same areas. At the Tikhaya Station on Franz-Joseph Land, temperatures below -40°C were not measured after 1929 whereas, except 1896, temperatures below -40°C were measured every winter.

Navigation records show that before the 1930s, considerable areas of

[1294] Kaufman, D. S., Ager, T. A., Anderson, N. J., Anderson, P. M., Andrews, J. T., Bartlein, B. J., Brubaker, L. B., Coats, L. L., Cwynar, L. C., Duvall, M. L., Dyke, A. S., Edwards, M. E., Eisner, W. R., Gajewski, K., Geirsdóttir, A., Hu, F. S., Jennings, A. E., Kaplan, M. R., Kerwin, M. W., Lozhkin, A. V., MacDonald, G. M., Miller, G. H., Mock, C. J., Oswald, W. W., Otto-Bliesner, B. L., Porincu, D. F., Rühland, K., Smol, J. P., Steig, E. J. and Wolfe, B. B. 2004: Holocene thermal maximum in the Western Arctic (0 to 180°W). *Quaternary Science Reviews* 23: 529-560.

[1295] http://psc.apl.washington.edu/publication/Arctic_Change/arctic.pdf

the Arctic were closed to ships, even icebreakers. Franz-Joseph Land was inaccessible (*Foka*, 1912) as was Novaya Zemlya (*Ermak*, 1901). Other ships were trapped in the ice and carried out of the Arctic by moving ice (*St Anna*, 1912). In the 1930s, ships were able to travel the North Sea Route, sail around Franz-Joseph Land (*Knipovich*, 1932) and Severnaya (*Sibiryak*, 1932). Many ships that were not icebreakers were able to make two return summer trips to Novaya Zemlya in the 1930s.

Some 12,000–10,000 years ago, warming was concentrated in northwest North America while it was still cool in northeast North America, principally due to the decaying Laurentide Ice Sheet. After the warming of Alaska and northwest Canada, northeast Canada warmed some 4000 years later. Again, this shows that what has been promoted as global climate is not global and that global warming is not just simply a result of humans adding $CO_2$ to the Earth's atmosphere.

There is a warming trend in Alaska but this may not be representative of the Arctic, as it has been influenced by the Pacific Decadal Oscillation event in 1976–1977.[1296] The northern Pacific Ocean is characterised by large sudden climatic shifts over 30-year cycles resulting from the Pacific Decadal Oscillation. There have been 11 such shifts since 1659 and, although Alaska is warming in contrast to the rest of the Arctic, the 1976–1977 Pacific Decadal Oscillation has probably given Alaska its latest temperature rise. Furthermore, there is a strong urban heat island effect in Alaska, even in small villages.[1297]

Greenland is warming. But what does this really mean? Nothing! Greenland temperature records comparing the current warming (1975–2000) with the previous warming (1920–1930) show that the current warming is not unprecedented.[1298] Furthermore, the measurements analysed were from Greenland coastal stations hence might not give the full picture. There are significant differences between the global temperature and the Greenland temperature records in the 1881–2005 period. This again suggests that "global warming" may not be global. Furthermore, while all the decadal averages of the post-1955 global temperature are higher (i.e. warmer climate) than the pre-1955 average, almost all post-1955 temperature averages at Greenland measuring stations are lower (i.e. colder climate) than the pre-

[1296] Gedaloff, Z. and Smith, D. J., 2001: Interdecadal climate variability and regime-scale shifts in Pacific North America. *Geophysical Research Letters* 28: 1515-1518.

[1297] Hinkel, K. M., Nelson, F. E., Klene, A. E. and Bell, J. H. 2003: The urban heat island in winter at Barrow, Alaska. *International Journal of Climatology* 23: 1889-1905.

[1298] Chylek, P., Dubey, M. K. and Lesins, G. 2006: Greenland warming of 1920-1930 and 1995-2000. *Geophysical Research Letters* 13: L11707, doi:10.1029/2006GL26510.

1955 temperature average. The Greenland data shows that the warming of 1920–1930 demonstrates that a high concentration of $CO_2$ and other greenhouse gases is not a necessary driver of warming. A general increase in solar activity[1299] since the 1990s and the sea surface temperature changes of tropical oceans[1300] may well be the contributing factors. The Greenland glacier acceleration observed during 1996–2005[1301] probably occurred previously in the 1920–1940 warming and the Medieval Warm period when temperatures on Greenland were far higher than today.[1302,1303] These were times of higher snowfall and greater accumulation of ice at the heads of the glaciers. And, to finish with the myth of a catastrophic collapse of the Greenland ice sheet, it appears that the Greenland ice sheet is growing.[1304] Ice thickness is increasing in the highlands of Greenland and this will result in an increase in flow rate that will one day reach the ice front as a surge.[1305]

There were dire predictions that in 2008, the Arctic Sea ice would melt and that, for the first time, the North Pole would be ice-free.[1306] If sea ice melted, far more solar energy would be absorbed by water rather than reflected by ice. The slighter warmer times over the last two decades of the 20th Century are similar to those in the early part of the 20th Century (1920–1940). However, some computer models predict that the Arctic will have no sea ice by the end of the 21st Century.[1307] The Arctic summer ice extent is largely determined by variable ocean and atmosphere currents such as the

[1299] Scafetta, N. and West, B. J. 2006: Phenomenological solar contribution to the 1900-2000 global surface warming. *Geophysical Research Letters* 33: 10.1029/2005GL025539.

[1300] Hoerling M. P., Hurrell, J. W. and Xu, T. 2001: Tropical origins for recent North Atlantic climate change. *Science* 292: 90-92.

[1301] Rignot, E. and Kanagaratnam, P. 2006: Changes in the velocity structure of the Greenland ice sheet. *Science* 311: 986-990.

[1302] Dahl-Jensen, D., Mosegaard, K., Gundestrup, N., Clow, G. D., Johnsen, S. J., Hansen, A. W. and Balling, N. 1998: Past temperatures directly from the Greenland ice sheet. *Science* 282: 268-271.

[1303] DeMenocal, P., Ortiz, J., Guilderson, T. and Sarnthein, M. 2000: Coherent high- and low-latitude climate variability during the Holocene warm period. *Science* 288: 2198-2202.

[1304] Zwally, H. J., Schultz, B., Abdalati, W., Abshire, J., Bentley, C., Brenner, A., Bufton, J., Dezio, J., Hancock, D., Harding, D., Herring, T., Minster, B., Quinn, K., Palm, S., Spinhirne, J. and Thomas, R. 2005: ICESA's laser measurements of polar ice, atmosphere, ocean and land. *Journal of Geodynamics* 34: 405-445.

[1305] Abdalati, W., Krabill, W., Frederick, E., Manizade, S., Martin, C., Sinntag, J., Swift, R., Thomas, R., Wright, W. and Yungel, J. 2001: Outlet glacial and marginal elevation changes: near coastal thinning of the Greenland ice sheet. *Journal of Geophysical Research* 106: 33729-33741.

[1306] http://news.nationalgeographic.com/news/2008/06/080620-north-pole.html

[1307] Johannessen, O. M., Bengtsson, L., Miles, M. W., Kuzmina, S. I., Semenov, V. A., Alekseev, G. V., Nagurnyi, A. P., Zakharov, V. F., Bobylev, L. P., Pettersson, L. H., Hasselmann, K. and Cattle, H. P. 2004: Arctic climate changes: observed and modelled temperature and sea-ice variability. *Tellus A* 56: 328-341.

Arctic Oscillation.[1308] Elsewhere in the Arctic there is a very different story. Monthly readings over the last 70 years from 37 Arctic and seven sub-Arctic stations show that the highest temperatures occurred in the 1930s. Even in the 1950s, the Arctic was warmer than fifty years later.[1309,1310] Data from 125 Arctic land stations and numerous drifting buoys shows a strong warming from 1917 to 1937, no warming since 1937 and even a possible slight cooling since 1937.[1311] Eight Danish weather stations show southwestern coastal Greenland has been cooling over the last 50 years.[1312] In addition, three sea-based stations show that sea surface temperatures in the Labrador Sea have fallen over the last 50 years.

In reality, these dire predictions about the Arctic sea ice were hopelessly wrong. On 11 August 2008, the area of the Arctic Ice was 30% greater than on 12 August 2007.[1313] Ice grew in almost every direction, with a large increase north of Siberia. The Northwest Passage has seen a significant increase in ice. Some of the islands in the Canadian Archipelago are surrounded by more ice than they were in the summer of 1980. Sea ice in the Arctic in 2007 and 2008 was some 2 million square kilometres less than in 1979–1980. NASA attributes this decrease in ice to winds pushing the sea ice in the direction of warm ocean currents.[1314] Warm water from the North Atlantic Ocean is flowing deeply into the Arctic Ocean[1315] and much of the prominent Arctic warming and cooling in Greenland during the last half of the 20th Century[1316] is due to natural changes, perhaps to multi-decadal oscillations like the Arctic Oscillation, the Pacific Decadal Oscillation and the El Niño.[1317] The resulting

---

[1308] http://www.jpl.nasa.gov/news.cfm?release=2007-131

[1309] Przybylak, R. 2000: Temporal and spatial variation of surface air temperature over the period of instrumental observations in the Arctic. *International Journal of Climatology* 20, 587-614.

[1310] Przybylak, R. 2002; Changes in seasonal and high-frequency air temperature variability in the Arctic from 1951 to 1990. *International Journal of Climatology* 22, 1017-1033.

[1311] Polyakov, I. V., Bekryaev, R. V., Bhatt, U. S., Colony, R. L., Maskstas, A. P. and Walsh, D. 2003: Variability and trends of air temperature and pressure in the Maritime Arctic 1875-2000. *Journal of Climate* 16, 2067-2077.

[1312] Hanna, E. and Capellan, J. 2003: Recent cooling in southern coastal Greenland and relation with the North Atlantic Oscillation. *Geophysical Research Letters* 30, 10.1029/2002GL015797.

[1313] http://www.theregister.co.uk/2008/08/15/goddard_arctic_ice_mystery/

[1314] http://www.emc.ncep.noaa.gov/research/cmb/sst_analysis/

[1315] Polyakov, I. V., Beszczynska, A., Carmack, E. C., Dmitreko, I. A., Fahrbach, E., Frolov, I. E., Gerdes, R., Hansen, E., Holfort, J., Ivanov, V. V., Johnson, M. A., Karcher, M., Kauker, F., Morison, J., Orvik, K. A. and Schauer, U. 2005: One more step toward a warmer Arctic. *Geophysical Research Letters* 32, doi.10.1029/2005GL023740.

[1316] Dickson, R. R., Osborn, T. J., Hurrell, J. W., Meincke, J., Blindheim, J., Adlandsvik, B., Vinje, T., Alekseev, G., Maslowski, W. and Cattle, H. 2000: The Arctic Ocean response to the North Atlantic Oscillation. *Journal of Climate* 15: 2671-2696.

[1317] Polyakov, I. V. and Johnson, M. A. 2000: Arctic decadal and interdecadal variability. *Geophysical Research Letters* 27: 4097-4100.

thin ice breaks up easily in storms, resulting in an ice retreat on the Canadian side. This retreat in ice is periodic and not permanent.[1318] During 2008, the Antarctic sea ice also increased in area, as it has been doing for the last 30 years.[1319] In some parts of the Arctic, temperature measurements are higher in communities rather than isolated areas because of the urban heat island effect.[1320]

It is not known how the submarine volcanoes along the Gakkel Ridge affect Arctic Ocean sea temperatures. This is where the ocean floor is being pulled apart at the slowest rate of all ocean ridges on Earth. It was expected that there would be little volcanic and geothermal activity but Nature had a surprise. Some 15 active geothermal fields have been identified.[1321] There are deep-water discharges of superheated volatiles, hot springs, lavas and products of large volcanic explosions.[1322] Large deep-water submarine explosive volcanism was only discovered in 2008 and we have no idea how such events have affected the Arctic. However, to have an explosion of a basalt volcano in more than 3 km water depth requires a basalt lava to have at least 13% of dissolved $CO_2$ and for a catastrophic release of $CO_2$ from the lava. The high-pressure cold Arctic waters would dissolve all the $CO_2$ released from the explosion. The large volume of basalt lava is then cooled by circulating Arctic water. This Arctic water is heated. Away from the Gakkal Ridge, there is other unseen volcanic and hot spring activity on the floor of the Arctic Ocean.[1323] One large unseen sub-Arctic Ocean or sub-glacial Antarctic volcano and associated geothermal activity could affect the world's climate by adding heat and $CO_2$ to cold high-pressure polar ocean

[1318] Polyakov, I. V., Bhatt, U. S., Colony, R., Walsh, D., Alekseev, G. V., Bekryaev R. V., Karklin, V. P., Yulin, A. V. and Johnson, M. A. 2003: Long term ice variability in Arctic marginal seas. *Journal of Climate* 16: 2078-2084.

[1319] http://arctic.atmos.uiuc.edu/cryosphere/IMAGES/current.365.south.jpg

[1320] Hinkel, K. M., Nelson, F. E., Klene, A. E. and Bell, J. H. 2003: The urban heat island in winter at Barrow, Alaska. *International Journal of Climatology* 23: 1889-1905.

[1321] Press Release, US National Science Foundation, 28th November 2001. "*Contrary to their expectations, scientists on a research cruise to the Arctic Ocean have found evidence that the Gakkal Ridge, the world's slowest mid-ocean ridge, may be volcanically active. A few years ago submarine exploration of the Arctic Ocean under the polar ice cap found some 15 large geothermal vents along the Arctic fracture zone, and evidence of the recent outflow of lava. We found more hydrothermal activity on this cruise than in 20 years of exploration on the mid-Atlantic Ridge', said Charles Langmuir, scientist from Lamont-Doherty Earth Observatory at Columbia University.*"

[1322] Sohn, R. A., Willis, C., Humphris, S., Shank, T. M., Singh, H., Edmonds, H. N., Kunz, C., Hedman, U., Helmke, E., Jakuba, M., Liljebladh, B., Linder, J., Murphy, C., Nakamora, K., Saato, T., Schlindwein, V., Stranne, C., Taisenfreund, M., Upschurch, L., Winsor, P., Jakobsson, M. and Soule, A. 2008: Explosive volcanism on the ultraslow-spreading Gakkal ridge, Arctic Ocean. *Nature* 453: 1236-1238.

[1323] Snow, J., Hellbrand, E., Jokat, W. and Muhe, R. 2001: Magmatism and hydrothermal activity in Lena Trough, Arctic Ocean. *Transactions of the American Geophysical Union* 82: 193.

waters. Changes in the Arctic waters, ice and volcanoes require observations over much larger time scales.

A drill hole in the Greenland ice sheet was to change the world. By looking at the heavy oxygen ($O^{18}$) and light oxygen ($O^{16}$) in ice over a 250,000-year time span, the temperature at the time of snow fall can be calculated.[1324] This research showed that not only were the well-known orbitally driven climate cycles of 90,000 years of cold and 10,000 years of warm recorded, but there were also cycles of 1500 ± 500 years. These 1500-year cycles were validated by ice drilling in 1987 in Antarctica and by advances and retreats in glaciers of the Arctic, Europe, the Americas, New Zealand and Antarctica. These 1500-year cyclical climate changes were also recorded from sediment cores in all oceans and major seas. Plants responded to these 1500-year climate cycles, as the 14,000 year pollen record from North America shows the cycles.[1325] Archaeologists have shown that humans moved up mountains during warming and down mountains during cooling.

Ice cores from Central Greenland show that climate in Greenland during the last interglacial period was characterised by a series of severe cold periods, which began extremely rapidly and lasted from decades to centuries. The last interglacial period was slighter warmer than the present interglacial period.[1326] Climate change due to natural reasons can therefore be very fast. If the Late 20th Century Warming is due to humans pumping $CO_2$ into the atmosphere, then why has it been warmer at times when there was no human industrialisation?

Computer simulations of Arctic warming and the melting of polar glaciers predict changes of uncertain sizes. Using these computer models, a look backwards at the last interglacial (130,000–116,000 years ago) indicates that the Greenland ice sheet and circum-Arctic ice fields contributed to a sea level rise of 2.2–3.4 metres.[1327] The computer simulations must be wrong as observational data shows a sea level rise of at least twice this figure. The average temperature during this 14,000-year interglacial was at least 5°C

[1324] Dansgaard, W., Johnsen, S. J., Clausen, H. B., Dahl-Jensen, D., Gundestrup, N., Hammer, C. H. and Oeschger, H. 1984: North Atlantic climate oscillations revealed by deep Greenland ice cores. In: Climate Processes and Sensitivity (ed. H. L. Hansen and T. Takahashi). *American Geophysical Union Monograph* 29: 288-298.

[1325] Viau, A. E., Gajewski, K., Fines, P., Atkinson, D. E. and Sawada, M. C. 2002: Widespread evidence of 1500 yr climatic variability in North America during the past 14,000 yr. *Geology* 30: 455-458.

[1326] GRIP members, 1993: Climate instability during the last interglacial period recorded in the GRIP ice core. *Nature* 364: 203-207.

[1327] Otto-Bliesner, B. L., Marshall, S. J., Overpeck, J. T., Miller, G. H., Hu, A. and CAPE Last Interglacial Project members. 2006: Simulating Arctic climate warmth and icefield retreat in the last interglaciation. *Science* 311: 1751-1753.

higher than at present. The margin of the Greenland ice sheet retreated rapidly during the first few thousand years of the Holocene about 10,000 to 8000 years ago. There was a short-lived cooling event approximately 8400 to 8000 years ago associated with a 5 to 7°C fall in mean annual air temperature over the centre of the ice sheet (the GH-8.2 Event). Short-lived warm and cold periods such as the Younger Dryas and GH-8.2 Event are poorly understood.[1328]

Ice cores from Greenland show that the temperature was warmer at 1000 AD. The same cores show two very cold periods at 1550 and 1850 during the Little Ice Age. Temperature was 0.7 to 0.9°C colder than at present. After the Little Ice Age, temperature increased to 1930 and then decreased until 1995 (the year the study ended).[1329] The Greenland continental climate is validated by sediment cores from a fjord in East Greenland that shows cooling after 1300 and very severe and variable climatic conditions from 1630 to 1900.[1330] These temperature changes may not seem great but temperature changes of up to 6°C over the last 8000 years have been recorded off the Alaskan coast by using floating single-celled organisms.[1331] These provide a good indication of sea surface temperature and sea-ice cover, both of which changed in the Little Ice Age. Vegetation studies also showed that the Alaskan land had temperature changes far greater than Greenland. Interglacial summer temperatures were 1 to 2°C higher than now and, in some locations, the summer temperature may have been as much as 5°C warmer.[1332] In northern Quebec, soil deformations from the freezing and thawing of water show that there was severe cold from 1500 to 1900 during the Little Ice Age.[1333]

The Arctic Climate Impact Assessment[1334] stated: "warming is predicted

[1328] Long, A. J., Roberts, D. H. and Dawson, S. 2006: Early Holocene history of the west Greenland Ice Sheet and the GH-8.2 event. *Journal of Quaternary Science Reviews* doi:10.1016/j. quascirev.20005.07.002.

[1329] Dahl-Jensen, D., Mosegaard, K., Gundestrup, N., Clow, G. D., Johnsen, S. J., Hansen, A. W. and Balling, N. 1998: Past temperatures directly from the Greenland ice sheet. *Science* 282: 268-271.

[1330] Jennings, A. E. and Weiner, N. J. 1996: Environmental change in eastern Greenland during the last 1,300 years: Evidence from foraminifera and lithofacies in Nansen Fjord, 68N. *The Holocene* 6: 171-191.

[1331] Darby, D., Bischof, J., Cutter, G., de Vernal, A., Hillaire-Marcel, C., Dwyer, G., McManus, J., Osterman, L., Polyak, L. and Poore, R. 2001: New record shows pronounced changes in Arctic Ocean circulation and climate. *EOS* 82: 601-607.

[1332] Muhs, D. R., Ager, T. A. and Begét, J. E. 2001: Vegetation and palaeoclimate of the last interglacial period, Central Alaska. *Quaternary Science Review* 20: 41-61.

[1333] Kasper, J. N. and Allard, M. 2001: Late Holocene climatic changes as detected by growth and decay of ice wedges on the southern shore of Hudson Straight, northern Quebec, Canada. *The Holocene* 11, 563-577.

[1334] Arctic Climate Impact Assessment, 2005: *Arctic Climate Impact Assessment – Special Report.* Cambridge University Press.

to enhance atmospheric moisture storage resulting in increased net precipitation".

This idea is supported by others.[1335,1336,1337] Trends in rainfall and snowfall over the six largest European drainage basins were calculated for the period 1936–1999. The annual total precipitation over the period decreased,[1338] in agreement with another study.[1339] What makes this study important is that some of the authors had previously argued that total precipitation increased and, on the basis of new data, the conclusions were reversed. This is science in action and demonstrates the uncertainty of science. The implications are profound. Either (a) the theoretical arguments and model predictions that suggest "net high-latitude precipitation increases in proportion to increases in mean hemispheric temperature" or (b) late 20th Century temperatures may not have been much warmer than those in the mid-1930s, or (c) both options may apply.

In the Medieval Warming (900-1300 AD) and the Holocene warming (8500 years ago to the present), temperatures in Greenland were much warmer than now and the ice cap did not disappear.[1340] Coral data shows no substantial sea level rise at these times. Furthermore, during the last interglacial period (130,000–116,000 years ago) when average global temperature was some 6°C warmer for at least 8000 years, Greenland ice remained frozen and preserved DNA from spiders and trees.[1341] This suggests that the Greenland Ice Sheet is far more stable than we are led to believe. The fact that we have ice cores

---

[1335] Peterson, B. J., Holmes, R. M., McClelland, J. W., Vorosmarty, C. J., Lammers, R. B., Shiklomanov, A. I., Shiklomanov, I. A. and Ramsdorff, S. 2002: Increasing river discharge to the Arctic Ocean. *Science* 298: 2171-2173.

[1336] Manabe, S. and Stouffer, R. J. 1994: Multiple-century response to a coupled ocean-atmosphere model to an increase of atmospheric carbon dioxide. *Journal of Climate* 7: 5-23.

[1337] Rahmstorf, S. and Ganopolski, A. 1999: Long-term global warming scenarios computed with efficient coupled climate model. *Climatic Change* 43: 353-367.

[1338] Rawlins, M. A., Willmott, C. J., Shiklomanov, A., Linder, E., Frolking, S., Lammers, R. B. and Vorosmarty, C. J. 2006: Evaluation of trends in derived snowfall and rainfall across Eurasia and linkages with discharges to the Arctic Ocean. *Geophysical Research Letters* 33: 10.1029/2005GL025231.

[1339] Berezovskaya, S., Yang, D. and Kane, D. L. 2004: Compatibility analysis of precipitation and runoff trends over the large Siberian watersheds. *Geophysical Research Letters* 31: 10.1029/2004GL021277.

[1340] Dahl-Jensen, D., Mosegaard, K., Gundestrup, N., Clow, G. D., Johnsen, S. J., Hansen, A. W. and Balling, N. 1998: Past temperatures directly from the Greenland ice sheet *Science* 282: 268-271.

[1341] Willerslev, E., Cappellini, E., Boomsma, W., Nielsen, R., Hebsgaard, M. B., Brand, T. B., Hofreiter, M., Bunce, M., Poinar, H. N., Dahl-Jensen, D., Johnsen, S., Steffensen, J. P., Bennike, O., Schwenninger, J.-L., Nathan, R., Armitage, S., de Hoog, C.-J., Alfimov, V., Christl, M., Beer, J., Muscheler, R., Barker, J., Sharp, M., Penkman, K. E. H., Haile, J., Taberlet, P., Gilbert, M. T. P., Casoli, A., Campani, E. and Collins, M. J. 2007: Ancient biomolecules from deep ice cores reveal a forested southern Greenland. *Science* 317: 111-114.

showing 100,000-year cycles of cooling and warming for the last 800,000 years further shows that in all this time, Greenland ice did not melt in warm interglacials.

And today? A continuous measurement of the Greenland ice sheet from satellites shows that the vast interior of Greenland above 1500 metres above sea level increases in height by 6.4 ± 0.2 cm per year. Below 1500 metres, the elevation decreases by 2.0 ± 0.9 cm per year in accord with the reported thinning of the ice sheet margins. Winter changes are linked to the North Atlantic ocean current changes.[1342] The melting of the Greenland ice sheet margin has increased during the period 1992–2006. The melt rate was even higher in the 1900s, 1930s, 1940s, 1950s and 1960s. This suggests that the current melting is just part of normal climate variability.[1343] The largest modern melting occurred in the 1920s and 1930s, concurrent with the warming of that period. Current changes of the Greenland ice sheet are smaller than changes observed during the 1920–1940 warm period.[1344]

It is hard to see how human-induced $CO_2$ emissions could possibly melt Arctic ice. Unless, of course, one ignores history and a large body of science.

## The Antarctic

Why Antarctica is covered by ice is a major geological problem. Equally problematic is the waxing and waning of the Antarctic ice sheet.[1345] Before Antarctica had ice, the continent was warm, wet and forested. Substantial Antarctic glaciation began at 34 Ma at the Eocene-Oligocene boundary. Although oxygen isotopes in floating organisms show that global temperature was steadily decreasing in the Eocene for more than 10 million years, the Eocene-Oligocene boundary at 33.8 Ma separates one of the warmest periods over the last 100 million years from one of the coldest periods.[1346,1347] This boundary is hard to understand because of the lack of deep sea evidence for

---

[1342] Johannessen, O. M., Khvorostovsky, K., Miles, M. W. and Bobylev, L. P. 2004: Recent ice-sheet growth in the interior of Greenland. *Science* 310: 1013-1016.

[1343] Chylek, P., McCabe, M., Dubey, M. K. and Dozier, J. 2007: Remote sensing of Greenland ice sheet using multispectral near-infra red and visible radiances. *Journal of Geophysical Research* 112: DS24S20, doi: 10.1029/2007 JD008742.

[1344] Chylek, P., Dubey, M. K. and Lesins, G., 2006: Greenland warming of 1920-1930 and 1995-2000. *Geophysical Research Letters* 13: L11707, doi:10.1029/2006GL26510.

[1345] Ingólfsson, O. 2004: Quaternary glacial and climate history of Antarctica. In: *Quaternary Glaciations – Extent and Chronology, Part III* (eds Ehlers, J. and Gibbard, P. L.), 3-43, Elsevier.

[1346] Zachos, J., Pagani, M., Sloan, I., Thomas, E. and Billups, K. 2001: Trends, rhythms, and aberrations in global climate 65 Ma to present. *Science* 292: 686-693

[1347] Pearson, P. N., Ditchfield, P. W., Singano, J., Harcourt-Brown, K. G., Nicholas, C. J., Olsson, R. K., Shackleton, N. J. and Hall, M. A. 2001: Warm tropical sea surface temperatures in the Late Cretaceous and Eocene epochs. *Nature* 413: 481-487.

this cooling. However, by integrating the chemistry of the shells of single-celled bottom-dwelling foraminifera with the chemistry of floating shells, the rapid increase in ice volume and the decrease in sea temperature of 2.5°C can be seen chemically.[1348] The chemical evidence suggests that sea level should have dropped 120 to 135 metres, yet the geological evidence shows that sea level dropped by 55 to 70 metres.[1349] This geological evidence includes the rising of the sea floor as the weight of the water has been removed. If sea level fall is calculated purely on the basis of removal of water into the Antarctic ice sheet, then sea level should have dropped by 82 to 105 metres. This is a very uncertain figure because the loading of ice on Antarctica would have resulting in the sinking of Antarctica and the rising of land elsewhere. Ice volume then expanded and sea level fell. The Antarctic ice sheet at 33.5 Ma was about 25% larger than the current Antarctic ice sheet. This work on the cooling of bottom water is in accord with other work which shows that the tropical sea surface temperature dropped by about 2.5°C at the same time.[1350]

During the Late Miocene (10–5 Ma), there was an expansion of Antarctic ice and the intensification of global wind-driven and atmospheric circulation established a cool strongly layered ocean. In the Pliocene (5–3 Ma), there were warm conditions. Compared to today, ocean temperature was 3°C higher, sea level was some 20 metres higher and atmospheric $CO_2$ was 30% higher.[1351] The Pliocene El Niño conditions were terminated at 2.67 Ma and the Northern Hemisphere ice sheet appeared. Concurrently, Central American volcanoes closed the seaway between the Pacific Ocean and the Caribbean and the Earth was bombarded by an additional dose of cosmic radiation. During these changes over the last 6 Ma, the Subtropical Convergent Zone over Australia has weakened and strengthened as well as migrating north and south of its current position on Australia's southern margin.

Ocean currents changed. By 800,000 years ago, the uplift of the Indonesian Archipelago and the onset of the 100,000-year climate cyclicity

[1348] Katz, M. E., Miller, K. G., Wright, J. D., Wade, B. S., Browning, J. V., Cramer, B. S. and Rosenthal, Y. 2008: Stepwise transition from the Eocene greenhouse to the Oligocene icehouse. *Nature Geoscience* 1: 329-334.

[1349] Nash, T. R., Wilson, G. S., Dunbar, G. B., Barrett, P. J. 2008: Constraining the amplitude of Late Oligocene bathymetric changes in western Ross Sea during orbitally-induced oscillations in the East Antarctic Ice Sheet: (2) implications for global sea-level changes. *Palaeoceanography, Palaeoclimatology, Palaeoecology* 260: 66-76.

[1350] Lear, C. H., Bailey, R., Pearson, P. N., Coxall, H. K. and Rosenthal, Y. 2008: Cooling and ice growth across the Eocene – Oligocene transition. *Geology* 36: 251-254.

[1351] Gallagher, S. J., Greenwood, D. R., Taylor, D., Smith, A. J., Wallace, M. W. and Holgate, G. R. 2003: The Pliocene climate and environmental evolution of southeastern Australia: evidence from the marine and terrestrial realm. *Palaeogeography, Palaeoclimatology, Palaeoecology* 193: 349-382.

again changed ocean currents. Tropical reefs developed further south. The modern ocean currents which transfer heat around the world are relatively recent and ephemeral in the geological record. A better place to look at what planet Earth would be like in a greenhouse is just before the latest glaciation.

The long-accepted explanation for the onset of the ice cover has been that it was due to the Antarctic Circum-Polar Current, which was initiated after Antarctica's neighbouring continental land masses drifted away.[1352] The Circum-Polar current was supposed to prevent warm subtropical waters reaching Antarctica. Changing the shape of continents and the sea floor changed ocean currents. These changes in ocean currents changed Antarctic climate.

Before the opening of the Drake Passage, continent shape had a profound effect on global climate. It still does. The complete Antarctic seaway could not operate until the South Tasman Rise had cleared the Oates Land coast of Antarctica at about 32 Ma.[1353] The Drake Passage possibly opened as late as the Early Miocene around 25 Ma, hence the current could not have been in existence 10 million years earlier.[1354] After the Drake Passage opened, Antarctica decreased in temperature by about 3°C and the Arctic increased by about 3°C.[1355] The opening of the Drake Passage and Oates Land changed the distribution of ocean currents[1356] providing a four-fold increase in the outflow of Antarctic bottom waters.[1357]

Some 5 million years ago, the Earth was warmer than now. There was no polar sea ice and there was no conveyor belt of ocean water circulation. As the Drake Passage widened, the Antarctic Circumpolar Current developed. Not only did Antarctica become isolated from warmer temperate waters but also wind stress caused vertical overturning with colder subsurface waters upwelling to the surface. The thermally isolated continent cooled, more land ice accumulated and sea ice developed in winter. Once sea ice develops, salts are expelled beneath the ice and water beneath the ice becomes more saline and denser. It sinks. This was the development of cold bottom water and a general cooling of the oceans. The formation of bottom water and the

---

[1352] Components of the ancient supercontinent Gondwana.

[1353] Lawver, L. A. and Gahagan, L. M. 2003: Evolution of Cenozoic seaways in the circum-Antarctic region. *Palaeogeography, Palaeoclimatology, Palaeoecology* 198: 11-37.

[1354] Scher, H. D. and Martin, E. E. 2006: Timing and climatic consequences of the opening of Drake Passage. *Science* 312: 428-430.

[1355] Toggweiler, J. R. and Bjornsson, H. 2000: Drake Passage and palaeoclimate. *Journal of Quaternary Science* 15: 319-328.

[1356] Toggweiler, J. R. and Samuels, B. 1995: Effect of Drake Passage on the global thermohaline circulation. *Deep Sea Research Part 1: Oceanographic Research Papers* 42: 477-500.

[1357] Mikolajewicz, U., Maier-Reimer, E., Crowley, T. J. and Kim, K.-Y. 1993: Effect of Drake and Panamanian gateways on the circulation of an ocean model. *Paleoceanography* 8: 409-426.

growth of ice changed the Earth's total energy and infra-red emissions in Antarctica exceeded solar absorption.

After nearly three million years of cooling, the oceans gradually filled with cold bottom water. Winter sea ice formed in the Arctic and cooling continued because infra-red radiation from the Arctic now exceeded solar absorption. Evaporation of water from the sea surface increases by 7.7% with each degree Celsius of sea surface temperature rise. Evaporation requires latent heat, and air from the tropics with potential energy is transferred to the poles. Upwelling in tropical areas reduces the sea surface temperature and this reduces the amount of energy that can be transferred to the poles resulting in polar cooling. It is this evaporation and transfer of energy that balances climate.

This agrees with other data that suggests that the Drake Passage was initially shallow and greatly altered the distribution of life in the Southern Ocean.[1358] Even now, the shallow Drake Passage does not allow deep ocean water to be carried to Antarctica.[1359] There needs to be another explanation for the onset of the Antarctic Ice Sheet. One may be a sudden drop in greenhouse gases sucked out of the atmosphere by the rising Himalayas. As rock decomposes to soil, $CO_2$ is removed from the atmosphere and fixed in soil. This soil is eroded and deposited as sediment. By this process, $CO_2$ is sequestered. However, the chronology of the evolution of the Himalayas shows that this mechanism is probably invalid. Although the Indo-Eurasia collision started at 50 Ma, the uplift of the Himalayas did not begin until 23 Ma.[1360] By then, Antarctica was covered by ice. This 23 million-year date coincides with the monsoonal weathering records from the South China Sea,[1361,1362] Bay of Bengal[1363] and Arabian Sea.[1364]

Let's just get our head around how cold it really is in Antarctica. Winter

---

[1358] Beu, A. G., Griffin, M. and Maxwell, P. A. 1997: Opening of the Drake Passage gateway and Late Miocene cooling reflected in Southern Ocean molluscan dispersal: evidence from New Zealand and Argentina. *Tectonophysics* 281: 83-97.

[1359] Seidov, D. and Maslin, M. 2001: Atlantic Ocean heat piracy and the bipolar climate see-saw during Heinrich and Dansgaard-Oeschger events. *Journal of Quaternary Science* 16: 321-328.

[1360] Clift, P. D., Hodges, K. V., Heslop, D., Hannigan, R., Long, H. V. & Calves, G. 2008: Correlation of Himalayan exhumation rates and Asian monsoon intensity. *Nature Geoscience* 1: 875-883.

[1361] Li, X., Wei, G., Shao, L., Liu, Y., Liang, X., Jian, Z., Sun, M. and Wang, P. 2003: Geochemical and Nd isotope variations in sediments of the South China Sea: a response to Cenozoic tectonism in SE Asia. *Earth and Planetary Science Letters* 211: 207-220.

[1362] Sun, X. and Wang, P. 2005: How old is the Asian monsoonal system. Palaeobotanical records from China. *Palaeogeography, Palaeoclimatology and Palaeoecology* 222: 181-222.

[1363] Clift, P. D. 2006: Controls on the erosion of Cenozoic Asia and the flux of clastic sediment to the ocean. *Earth and Planetary Science Letters* 241: 571-580.

[1364] Derry, L. A. and France-Lanord, C. N. 1996 Neogene Himalayan weathering history and river $^{87}Sr/^{86}Sr$: impact in the marine Sr record. *Earth and Planetary Science Letters* 142: 59-74.

temperatures on its high, cold interior plateau range from -40 to -65°C. The winds are intense. In summer, it warms with temperatures dipping to only -45°C and sometimes rising to -5°C. Even then, the ice reflects virtually all of the Sun's rays back out into space. If we raise the air temperature by 5°C, ice does not melt.

There are huge difficulties in determining whether the Antarctic ice sheet is expanding or contracting. It is claimed that the East Antarctic ice sheet may show signs of the Holocene warming and the West Antarctic ice sheet may be more sensitive to warming.[1365] This claim does not acknowledge that the East and West Antarctic ice sheets occupy basins, that ice flows uphill out of the basins and that such flow could not be due to a very slight change in global temperature.

However, the amount and pressure of water cannot explain movement along the base of the ice.[1366] With continental glaciers, there are highly variable daily and seasonal inputs of water along the base of the glacier. Sometimes water is rapidly released from under a glacier and floods occur. In the Kennicott Glacier of Alaska,[1367] high melt rates do not lead to rapid movement of ice and, in the Greenland Ice Sheet, large pulses of water through cold ice may explain the recent accelerations in movement of the Greenland Ice Sheet.[1368] These examples serve to show that the flow of glaciers is not from lubrication at the base of the ice sheet and that another mechanism must be operating.

However, the world's warming and cooling over the past 150 years have produced change in Antarctica. The huge East Antarctic ice sheet, which contains nearly 90% of the world's ice, has been thickening. European satellites measured the ice sheet's thickness 347 million times between 1992 and 2003 and found it was gaining about 45 billion tonnes of water per year because the planet has warmed enough for snow to fall in the coldest place on Earth.[1369] The Greenland ice sheet has also been thickening at its centre. Both ice sheets are growing at the middle and melting at the edge. This budget may leave us with a global warming sea level gain of about 1.8 mm per year or 10.2 cm per century. Al Gore exaggerated the impact of global

[1365] Remy, F. and Frezzotti, M. 2006: Antarctic ice sheet mass balance. *Comptes Rendus Geosciences* 338: 1084-1097.

[1366] Fountain, A. G. and Walder, J. S. 1998: Water flow through temperature glaciers. *Reviews in Geophysics* 36: 299-328.

[1367] Bartholomaus, T. C., Anderson, R. S. and Anderson, S. P. 2008: Response of glacier basal motion to transient water storage. *Nature Geoscience* 1: 33-37.

[1368] van der Veen, C. J. 2007: Fracture propagation as a means of rapidly transferring meltwater to the base of glaciers. *Geophysical Research Letters* 34: L01501.

[1369] Davis, C., Li, Y., McConnell, J. R., Frey, M. M. and Hanna, E. 2005: Snowfall-driven growth in East Antarctic ice sheet mitigates recent sea-level rise. *Science* 308: 1898-1901.

warming on Antarctica about 50-fold.

Ice sheets undergo marked changes in velocity and shape.[1370,1371] For example, in Antarctica the Whillans ice stream has slowed down[1372] and the Kamb ice stream stopped about 150 years ago.[1373] Although the sub-glacial supply and distribution has been proposed as the cause of the episodic increase and decrease in the rate of flow of glaciers, the shutdown process is debatable.[1374,1375] In many places, sub-glacial lakes exist. High sub-glacial heat flow and sub-glacial lakes may accelerate ice sheet flow but are not the reason for ice flow.

There is an active network of lakes linked to each other beneath two ice streams (Whillans and Mercer) draining the West Antarctic ice field.[1376] These lakes rise and fall producing a rise and fall of the surface ice by 9 metres in 14 places of area 120–500 square kilometres. We do not know the hydrostatic pressure on the lakes created by overlying ice. It is quite possible that the water in these lakes would have been squirted out long ago if the lakes had not been completely sealed by the ice. Before we make predictions about the future of the Antarctic ice sheet and sea level, we need to know far more about what is beneath the ice and why glacial speeds can vary so much.

Antarctica is a continent which does not disappoint. It is always full of surprises. Recent information from NASA shows that the total extent of ice in Antarctica is growing despite localised warming on the Antarctica Peninsula that has caused ice losses.[1377] Antarctica is an ice sheet sitting over an active plate boundary. Antarctica is rising. NASA data also shows that there is a hotspot centred directly over the Wilkins Ice Sheet.[1378] In 325 BC, a volcano erupted on the West Antarctic Ice Sheet. It is still active with a high explosive index. Volcanoes heat large volumes of rock. Hot springs

[1370] Conway, H., Catania, G., Raymond, C. F., Gades, A. M., Scambos, T. A. and Engelhardt, H. 2002: Switch of flow direction in an Antarctic ice stream. *Nature* 419: 465-467.

[1371] Ng, F. and Conway, H. 2004: Fast-flow signature in the stagnated Kamb Ice Stream, West Antarctica. *Geology* 32: 481-484.

[1372] Joughin, I., Tulaczyk, S., Bindschadler, R. and Price, S. F. 2002: Changes in west Antarctic ice stream velocities. Observation and analysis. *Journal of Geophysical Research* 107: doi:10.1029/2001JB001029.

[1373] Retzlaff, R. and Bentley, C. R. 1993: Timing of stagnation of ice stream C, West Antarctica from short-pulse-radar studies of buried surface crevasses. *Journal of Glaciology* 47: 533-561.

[1374] Anandakrishnan, S. and Alley, R. B. 1997: Stagnation of ice stream C, West Antarctica by water piracy. *Geophysical Research Letters* 234: 265-268.

[1375] Price, S. F., Bindschadler, R. A., Hulbe, C. L. and Joughin, I. R. 2001: Post-stagnation behaviour in the upstream regions of Ice Stream C, West Antarctica. *Journal of Glaciology* 47: 283-294.

[1376] Fricker, H. A., Scambos, T., Bindschadler, R. and Padman, L. 2007: An active sub-glacial water system in West Antarctica mapped from space. *Science* 315: 1544-1548.

[1377] http://earthobservatory.nasa.gov/Newsroom/NewImages/images.php3?img_id=17257

[1378] http://data.giss.nasa.gov/cgi-bin/gistemp/do_nmap.py?year_last=2007&month_last=09&sat=4&sst=1&type=trends&mean_gen=1212&year1=1951&year2=2004&base1=1=1951&base2=2006&radius=1200&pol=pol

and geysers are common and very hot gas is released. Volcanic heat creates melt water that lubricates the base of the ice sheet and may increase ice flow towards the sea. For example, Pine Island Glacier on the West Antarctic Ice Sheet is showing a rapid change.[1379]

Recently extinct and active sub-glacial volcanoes beneath the vast expanses of the West Antarctic Ice Sheet occur in Marie Byrd Land, Northern Victoria Land and Mt Early. Large portions of glacial bedrock derived from eruptions such as Mt Early at 16 Ma or recent eruptions such as at 207 ± 240 BC.[1380] Such eruptions provide a long-lived source of heat to the base of ice sheets, melt water flows towards the ocean in complex ice-constrained hydrological systems and fast-flowing ice streams may form. Such features are well known from Iceland where glaciers flow across an active volcanic spreading centre. Furthermore, the heat flux varies. For example, at the Grimsvotn Caldera, the heat flux varied from 1922 to 1991 by a factor of ten and peaked at the 1938 eruption.[1381] Although insignificant compared to the West Antarctic Ice Sheet, the maximum amount of melt water in Iceland is more than half the annual melt water produced from geothermal activity in the catchments of the Siple ice streams, Antarctica.[1382] Subglacial volcanoes on Earth (Iceland,[1383,1384,1385,1386,1387] Canada,[1388,1389,1390] USA,[1391]

[1379] http://www.sciencedaily.com/releases/2008/01/080120160720.htm

[1380] Corr, H. F. J. and Vaughan, D. G. 2008: A recent volcanic eruption beneath the West Antarctic ice sheet. *Nature Geoscience* 1: 122-125.

[1381] Bjornsson, H. and Guomundsson, M. T. 1993: Variations in the thermal output of the sub-glacial Grimsvotn Caldera, Iceland. *Geophysical Research Letters* 20: 2127-2130.

[1382] Joughin, I., Tulaczyk, S., MacAyeal, D. R. and Engelhardt, H. 2004: Melting and freezing beneath Ross ice streams, Antarctica. *Journal of Glaciology* 50: 96-108.

[1383] Nielsen, N. 1936: A volcano under an ice-cap, Vatnajökull, Iceland, 1934-1936. *Geographical Journal* 40: 6-23.

[1384] Jones, J. G. 1969: Intraglacial volcanoes of the Laugarvatn region, south-west Iceland I. *Journal of the Geological Society, London* 124: 197-211.

[1385] Jones, J. G. 1969: Intraglacial volcanoes of the Laugarvatn region, south-west Iceland II. *Journal of Geology* 78: 127-140.

[1386] Brandsdottir, B., Gudmundsson, O., Menke, W. H. and Olafsson, H. 1992: Preliminary results from a refraction profile across the Katla sub-glacial volcano, South Iceland; evidence for a shallow crustal magma chamber within a propagating rift zone. *EOS* 73: 277.

[1387] Gudmundsson, M. T., Sigmundsson, F. and Bjornsson, H. 1997: Ice-volcano interaction of the 1996 Gjalp sub-glacial eruption, Vatnajokull, Iceland. *Nature* 389: 954-957.

[1388] Allen, C. C., Jercinovic, M. J. and Allen, J. S. B. 1982: Subglacial volcanism in north-central British Columbia and Iceland. *Journal of Geology* 90: 699-715.

[1389] Jackson, L. E. 1989: Pleistocene sub-glacial volcanism near Fort Selkirk, Yukon Territory. *Geological Survey of Canada Paper* 89-1E: 251-256.

[1390] Dixon, J. E., Filiberto, J. R., Moore, J. G. and Hickson, C. J. 2002: Volatiles in basaltic glasses from a sub-glacial volcano in northern British Columbia (Canada): implications for ice sheet thickness and mantle volatiles. *Geological Society of London Special Publications* 202: 255-271.

[1391] Porter, S. C. 1987: Pleistocene sub-glacial eruptions on Mount Kea. In: *Volcanism in Hawaii* (eds Decker, R. W., Wright, T. T. and Stauffer, P. H.). US Geological Survey Professional Paper: 587-598.

Antarctica[1392,1393]) and Mars[1394,1395] have long been known. Not only is there a high geothermal gradient, there are constant emissions of very hot gases, hot springs and sporadic eruptions. Yet, every time an iceberg calves from an Antarctic glacier, it is promoted as being evidence of global warming. No other alternative is even considered and there is no questioning as to whether another mechanism might be operating. Very few in the public know that there are active sub-glacial volcanoes in Antarctica and that parts of the Antarctic continent are rising. Both processes destabilise the ice sheets and both processes could lead to a sudden catastrophic collapse of parts of the ice sheet.

Temperatures at and near the North and South Poles are lower now than they were in 1930. The International Geophysical Year of 1957 enabled the establishment of many additional measuring stations in Antarctica. Although public commentary alarms us about the warming of Antarctica, there are different stories in the scientific literature. Spatial analysis of Antarctic meteorological data demonstrates a net cooling on the Antarctic continent between 1966 and 2000.[1396] The measurements from 21 Antarctic surface measuring stations show an average annual continental decline of 0.008°C from 1978 to 1998. These measurements have been validated by satellite measurements since 1979 of infra-red radiation showing a decline in temperature in Antarctica of 0.42°C per decade.[1397] If Antarctica is cooling, then this should be reflected in an increase in the sea-ice season. And this is exactly what has been measured.[1398] Satellite measurements and drill holes show the Antarctic ice sheet is thickening, hence there must be more snow. There are also suggestions based on satellite measurements that the West Antarctic Ice Sheet is actually growing, rather than melting.[1399]

About two thirds of the West Antarctic Ice Sheet has already melted in

---

[1392] LeMasurier, W. E. 1976: Intraglacial volcanoes in Marie Byrd Land. *Antarctic Journal U.S.* 11: 269-270.

[1393] Behrendt, J. C., Blankenship, D. D., Damaske, D. and Cooper, A. K. 1995: Glacial removal of late Cenozoic sub-glacially emplaced volcanic edifices by the West Antarctic ice sheet. *Geology* 23: 1111-1114.

[1394] Hodges, C. A. and Moore, H. J. 1978: The sub-glacial birth of Olympus Mons and its aureoles. *Journal of Geophysical Research* 84: 8061-8074.

[1395] Allen, C. C. 1979: Volcano-ice interactions on Mars. *Journal of Geophysical Research* 84: 8048-8059.

[1396] Doran, P. T., Priscu, J. C., Lyons, W. B., Walsh, J. E., Fountain, A. G., McKnight, D. M., Moorhead, D. L., Virginia, R. A., Wall, D. H., Clow, G. D., Fritsen, C. H., McKay, C. P. and Parsons, A. N. 2002: Antarctic climate cooling and terrestrial ecosystem response. *Nature* 415, 517-520.

[1397] Comision, J. C. 2000: Variability and trends in Antarctic surface temperatures from *in situ* and satellite infra-red measurements. *Journal of Climate* 13: 1674-1696.

[1398] Watkins, A. B. and Simmonds, I. 2000: Current trends in Antarctic sea ice: The 1990s impact on a short climatology. *Journal of Climate* 13: 4441-4451.

[1399] Wingham 2005 (Univ College London) Earth Observation Summit Brussels Feb 2005.

post-glacial times and further melting would raise sea level. The temperature rise in West Antarctica, where the ice is growing, has coincided with cooling in South Antarctica. This serves to show the exciting complexities and contradictory findings that climate research in Antarctica yields. Observations from ice cores suggest that there has been an insignificant change in Antarctic snowfall since 1957, thereby suggesting that Antarctic snowfall is not mitigating sea level rise despite warming of the overlying atmosphere between 1976 and 1998.[1400] Flow measurements of ice in the Ross ice streams of Antarctica show that movement of ice has slowed or halted, allowing the ice to thicken by added snow.[1401] It was also shown that the harsh dry valleys of Antarctica, long considered the bellwether for global climate change, have grown noticeably cooler since the 1980s.[1402]

In East Antarctica, the use of chemicals created in the cosmos[1403] can be used to show the magnitude and timing of the East Antarctic ice sheet retreat since the peak of the last glaciation, some 20,000 years ago. And what was found? The ice sheet thickness has changed very little since the peak of the last glaciation, hence East Antarctica's role in sea level change is negligible.[1404]

Armadas of icebergs from the Antarctic ice sheet show pulses on millennial scales between 20,000 and 74,000 years ago during the last glaciation. These coincided with strong increases in North Atlantic deep-water production and short periods of warming in the North Atlantic during the last glaciation.[1405] These armadas of icebergs were not a result of warming, they occurred during very cold times.

The Ross Ice Shelf has disappeared and reformed many times[1406] due

[1400] Monaghan, A. J., Bromwich, D. H., Fogt, R. L., Wang, S.-H., Mayewski, P. A., Dixon, D. A., Elaykin, A., Frezzotti, M., Goodwin, I., Isaksson, E., Kaspari, S. D., Morgan, V. I., Oerter, H., van Ommen, T. D., van der Veen, C. J. and Wen, J. 2006: Insignificant change in Antarctic snowfall since the International Geophysical Year. *Science* 313: 827-831.

[1401] Joughin, I. and Tulaczyk, S. 2002: Positive mass balance of the Ross ice streams, West Antarctica. *Science* 295: 476-480.

[1402] Doran, P. T., McKay, C. P., Clow, G. D., Dana, G. L., Fountain, A. G., Nylen, T. and Lyons, W. B. 2002: Valley floor climate observations from McMurdo dry valleys, Antarctica, 1986-2000. *Journal of Geophysical Research* 107: doi:10.1029/2001JD002045.

[1403] $Be^{10}$ and $Al^{26}$

[1404] Mackintosh, A., White, D., Fink, D., Gore, D. B., Pickard, J. and Fanning, P. C. 2007: Exposure ages for mountain dipsticks from MacRobertson Land, East Antarctica, indicating little change in ice-sheet thickness since the Last Glacial Maximum. *Geology* 35: 551-554.

[1405] Kanfoush, S. L., Hodell, D. A., Charles, C. D., Guilderson, T. P., Mortyn, P. G. and Ninnemann, U. S. 2000: Millennial-scale instability of the Antarctic ice sheet during the last glaciation. *Science* 288: 1815-1819.

[1406] Putsey, C. J., Murray, J. W., Apopleby, P. and Evans, J. 2006: Ice shelf history from petrographic foraminiferal evidence, Northwest Antarctic Peninsula. *Quaternary Science Reviews* 25: 2357-2379.

to natural cycles. Over the past million years, temperatures on the Ross Ice Shelf were often 2° to 3°C warmer than now despite atmospheric $CO_2$ never rising above 300 ppmv, compared to today's 385 ppmv.[1407] The Larsen A and Larsen B Ice Shelves underwent widespread break-up about 5000 years ago. The break-up and regrowth probably took place over a few centuries and the maximum ice shelf limit may date only from the Little Ice Age, a few hundred years ago.[1408]

In 1998 the Larsen B Ice Shelf, on the eastern side of the Antarctic Peninsula, lost a large amount of ice. This happened again in March 2002. The media frenzy frightened people into believing that Antarctica was melting and there would be a global sea level rise. There was nothing abnormal about Antarctic temperature. Over five weeks, the 220-metre thick Larsen B Ice Shelf disintegrated into a large number of icebergs. Lateral crevasses up to 6 metres wide opened up. The calving of icebergs was at the end of summer after a season of 24 hours sunlight, warmer ocean currents and warmer winds. Lateral fracturing, the production of crevasses and the partial collapse of ice sheets is the norm, not the exception, and ice sheet loss is part of the glacial budget. This demonstrates what we already know: ice sheets are dynamic. In August 2002, only five months after the collapse of some of the Larsen B Ice Shelf, NASA reported that between 1979 and 1999, Antarctic sea ice had been increasing in all areas except the Antarctic Peninsula. The losses and gains of sea ice showed an increase in area of 2.6 million square kilometres.

Contrary to the horror predictions about global warming-induced mass wastage of the Antarctic ice sheet leading to sea level rise that destroys coastal lowland communities, there is contrary evidence.[1409] By using satellite altimeter echoes to determine the thickness of the Antarctic ice sheet from 1992 to 2003, it was shown that the ice sheet was growing by 27 ± 29 billion tonnes per year, sufficient to lower sea level by 0.08 mm per year as water is extracted from the oceans to produce accumulating snow. This study dealt with 85% of the East Antarctic ice sheet and 51% of the West Antarctic ice sheet accounting for 72% of the grounded ice sheet. It is the season for alarmist estimates to be revised. Estimated sea level rise from the melting of alpine valley glaciers and icecaps will only be 5 cm by 2100 AD,[1410] about half of previous estimates.

---

[1407] Joughlin, I. and Tulaczyk, S. 2002: Positive mass balance of the Ross Ice Streams, west Antarctica. *Science* 295: 476-480.

[1408] Pudsey, C. J. and Evans, J. 2001: First survey of Antarctic sub-ice shelf sediments reveals mid-Holocene ice shelf retreat. *Geology* 29: 789-790.

[1409] Wingham, D. J., Shepherd, A., Muirhead, A. and Marshall, G. 2006: Mass balance in the Antarctic ice sheet. *Philosophical Transactions of the Royal Society* A364, 1627-1635.

[1410] Raper, S. and Brathwaite, R. 2006: Low sea level rise projections from mountain glaciers and ice caps under global warming. *Nature* 439: 311-313.

The initial analyses of the Vostok ice core used samples spaced at intervals of hundreds of years. The initial conclusions were that high $CO_2$ in the atmosphere led to high air temperatures. However, with far more detailed measurements on the scale of decades over a 250,000-year ice core record and a correlation of a 35,000-year ice core record from Taylor Dome, it was shown that high air temperatures are followed some 400 to 1000 years later by a high atmospheric $CO_2$ content.[1411,1412] More recent work, using argon isotopes in Antarctic ice cores of just one temperature rise, shows that $CO_2$ increased 200 to 800 years after that particular temperature rise.[1413] During the last 420,000 years there have been massive temperature changes, and a rise in $CO_2$ concentration follows air temperature increase by about 800 years and it is only after a cooling event that $CO_2$ decreases. This is no surprise, as $CO_2$ is more soluble in cold water than warm water.

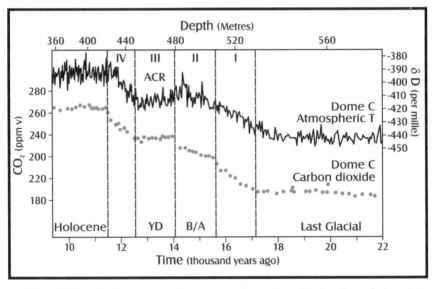

*Figure 28: Detailed determinations of temperature and atmospheric $CO_2$ from Dome C Antarctic ice core showing that at about 14,500 years ago temperature increased and at about 14,000 years ago atmospheric $CO_2$ increased. Other stepwise increases in $CO_2$ are unrelated to temperature. The hypothesis that human emissions of $CO_2$ create global warming is invalidated with this data which shows that a rise in atmospheric $CO_2$ follows temperature rise.*

[1411] Fischer, H., Wahlen, M., Smith, J., Mastroianni, D. and Deck, B. 1999: Ice core record of atmospheric $CO_2$ around the last three glacial terminations. *Science* 283: 1712-1714.

[1412] Mudelsee, M. 2001: The phase relations among atmospheric $CO_2$ content, temperature and global ice volume over the past 420 ka. *Quaternary Science Reviews* 20: 583-589.

[1413] Caillon, N., Severinghaus, J. P., Jouzel, J., Barnola, J.-M., Kang, J. and Lipenkov, V. Y. 2003: Timing of atmospheric $CO_2$ and Antarctic temperature changes across Termination III. *Science* 299: 1728-1731.

Ice drilling by the European Project for Ice Coring in Antarctica (EPICA) at Concordia Station, Dome C, Antarctica gives another look back in time over some 800,000 years. Eight icehouse-greenhouse cycles were recognised.[1414] Antarctic temperatures have been reconstructed[1415] as have the $CO_2$, methane and nitrous oxide content.[1416,1417]    More detailed measurements[1418,1419] concentrated on the history of $CO_2$ in ancient air trapped in the ice. Popular paradigms were destroyed. At 800,000 and 650,000 years ago, atmospheric $CO_2$ dropped below 180 ppmv yet temperature was unchanged. Temperature and $CO_2$ are not connected. Furthermore, there was a long-term trend in $CO_2$ which rose by 25 ppmv from 800,000 to 400,000 years ago and then fell by 15 ppmv thereafter. Again, a disconnection between temperature and $CO_2$. Even more intriguing was that the methane levels in trapped air changed from 100,000- to 20,000-year cycles. By contrast, temperature and $CO_2$ showed 100,000-year cycles.

Observations of the Antarctic ice sheet suggest that the East Antarctic Ice Sheet is nowadays more or less in balance, whereas the West Antarctic Ice Sheet exhibits some changes likely to be related to climate change and its negative balance. This conclusion, from altimetry surveying, suggests that

[1414] Wolff, E. W., Chappellaz, J., Fischer, H., Kull, C., Miller, H., Stocker, T. F. and Watson, A. J. 2004: The EPICA challenge to the Earth system modeling community. *EOS* 85: doi:10.1029/2004EO380003.

[1415] Augustin, L., Barbante, C., Barnes, P. R. F., Barnola, J.-M., Bigler, M., Castellano, M., Cattani, O., Chappelaz, J., Dahl-Jensen, D., Delmonte, B., Dreyfus, G., Durand, G., Falourd, S., Fischer, H., Flückiger, J., Hansson, M. E., Huybrechte, P., Jugie, G., Johnsen, S. J., Jouzel, J., Kaufmann, P., Kipfstuhl, J., Lambert, F., Lepinkov, V. Y., Littot, G. C., Longinelli, A., Lorrain, R., Maggi, V., Masson-Delmotte, V., Miller, H., Mulvaney, R., Oerlemans, J., Oerter, H., Orombelli, G., Parennin, F., Peel, D. A., Petit, J.-R., Reynaud, D., Ritz, C., Ruth, U., Schwander, J., Siegenthaler, U., Souchez, R., Stauffer, B., Peder Steffensen, J., Stenni, B., Stocker, T. F., Tabacco, I. E., Udisti, R., van de Wal, R. S. W., can den Broeke, M., Weiss, J., Wilhelms, F., Winther, J.-G., Wolff, E. W. and Zucchelli, M. 2004: Eight glacial cycles from Antarctic ice. *Nature* 429: 623-628.

[1416] Siegenthaler, U., Stocker, T. F., Monnin, E., Lüthi, D., Schwander, J., Stauffer, B., Raynaud, D., Barnola, J.-M., Fischer, H., Masson-Delmotte, V. and Jouzel, J. 2005: Stable carbon cycle-climate relationship during the Late Pleistocene. *Science* 310: 1313-1317.

[1417] Spahni, R., Chappellaz, J., Stocker, T. F., Loulergue, L., Hausammann, G., Kawamura, K., Flückiger, J., Schwander, J., Raynaud, D., Masson-Delmotte, V. and Jouzel, J. 2005: Atmospheric methane and nitrous oxide of the Late Pleistocene from Antarctic ice cores. *Science* 310: 1317-1321.

[1418] Lüthi, D., Le Floch, M., Bereiter, B., Blunier, T., Barnola, J.-M., Siegenthaler, U., Raynaud, D., Jouzel, J., Fischer, H., Kawamura, K. and Stocker, T. F. 2008: High-resolution carbon dioxide concentration record 650,000-800,000 years before present. *Nature* 453: 379-383.

[1419] Loulergue, L., Schilt, A., Spahni, R., Masson-Delmotte, V., Blunier, T., Lemieux, B., Barnola, J.-M., Raynaud, D., Stocker, T. F. and Chappellaz, J. 2008: Orbital and millennial-scale features of atmospheric $CH_4$ over the past 800,000 years. *Nature* 453: 383-386.

there is no evidence for the dramatic melting of the Antarctic ice sheet.[1420] The West Antarctic Ice Sheet has a significant decrease over the last 2.5 Ma and there are interannular variations of both the Antarctic and Greenland ice sheets.[1421] Changes in the evolution of the Antarctic ice sheet are only to be expected. After all, everything else on Earth is evolving

There are large uncertainties regarding Antarctica's current and future contribution to sea level change. Although warming may increase snowfall in the continent's interior,[1422] this warming may also accelerate glacier discharge at the coast where there may be warmer air and water.[1423] The use of radar has shown that in East Antarctica, glacier losses (Wilkes Land) and gains (Filchner and Ross Ice Shelves) combine to show ice losses equal ice gains.[1424] In West Antarctica, there are widespread ice losses along the Bellingshausen and Amundsen Seas which increased by about 60% in a decade whereas in the Peninsula losses increased by 140% in the same decade. Losses are concentrated along narrow channels occupied by disgorging glaciers and are caused by modern and ancient glacier acceleration. These changes in glacier flow rate have a major impact on the amount of ice in Antarctica.

As with the Arctic, the pushing of water by wind into Antarctic waters changes temperature. A period of warming in West Antarctica in the late 1980s and early 1990s followed the delivery of Circumpolar Deep Water to the inner continental shelf via a submarine trough. Variations in the influx of water are related to seasonal and decadal wind changes.[1425]

The Antarctic warms and cools. There have been 200- to 300-year cycles recognised from fossils[1426] and ice core measurements[1427] and these cycles are

[1420] Rémy, F. and Frezzotti, M. 2007: Antarctic ice sheet mass balance. *Comptes Rendus Geosciences* 338: 1084-1097.

[1421] Ramillien, G., Lombard, A., Cazenave, A., Ivins, E. R., Llubes, M., Remy, F. and Blancale, R. 2006: Interannual variations of the mass balance of the Antarctica and Greenland ice sheets from GRACE. *Global and Planetary Change* 53: 198-208.

[1422] Davis, C. H., Yi, Y., McConnell, J. R., Frey, M. M. and Hanna, E. 2005: Snowfall-driven growth in East Antarctica Ice Sheet mitigates recent sea level rise. *Science* 308: 1898-1901.

[1423] Rignot, E. 2006: Changes in ice dynamics and mass balance of the Antarctic ice sheet. *Philosophical Transactions of the Royal Society A* 364: 1637-1656.

[1424] Rignot, E., Bamber, J. L., van den Broeke, M., Davis, C., Li, Y., van de Berg, W. and van Meijgaard, E. 2008: Recent Antarctic ice mass loss from radar interferometry and regional climate modelling. *Nature Geoscience* 1: 106-110.

[1425] Jacobs, S. 2006: Observations of change in the Southern Ocean. *Philosophical Transactions of the Royal Society* 364A: 1657-1681.

[1426] Domack, E. W., Mashiotta, T. A., Burkley, L. A. and Ishman, S. E. 1993: 300 year cyclicity in organic matter preservation in Antarctic fjord sediments. In: The Antarctic paleoenvironment: a perspective on global change (eds Kennett, J. P. and Warnke, D. A.). *American Geophysical Union Antarctic Research Series* 60: 265-272.

[1427] Jones, P. D., Marsh, R., Wigley, T. M. L. and Peel, D. A. 1993: Decadal timescale links between Antarctic Peninsula ice-core oxygen-18, deuterium and temperature. *The Holocene* 3: 14-26.

linked to solar cycles which are seen by linked changes in the atmosphere, oceans and life.[1428] An indication of the fluctuating climate is the temperature-driven migration of penguin rookeries.[1429] The Arctic and Antarctic are out of phase.[1430] Antarctic temperatures increase while there is cooling in Greenland on millennial scales.[1431] These variations in polar climate are best recorded during glacial times.[1432] This shows that global climate changes are not quite global. Warming in Antarctica and the circumpolar current leads to changes in the North Atlantic some 1500 years later.[1433] This is supported by measurements that show that over the last 90,000 years there has been a correlation between methane in Greenland Ice Sheet Project 2 (GRIP2) and Byrd ice cores.[1434] Comparison of temperatures since the peak of the last glaciation in Greenland[1435] and Antarctica[1436] confirm this lack of synchronous climate change. In the past as climates changed, Antarctica was 1000–2000 years ahead of Greenland.[1437] The origin of this polar see-saw is unknown. It may be driven by changes in high-latitude or near-equatorial sea surface features. Other models suggest it is driven by a deep ocean conveyor because North Atlantic deep water cannot enter the Southern Ocean at depths shallower than the bottom of the Drake Passage. Hundreds of years are required to warm high-latitude areas and the see-saw operates like a pendulum between the two poles.[1438]

---

[1428] Leventer, A., Domack, E. W., Ishman, S. E., Brachfield, S., McClennan, C. E. and Manly, P. 1996: Productivity cycles of 200-300 years in the Antarctic Peninsula region: understanding linkages among the sun, atmosphere, ocean and biota. *Geological Society of America Bulletin* 108: 1626-1644.

[1429] Baroni, C. and Orombelli, G. 1994: Abandoned penguin rookeries as Holocene paleoclimate indicators in Antarctica. *Geology* 22: 23-26.

[1430] Rosqvist, G. C. and Schuber, P. 2003: Millennial-scale climate changes on South Georgia, Southern Ocean. *Quaternary Research* 59: 470-475.

[1431] Shackleton, N. 2001: Climate change across the hemispheres. *Science* 291: 58-59.

[1432] Blunier, T., Chappelaz, J., Schwander, J., Dällenbach, A., Stauffer, B., Stocker, T. F., Raynaud, D., Jouzel, J., Clausen, H. B., Hammer, C. U. and Johnsen, S. J. 1998: Asynchrony of Antarctic and Greenland climate change during the last glacial period. *Nature* 394: 739-743.

[1433] Leuschner, D. C. and Sirocko, F. 2000: The low-latitude monsoon climate during Dansgaard-Oeschger cycles and Heinrich Events. *Quaternary Science Reviews* 19: 243-254.

[1434] Blunier, T. and Brook, E. J. 2001: Timing of millennial-scale climate change in Antarctica and Greenland during the last glacial period. *Science* 291: 109-112.

[1435] Humlum, O. 1999: Late Holocene climate in central west Greenland: Meteorological data and rock glacier isotope evidence. *The Holocene* 9: 581-594.

[1436] Ingólfsson, O., Hjort, C. and Humlum, O. 2003: Glacial and climate history of the Antarctic Peninsula since the last glacial maximum. *Arctic, Antarctic and Alpine Research* 35: 175-186.

[1437] Wunsch, C. 2003: Greenland-Antarctic phase relations and millennial time scale climate fluctuations in the Greenland ice-cores. *Quaternary Science Reviews* 22: 1631-1646.

[1438] Seidov, D. and Maslin, M. 2001: Atlantic ocean heat piracy and the bipolar climate see-saw during Heinrich and Dansgaard-Oeschger events. *Journal of Quaternary Science* 16: 321-328.

## Alpine valley glaciers

In Europe, it is the tradition to go on a Sunday walk (*Spaziergang*). In 2003, Frau Ursula Leuenberger and her husband were walking in the mountains high above Thun (Switzerland) at the foot of the ice at Schnidejoch Pass. The ice had retreated slightly in the long hot summer of 2003 and Frau Leuenberger found a birch bark archer's quiver some 4700 years old. Later archaeological work discovered hundreds of relics from the Neolithic Era, the Bronze Age and Roman times, even part of a shoe from a time at the end of the Medieval Warming. The Schnidejoch Pass was a long-forgotten short cut across the Alps. At times it must have been ice-free, at other times advancing ice covered the archaeological relics. It is clear that alpine valley glaciers waxed and waned, that the Schnidejoch Pass was a useful trade route in warm times and that climate change had a significant effect on trade, travel and the lives of Europeans. The various opening times of the Schnidejoch Pass are not exactly coincidental with warm times because, although temperature may have increased in previous warmings, ice needs a large amount of heat to melt and there is a lag between warming and the melting of ice.

In Italy, the Ghiacciaio del Calderone glacier in the Apennine Mountains has lost half its mass since 1794 with a slow loss from 1794 to 1884 and then rapid melting until 1990.[1439] This ice loss at Ghiacciaio del Calderone took place in the Little Ice Age and hence could not possibly be related to modern global warming derived from human emissions of $CO_2$. Other European glaciers, such as Mont Blanc, just refuse to melt.[1440]

The good news is that the alpine valley glaciers are not retreating. Measurements of retreats and advances from glaciers in the period 1946–1995 for 246 glaciers show that there is no sign of any recent global trend towards increased glacier melting.[1441] Measurements were made in Western Europe, North America and the former USSR. There were fewer measurements in other parts of the world. The data is obtained from stakes and snow pits and this data will be replaced with geodetic and remote sensing techniques.

Glaciers retreat and advance. They are constantly melting. That's what they do. Glacial retreat or advance may be a lagging indicator of temperature

---

[1439] D'Orefice, M., Pecci, M., Smiragalia, C. and Ventura, R. 2000: Retreat of Mediterranean glaciers since the Little Ice Age: Case study of Ghiacciaio del Calderone (Central Apennines, Italy). *Arctic, Antarctic and Alpine Research* 32: 197-201.

[1440] Vincent, C., Le Meur, U., Six, D., Funk, M., Hoelzle, M. and Preunkert, S. 2007:Very high-elevation Mont Blanc glaciated areas not affected by the 20th Century climate change. *Journal of Geophysical Research* 112: doi: 10.1029/2006JD007407.

[1441] Braithwaite, R. J. 2002: Glacier mass balance: the first 50 years of international monitoring. *Progress in Physical Geography* 26: 76-95.

change. However, the growth and retreat of glaciers is not necessarily a simple indicator of modern temperature change. Glaciers gain ice during increased precipitation, which can occur during warmer times, and glaciers can calve icebergs as a result of plastic flow of ice. This plastic flow takes place in cold and warm times and results from processes that started thousands of years before. Glacier response to climate change varies over time depending upon their shape, location and elevation. They also respond to changes in precipitation, humidity and cloudiness. Glacier modelling suggests that with rapid warming (0.04°C per year) most alpine valley glaciers would disappear after a hundred years. However, a slower warming (0.01°C per year) with the likely associated increase in precipitation, would mean glaciers would only decrease in volume by 10 to 20% after a hundred years.[1442]

The major alpine valley glaciers of the world all started to shrink around 1850, half of them stopped shrinking around 1940 and some even started to grow after 1940. The Arctic ice field is shrinking and thinning. In the Arctic, there are 18 glaciers with a long observation history. More than 80% of them have lost mass since the Little Ice Age.

The Furtwängler Glacier on Mt Kilimanjaro (5895 metres altitude) has become the icon of human-induced global warming, having shrunk from more than 12 square km to 2.2 square km since 1880.[1443] The Furtwängler Glacier is a remnant of a bigger ice sheet and has 40-metre vertical ice cliffs at its termination. Glacial advance and retreat has a lag factor and we need to be mindful that the Little Ice Age ended in 1850. The shrinking could not be due to 20th Century human emissions of $CO_2$. It is due to the mountain air becoming drier since 1880. Furthermore, it is well known that solar radiation and sublimation, not air temperature, are the principal factors for the loss of ice from tropical glaciers. Blades of ice up to 2 metres high are formed by sublimation driven by direct solar radiation and the direction of the blade operates like a compass because it indicates the position of the Sun at noon.

At Kilimanjaro, the dry air resulted in less precipitation and melting ice was not replaced by snow, thereby pushing the ice into disequilibrium, with the melting rate exceeding the precipitation rate. This was not a slow process, it was a rapid and drastic drop in atmospheric moisture at the end of the 19th

[1442] Oerlemans, J., Anderson, B., Hubbard, A., Huybrechts, P., Jóhannesson, T., Knap, W. H., Schmeits, M., Stroeven, A. P., van de Wal, R. S. W., Wallinga, J. and Zuo, Z. 1998: Modeling the response of glaciers to climate warming. *Climate Dynamics* 14: 267-274.

[1443] Mahaney, W. C. 1990: *Ice on the equator: Quaternary geology of Mount Kenya*. W. Caxton Ltd.

Century, possibly due to land clearing at lower altitude.[1444] Between 1953 and 1976, some 21% of the ice area disappeared. However, this was in the cooling period from 1940 to 1976. The retreat of ice slowed after 1979. Global warming has not contributed to the retreat of ice on Mt Kilimanjaro.

The history of ice on Mt Kilimanjaro shows that the ice comes and goes. A study of dust and oxygen isotopes in ice cores from Kilimanjaro[1445] shows that there were three periods of abrupt climate change (8500, 5200 and 4000 years ago). The African Humid Period from 11,000–4,000 years ago has a chemical fingerprint of warmer conditions and less aerosols (magnesium, calcium, sulphate and nitrate). Solar radiation was higher at this time because of the Earth's orbit (precession). Ice was also forming on Kilimanjaro at this time. During much of this period, tropical African lakes rose as much as 100 metres above present levels, with Lake Chad expanding from about 17,000 square kilometres to 330,000 square kilometres, the size of the modern Caspian Sea.[1446,1447,1448] Rainfall over most of East Africa was almost twice the current rainfall.[1449] Within the African Humid Period was a second humid period from 6500 to 5000 years ago when it was drier than before and wetter than now. Lake levels changed and vegetation changed in response to the change in climate.

During this climate change at about 5300 years ago, the hierarchical societies of the Nile Valley and Mesopotamia started and the Neolithic settlements of the inner desert of Arabia were abandoned.[1450] Some 4000 years ago, lake levels dropped as conditions became cooler, drier and windier. This is represented by a major dust layer in the Kilimanjaro ice. A period of intense drought started some 4000 years ago. This 300-year drought was experienced in northern and tropical Africa, the Middle East and western

---

[1444] Kaser, G., Hardy, D. R., Mölg, T., Bradley, R. S. and Hyera, T. M. 2004: Modern glacier retreat on Kilimanjaro as evidence of climate change: observations and facts. *International Journal of Climatology* 24: 329-339.

[1445] Thompson, L. G., Mosley-Thompson, E., Davis, M. E., Henderson, K. A., Brecher, H. H., Zagorodnov, V. S., Mashiotta, T. A., Lin, Ping-Nan, Mikhalenko, V. N., Hardy, D. R. and Beer, J. 2002: Kilimanjaro ice core records: Evidence of Holocene climate change in tropical Africa. *Science* 298: 589-593.

[1446] Street, F. A. and Grove, A. T. 1976. Environmental and climatic implications of late Quaternary lake-level fluctuations in Africa. *Nature* 261: 385-390.

[1447] Grove, A. T. and Warren, A. 1968: Quaternary landforms and climate on the South side of the Sahara. *Geographical Journal* 134: 194-208.

[1448] Gasse, F. 2000: Hydrological changes in the African tropics since the Last Glacial Maximum. *Quaternary Science Reviews* 19: 189-211.

[1449] Street-Perrott, F. A. and Perrott, R. A. 1990: Abrupt climate fluctuations in the tropics: the influence of Atlantic circulation. *Nature* 343: 607-612.

[1450] Sirocko, F., Sarnthein, M., Erlenkeuser, H., Lange, H., Arnold, M. and Duplessy, J. C. 1993: Century-scale events in monsoonal climate over the past 24,000 years. *Nature* 364: 322-324.

Asia.[1451,1452] Many civilisations collapsed[1453] and the drought was probably global as dust is recorded in ice cores from the Huascarán glacier in the Andes of northern Peru.[1454] The Kilimanjaro ice sheet grew and contracted in concert with large-scale coolings and warmings in Africa over the last 4000 years. One of the ice fields on Kilimanjaro formed during the Little Ice Age.

Contrary to the general presumption that glaciers all over the world are rapidly retreating, the 74 km Siachen glacier in the Karakoram Mountains of India advanced 700 metres from 1862 to 1909. This advance was neutralised by a faster retreat from 1929 to 1958. The glacier is now neither retreating nor advancing. Tropical glaciers of South America, Africa, Papua New Guinea and Irian Jaya reached their greatest extent during the Little Ice Age and have been retreating since the late 19th Century.[1455] The Quelccaya glacier in Peru has retreated quickly in recent years, Mt Kilimanjaro has retreated at least 80% since 1912, the ice cap of Mt Kenya has shrunk by 40% since 1963 and, in the last 30 years, Venezuela has lost four of its six glaciers. The 1930s and 1940s saw a rapid mass loss, melting slowed in 1970 and some glaciers even advanced. In the 1990s, there was rapid glacial retreat.

The Russian Arctic island of Novaya Zemlya had rapid glacial retreat before 1920, the retreat then slowed and, after 1950, many glaciers stopped retreating and many tidewater glaciers started to advance.[1456] Over the last 40 years, both the summer and winter temperatures of Novaya Zemlya have been lower than the previous 40 years. Many Scandinavian glaciers have been growing and glaciers in the Caucasus Mountains of Russia are at equilibrium.[1457] For example, Sweden's Storglaciaren has expanded over

---

[1451] Pachur, H-J. and Hoelzmann, P. 2000: Late Quaternary palaeoecology and palaeoclimates of the eastern Sahara *Journal of African Earth Sciences* 30: 929-939.

[1452] Gasse, F. and van Campo, E. 1994: Abrupt post-glacial climate events in West Asia and North Africa monsoon domains. *Earth and Planetary Science Letters* 126: 435-456.

[1453] Dalfes, H. N., Kukla, G. and Wiess, H. 1994: Third Millennium Climate Change and Old World Collapse. In: NATO ASI Series 1: *Global Environmental Change* 49: Springer-Verlag.

[1454] Thompson, L. G., Mosley-Thompson, E. and Henderson, K. A. 2000: Ice-core palaeoclimate records in tropical South America since the Last Glacial Maximum. *Journal of Quaternary Science 15*: 377-394.

[1455] Kaser, G. 1998: A review of the modern fluctuations of tropical glaciers. *Global and Planetary Change* 22: 93-103.

[1456] Zeeberg, J. J. and Forman, S. L. 2001: Changes in glacial extent on North Novaya Zemlya in the twentieth century. *Holocene* 11: 161-175.

[1457] Caseldine, C. J. 1985: The extent of some glaciers in northern Iceland during the Little Ice Age and the nature of recent deglaciation. *The Geographical Journal* 151: 215-227.

the last 40 years[1458] and, in the 1990s, it increased in mass.[1459] The glaciers in the western part of Norway have been advancing for much of the last two decades.[1460]

Four glaciers in northern Iceland had maxima at 1868, 1885, 1898 and 1917. Two of the glaciers keep retreating until 1985 (the time when observations ceased) and the other two of the glaciers periodically re-advanced when temperature was below 8°C. The Sólheimajökull glacier in southern Iceland has advanced and retreated repeatedly during the last 300 years resulting from a combination of cooling and increased precipitation.[1461] Sólheimajökull retreated due to widespread Arctic warming between 1920 and 1940, while its advances since 1970 are in response to cooling and increased precipitation. In the 18th Century, Iceland was warm, comparable to the 20th Century, while the coldest period in recent history was the 1780s. Glacier advances were at about 3000 BC, 1100 BC, 600 AD, 900 AD and 1300 AD resulting from cooling of 1–2°C. These correlate with worldwide glacier expansion, atmospheric changes in Greenland and changes in ocean circulation north of Iceland. Glacier expansion follows periods when sea ice is extensive in the Greenland and Barents Seas.[1462]

Glacier retreat and advance is not just driven by temperature, humidity and ice dynamics. In the Southern Alps of New Zealand, the Waiho Loop moraine was interpreted as a mass of debris left behind by retreating ice thereby suggesting that the Younger Dryas did not occur in New Zealand.[1463] Other studies suggested glacial advance in New Zealand during the Younger Dryas.[1464] Mountain terrains commonly have catastrophic slope failure producing landslides.[1465] These landslides can block valleys and create dams,

---

[1458] Brathwaite, R. J. 2002: Glacier mass balance: The first 50 years of international monitoring. *Progress in Physical Geography* 26: 76-95.

[1459] Brathwaite, R. J. and Zhang, Y. 2000: Relationships between interannual variability of glacial mass balance and climate. *Journal of Glaciology* 45: 456-462.

[1460] Wangensteen, B., Tønsberg, O. M., Kääb, A., Eiken, T. and Haghen, J. O. 2006: Surface elevation change and high resolution surface velocities for advancing outlets of Jostedalsbreen. *Geografiska Annaler* A 88: 55-74.

[1461] Mackintosh A. N., Dugmore, A. J. and Hubbard, A. L. 1997: Holocene climate changes in Iceland: evidence from modeling glacier length fluctuation at Solheimajokull. *Quaternary International* 91: 39-52.

[1462] Mackintosh, A. N., Dugmore, A. J. and Hubbard, A. L. 2002: Holocene climatic changes in Iceland: evidence from modelling glacier length fluctuations at Sólheimajökull. *Quaternary International* 91: 39-52.

[1463] Barrows, T. T., Lehman, S. J., Fifield, L. K. and DeDecker, P. 2007: Absence of cooling in New Zealand and the adjacent ocean during the Younger Dryas Chronozone. *Science* 318: 86-89.

[1464] Denton, G. H. and Hendy, C. H. 1994: Younger Dryas age advance of Franz Josef Glacier in the Southern Alps of New Zealand. *Science* 264: 1434-1437.

[1465] Hewitt, K. 1988: Catastrophic landslide deposits in the Karakoram Himalaya. *Science* 242: 64-67.

displace water from lakes and dams and cover ice sheets.[1466] In some areas, products of landslides have been misidentified as debris left behind by glaciers.[1467] In 1991, some 14 million cubic metres of rock collapsed from Mt Cook, covered the Tasman Glacier and triggered rapid glacial advance.[1468] Huge landslides in alpine New Zealand blanketed the Franz Josef Glacier and significantly reduced the surface melt rate, insulated the ice, triggered rapid glacial advance and piled up rock at the glacier's snout.[1469] Just because an alpine valley glacier rapidly advances does not mean that the climate is cooling.

There is no evidence that glacial retreat has been faster in the 20th Century when there has been an increase in $CO_2$ in the atmosphere.[1470] Some of us are not surprised that glaciers have retreated since the Little Ice Age. Arctic glacial retreats also took place in the Medieval Warming and, in the coldest times of the Little Ice Age, glaciers advanced in the early 15th Century, the mid 17th Century and the last half of the 19th Century.[1471] It is clear that alternating glacial advance and retreat is normal.

To state blandly that glaciers are melting because of human-induced global warming and $CO_2$ emissions is demonstrably false. Even if glaciers do retreat, they do not retreat forever, and they retreat for a diversity of reasons. We humans may worry about shrinking glaciers. A bigger worry is expanding glaciers.

The IPCC 2007 *Summary for Policymakers* stated: "Mountain glaciers and snow cover have declined on average in both hemispheres. Widespread decreases in glaciers and ice caps have contributed to sea level rise." There is no evidence to support this statement. Only contrary evidence.

---

[1466] Jarman, D. 2006: Large rock slope failures in the Highlands of Scotland: Characterisation, causes and spatial distribution. *Engineering Geology* 83: 161-182.

[1467] Hewitt, K. 1999: Quaternary moraines vs catastrophic rock avalanches in the Karakoram Himalaya, Northern Pakistan. *Quaternary Research* 51: 220-237.

[1468] Kirkbride, M. P. and Sugden, D. E. 1992: New Zealand loses its top. *Geographical Magazine* July 1992: 30-34.

[1469] Tovar, D. S., Shulmeister, J. and Davies, T. R. 2008: Evidence for a landslide origin of New Zealand's Waiho Loop moraine. *Nature Geoscience* 1: 524-526.

[1470] Dowdeswell, J. A., Hagen, J. O., Björnsson, H., Glazovsky, A. F., Harrison, W. D., Holmlund, P., Jania, J., Koerner, R. M., Lefauconnier, B., Ommanney, C. S. L. and Thomas, R. H. 1997: The mass balance of circum-Arctic glaciers and recent climate change. *Quaternary Research* 48: 1-14.

[1471] Calkin, P. E., Wiles, G. C. and Barclay, D. J. 2001: Holocene coastal glaciation of Alaska. *Quaternary Science Review* 20: 449-461.

# Sea ice

Sea ice covers 7% of the surface of the Earth. Sea ice distribution and thickness are crude indicators of the sea surface temperature and ocean currents. In the Arctic, there are good long-term records of sea ice because of the sea traffic and fishing.[1472] The Arctic was warmer than now between 1920 and 1940. This is reflected in the sea ice and the local climate.[1473] Changes in Arctic sea ice are correlated with changes in water temperature measured by using oxygen chemistry in fossils that were floating animals in Lake Baikal (Siberia), the world's largest freshwater lake.[1474]

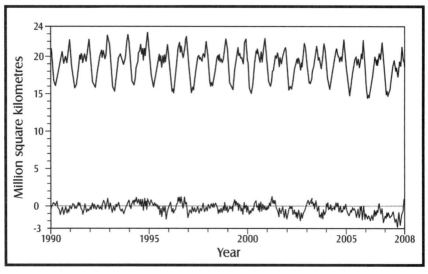

Figure 29: Area of global sea ice showing Arctic sea ice variations (lower graph) and cyclical changes in the total area of Antarctic sea ice (upper graph). There is little or no change in global sea ice associated with the Late 20th Century Warming.

Between 1921 and 1939, the area of sea ice in the Greenland Sea in April–August was 15–20% less than in 1898–1920.[1475] In the Barents Sea, a similar

[1472] Dickson, D. and Østerhus, S. 2007: One hundred years in the Norwegian Sea. Norwegian Journal of Geography 61: 56-75.

[1473] Benestad, R. E., Hanssen-Bauer, I., Skaugen, T. E. and Førland, E. J. 2002: Associations between sea-ice and the local climate on Svalbard. Norwegian Meteorological Institute Report 07/02: 1-7.

[1474] Mackay, A. W. 2007: The paleoclimatology of Lake Baikal: A diatom synthesis and prospectus. Earth Science Reviews 82: 181-215.

[1475] Karelin, D. B. 1953: Pervaya vysokoshirotnaya ekspeditsiya. In: Russkie moreplavateli (Ed. Lupach, V. S.) Voennoe izdatel'stvo Ministerstva oborony SSSR, 425-434 (from: Barr, W. 1984: The first Soviet high-latitude expedition, 205-216).

pattern was seen between 1920 and 1930 with 12% less sea ice than between 1898 and 1920.[1476] Since 1929, the southern part of the Kara Sea has been free of ice in September whereas in 1869–1928, the possibility of meeting ice there in September was 30%.[1477]

In the 19th Century, polar ice came very close to Iceland. During 1915–1940, except for 1929 when negligible amounts of polar ice were observed, there was no sea ice in the region. The *Fram* cruise[1478] (November 1892–August 1895) recorded a polar ice thickness of 655 cm (at 81°59'N 113°26'E) whereas the *Sedov* cruise (November 1937–August 1939, during the Arctic warming) recorded a thickness of 220 cm at a similar latitude (82°43'N, 121°30'E). Sea surface temperature measured by the *Fram* cruise did not exceed 1.13°C whereas on the *Sedko* cruise (1935 and 1938) sea surface temperatures reached 2.68 and 1.8°C respectively. *Sedko* was further north and east than the *Fram* in waters that are normally far colder than areas sailed by the *Fram*, as confirmed by the voyage of the RV *Akademik Shokalsky* in 1994.[1479]

Although there was more ice added from Greenland to the sea during the Arctic warming of 1920–1940, the ice melted due to the increased speed and temperature of the Norway and Spitsbergen currents and an increased wind velocity.[1480] During the Arctic warming of 1920–1940, cyclones moved northward and winds changed from cold easterlies to warmer southwest winds. In 1929 in West Greenland, no sea ice was attached to the land and the hunters were able to use kayaks throughout the whole winter.[1481]

Russian shipping records are a good indicator of the extent and intensity of Arctic sea ice. In 1901, the icebreaker *Ermak* failed to return from Novaya Zemlya, in 1912 *St Anna* in the Brusilov expedition was trapped in ice near Yamal and, in 1912, the *Foka* in the Seddov expedition could not reach Franz-Joseph Land. By contrast, in the Arctic warming of the 1930s, non-icebreaking ships could sail the North Sea route and could sail twice to and from Novaya Zemlya each summer. The *Sibiryak* sailed around Verernaya Zemlya (1932) and the *Knipovich* sailed around Franz-Joseph Land (1932).

---

[1476] Zubov, N. N. 1936: Ekspeditsiya "Sadko". *Sovetskaya Artika* 1: 28-50. (from: Barr, W. 1984: *The first Soviet high-latitude expedition*, 205-216).

[1477] Vise V. Yu. 1946: *Na "Sibiryakove" I "Litke" cherez ledovitye morya*. Izdatel'stvo Glavsevmorputi. (from: Barr, W. 1984: *The first Soviet high-latitude expedition*, 205-216).

[1478] Nansen, F. 1897: *Farthest North*. Westminster: Archibald Constable.

[1479] Joint Expedition 1995: Investigation of environmental radioactivity in waste dumping areas of the far eastern seas. *Results from the first Japanese-Korean-Russian Joint Expedition* 1994, 1-62.

[1480] Birkeland, B. J. 1930: Temperaturvariationen auf Spitsbergen. *Meteorologische Zeitscrift Juni 1930*: 234-236.

[1481] Humlum, O. 1999: Late-Holocene climate in central West Greenland: meteorological data and rock-glacier isotope evidence. *The Holocene* 9: 581-594.

Atlantic Ocean water entering the Arctic Basin was warmer and the lower boundary of the cold intermediate layer at 150–200 metres (beginning of 20th Century) rose to 75–100 metres in 1940–1945. During World War II, the reduced sea ice in the Arctic Ocean prompted the German naval high command to patrol the Kara Sea east of Novaya Zemlya to intercept convoys from North America to the Red Army.[1482] The pocket battleship *Admiral Scheer* had no ice protection for its exposed propellers even though it operated east of 100°E.[1483,1484] It sank a far larger and better protected ship, the *Alexander Sibiryakow*.[1485] A German freighter converted to an armed raider (*Komet*) sailed the northern route from the Atlantic to the Pacific Ocean in early summer 1940.[1486] The *Komet* quietly slipped through the Bering Strait and created havoc in the Pacific Ocean. This navigation route is not open today because of sea ice.

Melting of Northern Hemisphere sea ice occurs where ocean currents bring warmer waters from the south. Summer cloud cover in the region 60°N to 90°N has shifted substantially in recent years. It may be that the Southern Oscillation in the Pacific has an influence. It was positive in just 13 of the 48 months from January 2002 to December 2005 and on three of these occasions it was less than 1. Although we may not have been in El Niño conditions, we were heading in that direction. Perhaps those that claim human emissions of $CO_2$ drive climate change should demonstrate exactly how $CO_2$ drove these changes and why the change in cloud cover has only occurred since 1998.

Penguins can be a good proxy for Antarctic temperature. Records over more than 50 years show that Adélie penguins, cape petrels and other East Antarctic breeding birds arrive at their spring nesting locations now some nine days later than they did in the 1950s. This is in accord with the increasing length of the sea ice season.[1487,1488] Sea ice increased by 8% between 1978 and 2005. This was at a time when rapid global warming was claimed to have

---

[1482] Bellamy, C. 2007: *Absolute War. Soviet Russia in the Second World War*. Pan Macmillan..

[1483] Barr, W. 1975: Operation "Wunderland": Admiral Scheer in the Kara Sea, August 1942. *Polar Record* 17: 461-472.

[1484] Brennecke, J. and Krancke, T. 2001: *Schwerer Kreuzer Admiral Scheer*. Koehlers.

[1485] Beesley, P. 2006: *Very special intelligence: The story of the Admiralty's operation intelligence centre 1939-1945*. Naval Institute Press.

[1486] Hunt, W. R. 1975: *Arctic Passage: The turbulent history of the land and people of the Bering Sea, 1697-1975*. Scribner.

[1487] Barbraud, C. and Weimerskirsch, H. 2001a: Contrasting effects of the extent of sea-ice on the breeding performance of an Antarctic top predator, the Snow Petrel Pagodroma nivea. *Journal of Avian Biology* 32: 297-302

[1488] Barbraud, C. and Weimerskirsch, H. 2001b: Emperor penguins and climate change. *Nature* 411: 183-186

derived from human emissions of $CO_2$.

In 2008, a somewhat naïve and enthusiastic Englishman[1489] almost perished trying to paddle a kayak to the North Pole to highlight the effects of human induced global warming. He could only paddle to 960 km from the Pole. In 1893, Nansen was able to kayak to 800 km from the North Pole. The pathetic Pythonesque paddle in 2008 was to prove global warming had reduced the extent of sea ice. It demonstrated the exact opposite.

There have been suggestions that the Antarctic sea ice may reduce due to a human-driven temperature increase which would raise sea surface temperature.[1490] Once the sea ice area was reduced, there would be more solar energy absorbed by the oceans and less solar energy reflected by the ice. This has led to the suggestion that the polar areas, especially Antarctica, would experience the greatest change from global warming.[1491] These suggestions of the loss of Antarctic sea ice were in contrast to previous studies showing that, since the time of satellite measurements of sea ice, there has been no change to Antarctic sea ice.[1492,1493,1494,1495,1496] The Antarctic sea ice extent, open water area and ice area increased significantly during the period 1987–1996.[1497] Antarctic sea ice varies interannually[1498] and can be related to changes in the Southern Oscillation Index which appears to be one

[1489] Lewis Gordon Pugh.

[1490] Washington, W. M. and Meehl, G. A. 1989: Climate sensitivity due to increased $CO_2$: experiments and a coupled atmosphere and ocean general circulation model. *Climate Dynamics* 4: 1-38.

[1491] Fyfe, J. C., Boer, G. J. and Flato, G. M. 1999: The Arctic and Antarctic Oscillations and their projected changes under global warming. *Geophysical Research Letters* 26: 1601-1604.

[1492] Parkinson, C. L., Cosimo, J. C., Cavalieri, D. J., Gloersen, P. and Campbell, W. 1987: *Arctic sea ice 1973-1976*. NASA SP-489.

[1493] Gloersen, P. and Campbell, W. J. 1988: Recent variations in Arctic and Antarctic sea-ice covers. *Nature* 352: 33-36.

[1494] Gloersen, P., Campbell, W. J., Cavalieri, D. J., Comiso, J. C., Parkinson, C. L. and Zwally, H. J. 1992: *Arctic and Antarctic sea ice, 1978-1987: Satellite passive microwave observations and analysis*. NASA SP-511.

[1495] Johannessen, O. M., Bengtsson, L., Miles, M. W., Kuzmina, S. I., Semenov, V. A., Alekseev, G. V., Nagurnyi, A. P., Zakharov, V. F., Bobylev, L. P., Pettersson, L. H., Hasselmann, K. and Cattle, H. P. 2004: Arctic climate change: observed and modelled temperature and sea-ice variability. *Tellus A* 56: 328-341.

[1496] Bjørgo, E., Johannessen, O. M. and Miles, M. W. 1997: Analysis of merged SSMR-SSMI time series of Arctic and Antarctic sea ice parameters 1978-1995. *Geophysical Research Letters* 24: 413-416.

[1497] Watkins, A. B. and Simmonds, I. 2000: Current trends in Antarctic sea ice: The 1990s impact on a short climatology. *Journal of Climate* 13: 4441-4451.

[1498] Simmonds, I. and Jacka, T. H. 1995: Relationships between the interannual variability of Antarctic sea ice and the Southern Oscillation. *Journal of Climate* 8: 637-647.

of the factors driving climate change in the Antarctic.[1499] These may be due to changes in ocean currents or El Niño events, and decreases in the ice in the Weddell Sea could be related to a temperature increase. Extreme caution must be exercised using sea ice conditions as an indicator of climate trends and many local conditions change the extent and thickness of sea ice.

Even more extreme caution must be taken when computer simulations claim that there is human-induced global warming at the poles.[1500] We are a long way from understanding natural variability.

---

[1499] Smith, S. R. and Stearns, C. R. 1993: Antarctic climate anomalies surrounding the minimum in the Southern Oscillation Index. *Antarctic Research Series* 61: 149-174.

[1500] Gillett, N. P., Stone, D. A., Stott, P. A., Nozawa, T., Karpechko, A. Y., Hegerl, G. C., Whener, M. F. and Jones, P. D. 2008: Attribution of polar warming to human contribution. *Nature Geoscience* 1: 750-754.

# Chapter 6

# WATER

Question: Do human emissions of $CO_2$ create sea level rise?
Answer: No.

Question: Will the seas become acid?
Answer: No.

Question: Does sea level rise kill coral atolls?
Answer: No.

Question: Are humans forcing changes in ocean currents?
Answer: No.

*Sea level changes of over 600 metres have occurred in the past. Sea level rise creates biodiversity, sea level fall accelerates extinction. Sea level fall kills coral reefs. Not only does sea level rise and fall, the land level rises and falls, the volume of the ocean basins changes and the ocean basins change in shape and depth. Gravity also changes sea level.*

*Some parts of the world are sinking (e.g. eastern England, the Netherlands), others are rising (e.g. Scandinavia, Scotland) and others are not changing (e.g. northwest Alaska). These combined sea, land and ocean basin changes make accurate measurements of sea level change very difficult.*

*Since the last glaciation, there has been an average sea level rise of 1 cm per year. During this time, sea level has fallen and risen at rates of over 2 cm per year. This rate of sea level change is far higher than the most catastrophic IPCC speculations. Sea level rise over the last 14,000 years since the last glaciation has not destroyed coral reefs or coral atolls. The post-glacial sea level rise has had the opposite effect: it has stimulated coral growth. Island atoll states are being inundated from sinking, not from sea level rise.*

*Sea levels have been rising since the last glaciation ended and reached a maximum 6000 years ago. There has been no noted acceleration of sea level rise amid the period of industrialisation.*

*The oceans store a vast amount of dissolved $CO_2$ and the amount of $CO_2$ in the air is correlated with global sea surface temperature. A huge but unknown amount of $CO_2$ is added to ocean water from submarine volcanoes. The $CO_2$ dissolved in ocean water is exchanged with and cycles through the atmosphere, life, soils and rocks. The oceans continually remove dissolved $CO_2$ by shell formation, limestone formation and chemical reactions with rocks and sediments. The more $CO_2$ dissolves in the oceans, the more $CO_2$ is removed. Water drives the carbon cycle.*

*If humans burned all the fossil fuels on Earth, the atmospheric $CO_2$ content would not even double. A very slight change to any one of a number of natural systems would swamp any $CO_2$ additions by humans to the atmosphere.*

*The oceans have been salty and alkaline since the beginning of time, even when temperature was higher and atmospheric $CO_2$ was at least 25 times the current value. This is because rocks on the land chemically react with air, water and micro-organisms to form soil and because submarine volcanic rocks and sediments chemically react with seawater. When we run out of rocks, the oceans will become acid.*

*The surface of the Earth is more than 70% water, and evaporation is the regulator of global climate. Evaporation and the increase in exchange of latent heat between the ocean and atmosphere increase nearly exponentially with surface temperature of water. An increased exchange of latent heat energy for a very small temperature rise (0.3°C) is more than enough to offset a doubling of atmospheric $CO_2$.*

*There have been no "tipping" points in the past when temperature and $CO_2$ were far higher than now because all systems involving $CO_2$ have natural upper and lower regulators.*

*The ocean currents transfer huge amounts of heat. They are driven by wind and this is, in turn, driven by the Earth's rotation. Changes to the shape of the shoreline, the Earth and the ocean floor can change currents. The cool dense bottom currents of the ocean show that the Earth is far cooler than in past times. For only 20% of time the planet has had ice and most of the time planet Earth has been a warm wet greenhouse planet.*

*The origin of El Niño is poorly understood. Despite El Niño events being one of the greatest transfers of surface energy on Earth, they cannot be predicted by computer models. El Niño events are not factored into models of future climate.*

*The oceans and the atmosphere are non-linear chaotic and turbulent systems from bottom to top. We try to understand such systems with incomplete computer models.*

*Nature does not play computer games.*

## Weird water

Water, as we have seen, is weird.[1501] If the atoms comprising water were not held together by hydrogen bonding, it would boil at about -30°C. Water ice should be denser than liquid water. If ice sank, lakes, seas and oceans would freeze from the bottom up. This would prevent ice melting and would eventually produce permanent ice on Earth. This ice would reflect radiation, planet Earth would not be able to escape from being an ice ball.

A lot of heat is required to break off molecules of ice at 0°C to form water at 0°C.[1502] Even more heat is needed to break off molecules of water at 100°C to form steam at 100°C.[1503] Much heat is needed to warm water and, once warm, it takes a long time to cool.[1504] Evaporation and precipitation of water control the upper limit to air temperature. These weird properties prevent a runaway warming or permanent freezing of the Earth. If water were not weird, no nutrients could enter cells and sustain life. If water were not weird, there would be no heat held in the atmosphere and oceans and the air temperature would be a balmy -18°C.

For every 1°C increase in sea surface temperature, 7% more water can

---

[1501] Water has the highest heat capacity of all solids and liquids (except $NH_3$)[prevents extreme ranges in temperature and allows huge heat transfer by water movements], the highest latent of fusion (except $NH_3$)[allows thermostatic effect of freezing], the highest latent heat of evaporation [allows heat and water transport in the atmosphere], temperature of maximum density decreases with increasing salinity [controls temperature distribution in oceans and vertical circulation in lakes], highest surface tension of all liquids (except Hg) [important for cell physiology], dissolves more substances and in greater quantity than any other liquid [important for cell nutrition], has the highest dielectric constant of all liquids [allows cell nutrition], has a very small electrolytic dissociation, has a high transparency [allows photosynthetic life to live in deeper water], highest heat conduction of all liquids and large infra-red and ultra-violet adsorption [responsible for at least 75% of the greenhouse effect].

[1502] Latent heat of fusion of ice 333.55 J/g (79.72 cal/g).

[1503] Latent heat of vapourisation of water (at 0°C) 2500 J/g (598 cal/g) and (at 100°C) 2260 J/cal (539 cal/g).

[1504] Specific heats (cal/g°C) are water 1.00, ice 0.50 and steam 0.47. By contrast wood is 0.12 and gold is 0.03.

dissolve in air.[1505] A 99-year record of hourly precipitation in Holland shows that on an hourly basis, when the air temperature is above 12°C then precipitation increases at a rate of 14% per degree temperature rise.[1506] Evaporation and rain buffer temperature on Earth because both involve an exchange of heat.

The world's oceans are dynamic. They carry heat, contain dissolved $CO_2$ and are rich in floating photosynthetic micro-organisms that remove $CO_2$ from the atmosphere and water. Floating organisms in the oceans also remove calcium carbonate to build shells. Shells accumulate as fossil-bearing sediments and limestones. Chemical reactions between seawater and submerged rocks remove $CO_2$ from the oceans. This is why the oceans stay alkaline. Ocean chemistry has changed slightly over time.[1507,1508] The fossil record provides information on coral reefs, use of $CO_2$ by shells, past temperatures and sea levels. Sea level has risen and fallen countless times over the history of planet Earth and these changes are due to a diversity of competing processes.

The oceans are four-dimensional (latitude, longitude, depth and time) complex masses which show multidecadal temperature trends. A recent slight warming is recorded in 37% of the samples taken in the top 50 metres of some oceans, although Southern Hemisphere oceans are poorly sampled.[1509] This may result from a diversity of processes. Both warming and cooling trends are recorded over the last 50 years.[1510]

Sea level is always changing. The land level is also always changing but at varying rates and opposing directions simultaneously in different areas. In the past, sea level has risen or fallen by up to 600 metres and the land level has risen or fallen by up to 10,000 metres. Only 6000 years ago sea level was 2 metres higher than now. During the last glaciation, sea level rose and fell by 130 metres every 100,000 years. Some past sea level changes were very fast,

---

[1505] Clausius-Clapeyon relation

[1506] Lenderink, G. and van Meijgaard, E. 2008: Increase in hourly precipitation extremes beyond expectations from temperature changes. *Nature Geoscience* 1: 511-514.

[1507] Veizer, J., Ala, D., Azmy, K., Bruckschen, P., Buhl, D., Bruhn, F., Carden, G. A. F., Diener, A., Ebneth, S., Godderis, Y., Jasper, T., Korte, C., Pawellek, F., Podlaha, O. G. and Strauss, H. 1999: $^{87}Sr/^{86}Sr$, $\partial^{13}C$ and $\partial^{18}O$ evolution of Phanerozoic seawater. *Chemical Geology* 161: 59-68.

[1508] Kasting, J. F., Howard, M. T., Wallmann, K., Veizer, J., Shields, G. and Jaffrés, J. 2006: Paleoclimates, ocean depth, and the oxygen isotopic composition of seawater. *Earth and Planetary Science Letters* 252: 82-93.

[1509] Harrison, D. E. and Carson, M. 2006: Is the upper ocean warming? A data analysis approach. *EOS Trans.* 87:52.

[1510] Harrison, D. E. and Carson, M. 2006: Is the world ocean warming? Upper ocean temperature trends: 1950-2000. *Journal of Physical Oceanography* 37: 174-187.

some were slow.[1511]

Global sea level has changed over geological time.[1512] Sea level change is related to the presence or absence of ice caps, the shape of continents, the shape of the sea floor and the temperature of the oceans. Some changes were dramatic, such as the sea level drop associated with the initial expansion of the Antarctic Ice Sheet at 37 Ma.[1513]

Deep in the human psyche is a fear of a rapid sea level rise. This may derive from the Noah's Flood story. It may also derive from the regular destruction of coastal populations. Why did the great civilisations of the past not arise in far more pleasant areas along the seaside? Sporadic asteroid impacts in ocean basins, volcanic explosions, the slumping of sediments from the continental shelf, earthquakes and the collapse of volcanoes into the ocean have produced tsunamis. Some of these tsunamis were hundreds of metres high. The catastrophic tsunami of 26 December 2004 that killed at least 250,000 people in Asia was relatively small.

## The great flood

During an exceptionally cold period 8500 to 8000 years ago, the Anatolian highland people moved to a lower altitude.[1514] The Anatolian highlands were deserted and populations shifted to a 160,000 square kilometre basin with a warmer wetter climate. This basin is now occupied by the Black Sea. This protected basin was serviced by melt water rivers (Don, Dnieper, Danube) and comprised two large freshwater lakes and fertile plains. It became the breadbasket of the ancient world.

Almost 25% of the floor of the modern Black Sea is flat and less than 100 metres below sea level. During the last glaciation, the Sakarya River drained the basin into the Mediterranean Sea via the Gulf of Izmit and the Sea of Marmara. This kept the Marmara Sea out of the basin. However, the post-glacial sea level rise meant that the Marmara Sea level was about 100 metres higher than the floor of the Black Sea basin. At that time, there was no Bosphorus, just a low valley with a rock outcrop at its headwaters that protected the basin from inundation by the Marmara Sea.

[1511] Shakleton, N. J., Lamb, H. H., Worssam, B. C., Hodgson, J. M., Lord, A. R., Shotten, F. W., Schove, D. J. and Cooper, L. H. N. 1977: The oxygen isotope stratigraphic record of the Late Pleistocene. *Philosophical Transactions of the Royal Society of London B* 280: 169-182.

[1512] Miller, K. G., Kominz, M. A., Browning, J. V., Wright, J. D., Mountain, G. S., Katz, M. E., Sugarman, P. J., Cramer, B. S., Christie-Blick, N. and Pekar, S. F. 2005: The Phanerozoic record of global sea-level change. *Science* 310: 1293-1298.

[1513] Shevenell, A. E., Kennett, J. P. and Lea, D. W. 2004: Middle Miocene Southern Ocean cooling and Antarctic cryosphere expansion. *Science* 305: 1766-1770.

[1514] Wilson, I. 2001: *Before the flood*. Orion.

However, the area is at a plate boundary where Africa is colliding with Europe, and the North Anatolian Fault regularly moves. The last movement on 17 August 1999 caused 20,000 fatalities. Movement along the North Anatolian Fault 7600 years ago resulted in the breaking of rocks and a rush of water down from the Marmara Sea into the Black Sea basin. This carved out the Bosphorus and filled the basin to form the Black Sea. This process of forming the Black Sea took no longer than two years. Seawater poured into the basin with the force of 200 Niagara Falls, sea level rose by 15 cm per day and the shoreline advanced kilometres per day.[1515]

Marine sediments deposited from turbulent waters were deposited on fertile soils, dense salty water filled the bottom of the Black Sea displacing fresher water to the surface, and the bottom waters of the Black Sea became oxygen-poor. This was fortunate as it allowed the preservation of wooden village structures that were built on the shore of the former freshwater lakes.

This must have been a terrifying event. People and livestock perished, populations dispersed and the survivors carried with them their language, culture and knowledge of agriculture, animal husbandry, craftsmanship and metallurgy. It is no surprise that many cultures have myths about a great flood. Post-glacial sea level rise of 130 metres over the last 14,000 years, floods and catastrophic events such as the inundation of the Black Sea basin were passed down as stories whereas the comings and goings of daily life were not.

Drill cores from coral reefs from the same period show that the great flood that formed the Black Sea was not a global event.[1516,1517] Attempts to interpret the Black Sea flooding event were passed down from generation to generation. These appeared as culturally and linguistically transposed stories. They appeared as a fragmentary Sumerian story written some 2000 years later at 3400 BC, the Mesopotamian myth of Athrahasis, the Babylonian tale of Ut-napishtim in the Epic of Gilgamesh, the Greek stories of Deucalion and Pyrrha, the story of Dardanus, and the biblical story of Noah and the great flood.

These changes to what was considered a static world were so rapid and

---

[1515] Ryan, W. and Pitman, W. 1998: *Noah's Flood, the new scientific discoveries about the event that changed history.* Simon and Schuster.

[1516] Lighty, R. G., Macintyre, I. G. and Stuckenrath, R. 1982: *Acropora palmata* reef framework: A reliable indicator of sea level in the western Atlantic for the past 10,000 years. *Coral Reefs* 1: 125-130.

[1517] Ludwig, K. R., Muhs, D. R., Simmons, K. R., Halley, R. B. and Shinn, E. A. 1996: Sea-level records at approximately 80 ka from tectonically stable platforms; Florida and Bermuda. *Geology* 24: 211-224.

incomprehensible that the great flood to form the Black Sea was interpreted as the action of a god who flooded the known world in order to rid it of evil. So too with the modern world, which many interpret as static. The slightest change in Nature is viewed as a message that we humans are changing the climate, that this is evil and that we must rid the world of this evil. To many, it is incomprehensible that Nature can change the planet or that humans are an insignificant short-lived recent terrestrial vertebrate living on a planet where natural forces are many orders of magnitude greater than any human force. The amount and rate of measured modern changes in temperature and sea level are far slower than post-glacial processes. This is not in accord with the beliefs of many who claim that humans are driving global warming and sea level rise.

## Sea level

A dynamic Earth, climate change and associated sea level rises and falls have been well recorded in the geological literature since the times of Hutton in the late 18th Century.[1518] Since the last great interglacial 116,000 years ago, many sea level rises and falls have been measured.[1519] The Earth is constantly changing and it is only now in the modern industrial age that the general public has become aware that there are constant changes in temperature, sea level, ice and life on Earth. The planet is never static. It evolves and changes constantly. This dynamism does not necessarily mean that humans are driving the changes. In Australia, mineral sand deposits on old beaches 150 metres above the modern sea level are mined more than 500 kilometres from the coast (e.g. Ginko, Pooncarie). This shows that very large sea and land level rises and falls have occurred in the recent past. Another example is the raised coral terraces in the Huon Peninsula, Papua New Guinea.[1520] These changes are unrelated to any human activity.

Sea level rise creates many new shallow water environments.[1521] The geological record shows that there is an increasing biodiversity when there is a sea level rise. If sea level falls, an increased extinction of life may result.

---

[1518] Brooks, C. E. P. 1926: *Climate through the ages*. R. V. Coleman.

[1519] Dansgaard, W. and Oeschger, H., 1989: Past environmental long-term records from the Arctic. In: *The environmental record in glaciers and ice sheets*. (eds Oeschger H. and Langway, C. C.) 287-318, Wiley.

[1520] Chappell, J. 1974: Geology of coral terraces, Huon Peninsula, New Guinea: A study of Quaternary tectonic movements and sea-level changes. *Bulletin of the Geological Society of America* 85: 553-570.

[1521] Solé, R. V. and Newman, M. 2002: Extinctions and biodiversity in the fossil record. In: *The Earth system: biological and ecological dimensions of global environmental change* (eds Mooney, H. A. and Canadell, J. G.), JohnWiley, 297-301.

In the Neoproterozoic glaciation, sea level fell so much that there was no continental shelf.[1522] Sea level change in this glaciation was at least 12 times as great as the sea level changes associated with the latest glaciation. Sea level dropped 1500 metres due to capture of water as huge continental ice sheets and continents sank under the load of kilometres of ice. One change promotes another. The relative rise of coastal areas under a thinner ice sheet raised sea level, the lack of weight on the sea floor allowed sea level to rise and the gravitational pull of large ice sheets raised sea level by about 100 metres. Even today, the gravitational pull of mountains near the sea (e.g. the Andes) and large ice sheets result in a local sea level rise. The total sea level change associated with the Neoproterozoic glaciation was a fall of about 650 metres. A sudden change from glaciation to warm conditions happened over a few centuries. The seas may have been slightly acid. The very cold climate glacial rocks were immediately overlaid by carbonate rocks that formed over a few thousand years in shallow water at a temperature of at least 40°C.[1523] At that time, atmospheric $CO_2$ was more than 10% and up to 35% compared to the 0.0385% of today. During the formation of the carbonate rocks, the oceans were alkaline.

Modern global sea level changes are exceptionally difficult to determine. The earliest measurement sites were measuring sticks attached to piers. In the middle of the 19th Century, tide gauges using floats in stilling wells were installed. These buffered the effects of waves. Modern gauges use echo sounding and transmit data by satellite for measurement in real time. Over periods of a century or more, tide gauges need to be maintained, repaired, moved and upgraded, as do the piers. This often does not happen. Sea level measurements from tide gauges are made from a sparse network of coastal stations, many of which are in geologically unstable places. Precise satellite measurements give sea level rise as half that measured from tidal stations.[1524] Corrected data for a large part of the globe shows a rise of 1.8 mm per annum from 1900 to 1980[1525] and this is in accord with measurements from corals and other proxies for the past 3000 years. Historical records show no acceleration of sea level rise in the 20th Century.[1526] During the warming

---

[1522] Christie-Blick, N., Dyson, I. A. and von der Borch, C. C. 1995: Sequence stratigraphy and the interpretation of Neoproterozoic earth history. *Precambrian Research* 73: 3-26.

[1523] Hoffman, P. F. and Schrag, D. P. 2002: The snowball Earth hypothesis: testing the limits of global change. *Terra Nova* 14: 129-155.

[1524] Cabanes, C., Cazenave, A. and Le Provost, C. 2001: Sea level rise during the past 40 years determined from satellite and in situ observations. *Science* 294: 840-842.

[1525] Trupin, A. and Wahr, J. 1990: Spectroscopic analysis of global tide gauge sea level data. *Geophysical Journal International* 100: 441-453.

[1526] Douglas, P. C. 1992: Global sea level acceleration. *Journal of Geophysical Research* 97: 12699-12706.

from 1920 to 1940, sea level rise actually stopped.[1527]

Accurate determination of sea level changes from tide gauges is fraught with difficulty. Many piers slowly sink and the geographic location of tide gauges makes constant measurement unreliable. For example, at Port Adelaide (Australia), the tidal measuring station has been sinking, thereby recording a sea level rise.[1528]

Tide gauges need a consistency of measurement over a long period by having regular high-precision surveying of the gauge position and frequent calibration of the gauge. This just does not happen. Furthermore, tide gauges only make local measurements. In many tectonically unstable parts of the world, often very close to each other, the land has risen (e.g. Ephesus and Troy, Turkey)[1529] or fallen simultaneously (e.g. Lydia, Turkey).[1530]  In biblical times,[1531] Ephesus was a port city. Strabo recorded that King Attalus Philadelphus of Bergama built a breakwater to protect Ephesus and siltation combined with a land rise destroyed the port.[1532] The Romans tried to rebuild the port and failed. Ephesus is now 24 km inland and some 5 metres above sea level. Lydia, the birthplace of coinage, is now metres below sea level. Plate tectonics, rebound of land after loading (with ice, soil, sediment and water), volcanism, sediment compaction and extraction of fluids all affect tide gauge measurements. Vertical movement of land is also variable. Another example of tectonic instability is the Vanuabalavu island group of northeast Fiji, where some individual islands have sunk while the whole island group has actually risen.[1533]

Sea level changes are a result of competing forces. When most of the Northern Hemisphere north of 50°N was covered in ice during the last glaciation from 116,000 to 14,000 years ago, landmasses sank under the weight of ice. Now that the ice has melted, the landmasses are rising. Rocks are slightly plastic and, if a force of loading or unloading of ice is applied over time they bend. If a force is applied suddenly, they break. Earthquakes and earth tremors occur when rocks break, and areas that were covered

[1527] Singer, S. F. 1997: *Hot talk, cold science. Global warming's unfinished debate*. The Independent Institute.

[1528] Belperio, A. P. 1993: Land subsidence and sea level rise in the Port Adelaide estuary: implications for monitoring the greenhouse effect. *Australian Journal of Earth Sciences* 40: 359-368.

[1529] Kraft, J. C., Aschenbrenner, S. E. and Rapp, G. 1978: Paleogeographic reconstructions of coastal Aegean archaeological sites. *Science* 195: 941-947.

[1530] Kuniholm, P. I. 1990: Archaeological record: evidence and non-evidence for climate change. *Philosophical Transactions of the Royal Society of London* A330: 645-655.

[1531] Acts 19: 1-7.

[1532] Murphy-O'Connor, J. 2008: *St Paul's Ephesus: Texts and Archaeology*. Liturgical Press.

[1533] Nunn, P. D., Ollier, C., Hope, G., Rodda, P., Omura, A. and Peltier, W. R. 2002: Late Quaternary sea-level and tectonic changes in northeast Fiji. *Marine Geology* 187: 299-311.

by ice during the last glaciation have common earth tremors. At present, adjacent lands are both rising (e.g. Scandinavia, Scotland, Canada) and sinking simultaneously (e.g. Holland, northwest Denmark, southeast England). The floor of the North Sea is also sinking. This counterbalancing of land level rise and post-glacial sea level rise means that relative sea level changes may be small.

The example of post-glacial land rise in Scandinavia is well known.[1534,1535] The 12th Century castle of Turku (Finland)[1536] was built on an island. It is now connected to the mainland as a result of post-glacial land level rise. Tide gauges in the port of Turku are a guide to local uplift, not to sea level changes. Stockholm is no longer an island and is rising at 1 cm per year. Post-glacial uplift in Scandinavia results in subsidence of the Netherlands, Denmark and northwest Germany. Is this a case for an international court?

Evidence is presented in the scientific literature that, as a result of sea level rise, the additional weight of seawater causes uplift in the neighbouring land.[1537,1538,1539,1540,1541] The loading of the continental shelf with water from melting of the ice sheets causes the land to rise and is dependent upon the thickness of brittle rocks and the plasticity of the Earth's mantle to depths of at least 670 km.[1542] If the land rises, sea level may actually fall.

The post-glacial sea level records from northeast Ireland between 21,000 and 11,000 years ago show that there were many rises and falls in sea level in response to ice unloading.[1543] A plot of sea level change is like a sawtooth.

---

[1534] Lambeck, K. and Chappell, J. 2001: Sea level change through the last glacial cycle. *Science* 292: 679-686.

[1535] Clemmensen, L. B. and Andersen, C. 1998: Late Holocene deflation of beach deposits, Skagen Odde, Denmark. *Geological Society of Denmark Bulletin* 44: 187-188.

[1536] Frost, H. 2007: Some out-of-the-way European maritime museums and developments. *The International Journal of Nautical Archaeology* 4: 143-13.

[1537] Bloom, A. L. 1967: Pleistocene shorelines: A new test of isostasy. *Bulletin of the Geological Society of America* 78: 1477-1494.

[1538] Walcott, R. I. 1972: Late Quaternary vertical movements in eastern America: quantitative evidence of glacio-isostatic rebound. *Reviews in Geophysics and Space Physics* 10: 849-884.

[1539] Chappell, J., Rhodes, E. G., Thom, B. G. and Wallensky, E. P. 1982: Hydro-isostasy and the sea-level isobase of 5500 B.P. in north Queensland, Australia. *Marine Geology* 49: 81-90.

[1540] Gibb, J. G. 1986: A New Zealand regional Holocene eustatic sea-level curve and its application for determination of vertical tectonic movements. *Bulletin of the Royal Society of New Zealand* 24: 377-395.

[1541] Nakada, M. 1986: Holocene sea levels in ocean islands: implications for the rheological structure of the Earth's mantle. *Tectonophysics* 121: 263-276.

[1542] Nakada, M. and Lambeck, K. 2008: Late Pleistocene and Holocene sea-level change in the Australian region and mantle rheology. *Geophysical Journal International* 96: 497-517.

[1543] McCabe, A. M., Cooper, J. A. G. and Kelley, J. T. 2007: Relative sea-level changes from NE Ireland during the last glacial termination. *Journal of the Geological Society, London* 164: 1059-1063.

It shows very strong initial uplift (21,000–19,000 years ago), ice loading, land sinking and a relative sea level rise (19,000–17,500 years ago and 17,000–14,000 years ago) and a catastrophic ice loss, rapid uplift of the land and a relative sea level fall (14,500–13,000 years ago). These very rapid fluctuations in ice, sea level and land level are not considered in geophysical and climate models.[1544,1545]

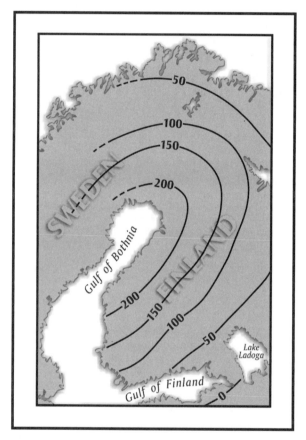

*Figure 30: Post-glacial uplift of Fennoscandia (in metres) over the last 9000 years. The present rate of uplift is 1 cm per year.[1546] Such uplift makes the determination of sea level changes very difficult.*

[1544] Lambeck, K. 1996: Late Devensian and Holocene shorelines of the British Isles and North Sea from models of glacio-hydrostatic rebound. *Journal of the Geological Society, London* 153: 437-448.

[1545] Shennan, I., Bradley, S., Milne, G., Brooks, A., Basset, A. and Hamilton, S. 2006: Relative sea-level changes, glacial isostatic modelling and ice sheet reconstructions from the British Isles since the last glacial maximum. *Journal of Quaternary Science* 21: 585-599.

[1546] Niskanen, E. 1939: On the upheaval of land in Fennsoscandia. *Isostatic Institution International Association of Geodesy Publication* 6, Helsinki.

The Netherlands is sinking.[1547] For more than 1000 years, the Dutch have been building dykes, pumping out water with windmills and suffering inundation when storms coincide with high tides. At present, 40% of the Netherlands is below sea level and more than 2000 km of dykes protect the country from inundation by the North Sea. Since 1200 AD, 580,000 hectares of agricultural land have been lost to the North Sea.

On 14 December 1287, St Lucia's flood killed between 50,000 and 80,000 people and the Zuider Zee was formed by the scouring away of peat and the incursion of the North Sea. In January 1362, the *Grote Mandrenke* (Great Drowning of Men) was a combination of hurricane force winds and a storm surge. Some 25,000 people were killed and in Slesvig (Denmark), 60 parishes totally disappeared. A storm on 18 November 1421 destroyed 65 villages and 10,000 perished.[1548] After another great storm in 1916, a major construction program in 1918 reclaimed 400,000 hectares of the Zuider Zee into arable land, some 10% of the country's agricultural land, reduced the shoreline by 300 km and pushed the old Zuider Zee 85 km southwards.[1549]

Many areas are currently sinking due to the extraction of groundwater (e.g. Bangkok, Mexico City, Denver) and petroleum (e.g. Texas Gulf coast). Submergence rates of about 11 mm per year occur at Galveston Bay (Texas) resulting from water extraction, petroleum extraction and subsidence.[1550] Subsidence increases the risk of inundation from hurricane activity, as was shown by Hurricane Katrina. The whole of the Texas Gulf area is subsiding. In the three years before the flooding associated with Hurricane Katrina devastated New Orleans in August 2005, the city and surrounding area had undergone rapid subsidence of about one metre.

Tide gauges at London show a sea level rise. At the end of the last glaciation 14,000 years ago, sea level rose. A river valley was filled with water. This valley is now the English Channel. What was a small freshwater river (Thames River) was inundated and became tidal with marshes in a wide open valley. The Romans built Londonium on the banks of the Thames River but had to move it onto the river terraces because storms coinciding with spring

---

[1547] Eitner, V. 1996: Geomorphological response of the East Frisian barrier islands to sea-level rise: an investigation of past and future evolution. *Geomorphology* 15: 57-65.

[1548] Van Baars, S. 2007: *The causes and mechanisms of historical dike failures in the Netherlands.* http://www.geo.citg.tudelft.nl/vanbaars/research/dikes/historicaloverview.pd

[1549] Van Lier, H. N. and Steiner, F. R. 1982: Review of the Zuiderzee reclamation works: An example of Dutch physical planning. *Landscape Plan* 9: 35-59.

[1550] Sharp, J. M. and Germiat, S. J. 1990: Risk assessment and causes of subsidence and inundation along the Texas Gulf Coast. In: *Greenhouse effect, sea level and drought* (eds Paepe R., Fairbridge, R. W. and Jelgersma, S.) 395-414, Kluwer.

tides and wind-generated surges caused inundation.[1551] The Romans and later Londoners did not know that all of eastern England was sinking. The post-glacial rise of Scotland is causing a compensatory fall in eastern England and parts of the east coast have sunk 6 metres in the last 6500 years.[1552,1553] Scotland is rising as eastern England is sinking (especially East Anglia). These events of inundation are also well recorded after Roman times.[1554,1555,1556,1557]

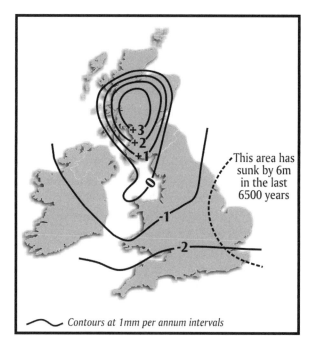

This area has sunk by 6m in the last 6500 years

Contours at 1mm per annum intervals

*Figure 31: The rise of Scotland and the sinking of England (especially eastern England) resulting from the loss of ice (unloading). Unloading started with the loss of ice 14,000 years ago, and land rise lags behind ice loss.*

[1551] Dodson, A. T. and Dines, J. S. 1929: Report on Thames floods and meteorological conditions associated with high tides in the Thames. *Geophysical Memoirs* 47: 1-39.

[1552] Dunning, F. W., Mercer, I. F., Owen, M. P., Roberts, R. H. and Lambert, J. L. M. 1978: *Britain before man*. HMSO.

[1553] Clayton, K. M. 1990: Sea-level rise and coastal defences in the UK. *Quarterly Journal of Engineering Geology* 23: 283-287.

[1554] Savage, A. 1995: The Anglo-Saxon Chronicles. *Crescent Books.*

[1555] Paris, Matthew 1236: *Chronica Majora.*

[1556] Stow, John 1580: *The Chronicles of England from Brute unto this Present Yeare of Christ, London.*

[1557] Pepys, Samuel 1666: *The diary of Samuel Pepys.* Cassell.

A stone bridge replaced the initial wooden London Bridge in the 12th Century. Subsidence of the bridge foundations led to the children's nursery rhyme "London Bridge is Falling Down".[1558] The extraction of groundwater from gravels and sands on the banks of the Thames has resulted in compaction which has exacerbated the sinking of London.[1559] After 2000 years of river flooding and inundation from the North Sea, the Thames Barrier was built and opened in 1984.[1560] The Thames Barrier is designed to protect London until 2030. London is still sinking and there is nothing we can do about it. London will probably rise during the next inevitable glaciation.

Sea level and land level changes force shorelines to advance and retreat. A rather depressing book documents the loss of towns to the sea by shoreline changes.[1561] Around The Wash (Yarmouth and Lowestoft), the coast is retreating.[1562] The Romans tried and failed to stop coastal erosion here and the shoreline has retreated 1.8 metres per year since Roman times.[1563] Cliffs in other areas have been undercut (e.g. Lyme Regis, Dorset) but elsewhere the coast is advancing (e.g. headland growth of 12 metres per year on the Yorkshire coast).[1564]

One has to be extremely optimistic to use tide gauge measurements to determine long-term sea level changes. The chance of long-term consistently accurate measurements is remote. Most of the reliable long-term tide gauge records come from Europe and the USA. For example, the annual mean relative sea level rise at New York City from 1893 to 1995 was 2.9 mm for the entire 103-year period. This is a relative sea level rise as sea level could have risen, land level could have fallen or both. It is well known that increasing urbanisation leads to land subsidence. New York is no exception. It also depends on the way the data is viewed. Blocks of 20-year intervals vary widely with sea level rise from 0 to 6 mm per year and blocks of 40-year intervals vary from 0.9 to 3.5 mm per year.[1565]

---

[1558] Wilson, G. and Grace, H. 1942: The settlement of London due to underdrainage of the London Clay. *Journal of the Institution of Civil Engineers* 19: 100-127.

[1559] Poland, J. F. and Davis, G. H. 1969: Land subsidence due to withdrawals of fluids. *Reviews in Engineering Geology* 2: 187-269.

[1560] Gilbert, S. and Horner, R. W. 1984: *The Thames Barrier*. Thomas Telford.

[1561] Sheppard, T. 1912. *The lost towns of the Yorkshire coast and other chapters bearing upon the geography of the district*. A. Brown & Sons.

[1562] Green, C. and Hutchinson, J. N. 1965: Relative land and sea levels at Great Yarmouth, Norfolk. *The Geographical Journal* 131: 86-90.

[1563] Green, C. 1961. East Anglian coastline levels since the Roman times. *Antiquity* 35: 21–28 and 155–156.

[1564] Lee, E. M., Hall, W. J. and Meadowcroft, I. C. 2001: Coastal cliff recession: the use of probabilistic prediction methods. *Geomorphology* 40: 253-269.

[1565] Douglas, B. C. and Peltier, W. R. 2002: The puzzle of global sea-level rise. *Physics Today* March 2002: 1-6.

Sea level is also subject to low-frequency variations that are recorded over large ocean regions for several decades or more. Relative sea level worldwide has an inter-annual to inter-decadal variation whose amplitude is larger than the overall rate of sea level rise over periods of decades. Only recording over long periods can accurately yield the underlying trend. Other causes of sea level changes are the 18.6-year lunar nodal cycle, El Niño events, earthquakes, volcanoes, ocean floor subsidence, high wind events and foul weather associated with atmospheric low-pressure cells.

Over a period of 18.6 years, the Moon's trajectory moves further north over the equator and then back again due to the Sun's gravity. The year 2006 was a maximum year for the 18.6-year lunar nodal tide, during which time tidal currents brought large volumes of warm water into the Arctic. As a result, the Arctic Ocean heated and the summer sea ice melting rate increased. The entire west coast of Greenland has been affected by the arrival of masses of warm water from the Irminger Sea near Iceland and some glaciers suddenly started to thin.[1566] However, the ice melting rate was not nearly as rapid as in the 1930s. The mass media promoted the idea that the decreased ice was proof of human-induced global warming. Extreme ocean tides produce changes in sea surface temperature at 90-day intervals based on the alignment of the Earth, Sun and Moon within the 18.6-year lunar nodal cycle. Extreme tides increase the vertical mixing of seawater thereby causing episodic cooling of sea surface water.[1567]

Besides a catastrophic tsunami that destroyed cities and drowned thousands from the Nile Delta to Dubrovnik, the 365 AD earthquake lifted western Crete 10 metres above sea level.[1568] The most severe shoreline changes are from local and regional-scale processes (e.g. cyclones,[1569] tsunamis[1570]) coinciding with the 18.6-year tidal cycle.[1571] On the coast of French Guiana,

[1566] Holland, D. M., Thomas, R. H., de Yong, B., Ribergaard, M. H. and Lyberth, B. 2008: Acceleration of Jakobshavn Isbræ triggered by warm subsurface ocean waters. *Nature Geoscience* 1: 659-664.

[1567] Keeling, C. D. and Whorf, T. P. 1997: Possible forcing of global temperature by the oceanic tides. *Proceedings of the National Academy of Sciences* 94: 8321-8328.

[1568] Shaw, B., Ambraseys, N. N., England, P. C., Floyd, M. A., Gorman, G. C., Higham, T. F. G., Jackson, J. A., Nocquet, J.-M., Pain, C. C. and Piggott, M. D. 2008: Eastern Mediterranean tectonics and tsunami hazard from the AD 365 earthquake. *Nature Geoscience* 1: 268-276.

[1569] Ericson, J. P., Vorosmarty, C. J., Dingham, S. L., Ward, L. G. and Meybeck, M. 2006: Effective sea-level rise in deltas: Causes of change and human dimension impacts. *Global and Planetary Change* 50: 63-82.

[1570] Borreo, J. C. 2005: Field and satellite imagery of tsunami effects in Banda Aceh. *Science* 308: 1596.

[1571] Wells, J. T. and Coleman, J. M. 1981: Periodic mudflat progradation, northeastern coast of South America: a hypothesis. *Journal of Sedimentary Petrology* 51: 1069-1075.

this could lead to a 6 cm sea level rise and a 90 metre shoreline retreat.[1572]

A French study of four decades of data from tide gauge stations and satellites showed that modern global sea level rise had been over-estimated.[1573] The island of Bermuda has been used as a deep-water example to show the variability of sea level and the trend of sea level rise. Between 1955 and 1998, tide gauges showed that relative sea level rose at 0.67 mm per year but the trend for the entire record (1933–1998) is three times this rate. Bermuda is an island consisting of a coral atoll capping an extinct volcano and is surrounded by water more than 3 km deep. It is more than likely that the relative sea level rise is measuring the rate of sinking of the volcano.

A change in sea level may initiate volcanic eruptions.[1574] During the last 2.67 Ma of glaciations and interglacials with rapidly rising and falling sea level, an increase in explosive volcanicity is recorded. Explosive volcanoes may correlate with changes in sea level[1575] and with changes in climate.[1576] Numerous submarine plumes of molten rock occur in the crust and mantle beneath the ocean floor. Examples are the Caribbean-Columbian Plateau, Kerguelen Plateau, Ontong Java Plateau, Manihiki Plateau and the Hikurangi Plateau. These generate broad domal uplift which can raise the ocean floor by 500 to 1000 metres over an area 1000 km wide.[1577] Not only does this have a profound effect on ocean currents and the addition of monstrous amounts of heat to the oceans, but it creates a sea level rise because a huge volume of seawater is displaced.[1578]

A study of the coastal salt marsh in the northeast USA did not record the Medieval Warming and the Little Ice Age yet showed a sea level rise since the late 19th Century.[1579] Warm periods did not correlate with sea level rise. Local

[1572] Gratiot, N., Anthony, E. J., Gardel, A., Gaucherel, C., Proisy, C. and Wells, J. T. 2008: Significant contribution of the 18.6 year tidal cycle to regional coastal changes. *Nature Geoscience* 1: 169-172.

[1573] Cabanes, C., Cazenave, A. and le Provost, C. 2001: Sea level rise during the past 40 years determined from satellite and in situ observations. *Science* 294: 840-842

[1574] McGuire, W. J., Howarth, R. J., Firth, C. R., Solow, A. R., Pullen, A. D., Saunders, S. J., Stewart, I. S. and Vita-Finzi, C. 1997: Correlation between the rate of sea-level change and frequency of explosive volcanism in the Mediterranean. *Nature* 399: 473-476.

[1575] McGuire, W. J. 2008: Changing sea levels and erupting volcanoes: cause and effect? *Geology Today* 8: 141-144.

[1576] Rampino, M. R., Self, S. and Fairbridge, R. W. 1979: Can rapid climate change cause volcanic eruptions? *Science* 206: 826-829.

[1577] Peate, I. U. and Bryan, S. E. 2008: Re-evaluating plume-induced uplift in the Emeishan large igneous province. *Nature Geoscience* 1: 625-629.

[1578] Peate, I. U., Larsen, M. and Lesher, C. E. 2003: The transition from sedimentation to flood volcanism in the Kangerlussuaq Basin, East Greenland: Basaltic pyroclastic volcanism during initial Palaeogene continental break-up. *Journal of the Geological Society, London* 160: 759-772.

[1579] Varekamp, J. C., Thomas, E. and van de Plassche, O. 2007: Relative sea-level rise and climate change over the last 1500 years. *Terra Nova* 4: 292-304.

factors seem to be driving land level and sea level changes and are unrelated to global changes. Various time scales were tested using carbon dating of the last 6000 years of organic-rich sediments and peat associated with estuarine sediments and shallow marshes along stable or apparently subsiding coasts. Another scale was the last 1000 years using peat and bottom-dwelling foraminifera from coastal salt marshes. The last scale used was the historic record of the last few hundred years based on tide gauge measurements. The conclusion was that the historic record was "a continuation of the past rather than a perturbation", that there was no indication of a pronounced temperature rise in the 20th Century and there is no relationship between atmospheric $CO_2$ and temperature.

The Isthmus of Corinth shows a 350,000-year sequence of fluctuations between lake, sea and land.[1580] If this area was used to calculate sea level rise, then relative sea level changes of 0.23 to 1.4 mm per year can be calculated, yet they have no bearing whatsoever on global sea level change. Again, this records a local feature, in this case tectonic movement as the Peloponnese Peninsula extends away from the mainland.

Melting of ice sheets and glaciers, expansion of seawater with warming and addition of new molten rocks to the sea floor create a sea level rise whereas the addition of snow to ice sheets, changes in the shape of the ocean floor, changes in landmass shapes and opening or closing of seaways can create a sea level fall.[1581] Thermal expansion of the oceans is not well known because of the lack of sampling in the Southern Hemisphere oceans at all depths and of the abyssal plains in both hemispheres. However, only surface water is affected by thermal expansion, response times can be very rapid, and sea level changes amount to only a few millimetres.[1582]

El Niño events can cause sea level rise by up to half a metre.[1583] The 1982–1983 El Niño raised sea level 0.35 metres along the west coast of the USA because there were no easterly winds to force the water back into the

[1580]  Kershaw, S., Guo, L. and Braga, J. C. 2005: A Holocene coral-algal reef at Mavra Litharia, Gulf of Corinth, Greece: structure, history and applications in relative sea-level change. *Marine Geology* 215: 171-192.

[1581]  Hogan, A. W. and Gow, A. J. 1997: Occurrence frequency of thickness of annual snow accumulation layers at South Pole. *Journal of Geophysical Research* 102: 021-014.

[1582]  Warrick, R. A. and Farmer, G. 1990: The greenhouse effect, climate change and rising sea level: implications for development. *Transactions of the Institute British Geographers* 15: 5-20.

[1583]  Meyers, G. 1996: Variation of Indonesian flowthrough and the El Niño:Southern Oscillation: Pacific low-latitude western boundary currents and the Indonesian flowthrough. *Journal of Geophysical Research* 101: 12255-12264.

west Pacific.[1584,1585,1586]  An increase in water load causes the ocean floor to sink. Even water storage dams show that the land beneath the water rises and falls, depending upon water level. Consequently, most large dams have a seismic monitoring network to measure land movement associated with the loading and unloading of dam water. Furthermore, if plate tectonics deepens ocean trenches and the medial rifts of mid ocean ridges, then sea level falls.

As continents undergo weathering to form soils, erosion removes soils and the continental masses gravitationally rise producing a relative sea level fall. This occurs in many alpine areas. Eroded material is deposited as sediment, most commonly on deltas, beaches and the continental shelf. The additional weight of material causes sinking of the continental shelf and near shore areas. For example, the Mississippi Delta is subsiding because of sediment compaction.[1587] This produces a relative sea level rise. These processes, although well known, cannot be quantified accurately at present. Other low-lying areas such as in southeast Asia have undergone considerable land exposure and inundation resulting in shoreline and river system changes during fluctuating climates.[1588] By contrast and contrary to popular belief, Bangladesh is actually growing because of the huge amount of sediment deposited in the Ganges delta.

Measurements of sea level change by satellites have increased accuracy, but the cause of any measured sea level change is no closer to resolution. Altimetric satellites such as TOPEX/Poseidon (launched 1992), Jason 1 (2001) and GRACE (2002) can measure the absolute sea level change for the whole planet. These satellites involve a paired satellite tracking system that measures the time dependence of the Earth's gravitational field and can measure global sea level with extraordinary accuracy. Satellite measurements need correction using a geophysical model that utilises the fact that the crust of the Earth is both elastic and brittle. Satellite-borne radar altimeters on TOPEX/Poseidon showed that, after correction, the average global sea level rise was in the order of 2.4 mm per year, a value close to that of corrected

[1584] Wyrtki, K. 1975: El Niño: The dynamic response of the equatorial Pacific Ocean to atmospheric forcing. *Journal of Physical Oceanography* 5: 572-584.

[1585] Harrison, D. E. and Crane, M. A. 1984: Changes in the Pacific during the 1982-83 El Niño event. *Oceanus* 27: 21-28.

[1586] Komar, P. D. 1986: El Niño and erosion on the coast of Oregon. *Shore and Beach* 54: 3-12.

[1587] Törnqvist, T. E., Wallace, D. J., Storms, J. E. A., Wallinga, J., van Dam, R. L., Blaauw, M., Derksen, M. S., Klerks, C. J. W., Meijneken, C. and Snijders, E. M. A. 2008: Mississippi Delta subsidence primarily caused by compaction of Holocene strata. *Nature Geoscience* 1: 173-176.

[1588] Voris, H. K. 2001: Maps of Pleistocene sea levels in Southeast Asia: shorelines, river systems and time durations. *Journal of Biogeography* 27: 1153-1167.

tidal gauges.[1589] Furthermore, these satellite measurements show rises and falls that vary on a local basis. For example, the Pacific Ocean floor around Tuvalu is sinking, giving the appearance of a sea level rise. For the last 20 years, Tuvalu has been the symbol of sea level inundation. Tuvalu is still there. It has not been inundated. Coral atolls grow upwards in response to relative sea level rise.

Unless the geology of the crust beneath our feet is very well known and the ice loading over the last 5 million years of glaciation is accurately known, the data on the rise or fall of land is somewhat speculative. It is incredibly difficult to measure whether sea level is rising or falling. However, we can measure this change after the event, and over time, satellites will determine long-term absolute global sea level changes with respect to the Earth's centre of mass. However, satellites have not measured sea level for very long. Calculations involving the size, shape and gravity of the Earth have shown that the global land rise and fall adjustments actually create a bias in the TOPEX/Poseidon and GRACE measurements.[1590]

The accuracy of satellite determination of sea level change cannot be tested by ground observations. Attempts to validate the satellite measurements of 2.8 ± 0.4 mm per year[1591,1592] produce a global average rise of 1.6 mm per year (16 cm per century) for the period 1993–2004. This is about 60% of the satellite estimate. Some 70% of the 1.6 mm per year is from the addition of fresh water to the oceans.[1593] More recent studies using the GRACE data show the rise was 2.5 mm per year for 2003–2005.[1594] This is some 20% less than the IPCC's projected sea level rise for 1993–2003 of 3.1 mm per year and confirms an earlier study.[1595] It seems that sea level has a large variability

[1589] Douglas, B. C., Kearney, M. S. and Leatherman, S. P. 2001: *Sea level rise. History and consequences*. Academic Press.

[1590] Wu, P. and Peltier, W. R. 1984: Pleistocene deglaciation and the Earth's rotation: a new analysis. *Geophysical Journal of the Royal Astronomical Society* 76: 202-242.

[1591] Leuliette, E. W., Nerem, R. S. and Mitchum, G. T. 2004: Results of TOPEX/Poseidon and Jason-1 calibration to construct a continuous record of mean sea level. *Marine Geology* 27: 79-94.

[1592] Cazenave, A. and Nerem, R. S. 2004: Present-day sea level change: Observations and causes. *Review of Geophysics* 42: doi:10.1029/2003RG000139.

[1593] Wunsch, C., Ponte, R. M. and Heimbach, P. 2007: Decadal trends in sea level patterns: 1993-2004. *Journal of Climate* 20: 5889-5911.

[1594] Cazenave, A., Dominh, K., Guinehut, S., Berthier, E., Llovel, W., Ramillien, G., Ablain, M. and Larnicol, G. 2008: Sea level budget over 2003-2008: A reevaluation from GRACE space gravimetry, satellite altimetry and Argo. *Global and Planetary Change* doi:10.1016/j.gloplach.2008.10.1004.

[1595] Willis, J. K., Chambers, D. K. and Nerem, R. S. 2008: Assessing the globally averaged sea level budget on seasonal to interannual timescales. *Journal of Geophysical Research* 113: C06015, doi: 10.1029/2007JC004517.

and is poorly understood, hence observations over a short period must be interpreted with great caution.

GPS measurements from many sites show that sea level rise is far less than predicted and in the first six years of the 21st Century was 1.35 ± 0.34 mm per year.[1596] This figure is less than the most widely quoted figures from tide gauge measurements (1.84 ± 0.35 mm per year)[1597,1598] and is more in accord with the 1.40 mm per year figure[1599] which ascribed 1.0 mm per year to the melting of global land ice reservoirs and 0.4 mm per year from thermal expansion of the oceans.[1600] The lack of correlation between various estimates[1601] is now showing that sea level rise is much slower than previously thought and far lower than any of the catastrophist models.

Studies of modern sea level changes provide results contrary to popular concern about a sea level rise. Ice shed from the giant ice sheets covering Antarctica and Greenland is responsible for only 12% of the current sea level rise.[1602] There is a major obstacle in predicting the future of ice sheets: we do not know what is going on beneath them.[1603] It is claimed that the rate of sea level rise has been accelerating in tandem with the rate of rise of the air's $CO_2$ concentration and temperature.[1604] This claim now has been tested[1605] and shown to be unfounded.

Warming does not necessarily mean that ice melts. Antarctica became isolated by a circumpolar current as it separated from South America at

[1596] Woppelmann, G., Miguez, B. M., Bouin, M.-N. and Altamini, Z. 2007: Geocentric sea-level trend estimates from GPS analyses at relevant tide gauges world wide. *Global and Planetary Change* 57: 396-406.

[1597] Douglas, B. C. 1997: Global sea level rise. *Journal of Geophysical Research* 96: 6981-6992.

[1598] Douglas, B. C. 2001: Sea level change in the era of the recording tide gauge. In: *Sea level rise: History and consequences* (eds Douglas, B., Keraney, M. and Leatherman, S.), Academic Press, 37-64.

[1599] Mitrovica, J. X., Wahr, J., Matsuyama, I., Paulson, A. and Tamisea, M. E. 2006: Reanalysis of ancient eclipse, astronomic and geodetic data: a possible route to resolving the enigma of global sea level rise. *Earth and Planetary Science Letters* 243: 390-399.

[1600] Antonov, J. I., Levitus, S. and Boyer, T. P. 2005: Themosteric sea level rise: 1955-2003. *Geophysical Research Letters* 32: doi: 10.1029/2005GL023112.

[1601] Munk, W. 2002: Twentieth century sea level: an enigma. *Proceedings of the National Academy of Science* 99: 6550-6555.

[1602] Shepherd, A. and Wingham, D. 2007: Recent sea-level contributions of the Antarctic and Greenland ice sheets. *Science* 315: 1529-1532.

[1603] Vaughan, D. and Arthern, R. 2007: Why it is hard to predict the future of ice sheets. *Science* 315: 1503-1504.

[1604] Mann, M. E., Bradley, R. S. and Hughes, M. K. 1999: Northern Hemisphere temperatures during the past millennium: Inferences, uncertainties, and limitations. *Geophysical Research Letters* 26: 759-762.

[1605] Larsen, C. E. and Clark, I. 2006: A search for scale in sea level studies. *Journal of Coastal Research* 22: 788-800.

37 Ma.[1606] This effectively refrigerated Antarctica.[1607] There is a popular fear today that the polar ice caps will melt and, as a result, sea level will inundate coastal populations. Ice core drilling in both Greenland and Antarctica has given a continuous record of some 800,000 years of past glacials and interglacials. In past interglacials when the temperature was at least 5°C warmer than now for about 10,000 years, the polar ice caps did not completely melt. Sea level has been rising and falling by about 130 metres over the past 800,000 years of alternating glacials and interglacials. The ice sheets could not have melted, otherwise there would be no old ice to drill. Well before Greenland was covered with ice, Antarctica had a thick ice sheet which remained during a very warm period of 4 million years (17–13 Ma).[1608] This was followed by rapid cooling.[1609] Ice volume trends suggest that this cooling was driven by the Earth's orbit and was unrelated to $CO_2$.

The most alarmist predictions of sea level rise from ice sheet melting caused by global warming need to be substantially scaled back. Ice persisted for much longer on Earth when the Earth was much hotter than today. Al Gore's Oscar-winning movie predicted that sea level would increase by 6 metres in the near future, while other predictions are that the rise would be 2 to 4 metres. These predictions are made in the absence of information from the past and do not consider all the other processes involved in sea level changes. Frightening seven-second Hollywood grabs are not the way science operates. This is the way Hollywood works and, like other products from Hollywood, it can only be regarded as fantasy.

Other scientists suggest that because we do not understand how the ice sheets behave, then we cannot make accurate predictions. Instead, they calculate how fast glaciers must move to dump ice into the sea to create a 2-metre sea level rise. The rate of ice movement required is far faster than measured ice sheet movement. If there were to be a 2-metre sea level rise, then Greenland glaciers would have to increase their speed to 48 kilometres per year and stay at that speed for the next hundred years.[1610] The maximum measured speed of a glacier today is 15 kilometres per year.

[1606] Cunningham, W. D., Dalziel, I. W. D., Lee, T.-L. and Lawver, L. A. 1995: Southernmost South America-Antarctic Peninsula relative plate motions since 84 Ma: Implications for the tectonic evolution of the Scotia Arc region. *Journal of Geophysical Research* 100: 827-8266.

[1607] Rintoul, S. R., Hughes, C. W. and Olbers, D. 2001: The Antarctic circumpolar current system. *Ocean Circulation and Climate International Geophysics Series* 77: 271-302.

[1608] Zachos, J., Pagani, M., Sloan, L., Thomas, E. and Billups, K. 2001: Trends, rhythms and aberrations in global climate 65 Ma to present. *Science* 292: 686-693.

[1609] Shevenell, A. E., Kennett, J. P. and Lea, D. W. 2004: Middle Miocene Southern Ocean cooling and Antarctic cryosphere expansion. *Science* 305: 1766-1770.

[1610] Pfeffer, T., Harper, J. and O'Neel, S. 2008: Kinematic constraints on glacier contribution to 21st-Century sea-level rise. *Science* 321: 1340-1343.

Sea level changes over the last 20 years are not particularly unusual. Sea level change was larger in the early part of the last century (2.03 ± 0.35 mm per year, 1904–1953) in comparison to the latter part (1.45 ± 0.34 mm per year, 1954–2003). The highest decadal rate of rise occurred in the decade centred on 1980 (5.31 mm per year) with a fall in the decade centred on 1964 (1.49 mm per year). Over the entire century the mean rate of change was 1.74 ± 0.16 mm per year,[1611] far less than any figure promoted by the IPCC.

Another model for present-day sea level rise suggests components are thermal expansion (+28.8 cm rise by 2100 AD), melting of alpine valley glaciers (+10.6 cm), Greenland ice melting (2.4 cm) and Antarctic snow accumulation (-7.4 cm).[1612] The projections of sea level rise due to melting of alpine valley glaciers (4.6 cm) and ice caps (5.1 cm) are about half of other projections and far less than the IPCC speculations. The greatest uncertainty about calculating sea level rise from the melting of the polar ice sheets is the lack of understanding of the processes affecting changes in the ice and the polar climate.[1613] None of these calculations incorporate the rise and fall of land, the rise and fall of the sea floor, the subsidence of deltas and coastal areas, the see-saw effect of loading and unloading and the various meteorological and gravity effects on sea level.

However, as the oceans and air warm, more water evaporates from the oceans and this moisture deposits as snow on polar ice caps and glaciers. Unless there is a sustained local temperature to induce long-term local melting, then global warming would actually force a growth of the glaciers and ice caps. This conclusion is supported by studies on the melting of Greenland's ice sheet which show that a 1°C temperature rise would increase sea level by 0.03 to 0.77 mm per year and in Antarctica would decrease sea level by 0.2 to 0.7 mm per year resulting from increased precipitation adding to the ice cap.[1614] If continental ice masses melt, it does not necessarily mean that sea level will rise. Land that was once covered by ice will rise and the loading of additional water into the oceans will make the ocean floor sink.

The 130-metre sea level rise over the last 14,000 years means that much of the West Antarctic Ice Sheet is now not underpinned by land. Already, two thirds of the West Antarctic Ice Sheet has collapsed into the ocean. The

[1611] Holgate, S. J. 2007: On the decadal rates of sea level change during the twentieth century. *Geophysical Research Letters* 34, L01602, doi: 10.1029/2006GL028492.

[1612] Raper, S. C. B. and Braithwaite, R. J. 2006: Low sea level rise projections from mountain glaciers and ice caps under global warming. *Nature* 439: 311-313.

[1613] Rémy, F. and Frezzotti, M. 2006: Antarctic ice sheet mass balance. *Comptes Rendus Geosciences* 338: 1084-1097.

[1614] Reeh, N. 1999: Mass balance of the Greenland ice sheet: Can modern observation methods reduce the uncertainty. *Geografiska Annaler* 81A: 735-742.

other third would require some 7000 years of similar melting which would create a 7-metre sea level rise.[1615] This is in fact a very slow rate of sea level change. This conclusion is supported by another study that showed glacial melting due to higher 20th Century temperature can only account for a 10 cm sea level change per century.[1616] Since 800 AD, sea level changes due to the Dark Ages, Medieval Warming and Little Ice Age have been within 0 ± 1.5 mm per year with an average close to zero.[1617] Sea level has essentially not changed in some regions. The coastline of the Chukchi Sea off northwest Alaska is geologically stable and sea level has risen by 0.025 cm per year over the last 6000 years. Sea level measurement involves huge uncertainties and we really do not know why the level can rise or fall at fast rates.[1618]

Sea level changes and the rate of heating of ocean water are not a simple matter of cause and effect. If sea level rise was only from thermal expansion of the oceans and from melting glaciers, then it would take a long time to heat the oceans. If indeed we are experiencing a sea level rise at all, it is likely to be due to the Medieval or Roman Warmings or rebound from the Little Ice Age and is certainly not due to contemporary processes. Sea level in the central Mediterranean region during Roman times was about 1.4 metres lower than now.[1619] Melting of sea ice does not change sea level.[1620] Only the melting of terrestrial ice can raise sea level. Again, this is a process that takes a long time.

There is a suggestion that sea level will fall and not rise. Oceans are getting deeper and sea level has fallen some 170 metres since 80 Ma and this may be followed by a further 120-metre sea level fall over the next 80 million years.[1621] Using an isotope chemistry approach, it was also shown that over a 550 million-year period, sea level has actually been falling as a result of the deepening of the oceans.[1622] Between 542 and 251 Ma,[1623] at least 172

---

[1615] Stone, J., Balco, G. A., Sugden, D. E., Caffee, M. W., Sass, L. C. III, Cowdery, S. G. and Siddoway, C. 2003: Holocene deglaciation of Marie Byrd Land, West Antarctica. *Science* 299, 99-102.

[1616] Munk, W. 2003: Ocean freshening, sea level rising. *Science* 300, 2014-2043.

[1617] Ekman, M. 1999: Climate changes detected through the world's longest sea level series. *Global and Planetary Change* 21, 1215-1224.

[1618] Holgate, S. J. 2007: On the decadal rates of sea level change during the twentieth Century. *Geophysical Research Letters* 34: L01602, doi: 10.1029/2006GL028492.

[1619] Lambeck, K., Anzidel, M., Antonioli, F., Benini, A. and Esposito, A. 2004: Sea level in Roman time in the Central Mediterranean and implications for recent change. *Earth and Planetary Science Letters* 224: 563-575.

[1620] Archimedes Principle.

[1621] Müller, D., Sdrolias, M., Gaina, C., Steinberger, B. and Heine, C. 2008: Long-term sea-level fluctuations driven by ocean basin dynamics. *Science* 319: 1357-1362.

[1622] Kasting, J. F., Howard, M. T., Wallmann, K., Veizer, J., Shields, G. and Jaffrés, J. 2006: Paleoclimates, ocean depth, and the oxygen isotopic composition of seawater. *Earth and Planetary Science Letters* 252: 82-93.

[1623] Palaeozoic Era.

sea level cycles have been recorded varying in magnitude from 10 to about 125 metres.[1624] Over time, the constant rise and fall of sea levels have left cycles of sedimentary rocks that can be used to show if the shoreline was advancing or retreating over land.[1625]

Although there is concern that the Pacific island atoll states may be the victims of sea level rise, in reality it is because the ocean floor and atolls are sinking. This long-term sea level fall is in agreement with other research.[1626,1627] Geologists are always reading the rocks to determine sea level changes and palaeoenvironments. The long-term rises and falls of sea level are vital for understanding marine basins prospective for oil exploration.[1628,1629]

In 1990, the IPCC predicted that human-induced warming would result in a sea level rise by 2100 of 66 cm with a potential range of 30 to 100 cm.[1630] In 1996, the IPCC predicted a sea level rise by 2100 of 49 cm with a range of 13 to 94 cm. In 2001, the same organisation predicted that the sea level rise would be 9 to 88 cm[1631] and in 2007, a range of 18 to 59 cm was predicted. At this rate of changing of predictions, it appears that we only have to wait a few more years and the IPCC sea level rise prediction will be zero! If the IPCC had only looked at geological history, then their predictions would be very different.

The International Union for Quaternary Research is almost 80 years old and deals with the last 2 million years of environmental and climate change. The former president of their Sea Level Commission states that no regular trend in sea level is evident over the last 300 years and satellite telemetry shows virtually no change over the last decade.[1632] The IPCC suggests that

[1624] Haq, B. U. and Schutter, S. R. 2008: A chronology of Paleozoic sea-level changes. *Science* 322: 64-68.

[1625] Vail, P. R. and Mitchum, R. M. 1979: Global cycles of relative changes of sea level from seismic stratigraphy. *American Association of Petroleum Geologists Memoir* 29: 469-472.

1626 Pitman, W. C. III 1978: Relationship between eustacy and stratigraphic sequences of passive margins. *Geological Society of America Bulletin* 89: 1389-1403.

[1627] Xu, X. Q., Lithgow-Bertelloni, C. and Conrad, C. P. 2006: Global reconstructions of Cenozoic seafloor ages: Implications for bathymetry and sea level *Earth and Planetary Science Letters* 243: 552-564.

[1628] Watts, A. B. and Thorne, J. A. 1984: Tectonics, global changes in sea level and their relationship to stratigraphic sequences at the US Atlantic continental margin. *Marine and Petroleum Geology* 1: 319-339.

[1629] Haq, B. U., Hardenbol, J. and Vail, P. R. 1987: Chronology of fluctuating sea levels since the Triassic. *Science* 235: 1156-1167.

[1630] Warwick, R. A. and Oerlemans, J. 1990: Sea level rise. In *Climate Change, the IPCC Assessment*, (eds Houghton, J. H., Jenkins, G. J. and Ephron, J. J.), Cambridge University Press.

[1631] Intergovernmental Panel on Climate Change, 2001: *Third Assessment Report*. Cambridge University Press.

[1632] Morner, N. A. 2004: Estimating future sea level changes from past records. *Global and Planetary Change* 40: 49-54.

sea level rise will be in the range of 0.09 to 0.88 metres between 1990 and 2100 whereas the Sea Level Commission states that sea level rise will be 10 ± 10 cm over the same period, thereby suggesting that a sea level rise cannot be predicted. The Environmental Protection Agency of the USA predicts a 50% chance that sea level will rise 45 cm by 2100 and a 1% chance that sea level will rise 110 cm by 2100.[1633] If we use the past to try to understand the present and future, we have no reason to expect such a significant change of sea level in the near future.

During the last interglacial period (130,000–116,000 years ago), global mean surface sea temperature was at least 2°C warmer than at present[1634] and mean sea level was 4 to 6 metres higher than at present.[1635,1636,1637] These figures have been used to analyse the effects of a modern interglacial sea level rise.[1638] At 123,500 years ago, the rate of sea level rise was 1.5 to 2.5 metres per century whereas at 119,000 years ago, the rate of sea level fall was 1.3 to 1.6 metres per century. Over the duration of the current interglacial, the average sea level rise over the last 14,000 years has been 1.0 metre per century. This is slightly more than the worst-case scenario sea level rise predicted by the IPCC. If sea level did rise 1 metre by 2100 AD, then this is exactly what would be expected at the current post-glacial sea level rise rate. The IPCC's own computer models show that their predicted sea level rise is probably not due to increased additions of $CO_2$ by humans into the atmosphere but rather is due to the continuing post-glacial sea level rise.

The sea level rise of 1.0 metre per century over the last 14,000 years must be placed in context. Most of this sea level rise was from 14,000 to 8000 years ago. By 8000 years ago, sea level was 3 metres lower than at present and sea level attained its current position by 7700 years ago. This means that sea level rose by 2 metres a century during that period. Sea level continued to rise

[1633] U.S. Environmental Protection Agency: *Global warming – climate, sea level.* <http://yosemite.epa.gov/oar/globalwarming/nsf/content/ClimateFutureClimateSea.level.html>

[1634] Otto-Bliesner, B. L., Marshall, S. J., Overpeck, J. T., Miller, G. H., Hu, A. and CAPE Last Interglacial Project members 2006: Simulating Arctic climate warmth and icefield retreat in the last interglaciation. *Science* 311: 1751-1753.

[1635] McCulloch, M. T. and Esat, T. 2000: The coral record of last interglacial sea levels and sea surface temperatures. *Chemical Geology* 169: 107-129.

[1636] Stirling, C. H., Esat, T. M., Lambeck, K. and McCulloch, M. T. 1998: Timing and duration of the Last Interglacial: Evidence for a restricted interval of widespread coral growth. *Earth and Planetary Science Letters* 160: 745-762.

[1637] Neumann, A. C. and Hearty, P. J. 1996: Rapid sea-level changes at the close of the last interglacial period: $^{234}U$-$^{230}Th$ data from fossil coral reefs in the Bahamas. *Bulletin of the Geological Society of America* 103: 82-97.

[1638] Rohling, E. J., Grant, K., Hemleben, Ch., Siddall, M., Hoogakker, B. A. A., Bolshaw, M. and Kucera, M. 2008: High rates of sea-level rise during the last interglacial period. *Nature Geoscience* 1: 38-42.

and, by 7400 years ago, sea level was at least 1.5 metres higher than at present and this was followed by a high static sea level that lasted until about 3000 or 2000 years ago.[1639] Sea level has dropped over the last 3000 years.

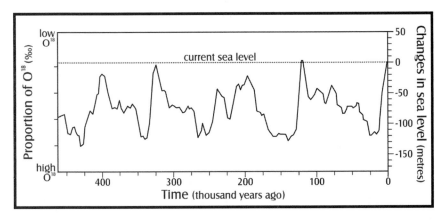

*Figure 32: Global sea level over the last 450,000 years calculated from oxygen isotope proxies[1640] showing high sea levels in interglacials and low sea levels during glaciation. During glaciation, sea level fluctuates greatly and sea level during the last interglacial was some 7 metres higher than at present. Modern sea level is both slightly higher and lower than previous interglacials. There is nothing extraordinary about the modern sea level or modern sea level change.*

In the geological literature, it has been known for more than two centuries that sea level on a local and global scale rapidly rises and falls for a great diversity of reasons. Over the last 6000 years, sea level rises and falls of 2 to 4 metres over periods of several decades are common.[1641,1642,1643] Changes were both global and regional[1644] and comparison of widespread regions shows

[1639] Sloss, C. R., Murray-Wallace, C. V. and Jones, B. G. 2007: Holocene sea-level change on the southeast coast of Australia: a review. *The Holocene* 17: 999-1014.

[1640] Imbrie, J and Imbrie, K. P. 1979: *Ice Ages*. Elsevier.

[1641] Fairbridge, R. W. 1958: Dating the latest movements in the Quaternary sea level. *New York Academy of Science Transactions* 20: 471-482.

[1642] Fairbridge, R. W. 1960: The changing level of the sea. *Scientific American* 202: 70-79.

[1643] Fairbridge, R. W. 1961: Eustatic changes in sea level. In: *Physics and Chemistry of the Earth, Vol 4* (eds Ahrens, L. H., Rankama, K., Press, F. and Runcorn, S. K.) Pergamon Press, 99-185.

[1644] Baker R. G. V., Haworth, R. J. and Flood, P. G. 2001: Warmer or cooler late Holocene marine palaeoenvironments?: Interpreting Southeast Australian and Brazilian sea-level changes using fixed biological indicators and their $\partial^{18}O$ composition. *Palaeogeography, Palaeoclimatology, Palaeoecology* 168: 249-272.

only synchronous rapid sea level change.[1645] Elsewhere, studies of adjacent islands show that some are rising and others are sinking.[1646] Climate change modelling requires a stable system upon which the variable of man-made $CO_2$ can be placed. The history of sea level changes shows that surface processes are unstable and that atmospheric $CO_2$ is not a destabilising force.[1647]

## Coral atolls

Coral reef records show that when the rest of the world was experiencing warm weather, the Pacific was cooler and *vice versa*. And, during a period of cool weather elsewhere in the world, the Pacific was warm and stormy.[1648] During the Medieval Warm Period conditions in the tropical Pacific were cool and possibly dry. Similarly, during the Little Ice Age, the central Pacific was comparatively warm and wet with stormy conditions more common. This suggests that warming and cooling events may not be global and that corals are far more resilient than some may think. Most corals, but not all, thrive in warm water.[1649]

Most coral species present in the Great Barrier Reef (Australia) and similar climatic ecologies are also present in areas with much warmer water.[1650] The Great Barrier Reef has grown throughout a time of rapidly rising sea level, high turbidity and rapidly rising temperature since the last ice age.[1651] If sea surface temperature increases and sea level rises, the reef will keep pace as it has done in the past. This is contrary to the alarmist press view. Over the last 500 million years, corals have survived in much warmer and much cooler waters than the modern Great Barrier Reef. Coral tissue thickness, one of the indicators of coral health, is greater for corals in warmer water.[1652] Some

[1645] Baker, R. G. V., Haworth, R. J. and Flood. P. G. 2001: Inter-tidal fixed indicators of former Holocene sea levels in Australia: a summary of sites and a review of methods and models. *Quaternary International* 83-85: 257-273.

[1646] Nunn, P. D., Ollier, C., Hope, G., Rodda, P., Omura, A. and Peltier, W. R. 2002: Late Quaternary sea-level and tectonic changes in northeast Fiji. *Marine Geology* 187: 299-311.

[1647] Baker, R. G., Haworth, J. and Flood, P. G. 2005: An oscillating Holocene sea-level? Revisiting Rottnest Island, Western Australia and the Fairbridge eustatic hypothesis. *Journal of Coastal Research* 42: 3-14.

[1648] Allen, M. 2006: New ideas about late Holocene climate variability in the central Pacific. *Current Anthropology* 47: 3.

[1649] Lough, J. M. and Barnes, D. J. 2000: Environmental controls on growth of the massive coral *Porites. Journal of Experimental Marine Biology and Ecology* 245: 225-243.

[1650] LaJeunesse, T. C., Loh, W. K. H., van Woesik, R., Hoegh-Guldberg, O., Schmidt, G. W. and Fitt, W. K. 2003: Low symbiont diversity in southern Great Barrier Reef corals, relative to those of the Caribbean. *Limnology and Oceanography* 48: 2046-2054.

[1651] Johnson, D. 2004: *The geology of Australia.* Cambridge University Press.

[1652] Barnes, D. J. and Lough, J. M. 1992: Systematic variations in the depth of skeleton occupied by coral tissue in massive colonies of *Porites* from the Great Barrier Reef. *Journal of Experimental Marine Biology and Ecology* 159: 113-128.

of the highest thicknesses of tissues are found on reefs around Papua New Guinea where the water is far warmer that the Great Barrier Reef.

Over geological time, coral reefs have been bleached and almost all those that did bleach have since recovered. Bleaching is not exclusive to the last 25 years. It occurs in summer when there is a combination of low cloud cover and light winds. This drives water temperatures up by a couple of degrees. Over the last 25 years, water temperatures along the Great Barrier Reef have not increased by more than 1°C and the reported increase in bleaching may well reflect the number of scientists and environmentalists now observing this phenomenon.

Coral bleaching has been described as a vivid demonstration of climate change in action[1653] but again, this media view is contrary to much published science. Growth rate records for massive corals show a small but significant increase over the last 100 years. This is related to the small but significant temperature increase that has occurred in the post Little Ice Age times from the late 19th Century to the present. This is not surprising as most corals like it warm.

Some coral species are killed by unusually elevated temperatures. These are not the long-lived massive corals but rather the plate and staghorn corals. These more specialist corals have the philosophy of a weed (i.e. live fast and die young). The massive corals are like a forest giant tree and live for hundreds of years, hence can withstand the more extreme conditions of high temperature and cyclones that temporarily destroy their frail and fast-growing cousins. Even so, the frail corals have a trick up their sleeve for surviving higher temperature conditions. They replace their embedded symbiotic organisms[1654] with a new resident more suitable to the changed conditions.[1655] If sea temperature rises, corals just migrate.

In 1830, Charles Lyell[1656] suggested coral atolls grow on top of sinking volcanoes, a concept validated a few years later by Charles Darwin.[1657] Darwin received the first volume of Lyell's book in 1830 and took it with him on HMS *Beagle* while the two later volumes were sent to him *en route*. He became fascinated with coral reefs and suggested that the common circular ring shape

---

[1653] http://news.bbc.co.uk/1/hi/sci/tech/4772715.stm

[1654] Zooxanthellae

[1655] Kuehl, M., Cohen, Y., Dalsgaard, T., Jorgensen, B. B. and Revsbech, N. P. 1995: Microenvironment and photosynthesis of zooxanthellae in scleractinian corals studied with microsensors for O sub(2), pH and light. *Marine Ecology Progress Series* 117: 159-172.

[1656] Lyell, C. 1830: *Principles of geology, being an attempt to explain the former changes of the Earth's surface, by reference to causes now in operation.* John Murray.

[1657] Darwin, C. 1837: On certain areas of elevation and subsidence in the Pacific and Indian Oceans, as deduced from the study of coral formations. *Proceedings of the Geological Society of London* 2: 552-554.

of atolls was due to formation around the circular rim of ancient sunken volcanoes. Darwin's voyage allowed him to view many volcanic islands, coral reefs and atolls.

Darwin suggested that the coralline limestone on the atoll could be up to about 180 metres thick. In his 1842 book *The Structure and Distribution of Coral Reefs*, Darwin[1658] showed that volcanoes were at various heights above the sea floor. If sea level falls or a volcano rises from the sea floor, coral attaches itself to the volcano, only to be killed upon later exposure to air. Vanuatu, for example, has many dead coral reefs well above sea level due to the rise of a local volcano,[1659,1660,1661] a feature that is seen in many parts of the world (e.g. Indonesia,[1662] Papua New Guinea,[1663] Barbados[1664]).

As sea level rises, mid-oceanic volcanoes become inundated and coral progressively grows vertically and horizontally. Coral normally grows to maximum depths of 40 metres. Coral growing on sinking volcanoes produces coral atolls. Sea level rise or volcanic island sinking produces coral atolls, it does not destroy them. Darwin showed this in 1842.

In the late 19th and early 20th Century, Professor Sir T. W. Edgeworth David drilled atolls to test Darwin's theory. In 1897, David commenced a drilling program on the island of Funafuti (now part of Tuvalu). After a number of attempts over the next decade, the drilling reached about 340 metres and was still in coralline limestone. David argued that Darwin's theory of coral reefs had been validated. In 1934, Japanese scientists drilled to 431.7 metres on the island of Kita Daito, east of Okinawa. They too drilled coralline limestone and did not reach a basalt volcano.

After World War II, there was interest in using atolls as a place to test atomic bombs and both Bikini and Enewetak atolls in the South Pacific were drilled. In 1952, a basalt volcano was reached at a depth of more than 1300

[1658] Darwin, C. 1842: *The Structure and Distribution of Coral Reefs*. D. Appleton and Co.

[1659] Neef, G. and Veeh, H. H. 1977: Uranium series ages and late Quaternary uplift in the New Hebrides. *Nature* 269: 682-683.

[1660] Neef, G. and Hendry, C. 1988: Late Pleistocene-Holocene acceleration of uplift rate in southwest Erromango Island, Southern Vanuatu, South Pacific: relation to the growth of the Vanuatuan mid sedimentary basin. *Journal of Geology* 96: osti: 7036330.

[1661] Cabioch, G. and Ayliffe, L. K. 2001: Raised coral terraces at Malakula, Vanuatu, Southwest Pacific, indicate high sea level during Marine Isotope Stage 3a. *Quaternary Research* 56: 357-365.

[1662] Bard, E., Jouannic, C., Hamelin, B., Pirazzoli, P., Arnold, M., Faure, G., Sumosusatro, P. and Syaefudin 1996: Pleistocene sea levels and tectonic uplift based on dating of corals from Sumba, Indonesia. *Geophysical Research Letters* 23: 1473-1476.

[1663] Chappell, J., Omura, A., McCulloch, M., Pandolfi, J., Ota, Y. and Pilians, B. 1996: Reconciliation of late Quaternary sea levels derived from coral terraces at Huon Peninsula with deep sea oxygen isotope records. *Earth and Planetary Science Letters* 80: 241-251.

[1664] Bender, M. L., Fairbanks, R. G., Taylor, F. W., Matthews, R. K., Goddard, J. G. and Broecker, W. S. 1979: Uranium-series dating of the Pleistocene reef tracts of Barbados, West Indes. *Geological Society of America Bulletin* 90: 577-594.

metres in two drill holes.[1665] Despite a combination of rapid subsidence of oceanic basalt volcanoes and a 130-metre sea level rise over the last 14,000 years, coral reefs in the South Pacific were not killed. They just kept growing. Darwin's coral atoll theory has now been validated by more than 170 years of independent interdisciplinary science.

A popular concern is that the Pacific atoll island nation Tuvalu is being inundated as a result of modern sea level rise. It is suggested in the popular press that this sea level rise is because the addition of $CO_2$ to the atmosphere by humans is creating global warming resulting in ice melting and sea level rise. Maybe the sporadic tidal inundation of Tuvalu by the sea is related to something else, such as the blasting out of a wartime harbour, the quarrying of huge amounts of coralline limestone for the 2 km World War II airstrip, the quarrying of coralline limestone for roads, subsidence due to extraction of underground fresh water, subsidence due to the compaction of coralline sand and subsidence due to traffic? Over-pumping of ground water has drawn seawater into aquifers. Tuvalu's problems have nothing to do with global climate change. They relate to the factors above and the fact that the floor of the Pacific Ocean is sinking. At times, coral-atoll-capped submarine volcanoes sink at such an extraordinary rate that the upward growth of coral cannot keep pace. This results in coral-capped guyots, a common topographic feature on the floor of the Pacific Ocean.

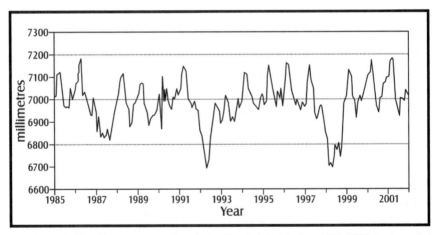

*Figure 33: Monthly measurements of sea level at Funafuti, Tuvalu, from 1985–2002 showing cyclical relative sea level changes but no absolute sea level change.[1666] There is no evidence to suggest that Tuvalu is being inundated by modern sea level change.*

[1665] Branagan, D. 2005: *T. W. Edgeworth David: A Life*. National Library of Australia.
[1666] USLC (Record $A_m$).

Similar concerns have been expressed in the popular press about other atoll island nations. The Maldives, we are told, are about to be inundated as a result of sea level rise caused by the addition of human-induced $CO_2$ into the atmosphere. Well, are they? The Maldives, a collection of 1200 islands in the Indian Ocean, are only 1 to 2 metres above the current sea level. Analysis of three islands in South Maalhosmadulu Atoll shows that the islands formed on a foundation of lagoonal sediments between 5500 and 4500 years ago when the reef surface was as much as 2.5 metres below sea level.[1667] Since then, sea level has risen. The islands reached their current dimensions 4000 years ago during a 1500-year period of rapid accumulation. Over the last 4000 years, the peripheral ridge has been subject to seasonal and longer-term shoreline changes while the outer reef has grown upward, reducing the energy window and confining the islands. Some 3900 years ago the ocean was 1.1 to 1.2 metres above the modern sea level following a very warm climate. The Indian Ocean at the Maldives was 0.1 to 0.2 metres higher 2700 years ago than at present and during the Medieval Warming, the sea level was 0.5 to 0.6 metres higher than at present.

Since 1970, sea level at the Maldives has fallen by 0.2 to 0.3 metres, and between 1900 and 1970, sea level was higher than at present.[1668] If there is significant Late 20th Century Warming and associated sea level rise, it will just reactivate the upward reef growth. This is in accord with the work of Darwin more than 150 years ago. Contrary to popular commentaries on the fragility of reef systems, atolls are very resilient and grow during periods of sea level rise. Sea level has been rising and falling as a result of short-term local events. In the light of this, it is hard to conclude that the Maldives will become inundated by the Indian Ocean as a result of modern temperature changes. Sea level changes in modern times are tiny compared to sea level changes associated with periods of change between ice ages and interglacials.

Drilling of coral reefs in Florida and the Bahamas and dating of the corals shows a 20-metre rise in sea level 17,000–12,500 years ago.[1669,1670] A rapid rise of 24 metres 12,000–11,000 years ago was followed by a decreased rate of sea level rise 11,000–10,500 years ago. Sea level rose by 17 metres 8500–6500

[1667] Kench, P. S., McLean, R. F. and Nicol, S. L. 2005: New model of reef-island evolution: Maldives, Indian Ocean. *Geology* 33: 145-148.

[1668] Morner, N.-A., Tooley, M. and Possnert, G. 2004: New perspectives for the future of the Maldives. *Global and Planetary Change* 40: 177-182.

[1669] Digerfeldt, G. and Hendry, M. D. 1987: An 8,000 year Holocene sea-level record from Jamaica: Implications for interpretation of Caribbean reef and coastal history. *Coral Reefs* 5: 165-169.

[1670] Neimann, A. C. and Hearty, P. J. 1996: Rapid sea-level changes at the close of the last interglacial (Substage 5e) recorded in Bahamian island geology. *Geology* 24: 775-778.

years ago and during this time, there was a rapid 6.5-metre sea level rise 7600–7200 years ago. This rapid rise was probably due to the surging of the unstable West Antarctic Ice Sheet and overall changes in the amount of marine ice 8000–6000 years ago. Between 6500 and 5000 years ago, sea level rose 3.7 metres and another 7.5 metres over the last 5000 years. The rise is highly variable. Throughout post-glacial times, sea level rose by 9.5 mm per year and, about 9000 years ago, slowed to 1.25 mm per year. The maximum rate of sea level rise was 45 mm per year between 7600 and 7200 years ago. These changes in Florida and the Bahamas are a combination of local and global sea level changes and differ from local and global sea level studies in other areas.[1671]

The numerous sea level rises and very rapid rates of sea level rise did not kill coral reefs. When sea level rose quickly (up to 45 mm per year) during post-glacial ice sheet melting and breaching of melt water dams, coral reefs were drowned but did not die.[1672] A complete continuous record of shallow water reef growth during post-glacial times (14,000 years to now) shows that coral reefs were not killed by sea level rises that were far faster than anything recorded today.[1673,1674]

## Dissolved $CO_2$ in seawater

Oceans have a vast storage capacity for $CO_2$, far more than the atmosphere or plants. The oceans contain about 80 times as much $CO_2$ as the atmosphere and provide a better understanding of surface $CO_2$ than the atmosphere. Only 0.001% of all carbon in the atmosphere-ocean-upper crust system is in the atmosphere. The oceans continually remove $CO_2$ to form carbonate sediments which become carbonate rocks. Carbonate rocks contain 40,000 times more $CO_2$ than the atmosphere. Soils contain more carbon than all living matter. The surface area of the oceans, some 70% of the surface area of planet Earth, provides a large surface area for adsorption of $CO_2$. The wind also pumps $CO_2$ into seawater.[1675] The solubility of $CO_2$ in water is

[1671] Sloss, C. R., Murray-Wallace, C. V. and Jones, B. G. 2007: Holocene sea-level change on the southeast coast of Australia: a review. *The Holocene* 17: 999-1014.

[1672] Toscano, M. A. and Lundberg, J. 1998: Early Holocene sea-level record from submerged fossil reefs on the southeast Florida margin. *Geology* 26: 255-258.

[1673] Toscano, M. A. and Lundberg, J. 1999: Submerged Late Pleistocene reefs on the tectonically-stable S.E. Florida margin: high-precision geochronology, stratigraphy, resolution of Substage 5a sea-level elevation, and orbital forcing. *Quaternary Science Reviews* 18: 753-767.

[1674] Ludwig, K. R., Muhs, D. R., Simmons, K. R., Halley, R. B. and Shinn, E. A. 1996: Sea-level records at approximately 80 ka from tectonically stable platforms; Florida and Bermuda. *Geology* 24: 211-214.

[1675] Smith, S. D. and Jones, E. P. 1985: Evidence of wind-pumping of air-sea gas exchange based on direct measurements of $CO_2$ fluxes. *Journal of Geophysical Research* 90: 869-875.

an inverse solubility. With many substances (e.g. salt, sugar), the warmer the water the more solid can dissolve in the water. With $CO_2$, it is the opposite. The colder the water, the more $CO_2$ can dissolve and the higher the pressure, the more $CO_2$ can dissolve.

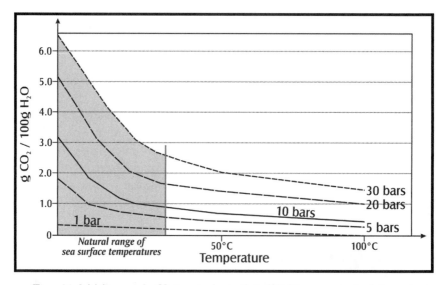

*Figure 34: Solubility curves for $CO_2$ in water showing that cold high-pressure water (i.e. bottom waters in the ocean) dissolve huge amounts of $CO_2$. These bottom waters are undersaturated in $CO_2$ hence can dissolve the monstrous amounts of $CO_2$ emitted by submarine volcanoes whereas the warm surface waters are nearly saturated in $CO_2$. The oceans have a huge capacity to absorb and emit $CO_2$ and each small increment of ocean temperature and pressure has a major effect on $CO_2$ solubility.*

The long-term global average sea surface temperature is 15 °C. At 15°C at sea level, seawater can dissolve its own volume of $CO_2$. At 10°C, seawater absorbs 19% more than its own volume of $CO_2$, while at 20°C it absorbs 12% less than its own volume of $CO_2$.[1676] This is supported by the very tight correlation between the global average sea surface temperature and the amount of $CO_2$ in the air.

In polar areas, the cold surface water adsorbs more $CO_2$ than elsewhere. This water is carried to the tropics in cold dense bottom currents, and upwells to the surface releasing $CO_2$ as it warms. About 70% of ocean degassing occurs by this process (thermally driven solubility pump), and the other 30%

---

[1676] Endersbee, L. 2008: Carbon dioxide and the oceans. *Focus* 151: 20-21.

is degassed by life (biological pump).[1677] The role of subsurface bacteria is unknown and not factored into such figures. If these biological processes were removed, the level of atmospheric $CO_2$ would increase five-fold.[1678] A slight variation in the amount of floating organisms in the oceans or bacteria beneath our feet could account for variations far greater than all the human input of $CO_2$ into the atmosphere.

At a depth of 10 metres, seawater can dissolve twice its own volume of $CO_2$, as the amount that can dissolve in seawater increases with decreasing temperature and increasing pressure. Cold seawater under high pressure at the bottom of the oceans contains a huge amount of $CO_2$ and can dissolve much more. If water rises to the surface, $CO_2$ is released. We see the same process with aerated drinks. When the pressure is released (i.e. opening the bottle or can), $CO_2$ bubbles appear and continue to rise as the drink warms to room temperature.

Exchange of $CO_2$ between the atmosphere and ocean is well known.[1679,1680] An upper limit on how much the $CO_2$ concentration in the atmosphere will rise if all available fossil fuel is burned can be calculated. In order to permanently double the current level of $CO_2$ in the atmosphere and keep the oceans and atmosphere balanced, the atmosphere needs to be supplied with 51 times the present amount of atmospheric $CO_2$. The total amount of carbon in known fossil fuel could only produce 11 times the amount of $CO_2$ in the atmosphere.[1681] Unless we change the fundamental laws of chemistry and change the way in which oceans work, humans do not have enough fossil fuel on Earth to permanently double the amount of $CO_2$ in the atmosphere. If humans burned all the available fossil fuels over the next 300 years, there would be up to 15 turnovers of $CO_2$ between the oceans and atmosphere and all the additional $CO_2$ would be consumed by ocean life and precipitated as calcium carbonate in sea floor sediments.[1682]

Burning of fossil fuels adds $CO_2$ to the atmosphere and, in turn, to the oceans. Fossil fuel contains no $C^{14}$ (derived from cosmic radiation and nuclear bombs) and hence the increase in the $C^{13}$ and $C^{12}$ of seawater has been

---

[1677] Volk, T. and Liu, Z. 1988: Controls of $CO_2$ sources and sinks in the Earth scale surface ocean: temperature and nutrients. *Global Biogeochemical Cycles* 2: 73-89.

[1678] Eriksson, E. 1963: Possible fluctuations in atmospheric carbon dioxide due to changes in the properties of the sea. *Journal of Geophysical Research* 68: 3871-3876.

[1679] Revelle, R. and Suess, H. E. 1957: Carbon dioxide exchange between atmosphere and ocean and the question of an increase of atmospheric $CO_2$ during the past decades. *Tellus* 9: 18-27.

[1680] Skirrow, G. 1975: The dissolved gases – carbon dioxide. In: *Chemical Oceanography, Vol. 2*, 2nd Ed. (eds Riley, J. P. and Skirrow, G.), *Academic Press*.

[1681] Jaworowski, Z., Segalstad, T. V. and Hisddal, V. 1992: Atmospheric $CO_2$ and global warming. A critical view, Second Revised Edition. *Norsk Polarinstitutt Meddelelser* 119: 1-76.

[1682] Abelson, P. H. 1990: Uncertainties about global warming. *Science* 247: 1529.

used to calculate the addition of $CO_2$ derived from coal and oil burning.[1683] This calculation ignores the contribution of $CO_2$ from other sources such as bacteria in rocks, soil bacteria, volcanoes, floating micro-organisms and burning of wood, grass, stubble and dung. Even if these other sources are ignored, $CO_2$ of fossil fuel origin is confined to the near surface of the oceans and totals about 3% of the $CO_2$ of that surface water.[1684]

Calculation of how much $CO_2$ is dissolved in the oceans at high latitudes is fraught with difficulty. The $CO_2$ outgassing from tropical oceans plus $CO_2$ of human origin equals the amount of $CO_2$ dissolved in the polar oceans. Outgassing occurs in the Southern Ocean between 44°S and 59°S and strongly in the tropics, with vigorous uptake at mid latitudes and in the far Northern Hemisphere.[1685]

Some 85% of the planet's volcanoes are submarine and account for 75% of the heat transferred to the surface from molten rocks.[1686] The material for these volcanoes, including $CO_2$, derives from the mantle of the Earth. They occur along mid ocean ridges and involve processes that have been taking place for billions of years.[1687] Molten rock commonly has a very high content of dissolved gases such as $H_2O$ and $CO_2$.[1688,1689] When molten rocks rise and cool, they release monstrous amounts of $CO_2$ and other gases.[1690] In the past this gas release provided the initial high $CO_2$ content of the Earth's atmosphere before plants and animals with hard parts were abundant.[1691] Terrestrial volcanoes explode because they suddenly release gas whereas the

[1683] Key, R. 2006: The dangers of ocean acidification. Scientific American March 2006, 58-65.

[1684] Jaworowski, Z., Segalstad, T. V. and Hisddal, V. 1992: Atmospheric $CO_2$ and global warming. A critical view, Second Revised Edition. Norsk Polarinstitutt Meddelelser 119: 1-76.

[1685] Mikaloff Feltcher, S. E., Gruber, N., Jacobson, A. R., Gloor, M., Doney, S. C., Dutkiewicz, S., Gerber, M., Follows, M., Joos, F., Lindsay, K., Menemenlis, D., Mouchet, A., Müller, S. A. and Sariento, J. L. 2007: Inverse estimates of the oceanic sources and sinks of natural $CO_2$ and the implied ocean carbon transport. Global Biogeochemical Cycles 21: GB1010, doi:10.1029/2006GB002751.

[1686] Crisp, J. A. 1984: Rates of magma emplacement and volcanic output. Journal of Volcanology and Geothermal Research 20: 177-211.

[1687] Lagabrielle, Y., Goslin, J., Martin, H., Thirot, J. L. and Auzende, J. M. 1997: Multiple active spreading centers in the hot North Fiji Basin (Southwest Pacific): A possible model for Archaean seafloor dynamics. Earth and Planetary Science Letters 149: 1-14.

[1688] Stolper, E. and Holloway, J. R. 1988: Experimental determination of the solubility of carbon dioxide in molten basalt at low pressure. Earth and Planetary Science Letters 87: 397-408.

[1689] Shilobreyeva, S. N. and Kadik, A. A. 1990: Solubility of $CO_2$ in magmatic melts at high temperatures and pressures. Geochemistry International 27: 31-41.

[1690] Marty, B. and Zimmermann, L. 1999: Volatiles (He, C, N, Ar) in mid-ocean ridge basalts: assessment of shallow-level fractionation and characterisation of source composition. Geochimica et Cosmochimica Acta 63: 3619-3633.

[1691] Bottinga, Y. and Javoy, M. 1989: MORB degassing: evolution of $CO_2$. Earth and Planetary Science Letters 95: 215-225.

weight of at least 3 km of water stops most submarine volcanoes exploding. In gas-rich volcanoes, the most abundant gas is $CO_2$.

The $CO_2$ from submarine volcanoes does not bubble up from the mid ocean ridges and enter the atmosphere, as it is dissolved by the cold bottom waters. By contrast, methane and helium do. The high-pressure cool bottom waters dissolve all the volcanic $CO_2$. Many mid ocean ridge lavas are supersaturated in $CO_2$[1692] and the release of $CO_2$ into bottom waters is a part of the normal sea floor spreading process,[1693,1694,1695] The total length of submarine spreading centres is about 64,000 km so the amount of $CO_2$ released is huge.[1696] Huge volumes of molten rock and an accumulation of gas, mainly $CO_2$,[1697] are present beneath these spreading centres.

Volcanic gases such as $CO_2$ escape from the molten rock before, during and after eruptions.[1698] Unless released $CO_2$ can be measured by instruments before, during and after a submarine volcanic eruption, then the amount is unknown. However, if submarine volcanic rocks suddenly freeze to a glass, then this traps some of the $CO_2$.[1699,1700] Although the molten rock may have lost most of its $CO_2$ during its rise and cooling, the trapped remainder permits estimation of a minimum figure initial content.[1701]

Not only are large amounts of $CO_2$ released from rising and cooling molten rocks at or beneath the sea floor at the mid ocean ridges, the associated

[1692] Jendrzejewski, N., Trull, T. W., Pineau, F. and Jovoy, M. 1997: Carbon solubility in mid-ocean ridge basaltic melt at low pressures (250-1950 bar). *Chemical Geology* 138: 81-92.

[1693] Gerlach, T. M. 1989: Degassing of carbon dioxide from basaltic magma at spreading centers, II. Mid-ocean ridge basalts. *Journal of Volcanology and Geothermal Research* 39: 221-232.

[1694] Jendrzejewski, N., Pinean, F. and Javoy, M. 1992: Water and carbon contents and isotopic compositions in Indian Ocean MORB. *EOS* 73: 352.

[1695] Dixon, J. E., Stolper, E. M. and Holloway, J. R. 1995: An experimental study of water and carbon dioxide solubilities in mid-ocean ridge basaltic liquids, Part II. Applications to degassing. *Journal of Petrology* 36: 1633-1646.

[1696] Pineau, F. and Javoy, M. 1994: Strong degassing at ridge crests: the behaviour of dissolved carbon and water in basaltic glasses at 14°N (M.A.R.). *Earth and Planetary Science Letters* 123: 179-198.

[1697] Hauri, E., Shimizu, N., Dieu, J. J. and Hart, S. R. 1993: Evidence for hot-spot related carbonatite metasomatism in the oceanic upper mantle. *Nature* 365: 221-227.

[1698] Kingsley, R. H. and Schilling, J.-G. 1995: Carbon in mid-Atlantic ridge basalt glasses from 28°N to 63°N: evidence for a carbon-enriched Azores mantle plume. *Earth and Planetary Science Letters* 129: 31-53.

[1699] Des Marais, D. J. and Moore, J. G. 1984: Carbon and its isotopes in mid-oceanic basaltic glasses. *Earth and Planetary Science Letters* 69: 43-57.

[1700] Dixon, J. E. and Stolper, E. M. 1995: An experimental study of water and carbon dioxide solubilities in mid-ocean ridge basaltic liquids. Part II: Applications to degassing. *Journal of Petrology* 36: 1633-1646.

[1701] Marty, B. and Tolstikhin, I. N. 1998: $CO_2$ fluxes from mid-ocean ridges, arcs, and plumes. *Chemical Geology* 145: 233-248.

hot springs also release small amounts of $CO_2$.[1702] Submarine hydrothermal hot springs are acidic, caused by the venting of $CO_2$ and sulphuric acid.[1703] The $CO_2$ content varies greatly[1704] and can be removed from hot springs by precipitation of carbonates in fluid-rock reactions. These springs release an estimated 0.3 to 1.2% of the annual input of $CO_2$ into the oceans.[1705] The amount released by gas vents in the volcanoes is unknown but the evidence suggests a much higher amount. Volcanoes add far more $CO_2$ to the oceans and atmosphere than humans.

Volcanically active areas, commonly rimming the oceans, where slabs of crust are pushed under adjacent slabs, also release large amounts of $CO_2$ from gas vents and hot springs.[1706] This process of recycling of the Earth's rocks has been happening for a very long time and has provided variable but large amounts of $CO_2$ to the Earth's atmosphere.[1707] Visible volcanoes represent only 15% of the Earth's total but are the only volcanoes accounted for in IPCC models. They are explosive due to the sudden expansion of trapped $H_2O$ and $CO_2$ in molten rocks.

When one slab of crust pushes underneath another, not only is $CO_2$ released from the resulting molten rocks but also from gas vents resulting from heating limey rocks.[1708,1709] The volume released from vents is hundreds of times greater than that released from molten rocks and the sudden expulsion can form craters 1000 metres in diameter.[1710] These $CO_2$ crater volcanoes[1711] are quite common globally, for example in the Arctic Ocean

[1702] Resing, J. A., Lupton, J. E., Feely, R. A. and Lilley, M. D. 2004: $CO_2$ and $^3$He in hydrothermal plumes: implications for mid-ocean ridge $CO_2$ flux. *Earth and Planetary Science Letters* 226: 449-464.

[1703] Mottl, M. J. and McConachy, T. F. 1990: Chemical processes in buoyant hydrothermal plumes on the East Pacific Rise near 21°N. *Geochimica et Cosmochimica Acta* 54: 1911-1927.

[1704] Sansone, F. J., Mottl, M. J., Olson, E. J., Wheat, C. G. and Lilley, M. D. 1998: $CO_2$-depleted fluids from mid-ocean ridge-flank hydrothermal springs. *Geochimica et Cosmochimica Acta* 62: 2247-2252.

[1705] LeQuéré, C. and Metzel, N. 2004: Chapter 12: Natural processes regulation the ocean uptake of $CO_2$. In: *SCOPE 62, The global carbon cycle: Integrating humans, climate, and the natural world* (eds Field, C. B. and Raupach, M. R.). Island Press, 243-256.

[1706] Hilton, D. R., Fischer, T. P., Hauri, E. and Shaw, A. M. 2006: Controls on the He-C systematics of the Izu-Bonin-Marianas (IBM) subduction zone. *Geochimica et Cosmochimica Acta* 70: doi:10.1016/j.gca.2006.06.507.

[1707] Yamamoto, J., Watanabe, M., Nozaki, Y. and Sano, Y. 2001: Helium and carbon isotopes in fluorites: implications for mantle carbon contribution in an ancient subduction zone. *Journal of Volcanology and Geothermal Research* 107: 19-26.

[1708] Kerrick, D. M. and Caldeira, K. 1998: Metamorphic $CO_2$ degassing from orogenic belts. *Chemical Geology* 145: 213-232.

[1709] Schuiling, R. D. 2005: *Our bubbling Earth*. Elsevier.

[1710] Fytikas, M. 1989: Updating of the geological and geothermal research on Milos Island. *Geothermics* 18: 485-496.

[1711] Maar.

where large craters have released massive amounts of heat and $CO_2$ into Arctic waters.[1712] Elsewhere in the Arctic, submarine volcanic and hot spring activity[1713] releases hot $CO_2$ which heats a large volume of rocks[1714] and even pools beneath the crust for later leakage into the atmosphere.[1715]

The Amazon River has a plume of low salinity water that stretches 3000 km into the tropical Atlantic Ocean which, when mixed with seawater, should emit $CO_2$ into the atmosphere. However, significant amounts of $CO_2$ are adsorbed from the atmosphere into this water. Nitrogen-fixing bacteria, reliant on the nutrients in the Amazon River runoff, change the ocean-air balance so that instead of emitting $CO_2$, the ocean adsorbs it.[1716] Carbon sequestration of 15 megatonnes per annum occurs in a region that was previously thought to emit $CO_2$ into the atmosphere. Nature continues to amaze us with little surprises.

Floating organisms extract dissolved $CO_2$ in ocean water and sunlight for photosynthesis.[1717] Not only is $CO_2$ plant food, it increases cell size.[1718] It is used by organisms to make carbonate shells and, when the floating organism dies, the shell sinks. In this way $CO_2$ is constantly removed from the oceans, estuaries and lakes. An increase in $CO_2$ and nutrients in seawater is the main factor that increases the growth rate of the floating organisms. Variations in sea surface temperature change the proportions of heavy and light oxygen isotopes in the shells[1719] and life preferentially accumulates the light isotope

---

[1712] Sohn, R. A., Willis, C., Humphris, S., Shank, T. M., Singh, H., Edmonds, H. N., Kunz, C., Hedman, U., Helmke, E., Jakuba, M., Liljebladh, B., Linder, J., Murphy, C., Nakamora, K., Saato, T., Schlindwein, V., Stranne, C., Taisenfreund, M., Upschurch, L., Winsor, P., Jakobsson, M. and Soule, A. 2008: Explosive volcanism on the ultraslow-spreading Gakkal ridge, Arctic Ocean. *Nature* 453: 1236-1238.

[1713] Snow, J., Hellebrand, E., Jokat, W. Muhe, R. 2001: Magmatic and hydrothermal activity in the Lena Trough, Arctic Ocean. *Transactions of the American Geophysical Union* 82: 193.

[1714] Schuiling, R. D. 2004: Thermal effects of massive $CO_2$ emissions associated with subducted volcanism. *Comptes Rendus Geosciences* 336: 1053-1059.

[1715] Schuiling, R. D. and Kreulen, R. 1979: Are thermal domes heated by $CO_2$-rich fluids from the mantle? *Earth and Planetary Science Letters* 43: 298-302.

[1716] Cooley, S. R., Coles, V. J., Subramanian, A. and Yager, P. L. 2007: Seasonal variations in the Amazon plume-related atmospheric carbon sink. *Global Biogeochemical Cycles* doi:10.1029/2006Gboo2831.

[1717] Arrigo, K. R., Robinson, D. H., Worthen, D. L., Dunbar, R. B., DiTullio, G. R., VanWoert, M. and Lizotte, M. P. 1999: Phytoplankton community structure and the drawdown of nutrients and $CO_2$ in the Southern Ocean. *Science* 283: 365-367.

[1718] Burkhardt, S., Riebesell, U. and Zondervan, I. 1999: Effect of growth rate, $CO_2$ concentration, and cell size on the stable carbon isotope fractionation in marine phytoplankton. *Geochimica et Cosmochimica Acta* 63: 3729-3741.

[1719] Wolf-Gladrow, D. A., Riebesell, U., Burkhardt, S. and Bijma, J. 2002: Direct effects of $CO_2$ concentration on growth and isotopic composition of marine plankton. *Tellus* 51: 461-476.

of carbon.[1720,1721] Fossilised floating organisms can therefore be used to track sea surface temperature changes as a proxy for climate change and can be used to calculate the $CO_2$ content of atmospheres in former times.

There is a hypothesis that the uplift of mountains changes the rate of weathering by removal of $CO_2$ from the atmosphere to produce a negative greenhouse effect. Increased mountain building increases the draw-down of $CO_2$. The uplift of the Himalayas is proposed as one of the climate drivers that triggered the latest glaciation.[1722,1723] The story goes that increased weathering adds more nutrients to the oceans. For example, the rapid uplift of the Himalayan-Tibetan Plateau led to an increased input of phosphorus to the oceans. This led to a blooming of algae in the oceans between 8 and 4 Ma coincidental with the intensification of the Indian-Asian monsoon. Not only did the uplift of mountains draw down $CO_2$ from the atmosphere, but algal blooming[1724] also depleted the atmospheric $CO_2$. This occurred in advance of glaciation.[1725]

Weathering is principally a process of addition of water and oxygen to rocks, and mountain building may not necessarily control climate through a negative greenhouse effect. For example, the deepest weathering profiles around the world are far older than the uplift of the Himalayas.[1726] Glaciation in the Northern Hemisphere started at about 2.67 Ma yet Himalayan mountain building began at 23 Ma, reached its peak at 15 Ma, remained high until 10.5 Ma, slowed gradually to 3.5 Ma and then started to increase again.[1727] Because weathering and erosion (the removal of weathered material) do not occur at the same time, the uplift of the Himalayan-Tibetan Plateau did not occur at the same time as climate change.[1728] Therefore, the hypothesis that mountain building triggers glaciation fails.

---

[1720] Peterson, B. J. and Fry, B. 1987: Stable isotopes in ecosystems studies. *Annual Review of Ecology and Systematics* 18: 293-320.

[1721] Descolas-Gros, C. and Fontungne, M. 2006: Stable carbon isotope fractionation by marine phytoplankton during photosynthesis. *Plant, Cell & Environment* 13: 207-218.

[1722] Ruddiman, W. F. and Kutzbach, J. E. 1991: Plateau uplift and climatic change. *Scientific American*, March 1991: 42-50.

[1723] Raymo, M. E. and Ruddiman, W. F. 1992: Tectonic forcing of late Cenozoic climate. *Nature* 359: 117-122.

[1724] Excess surface water nutrients or eutrophication, commonly by marine algae called coccolithophoridae.

[1725] Fillippelli, G. M. 2008: The global phosphorus cycle: Past, present and future. *Elements* 4: 89-95.

[1726] Ollier, C. D. and Pain, C. F. 1996: *Regolith, soils and landforms.* Wiley.

[1727] Clift, P. D., Hodges, K. V., Heslop, D., Hannigan, R., Long, H. V. and Calves, G. 2008: Correlation of Himalayan exhumation rates and Asian monsoon intensity. *Nature Geoscience* 1: 875-880.

[1728] Ollier, C. D. 2004: Mountain building and climate: Mechanisms and timing. *Geografia fisica e dinamica Quaternaria* 27: 139-149.

Most of the world's oceans are extremely depleted in iron. This is because the oceans are not acid and contain dissolved oxygen. Iron is a micronutrient for photosynthetic micro-organisms. Oceans are depleted in micronutrients and natural iron fertilisation induces algal blooms that extract $CO_2$ from the air and water. Such blooms are normally seen at the surface. However, the supply of nutrients from sediments derived from Antarctica has produced blooms at water depths of 3 km.[1729] Such observations indicate that estimates of the uptake of $CO_2$ by ocean waters can only be a minimum figure. During ice ages, forests and grassland turn to desert. The increased winds blow red iron-bearing desert dust into the oceans. This results in a blooming of micro-organisms in the oceans. The blossoming of these photosynthetic organisms withdraws even more $CO_2$ from the atmosphere.[1730] If $CO_2$ drives climate, then during an ice age red dust blown into the oceans would accelerate the removal of $CO_2$ into the atmosphere and the Earth would suffer from a runaway ice age. This has not happened. Clearly $CO_2$ is not the main driver of climate.

Increased human activity, principally from farming and livestock grazing, has produced up to 500% more dust from wind erosion of surface sediments in the USA.[1731] There has been a five-fold increase in dust into alpine ecosystems and this is stimulating the growth of photosynthetic organisms in the oceans, thereby removing $CO_2$ from the atmosphere. Much of North Africa, especially the Sahara, contributes iron-rich dust to the oceans.[1732]

## Acid oceans

Oceans have an acidity, measured as pH, of 7.9 to 8.2. This figure is higher than neutral (pH = 7) which means that the oceans are alkaline. The pH scale ranges from 0 to 14 – pH 6 is ten times more acid than pH 7 and pH 5 is a

[1729] Pollard, R. T., Salter, I., Sanders, R. J., Lucas, M. I., Moore C. M., Mills, R. A., Statham, P. J., Allen, J. T., Baker, A. R., Bakker, D. C. E., Charette, M. A., Fielding, S., Fones, G. R., French, M., Hickman, A. E., Holland, R. J., Hughes, J. A., Jickells, T. D., Lampitt, R. S., Morris, P. J., Nédélec, F. H., Nielsdóttir, M., Planquette, H., Popova, E. E., Poulton, A. J., Read, J. F., Seeyave, S., Smith, T., Stinchcombe, M., Taylor, S., Thomalla, S., Venables, H. J., Williamson, R. and Zubkov, M. V. 2009: Southern Ocean deep-water carbon export enhanced by natural iron fertilisation. *Nature* 457: 577-581.

[1730] de Baar, H. J. W., de Jong, J. T. M., Bakker, D. C. E., Löscher, B. M., Veth, C., Bathmann, U. and Smetacek, V. 1999: Importance of iron for plankton blooms and carbon dioxide drawdown in the Southern Ocean. *Nature* 373: 412-415.

[1731] Neff, J. C., Ballantyne, A. P., Farmer, G. L., Mahowald, N. M., Conroy, J. L., Landry, G. C., Overpeck, J. T., Painter, T. H., Lawrence, C. R. and Reynolds, R. L. 2008: Increasing aeolian dust deposition in the western United States linked to human activity. *Nature Geoscience* 1: 189-195.

[1732] Duce, R. A. and Tindale, N. W. 1999: Atmospheric transport of iron and its deposition in the ocean. *Limnology and Oceanography* 36: 1715-1726.

hundred times more acid than pH 7. To acidify seawater from pH 8 to pH 6, an extraordinarily large amount of acid is needed. Once there is acid present, sediments, rocks and shells become very reactive. These reactions destroy acid and the oceans return to their normal alkaline state.

The most alkaline waters in the oceans occur in the centre of ocean circulation patterns whereas less alkaline waters occur at sites of upwelling where deep ocean water is brought to the surface. This water is rich in nutrients. As a result photosynthetic micro-organisms thrive and become the bottom of the food chain for other abundant marine life.

If $CO_2$ dissolves in seawater, the oceans should become more acid.[1733] However, when $CO_2$ is dissolved in seawater, it is neutralised to bicarbonate by reacting with dissolved carbonate and borate in water and with calcium carbonate sediment covering much of the ocean floor. If the oceans became acid (pH <7), then it is argued that shells of marine organisms could dissolve.[1734] In some parts of the oceans, it is claimed that the pH has decreased since the Industrial Revolution by 0.1.[1735] Considering that the oceans have a pH of 7.9 to 8.2, this variation could be for a great diversity of reasons. Acidification is greatly exaggerated in the popular media as a potential environmental catastrophe[1736] and projections are made that by the end of the 21st Century, pH may have decreased by as much as 0.4.[1737] The geological record shows that shells do not dissolve, otherwise there would be no shelly fossils. The oceans are saturated with calcium carbonate to a depth of 4.8 km. If any more $CO_2$ were added to the oceans, then calcium carbonate would precipitate.[1738]

An additional small amount of $CO_2$ in the oceans would have precipitated gypsum from the oceans instead of calcium carbonate. This has not been found. If the $CO_2$ content were extremely high, dolomite would have

---

[1733] $CO_2 + H_2O \Leftrightarrow H_2CO_3$; $H_2CO_3 \Leftrightarrow H^+ + HCO_3^-$; $HCO_3^- \Leftrightarrow H^+ + CO_3^{2-}$. In the oceans at pH 7.9 to 8.2, $CO_2$ exists as dissolved gas (1%), $HCO_3^-$ (93%) and $CO_3^{2-}$ (8%). Calcium in seawater binds $CO_2$ into insoluble carbonates of calcium in shells, coral reefs and mineral precipitates ($Ca^{2+}_{[aq]} + CO_3^{2-}_{[aq]} \Leftrightarrow CaCO_3 \downarrow$). Furthermore, trapped seawater in sediments precipitates carbonate cement. By these processes $CO_2$ is removed from the atmosphere and stored in marine sediments as fossils, cement and rock. At present, $CaCO_3$ plankton shells can dissolve back into seawater at a depth of >4.8 km.

[1734] Calderia, K. and Wickett, M. E. 2003: Anthropogenic carbon & ocean pH. *Nature* 425: 365.

[1735] Haugan, P. M. and Drange, H. 1996: Effects of $CO_2$ on the ocean environment. *Energy, Conservation and Management* 37: 1019-1022.

[1736] Doney, S. C. 2006: The dangers of ocean acidification. *Scientific American*. March 2006, 58-65.

[1737] Andersson, A. J., Mackenzie, F. T. and Lerman, A. 2006: Coastal ocean $CO_2$-carbonic acid-carbonate sediment system of the Anthropocene. *Global Biological Cycles* 20: GBIS92.

[1738] Broeckner, W. S., Takahashi, T., Simpson, H. J. and Peng, T.-H. 1979: The fate of fossil fuel carbon dioxide and the global carbon budget. *Science* 206: 409-418.

precipitated from the oceans.[1739]   This has not been found except for immediately after Neoproterozoic glaciations. In the Neoproterozoic when atmospheric $CO_2$ was far greater than 1%, huge volumes of dolomite were precipitated in the Neoproterozoic seas. This means that the balance of $CO_2$ between the oceans and atmosphere we see today has not changed for thousands of millions of years.[1740]   This balance has not changed during times of intense sudden release of $CO_2$ from volcanoes. Increased volcanic production of $CO_2$ correlates well with increased sedimentation of calcium carbonate from the oceans.[1741]   This geological process that has taken place for billions of years is ignored in the computer climate models of the IPCC. This demonstrates once again that the models oversimplify reality due to data deprivation.

During times of coal formation, the atmospheric methane content rises slightly because peat is undergoing decay and attack by micro-organisms. The main coal formations were deposited when there was abundant atmospheric methane and $CO_2$. This was during cold times when micro-organisms were relatively inactive.[1742]   During the Permo-Carboniferous glaciation 300–260 Ma, the high atmospheric methane and $CO_2$ clearly did not have a strong warming effect. In cool marine conditions, methane gas in sediments bonds with water to form methane hydrate. The release of methane hydrate into the atmosphere can occur with earth tremors, submarine volcanic activity, meteorite and comet impact, unloading of sediment with sea level fall, submarine debris flows and warming of seawater.

At 55.8 Ma, there was a short period (about 10,000 years) of rapid warming[1743] resulting from a sudden release of methane hydrate from the ocean floor. Methane hydrate oxidised in the atmosphere to $CO_2$, and water vapour and the oceans were acidic for a short time.[1744,1745]   Understanding this period of warming and ocean acidity comes from the study of the

[1739] Gypsum $CaSO_4.2H_2O$, dolomite $CaMg(CO_3)_2$

[1740] Holland, H. D. 1984: *The chemical evolution of the atmosphere and oceans*. Princeton University Press.

[1741] Budyko, M. I., Ronov, A. B. and Yanshin, A. L. 1987: *History of the Earth's atmosphere*. Springer-Verlag.

[1742] Bartdorff, O., Wallmann, K., Latif, M. and Semenov, V. 2008: Phanerozoic evolution of atmospheric methane. *Global Biogeochemical Cycles* 22: GB1008, doi: 10.1029/2007BG002985.

[1743] Nunes, F. and Norris, R. D. 2006: Abrupt reversal of ocean overturning during the Palaeocene/Eocene warm period. *Nature* 439: doi:10.1038/nature04386.

[1744] Pearson, P. N. and Palmer, M. R. 1999: Middle Eocene seawater pH and atmospheric carbon dioxide concentrations. *Science* 284: 1824-1826.

[1745] Zachos, J. C., Röhl, U., Schellenberg, S. A., Sluijis, A., Hodell, D. A., Kelly, D. C., Thomas, E., Nicolo, M., Raffi, I., Lourens, L. J., McCarren, H. and Kroon, D. 2005: Rapid acidification of the ocean during the Paleocene-Eocene thermal maximum. *Science* 308: 1611-1615.

oxygen and carbon isotopes in calcium carbonate shells.[1746] In what was a catastrophic sudden change in warmth and acidity, these shells did not dissolve. They evolved.[1747] During this period of sudden warming, the depth at which calcium carbonate dissolved in the ocean moved from 4.8 to 2.7 km. A very large amount of $CO_2$ was dissolved in the oceans and was permanently sequestered by submarine weathering processes over a period of 100,000 years.[1748] Methane hydrate is not factored into IPCC models.

This short warming event at 55.8 Ma should have produced a significant sea level rise. Evidence from marine sediments from the New Jersey Shelf, the North Sea and the New Zealand Shelf shows that sea level started to rise 200,000 years before the event and peaked during the warming event. This sea level rise probably resulted from the establishment of a supervolcano (the North Atlantic Igneous Province) which reduced the volume of the oceanic basin.[1749] It is clear that warming events and sea level changes are not fully understood and hence computer modelling is simplistic.

Although rainwater is slightly acid (pH 5.6), by the time it runs over the surface and chemically reacts with minerals in soils and rocks, it enters the oceans as alkaline water.[1750,1751,1752] The salts transported down rivers result from rainwater reacting with rocks and they balance dissolved and atmospheric $CO_2$ by making river water alkaline and slightly saline.[1753] Soils contain far more $CO_2$ than the atmosphere, and during weathering release a huge amount of $CO_2$ which ends up in river systems.[1754] The total dissolved $CO_2$ in river systems depends upon the season, the position of the water in

[1746] Pak, D. K. and Miller, K. G. 1992: Paleocene to Eocene benthic foraminiferal isotopes and assemblages: implications for deepwater circulation. *Palaeoceanography* 7: 405-422.

[1747] Kelly, D. C., Bralower, T. J. and Zachos, J. C. 1998: Evolutionary consequences of the latest Paleocene thermal maximum from tropical planktonic foraminifera. *Palaeogeography, Palaeoclimatology and Palaeoecology* 141: 139-161.

[1748] Zachos, J. C., Röhl, U., Schellenberg, S. A., Sluijis, A., Hodell, D. A., Kelly, D. C., Thomas, E., Nicolo, M., Raffi, I., Lourens, L. J., McCarren, H. and Kroon, D. 2005: Rapid acidification of the ocean during the Paleocene-Eocene thermal maximum. *Science* 308: 1611-1615.

[1749] Courtillot, V., Jaupart, I., Manighetti, P., Tapponnier, P. and Besse, J. 1999: On causal links between flood basalts and continental breakup. *Earth and Planetary Science Letters* 166: 177-195.

[1750] Velbel, M. A. 1993: Temperature dependence of silicate weathering in nature: How strong a negative feedback on long-term accumulation of atmospheric $CO_2$ and global greenhouse warming? *Geology* 21:1059-1061.

[1751] Kump, L. R., Brantley, S. L. and Arthur, M. A. 2000: Chemical weathering, atmospheric $CO_2$ and climate. *Annual Review of Earth and Planetary Sciences* 28: 611-667.

[1752] Gaillardet, J., Dupré, B., Louvat, P. and Allègre, C. J. 1999: Global silicate weathering and $CO_2$ consumption rates deduced from the chemistry of large rivers. *Chemical Geology* 159: 3-30.

[1753] Karim, A. and Veizer, J. 2000: Weathering processes in the Indus River Basin: implications from riverine carbon, sulfur, oxygen, and strontium isotopes. *Chemical Geology* 170: 153-177.

[1754] Telmer, K. and Veizer, J. 1999: Carbon fluxes, $pCO_2$ and substrate weathering in a large northern river basin, Canada: carbon isotope perspectives. *Chemical Geology* 159: 61-86.

the river system and whether dissolved carbon has been converted to $CO_2$.[1755] This process of weathering has been removing $CO_2$ from the atmosphere and soils for billions of years and storing it in rocks.[1756,1757,1758] This removal of $CO_2$ does not trigger glaciation. The higher the temperature and $CO_2$ content, the quicker the removal of $CO_2$ by calcium carbonate precipitation.[1759] Over time, $CO_2$ uptake by soils, rocks, water and life has balanced the release of $CO_2$ into the atmosphere.[1760] This has resulted in the long-term stabilisation of the global surface temperature. Even if this geological stabilisation did not occur, there still would be long-term stabilisation by life.[1761]

Rainwater runs into and accumulates in freshwater lakes which are invariably slightly acid.[1762,1763] Yet lakes, especially those that are alkaline, commonly contain shells of floating organisms and macrofauna.[1764,1765,1766] When lakes are extremely acid, non-bacterial life dies.[1767] Freshwater lakes

[1755] Barth, J. A. C. and Veizer, J. 1999: Carbon cycle in St. Lawrence aquatic ecosystems at Cornwall (Ontario), Canada: seasonal and spatial variations. *Chemical Geology* 159: 107-128.

[1756] Berner, R. A., Lasagna, A. C. and Garrels, R. M. 1983: The carbonate-silicate geochemical cycle and its effect on atmospheric carbon dioxide over the past 100 million years. *American Journal of Science* 283: 641-683.

[1757] Raymo, M. E. and Ruddiman, W. F. 1992: Tectonic forcing of late Cenozoic climate. *Nature* 359: 117-122.

[1758] $CO_2 + H_2O \Leftrightarrow H_2CO_3$; $H_2CO_3 \Leftrightarrow H^+ + HCO_3^-$; $2Ca^{2+} + 2HCO_3^- \Leftrightarrow 2CaCO_3 + 2H^+$; $2KAlSi_3O_8 + 2H^+ + H_2O \Leftrightarrow Al_2Si_2O_5(OH)_4 + 2K^+ + 4SiO_2$; $2NaAlSi_3O_8 + 2H^+ + H_2O \Leftrightarrow Al_2Si_2O_5(OH)_4 + 2Na^+ + 4SiO_2$; $CaAl_2Si_2O_8 + 2H^+ + H_2O \Leftrightarrow Al_2Si_2O_5(OH)_4 + Ca^{2+}$; $KAl_2AlSi_3O_{10}(OH)_2 + 3Si(OH)_4 + 10H^+ \Leftrightarrow 3Al^{3+} + K^+ + 6SiO_2 + 12H_2O$; $CO_2 + CaSiO_3 \Leftrightarrow CaCO_3 + SiO_2$; $CO_2 + FeSiO_3 \Leftrightarrow FeCO_3 + SiO_2$; $CO_2 + MgSiO_3 \Leftrightarrow MgCO_3 + SiO_2$

[1759] Walker, J. C. B., Hays, P. B. and Kasting, J. F. 1981: A negative feedback mechanism for the long term stabilization of the Earth's surface temperature. *Journal of Geophysical Research* 86: 9776-9782.

[1760] Berner, R. A. 1980: Global $CO_2$ degassing and the carbon cycle: comment on 'Cretaceous ocean crust at DSDP sites 417 and 418: carbon uptake from weathering vs loss by magmatic activity." *Geochimica et Cosmochimica Acta* 54: 2889.

[1761] Schwartzman, D. W. and Volk, T. 1989: Biotic enhancement of weathering and the habitability of Earth. *Nature* 311: 45-47.

[1762] Dermott, R., Kelso, J. R. M. and Douglas, A. 1986: The benthic fauna of 41 acid sensitive headwater lakes in North Central Ontario. *Water, Air and Soil Pollution* 28: 283-292.

[1763] Harvey, H. H. and McArdle, J. M. 2004: Composition of the benthos in relation to pH in the LaCloche lakes. *Water, Air and Soil Pollution 30*: 529-536.

[1764] Pip, E. 1987: Species richness of freshwater gastropod communities in central North America. *The Malacological Society of London* 53: 163-170.

[1765] Rintelen, T. von and Glaubrecht, M. 2003: New discoveries in old lakes: Three new species of *Tylomelania* Sarasin & Sarasin, 1897 (Gastropoda: Cerithioidea: Pachychilidae) from the Malili Lake system on Sulawesi, Indonesia. *The Malacological Society of London* 69: 3-17.

[1766] Bennike, O., Lemke, W. and Jensen, J. B. 1998: Fauna and flora in submarine early Holocene lake-marl deposits from the southwestern Baltic Sea. *The Holocene* 8: 353-358.

[1767] Nilssen, J. P. 1980: Acidification of a small watershed in southern Norway and some characteristics of acidic aquatic environments. *International Revue der Gesamten Hydrobiologie* 65: 177-207.

also have an excess of calcium[1768] and some lakes and restricted-circulation seas have oxygen-poor bottom waters (e.g. Black Sea). Apart from a few short sharp events,[1769] the oceans have remained alkaline for billions of years. Dissolved $CO_2$ is removed by precipitation of calcium carbonate minerals in shells, coral reefs, cement that binds mineral grains together and mineral deposits. Because the oceans have an excess of calcium, if more $CO_2$ is dissolved in the oceans, then more calcium carbonate is precipitated. While the oceans have an excess of calcium, they cannot become acid.

Acid enters the ocean from submarine hot springs so the surroundings are less alkaline than elsewhere. This is especially the case with hot springs that are close to the coast in populated areas where water clarity, sediment deposition, shelter from waves, runoff of rainwater and human influences (e.g. sewage) change alkalinity. In these cases, short periods of decreased alkalinity lead to a decrease in animals that graze on green algae and hence an increase in green algae.[1770]

A reaction between seawater and minerals on the ocean floor keeps the oceans alkaline.[1771] The floor of the oceans is covered with the volcanic rock basalt. This is a highly reactive rock, especially when it is glassy and fractured.[1772] Reactions between seawater and basalt make the oceans more alkaline, balance the acid added by hot springs and balance the addition of $CO_2$ to seawater. Over time, basalt-seawater reactions have controlled the atmosphere and seawater chemistry.[1773] Contact with ocean floor rocks, especially basalt, also removes $CO_2$ from seawater to form carbonates.[1774] A fine balance is maintained where micro-organisms (plants) consume $CO_2$ as plant food thereby increasing alkalinity whereas the decomposition of

---

[1768] For example, the weathering of limestone by acid rains produces calcium for accumulation in lakes and oceans, $CO_2 + CaCO_3 + H_2O \Leftrightarrow 2(HCO_3)^- + Ca^{2+}$

[1769] Lowenstein, T. K. and Demicco, R. V. 2006: Elevated Eocene atmospheric $CO_2$ and its subsequent decline. *Science* 313: 1928.

[1770] Hall-Spencer, J. M., Rodolfo-Metalpa, R., Martin, S., Ransome, S., Fine, M., Turner, S. M., Rowley, S. J., Tedesco, D. and Buia, M.-C. 2008: Volcanic carbon dioxide vents show ecosystem effects of ocean acidification. *Nature* 453: doi:10.1038/nature07051.

[1771] For example, the weathering of silicates such as pyroxenes consumes $CO_2$ and forms carbonates. The same reactions apply for olivines, a far more reactive family of minerals than the pyroxenes. $CO_2 + CaSiO_3 \Leftrightarrow CaCO_3 + SiO_2$; $CO_2 + FeSiO_3 \Leftrightarrow FeCO_3 + SiO_2$; $CO_2 + MgSiO_3 \Leftrightarrow MgCO_3 + SiO_2$

[1772] Feldspars are the most abundant minerals in terrestrial and submarine rocks and buffer acidity by reaction to form kaolinite. $2KAlSi_3O_8 + 2H+ + H_2O \Leftrightarrow Al_2Si_2O_5(OH)_4 + 2K^+ + 4SiO_2$; $2NaAlSi_3O_8 + 2H+ + H_2O \Leftrightarrow Al_2Si_2O_5(OH)_4 + 2Na^+ + 4SiO_2$; $CaAl_2Si_2O_8 + 2H^+ + H_2O \Leftrightarrow Al_2Si_2O_5(OH)_4 + Ca^{2+}$

[1773] Arvidson, R. S., Guidry, M. and Mackenzie, F. T. 2005: The control of Phanerozoic atmosphere and seawater composition by basalt-seawater exchange reactions. *Journal of Geochemical Exploration* 88: 412-415.

[1774] $Ca^{2+} + H_2O + CO_2 \Leftrightarrow CaCO_3 + 2H^+$; $H^+ + (OH)^- \Leftrightarrow H_2O$

organisms increases acidity. The more $CO_2$ in the atmosphere, the more micro-organisms thrive in the oceans.

These mineral and biological processes have taken place for billions of years. Reactions with modern and ancient seafloor basalts show an acidity balance even when the $CO_2$ in the atmosphere was 25 times the current amount. The same water-rock chemical reactions that have kept the oceans saline have also kept them alkaline. If the oceans were becoming acid, then they should also become less saline. This we don't see.

Despite huge changes in atmospheric $CO_2$ over the last few hundred million years, average global temperature has not changed significantly, oceans have not become acid and there has been no runaway greenhouse.[1775] Fossil shells, algal reefs and coral reefs in ancient rocks show maintenance of alkalinity even when atmospheric $CO_2$ and temperature were far higher than now. Plants also thrived at those times.[1776]

In fact, the higher the atmospheric $CO_2$ in the past, the easier it has been to form shells. If the oceans were acid, shells would dissolve and the oceans would become alkaline. Furthermore, at these times the high temperature and high $CO_2$ content of the atmosphere are unrelated.[1777,1778] Geological history shows us that for the oceans (and land plants) to efficiently fix atmospheric $CO_2$ and store it in the rocks, the atmospheric $CO_2$ content needs to be far higher than at present.

The salinity of the oceans has been almost constant for billions of years[1779] with slight decreases in warmth and salinity through to the present.[1780] Acid rain leaches salts from rocks on the land and rivers transport them to accumulate in the oceans and be constantly recycled.[1781] By this process, acid rainwater is neutralised and the oceans remain alkaline.

[1775] Royer, D. L., Berner, R. A. and Park, J. 2007: Climate sensitivity constrained by $CO_2$ concentrations over the past 420 million years. *Nature* 446: 530-532.

[1776] Bice, K. L., Huber, B. T. and Norris, R. D. 2003: Extreme polar warmth during the Cretaceous greenhouse? Paradox of Turonian $\partial^{18}O$ record at Deep Sea Drilling Project Site 511. *Palaeoceanography* 18:1-11.

[1777] Veizer, J., Godderis, Y. and Francois, L. M. 2000: Evidence for decoupling of atmospheric $CO_2$ and global climate during the Phanerozoic eon. *Nature* 408: 698-701.

[1778] Donnadieu, Y., Pierehumbert, R., Jacob, R. and Fluteau, F. 2006: Cretaceous climate decoupled from $CO_2$ evolution. *Earth and Planetary Science Letters* 248: 426-437.

[1779] Hay, W. W., Wold, C. N., Soeding, E. and Floegel, S. 2001: Evolution of sediment fluxes and ocean salinity. In: *Geologic modeling and simulation: sedimentary systems* (eds Merriam, D. F. and Davis, J. C.), Kluwer, 163-167.

[1780] Knauth, L. P. 2005: Temperature and salinity history of the Precambrian ocean: implications for the course of microbial evolution. *Palaeogeography, Palaeoclimatology, Palaeoecology* 219: 53-69.

[1781] Rogers, J. J. W. 1996: A history of the continents in the past three billion years. *Journal of Geology* 104: 91-107.

Seawater evolves over time, measured by accurate determinations of isotopes in shells of known ages.[1782] Although the Earth's atmosphere has been far warmer than at present for most of geological time,[1783] changes in shell isotope chemistry may result from an increasing depth of the oceans since 500 Ma.[1784] Seawater evolution can be traced through changes in oxygen and strontium isotope ratios (driven by plate tectonics and the evolution of continents) and the concentrations of carbon and sulphur (driven by biological and chemical cycles). Since 550 Ma, seawater evolution has been unaffected by atmospheric $CO_2$, despite the concentration and temperature being far higher in the past.[1785] There is no reason why these tectonic, biogeochemical and geochemical cycles should change because we humans are now on Earth.

A popular catastrophist view is that as the climate warms, less and less $CO_2$ will be dissolved in the oceans and a "tipping point" will be reached when the Earth enters a runaway greenhouse. Another "tipping point" is that the oceans will become acid. Permanently. In fact, there is no such thing as a "tipping point" (or even a "precautionary principle") in science. The use of these words in the popular media and by political advocates immediately advertises non-scientific opinions. These views ignore geological history, natural $CO_2$ recycling, sequestration of $CO_2$ by shells and rocks and the logarithmic atmospheric $CO_2$-temperature relationship.

Computer simulations tell a different story from reality and indicate that the oceans will become acid.[1786,1787] Experiments with seawater are flawed because they are done in laboratories removed from the ocean floor rocks, sedimentation from continents and flow of river waters into the oceans. It is these real processes that have kept the oceans alkaline for billions of years. Laboratory experiments have to provide results in a short time to be

[1782] $\partial^{13}C$, $\partial^{18}O$ and $^{87}Sr/^{86}Sr$

[1783] Apart from the major glaciations, see *Ice*.

[1784] Kasting, J. F., Howard, M. T., Wallmann, K., Veizer, J., Shields, G. and Jaffrés, J. 2006: Paleoclimates, ocean depth, and the oxygen isotopic composition of seawater. *Earth and Planetary Science Letters* 252: 82-93.

[1785] Veizer, J., Ala, D., Azmy, K., Bruckschen, P., Buhl, D., Bruhn, F.,, Carden, G. A. F., Diener, A., Ebneth, S., Godderis, Y., Jasper, T., Korte, C., Pawellek, F., Podlaha, O. G. and Strauss, H. 1999: $^{86}Sr/^{87}Sr$, $\partial^{13}C$ and $\partial^{18}O$ evolution of Phanerozoic seawater. *Chemical Geology* 161: 59-88.

[1786] Caldeira, K. & Wickett, M. 2003: Anthropogenic carbon and ocean pH. *Nature* 425: 365.

[1787] Orr, J. C., Fabry, V. J., Aumont, O. , Bopp, L., Doney, S. C., Freely, R. A., Gnanadesikan, A., Gruber, N., Ishida, A., Joos, F., Key, R. M., Lindsay, K., Maier-Raimer, E., Matear, R., Monfray, P., Mouchet, A., Najjar, R. G., Plattner, G.-K., Rodgers, K. B., Sabine, C. L., Sarmiento, J. L., Schlitzer, R., Slater, R. D., Totterdell, I. J., Weirig, M.-F., Yamanaka, Y. and Yool, A. 2005: Anthropogenic ocean acidification over the twenty-first century and its impact on calcifying organisms. *Nature* 437: 681-686.

reported in scientific journals. Processes over geological time cannot be that easily replicated. Limited constrained experiments show that when increasing amounts of $CO_2$ were added to seawater, it became acid and dissolved shells. If a few handfuls of gravel, sediment and clay from the sea floor and some floating photosynthetic life had been added in the experiment to simulate real conditions, then the result would be completely different. Computer simulations that ignore observations and natural processes that have taken place over billions of years end up with a result unrelated to reality. Reality is written in rocks, not models based in incomplete information.

## Sea surface temperature

Oceans cover 71% of the surface of the Earth. The heat content of the ocean drives climate and the sea surface temperature essentially determines the surface temperature of the planet.[1788] There has been little if any global warming as diagnosed using ocean heat content.[1789] Measurement of sea surface temperature does not suffer from the limitations of land-based measurements such as the urban heat island and weather station placement, maintenance and correlation. Sea surface temperature measurements have other problems. Transferring heat from the atmosphere to the oceans is difficult because of the volume and the heat capacity of the atmosphere compared to the oceans. If human emissions have actually heated the atmosphere by anthropogenic gases, there is still not enough additional heat to warm the oceans. Global warming of the atmosphere derives from warming of the oceans which can be heated by a great diversity of natural processes.[1790]

We need to look back some millions of years to begin to understand modern sea surface temperature. Millions of years ago, the Earth was warmer yet there was polar ice. Sea surface temperature graded from the poles to the equator and the oceans had temperature layering, but no ocean floor currents carried cold polar waters to tropical areas to upwell. Wind and the Earth's rotation drove the horizontal circulation but not the vertical circulation in the ocean. Variations in the sea surface temperature were slow and the magnitude of the variation was limited.

---

[1788] Pielke, R. A. Snr 2003: Heat storage within the Earth system. *Bulletin of the American Meteorological Society* 84: 331-335.

[1789] Cazenave, A., Dominh, K., Guinehut, S., Berthier, E., Llovel, W., Ramillien, G., Ablain, M. and Larnicol, G. 2008: Sea level budget over 2003-2008: A reevaluation from GRACE space gravimetry, satellite altimetry and Argo. *Global and Planetary Change* (2008) doi: 10.1016/j.gloplach.2008.10.004.

[1790] Compo, G. P. and Sardeshmukh, P. D. 2008: Oceanic influences on recent continental warming. *Climate Dynamics* doi:10.1007/s00382-008-0448-9.

The Drake Passage deepened at 4 Ma. The Antarctic Circumpolar Current developed. Antarctica became colder, land ice grew and sea ice developed in winter. Antarctica became isolated, wind stress caused vertical overturning with cold subsurface waters upwelling to the surface. As sea ice increased, salt was expelled and sank from the underlying water so that bottom water became denser. This led to colder bottom water and a general cooling of the oceans. Cool bottom water and ice on the land caused an energy imbalance so that more solar energy was emitted into space than was adsorbed by water. This led to global cooling and the filling of the oceans with cool bottom water until about 1 Ma. Sea ice appeared in the Northern Hemisphere at about 2.67 Ma.

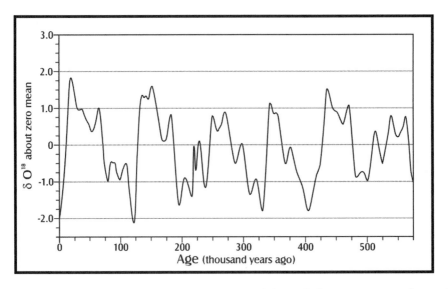

*Figure 35: Variations in oxygen isotope composition of bottom-dwelling micro-organisms showing variations in bottom ocean water temperature in response to six glacial cycles. Oxygen isotope data is a proxy for ocean temperature. During glaciation there are high values of $O^{18}$ which reflects the volume of ice on land. Note that the interglacials are much shorter than glaciations and that there is nothing unusual about the current interglacial.[1791]*

Sea floor sediments give an indication of the strengths of currents in former times. During the last ice age, currents were strong but weakened when the surface of the North Atlantic Ocean was freshened by melt

[1791] Bintanja, R., van de Wal, R. S. W. and Oerlemans, J. 2005: Modelled atmospheric temperatures and global sea levels over the past million years. *Nature* 437: 125-128.

waters from the north.[1792] Bottom currents weakened when the Greenland temperature was high. During the Younger Dryas currents weakened[1793] and during the Holocene warming, currents strengthened.[1794]

The oceans now have a thin surface layer warmed in the tropics and cooled in the middle and high latitudes by a net loss of energy to the atmosphere and space. The oceans transfer heat around the world. Cold dense polar water accumulates at the bottom. The climate system is very sensitive to the oceans because of the heat gains (tropics) and losses (high latitudes), evaporation and upwelling in the tropics which reduces the sea surface temperature and the amount of energy available to the atmosphere. These, in turn, reduce the amount of energy transport to higher latitudes resulting in polar cooling. Reduced upwelling results in polar warming. This amplification of cooling and warming at the poles relates to El Niño and La Niña events. The oceans contain some 22 times more heat than the atmosphere. They, not the atmosphere, control the surface heat balance. However, we are bombarded with media opinion telling us exactly the reverse and that increases in atmospheric temperature increase ocean temperature.

At about 1 Ma, external processes (orbital, solar, galactic) imposed a 100,000-year cycle on tropical ocean upwelling. At the zenith of the last ice age some 21,000 years ago, the tropical sea surface temperature was about 3°C cooler than now. The energy available to the atmosphere was about 20% less than now. This allowed snow to accumulate and form great ice sheets. As the tropical sea surface temperature increases, so does evaporation, which takes heat from the sea surface. The sea surface is heated by solar energy and infra-red energy from clouds and greenhouse gases. This provides an upper limit to sea surface temperature. As the tropical sea surface temperature increases, additional heat is lost due to infra-red emission and evaporation. A new sea surface temperature is reached when the $CO_2$ infra-red forcing is balanced by the energy loss from the surface. If the atmospheric $CO_2$ content is doubled, the sea surface temperature increases by about 0.3°C.

As long as the Earth spins on the polar axis, the ever-blowing trade winds and rotational momentum due to centrifugal force will drive ocean currents. The Brazil Current in the South Atlantic Ocean, the Pacific Ocean Japan Current and the Gulf Stream are good examples. Warm surface layers in the tropical Atlantic Ocean form the Equatorial Current. This is driven west by

[1792] Bianchi, G. and McCave, I. 1999: Holocene periodicity in North Atlantic climate and deep-ocean flow south of Iceland. *Nature* 397: 479-482.

[1793] Tarasov, L. and Peltier, W. R. 2005: Arctic freshwater forcing of the Younger Dryas cold reversal. *Nature* 435: 662-665.

[1794] Praetorius, S. K., McManus, J. F., Oppo, D. W. and Curry, W. B. 2008: Episodic reductions in bottom-water currents since the last ice age. *Nature Geoscience* 1: 449-452.

the trade winds and the Earth's rotation. The northern part of the Equatorial Current enters the Caribbean Sea and the Gulf of Mexico where surface waters are heated further and exit into the North Atlantic Ocean in a channel between Florida and the Bahamas. The Gulf Stream is a "river" in the warm water of the Florida Straits. It ends along the edge of the Arctic ice pack. These warm surface layers in the Caribbean Sea have no way out but through the narrow Florida Straits. Heading north is the only option. Sea surface temperatures fluctuate by up to 2°C.

Westerlies drive the Gulf Stream towards Europe. As they do, water piles up along the Norwegian coast and in the Arctic Ocean. The waters have nowhere to go, they are forced to sink along the pack ice boundary line and return southwards along the ocean floor. As this current flows almost along a meridian and the waters are overturned at high latitude, it is called the Atlantic meridional overturning circulation. The Labrador Current is the only other way that water can be released from the Arctic Ocean. This current carries icebergs south and plunges beneath the Gulf Stream near New York. In summer when the pack ice melts and ocean salinity drops, the Labrador Current should strengthen and the southward return of water along the sea floor should weaken. This is indeed the case.[1795]

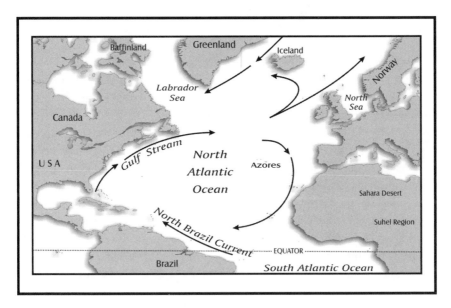

*Figure 36: North Atlantic Ocean currents showing the distribution of heat by the Gulf Stream from low latitude to high latitude areas.*

[1795] Kerr, R. A. 2006: Global climate change: False alarm: Atlantic conveyor belt hasn't slowed down after all. *Science* 314, 1064.

A similar situation exists in the Pacific Ocean where the shape of the coastline allows a clockwise circuit of the Japan Current to heat Alaska and then turn south to become the cold Alaska Current that cools California. There appears to be no deep-water return current as the coastal shape does not require it. Both trade winds and westerlies should fade when solar energy declines and, as a result, Europe gets a double dose in the reduction of warming. Although ocean temperature and density may drive deep currents, surface currents like the Gulf Stream are driven by the ever-present trade winds. In the tropical Pacific, El Niño influences climate. It also does over much of the globe. However, the history of sea surface temperature change in the Pacific is poorly known. In the El Junco Lake (San Cristóbal Island, Galápagos), fossilised floating animals[1796] were used as proxies in an attempt to determine the temperature changes of the tropical eastern Pacific.[1797] Lake levels increase during El Niño events. It was argued that the last 50 years was the warmest in the last 1200 years. The Medieval Warming and Little Ice Age are seen, and the temperatures deduced from fossilised floating animals in the Medieval Warming are slightly less than temperatures deduced from modern floating animals. This study did not determine what local features might influence El Junco Lake.

We all know that summers at the seaside are cooler and winters warmer than at an equivalent latitude inland. This is because the heat capacity of water is far higher than that of rocks and soils. The Sun's heat is stored in the upper few metres of the ocean. Heat is also lost from the upper few metres of the oceans. If the oceans are turbulent, they lose heat more quickly. Warm surface water is mixed to depths of tens of metres due to the wind so the heat is dispersed downwards. On the land, the summer heat reaches to a depth of only a metre or so in soil and rock.

Sea surface temperature is always changing. This can be shown by temperature measurements, oxygen isotope measurements or surface animals (e.g. floating shells, corals) or chemical changes in surface animals. Great care must be used with sea surface temperature measurements unless long-term variations and temperature cycles are accurately known.

---

[1796] Foraminifera.
[1797] Conroy, J. L., Restrepo, A., Overpeck, J. T., Steinitz-Kannan, M., Cole J. E., Bush, M. B. and Colinvaux, P. A. 2009: Unprecedented recent warming of surface temperatures in the eastern tropical Pacific Ocean. *Nature Geoscience* 2: 46-50.

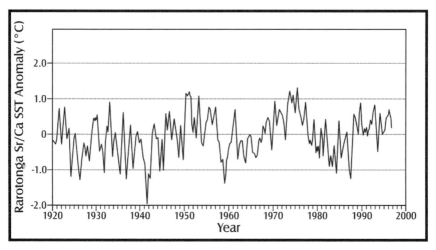

*Figure 37: South Pacific sea surface temperature at Rarotonga deduced from the chemistry of corals (Sr/ Ca proxy). Temperature changes of up to 3°C and minor temperature cycles superimposed on larger cycles are a feature of these measurements. There is no evidence that the sea surface temperature has increased with increasing human emissions of $CO_2$ or that dangerous warming is occurring.*

Sea surface temperature used to be measured in the major sea lanes of the world by passing ships. This was voluntary and only covered maritime ship routes. Almost all the surface of the oceans went unmeasured, especially in the Southern Hemisphere where about 80% is ocean and these oceans have far less shipping than the Northern Hemisphere.[1798] Until recently, the sea surface temperature was measured with a thermometer in a bucketful of seawater. These measurements had an accuracy of no better than ±0.5°C, which is highly inaccurate. The change from wooden to canvas buckets led to a "discontinuity" of temperature measurements and significant errors. Currents were also measured by similarly crude methods. A large proportion of historical sea surface temperature measurements were by unknown methods.

The last 30 years has seen more accurate and more widespread measurements of sea surface temperature and ocean currents. Since 1980, floating buoys have recorded the temperature of the oceans at a depth of 50 cm and this has been combined with the inlet temperature of ships' cooling water, which extracts considerably cooler seawater from a depth of about 10

[1798] Kent, E. C., Woodruff, S. D. and Berry, D. I. 2007: Metadata from WMO Publication No. 47 and an assessment of voluntary observing ship observation heights in ICOADS. *Journal of Oceanic and Atmospheric Technology* 24: 214-234.

metres. Hull sensors have also been used. The estimation of historical sea surface temperature trends from combining bucket measurements, floating buoys and ship inlet water temperature is meaningless. Yet it is this data that is being used to estimate changes in historical sea surface temperature. Current and sea surface temperatures older than 30 years are only crude measurements that should not be used in climate models. But they are.

There is great uncertainty about ocean warming which only measures changes since the 1950s. Large discrepancies are apparent between the various measuring methods.[1799] Since expandable bathythermograph (XBT) measurements are the largest proportion of the data set, this bias results in a significant world ocean warming artefact. If the bias is omitted, this reduces ocean heat content change since the 1950s by a factor of 0.62.[1800,1801]

Some 3000 ARGO scientific robots in the oceans have shown a sea surface temperature decrease. These do not assist with calibration of historical temperature measurements which are biased and cover only limited shipping routes. Furthermore, the historical sea surface temperature measurements do not correlate with satellite measurements. The multi-decadal models of the near surface show an increase in total heat from 1955 to 2003 followed by a dramatic loss of the average heat content of the upper oceans from 2003 to 2005.[1802,1803] This should be a wake-up call for computer climate models.

The current cooling of the North Atlantic waters may be due to a decrease in the mean annual heat from the Sun at middle and high northern latitudes during the last 11,000 years.[1804] The levels are now those of the Last Glacial Maximum. Perhaps the cooling of the northwest Atlantic slope waters is heralding the next inevitable glacial period.[1805] Although climate over the last 11,000 years is usually described as warm and stable, there have been significant large (e.g. Younger Dryas) and small variations (e.g. Medieval Warming, Little Ice Age). The cooling of slope waters east of the United States and Canada by 4–10°C during the Holocene probably resulted from

---

[1799] Expendable bathythermographs (XBT), bottle and CTD data. Most data is XBT and this XBT data is positively biased.

[1800] Gouretski, V. and Koltermann, K. P. 2007: How much is the ocean really warming? *Geophysical Research Letters* 34, L01610, doi: 10.1029/2006GL027834.

[1801] Ocean heat content increase (0-3000m) between 1957-1966 and 1986-1996 estimates of 12.8± >8 x $10^{22}$ Joules.

[1802] Lyman, J. M., Willis, J. K. and Johnson, G. C. 2006: Recent cooling of the upper ocean. *Geophysical Research Letters* 33: L18604, doi:10.1029/2006GL027033.

[1803] This equates to a global radiative imbalance of -1.0 ± 0.3 watts per square metre.

[1804] Seager, R., Battisti,. S., Yin, J., Gordon, N., Naik, N., Clement, A. C. and Cane, M. A. 2002: Is the Gulf Stream responsible for Europe's mild winters? *Quarterly Journal of the Royal Meteorological Society* 128: 2563-2586

[1805] Sachs, J. P. 2007: Cooling of Northwest Atlantic slope waters during the Holocene. *Geophysical Research Letters* 34: 10.1029/2006GL028495.

declining heat from the Sun, increasing convection in the Labrador Sea and a shift of the Gulf Stream towards the equator.

Once sea surface temperature is measured, then how is the data evaluated? If annual, five-month average or monthly averages are used, then different conclusions can be made. Annual averages show sea surface temperature rises whereas monthly averages do not.

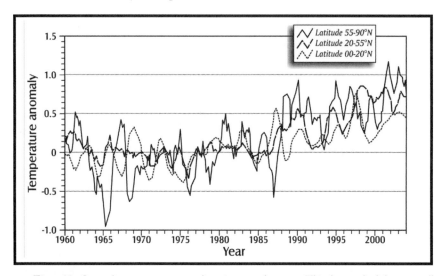

*Figure 38: Sea surface temperature anomalies using annual averages. This shows cyclical changes at all latitudes and an increase in sea surface temperature since 1985.*

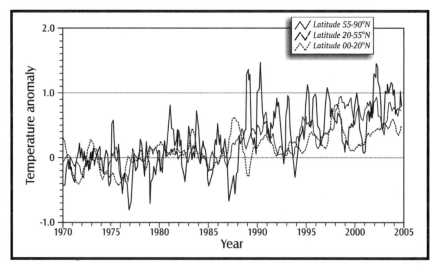

*Figure 39: Sea surface temperature anomalies using 5-month averages. This shows cyclical changes at all latitudes and a slight increase in sea surface temperature since 1988.*

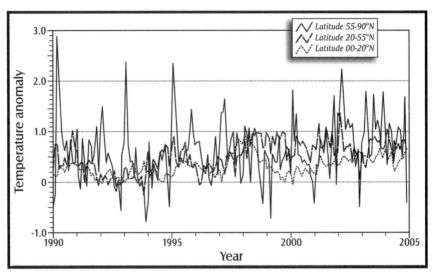

*Figure 40: Sea surface temperature anomalies using monthly averages. This shows cyclical changes at all latitudes and no increase in sea surface temperature. These three diagrams show that data can easily be manipulated to create a desired outcome.*

Records of the flow of the River Po in Italy show a relationship between rainfall and river discharge, as one would expect. For the last 200 years, wet and dry periods alternate in accord with solar cycles and show a relationship between the North Atlantic Oscillation and solar activity.[1806]

The Atlantic Ocean, especially the North Atlantic Ocean, has an influence on weather and climate in North America and Europe. Temperature-measuring buoys in the ice-free parts of the Atlantic Ocean between 10°N and 70°N from January 1999 to December 2005 showed that the trend in the heat content was decreasing or stable and the only warming appears north of 50°N.[1807] Furthermore, at low latitudes, the top 100 metres of water demonstrate warming but substantial cooling occurs from 100 to 1000 metres depth. If there was no measured warming in the lower latitude regions, then it is hard to reconcile that warming of the tropical Atlantic has led to an increase in hurricane activity.

In the sub-polar North Atlantic Ocean, circulation of deep convection gyres of the Labrador and Irminger Seas can suddenly change, as they did in

[1806] Zanchettin, D., Rubino, A., Traverso, P. and Tomasino, M. 2008: Impact of variations in solar activity on hydrological decadal patterns in northern Italy. *Journal of Geophysical Research* 113:, D12102, doi:10.1029/2007Jd009157.

[1807] Ivchenko, V. O., Wells, N. C., and Aleynik, D. L. 2006: Anomaly of heat content in the northern Atlantic in the last 7 years: Is the ocean warming or cooling? *Geophysical Research Letters* 33, L22606.

2007–2008. This results in the mixing of surface water with deep water and the absorption of $CO_2$ from the atmosphere. Changes in Northern Hemisphere air temperature, storm tracks, influxed fresh water in the Labrador Sea and the distribution of pack ice all contributed to an increased heat loss from water to air. Once the water was sufficiently cold enough, deep convection started. This process is so complex that it is difficult to predict when it might occur again.[1808]

At times during the last ice age and at the end of the ice age, the North Atlantic Ocean meridional overturning circulation was much weaker than now or perhaps did not exist at all. Climate models predict that the ocean's circulation will weaken in response to global warming, but the warming at the end of the last ice age suggests a different outcome.[1809] By using a chemical proxy in corals from Bermuda from 1781–1999, it is concluded that circulation of the North Atlantic Ocean had showed increased variability towards the end of the Little Ice Age (1800–1850).[1810]

Westerlies and the trade winds predominate in middle latitudes of both the Northern and Southern Hemispheres. Over the last 40 years, wind patterns have been migrating polewards. As the westerlies move, the ocean's circulation increases, releasing more $CO_2$ from the deep ocean. This leads to more warming and even stronger circulation in a feedback loop strong enough to push Earth out of an ice age. Previous models have placed the westerlies in the wrong place, thereby making predictions erroneous from the start. Early models suggested that ocean circulation was controlled by wind at the surface and buoyancy for deeper circulation. If fresh water were added to the oceans as the Earth warms, this would lead to less movement of the oceans. This model is invalid and current data suggests that wind has the major effect on ocean circulation. In glacial times, climatic belts do not simply compress concertina-wise towards the equator. For example, in Australia the westerly pattern moved north and was concentrated in a narrower band. Winds were stronger than ever.

There was a suggestion that the Atlantic meridional overturning circulation has slowed by 30% between 1957 and 2004.[1811] The North Atlantic is

[1808] Våge, K., Pickart, R. S., Thierry, V., Reverdin, G., Lee, C. M., Petrie, B., Agnew, T. A., Wong, A. and Ribergaard, M. H. 2009: Surprising return of deep convection to the subpolar North Atlantic Ocean in winter 2007-2008. *Nature Geoscience* 2: 67-72.

[1809] Toggweiler, J. R. and Russell, J. 2008: Ocean circulation in a warming climate. *Nature* 451: 286-288.

[1810] Goodkin, N. F., Hughen, K. A., Doney, S. D. and Curry, W. B. 2008: Increased multidecadal variability of the North Atlantic Oscillation since 1781. *Nature Geoscience* 1: 844-848.

[1811] Bryden, H. L., Longworth, H. R. and Cunningham, S. A. 2006: Slowing of the Atlantic meridional overturning circulation at 25°N. *Nature* 438: 655-657.

dominated by the Gulf Stream. At the latitude of New York it divides, and some water swirls towards the south in a surface current whereas the rest continues north, helping to warm Europe by 5–10°C. The north-south heat flow was measured using a set of instruments strung out across the Atlantic from the Bahamas to the Canary Islands. Data retrieval in 1957, 1981 and 1992 showed no change but those in 2004 showed that a 30% reduction had occurred between 1992 and 1998. If the Atlantic meridional overturning circulation had shut down, then Europe would refrigerate. The only way to stop the Atlantic meridional overturning circulation is to stop wind, stop the Earth's rotation or stop both. However, the media went into alarmist overdrive and exaggerated predictions were made about Europe cooling by up to 4°C. This was, of course, attributed to human-induced global warming resulting from $CO_2$ emissions. However, there was an incorrect treatment of measurement errors,[1812] suggesting that *Nature* rushed the publication of the relevant paper. It is a mystery why the editors and reviewers of the paper missed what is the first thing that people learn in school physics: errors of measurement. The infamous "hockey stick" of Mann *et al.* had similar sloppy editing.

This doomsday prediction was destroyed a year or so later – this time, without the blaze of publicity. The Atlantic meridional overturning circulation was measured from September 2004 to September 2005 with a picket fence of instruments east of Abaco Island (Bahamas) at 26.5°N. The picket fence allowed measurement of millions of cubic metres of water per second for a year and showed that there was no evidence at all for a 30% reduction in circulation.[1813] This conclusion was validated by independent measurements of the Deep Western Boundary Current east of the Grand Banks between 1999 and 2005.[1814] The lags in response of the North Atlantic Oscillation and the variation in the Labrador Sea convection are such that any human-

---

[1812] According to Bryden *et al.,* the 1957 transport in a layer shallower than 1000 m was 22.9 ± 6 Sverdrups (1Sv = 106 m³/s) compared with the transport in 2004 of 14.8 ± 6 Sv. The order of accuracy, ± 6 Sv, is an uncorrelated error of each measurement. Bryden *et al.* subtract the two numbers and present the result as 8.1 ± 6 Sv (instead of 8.1 ± 12 or 8.1 ± 8.5 Sv, depending on the type of errors). This is an incorrect result. The observed 8.1 Sv is well within the uncertainty of measurement hence the conclusion is invalid. The errors were repeated in a *Physics Today* article (April 2006, p. 26) and were exposed by Petr Chylek of Los Alamos Laboratories in *Physics Today* in 2007.

[1813] Meinen, C. S., Baringer, M. O. and Garzoli, S. L. 2006: Variability in Deep Western Boundary Current transports: Preliminary results from 26.5°N in the Atlantic. *Geophysical Research Letters* 33: L17610, doi.1029/2006GL026965.

[1814] Schott, F. A., Fischer, J., Dengler, M. and Zantopp, R. 2006: Variability of the Deep Western Boundary Current east of the Grand Banks. *Geophysical Research Letters* 33: L21S07, doi:10.1029/2006GL026563.

induced weakening of the Atlantic meridional overturning circulation will remain well within the natural variability.[1815]

Myths are passed down through the generations. One of them is that the Gulf Stream (the Atlantic meridional overturning circulation) is responsible for Europe's mild winters.[1816] The Gulf Stream has its source in the Gulf of Mexico and flows northeast along the USA coast. Myths often rest on a strand of truth. Today's ocean circulation systems deliver an enormous amount of tropical heat to the North Atlantic. During winter, this heat is released to the overlying easterly air masses, thereby making European winters mild.[1817] Mild compared with what? This view of the Gulf Stream is both correct and misleading. Western coastal North America also has mild winters but for a different reason. The Pacific Ocean equivalent of the Gulf Stream is the Kushiro Current. It flows north along the coast of Asia and, east of Japan, it heads due east towards California and Oregon. No heat is carried northward. Hence ocean heat transport cannot explain the vast difference in winter climate between the Pacific Northwest (e.g. Vancouver) and Vladivostok at the same latitude in East Asia.

The shape of the ocean floor is constantly changing. Not only does this change sea level but it also changes the transport of water in deep ocean currents and later mixing of deep with shallow water.[1818] The magnitude of transport, location and mixing of deep ocean water is very poorly known. Oceans contain 22 times more heat than the atmosphere, so change in ocean currents drives change in atmospheric temperature. Those creating alarm tell us the opposite; that a warming atmosphere is warming the oceans.

## El Niño

The El Niño-Southern Oscillation, commonly referred to as just El Niño (the little boy or the Christ child), is an ocean-atmosphere interaction.[1819] El Niño most commonly occurs in late December, lasts for a month or so and results from an intrusion of warm ocean waters replacing normally cold waters in the eastern Pacific. Torrential rains in upland Ecuador, loss of anchovy from

[1815] Latif, M., Collins, M., Pohlman, H. and Keenlyside, N. 2006: A review of predictability studies of Atlantic sector climate on decadal time scales. *Journal of Climate* 19: 5971-5987.

[1816] Seager, R., Battisti,. S., Yin, J., Gordon, N., Naik, N., Clement, A. C. and Cane, M. A. 2002: Is the Gulf Stream responsible for Europe's mild winters? *Quarterly Journal of the Royal Meteorological Society* 128: 2563-2586.

[1817] Broecker, W. S. 1997: Thermohaline circulation, the Achilles heel of our climate system: Will man-made $CO_2$ upset the climate balance. *Science* 278: 1582-1588.

[1818] MacKinnon, J. A., Johnston, T. M. S. and Pinkel, R. 2008: Strong transport and mixing of deep water through the Southwest Indian Ridge. *Nature Geoscience* 1: 755-758.

[1819] Trenberth, K. E. 1997: The definition of El Niño. *Bulletin of the American Meteorological Society* 78: 2771-2777.

the Peruvian fisheries and drought in Brazil result. However, the effect is far more widespread.[1820]

In India in the 1920s Sir Gilbert Walker studied the variations in the Southwest Monsoon. Failure of the monsoon led to crop failure, famine and widespread depopulation. Walker used weather stations in Darwin (Australia) and Tahiti, which had 100 years of records and showed a see-saw correlation (hence the Southern Oscillation). When the average pressure of one was abnormally high, the pressure of the other was abnormally low. High pressure in Tahiti also correlated with strong easterly winds extending across the Pacific Ocean providing strong upwelling along the west coast of South America. These winds dragged surface waters to the west of the Pacific Ocean decreasing sea level in the eastern Pacific by 0.5 metres and provoking strong upwelling along the west coast of South America.[1821] When the pressure dropped in Tahiti and rose in Darwin, the trade winds weakened and the warm air that was in the west of the Pacific Basin drifted eastward to the central and eastern Pacific. Westerly winds gave torrential rains in the central Pacific and drought in Australia and Indonesia.

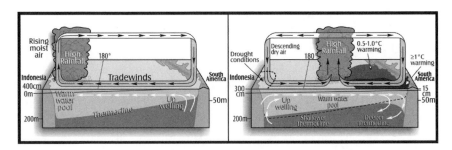

*Figure 41: The shift of the El Niño-Southern Oscillation across the South Pacific Ocean.*

The El Niño-Southern Oscillation phenomenon is a see-saw wave effect that extends around the globe.[1822] A weakening of the Equatorial Trade

[1820] Jianhua, J. and Sligo, J. 1995: The Asian summer monsoon and ENSO. *Quarterly Journal of the Royal Meteorological Society* 121: 1133-1168.

[1821] Meyers, G. 1996: Variation of Indonesian flowthrough and the El Niño:Southern Oscillation: Pacific low-latitude western boundary currents and the Indonesian flowthrough. *Journal of Geophysical Research* 101: 12255-12264.

[1822] Anyamba, A. and Eastman, J. R. 1996: Interannual variability of NDVI over Africa and its relation to El Niño/Southern Oscillation. *International Journal of Remote Sensing* 17: 2533-2548.

Wind easterlies marks the onset of an El Niño-Southern Oscillation event. These winds arise from the warm high-pressure systems at latitudes of 30–40 degrees in both the South and North Pacific.[1823] It is not known what causes a weakening of these highs.

The Southern Oscillation reflects the seasonal fluctuations in air pressure difference between Darwin and Tahiti.[1824] The El Niño is a 3–8-year variability in weather in the Pacific and Indian Oceans, and the Atlantic Ocean lags 12–18 months behind. El Niño (and La Niña, the little girl) are defined as sea surface temperature anomalies of more than 0.5°C across the central tropical Pacific Ocean for more than 5 months. El Niño lasts for 1 to 2 years. In the Pacific, equatorial winds pool warm water in the western Pacific and cold bottom water upwells along the South American coast. During El Niño, warm water pools near the South American coast and, because upwelling is reduced, the South Pacific water warms further. Trade winds in the South Pacific weaken or head east, there is high rainfall in the eastern Pacific and drought in the western Pacific. Pollen extracted from the Sajama Ice Cap, located in the western Bolivian Altiplano, shows significant variations in the amount and type of pollen over a long time and records these events.[1825] Drill core from lakes in the Ecuadorian Andes shows that El Niño has been a feature of the climate for at least 12,000 years.[1826] The frequency of El Niño events has not changed for at least 400 years.[1827]

Although El Niño and La Niña events have profound effects in the tropics, they also have a more significant effect on global temperature anomalies than $CO_2$, for example the 1998 event. Since satellite measurements of temperature started in 1978, variations in global temperature are shown to be due to climate effects in the Northern Hemisphere. Such changes cannot therefore be attributed to $CO_2$ because the increase in atmospheric $CO_2$ is global.[1828] This is contrary to the computer generated models of the IPCC, who state:

---

[1823] Rasmusson, E. M. and Carpenter, T. H. 1982: Variations in tropical seas surface temperature and surface wind fields associated with the Southern Oscillation/El Niño. *Monthly Weather Review* 110: 354-384.

[1824] Philander, S. G. 1990: *El Niño, La Niña and the Southern Oscillation*. Academic Press.

[1825] Liu, K.-B., Reese, C. A. and Thompson, L. G. 2007: A potential pollen proxy for ENSO derived from the Sajama ice core. *Geophysical Research Letters* 34: doi 10.1029/2006GL029018.

[1826] Moy, C. M., Seltzer, G. O., Rodbell, D. T. and Anderson, D. M. 2002: Variation of El Niño/Southern Oscillation activity at millennial timescales during the Holocene epoch. *Nature* 420: 162-165.

[1827] Verdon, D. C. and Franks, S. W. 2006: Long-term behaviour of ENSO: Interactions with the PDO over the past 400 years inferred from paleoclimate records. *Geophysical Research Letters* 33: L06712, doi:10.1029/2005GL025052.

[1828] Douglas, D. and Christy, J. 2008: Limits on $CO_2$ climate forcing from recent temperature data of Earth. *Energy and Environment* Arvix preprint arXiv 0809.0581.

Most of the observed increase in global average temperatures since the mid-20th century is very likely due to the observed increase in anthropogenic greenhouse gas concentrations, mainly carbon dioxide.

The variations show that during El Niño years, the weather on the Altiplano is warmer and drier. During La Niña, warm water pools further west than normal, forcing increased upwelling along the west coast of South America. Indonesia and Australia have increased rainfall, cyclones and thunderstorms, the Pacific Ocean off South America becomes cold and Atlantic cyclone activity increases. A significant amount of the sea surface temperature in the Indian Ocean is due to El Niño, as are heavy rains in east Africa and drought in Indonesia.[1829] A similar process rarely takes place in the Atlantic Ocean, where eastern Brazil becomes cooler and drier while waters in Africa's Gulf of Guinea become warmer. The tropical Atlantic sea surface temperature can be correlated with El Niño.[1830]

In Australia, 48 of the last 144 years have been drought years in some part of the continent. Drought is driven by El Niño whereas La Niña brings rain. The 2002 El Niño created drought in eastern Australia, it returned in 2006 (as did drought) and there was a mild La Niña in 2007 bringing some rain. There was not enough rain to ease the drought, especially in the breadbasket of Australia (Murray-Darling Basin). The nine-year absence of the rain-bringing La Niña is not uncommon and in the past La Niña has been absent for up to 15 years. Water consumption in the Murray-Darling Basin has been mismanaged for decades. Drought is part of life in the inland of Australia. Some argue, in a breathtaking leap of doctrinal faith, that the drought results from human-induced climate change. The current Prime Minister of Australia has actually predicted that rainfall will decrease over the coming decades, despite the unpredictability of El Niño and La Niña events. Does he know some science that has not yet been discovered? Chronic mismanagement combined with an El Niño-driven drought are why the rivers run dry, not $CO_2$.

Normally, most of the Sun's energy over land evaporates water from soils. The conversion of water into water vapour absorbs energy from the soil. The soils are kept cool. During drought, soil moisture is low, evaporation decreases, soil cooling decreases, solar energy heats the soil surface and air temperature rises. The higher air temperatures during drought are not the

[1829] Saji, N. N., Goswami, B. N., Vinayachandran, P. N. and Yamagata, T. 1999: The dipole mode of the tropical Indian Ocean. *Nature* 401: 360-363.

[1830] Enfield, D. B. and Mayer, D. A. 1997: Tropical Atlantic sea surface temperature variability and its relation to El Niño-Southern Oscillation. *Journal of Geophysical Research* 102: 929-945.

cause of the evaporation, they are the result of the lack of evaporation. El Niño in eastern Australia delivers less rainfall and also less cloud cover. Fewer clouds allow far more solar energy to reach the soil surface. The increase in solar energy hitting the soil is far greater than the trivial increase in radiant energy caused by an increase in $CO_2$.

Marine sediments provide a window into past El Niño events. A sheltered basin on the edge of the Peruvian Shelf 80 km west of Lima has given a 20,000-year history of El Niño. The most prominent feature of the whole record was a dramatic depression of El Niño activity between 5600 and 6000 years ago. This coincided with the major warmth of the Holocene Climate Optimum. El Niño activity was also weak between 800 and 1250 AD,[1831] a period dominated by the Medieval Warming. Over the last 1000 years, there was the opposite trend. In the late 13th and early 17th Centuries, temperatures in the Northern Hemisphere were rather cool but El Niño activity was high. During the 19th Century when the planet was warming after the Little Ice Age, El Niño activity started to decline. Higher temperatures were expected to be associated with higher El Niño activity. Such is expected as these events create a spike in the global air temperature, dramatically demonstrated by the 1997–1998 El Niño that produced the highest mean annual temperature of the complete satellite record.[1832]

Recent El Niño events took place in September 2006–early 2007, 2004–2005, 2002–2003, 1997–1998, 1994, 1993, 1991–1992, 1986–1987 and 1976.[1833,1834] Major El Niño events occurred at 1790–1793, 1828, 1876–1877, 1891, 1925–1926, 1982–1983 and 1997–1998.[1835] The 1997–1998 event received special attention as air temperature was warmed by 1.5°C (compared to the normal El Niño temperature rise of 0.3°C). A weak La Niña ensued in early 2008, modest La Niña events in 2000–2001, 1999–2000 and 1995, and an intense La Niña in 1988–1989.[1836]

Although El Niño is the largest interannual climate signal in the tropics, it is not just felt in the Southern Hemisphere. European climate responds

[1831] Rein, B., Luckge, A. and Sirocko, F. 2004: A major Holocene ENSO anomaly during the Medieval period. *Geophysical Research Letters* 31: doi.10.1029/2004GL020161.

[1832] Rein, B., Luckge, A., Reinhardt, L., Sirocko, F., Wolf, A. and Dullo, W.-C. 2005: El Niño variability off Peru during the last 20,000 years. *Palaeoceanography* 20: doi.10.1029/2004PA001099.

[1833] Guilderson, T. P. and Schrag, D. P. 1998: Abrupt shift in subsurface temperatures in the tropical Pacific associated with changes in El Niño. *Science* 281: 240-243.

[1834] Trenbath, K. E. and Hoar, T. J. 1996: The 1990-1995 El Niño-Southern Oscillation event: longest on record. *Geophysical Research Letters* 23: 57-60.

[1835] Caviedes, C. N. 2001: *El Niño in history: Storming through the ages*. The University of Florida Press.

[1836] Wang, B. 1995: Interdecadal changes in El Niño onset in the last four decades. *Journal of Climate* 8: 267-285.

to El Niño in late winter with colder conditions in northern Europe and mild conditions in southern Europe. This may be because of a connection to Europe from the Pacific via the stratosphere.[1837] The interplay between El Niño, the Asian Monsoon and the Indian Ocean drives climate extremes in and around the Indian Ocean.[1838,1839] Historical and proxy[1840] records show changes in the El Niño and Asian Monsoon over recent decades.[1841] Coral proxies show that, since 1846, there has been an increase in seasonal upwelling in the Indian Ocean[1842] and such upwelling has historically been associated with El Niño.[1843]

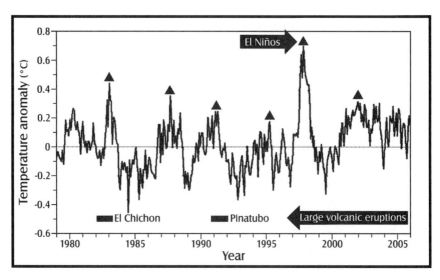

*Figure 42: Rises and falls in temperature over the last 30 years showing the influence of warming by El Niño events and cooling volcanic events. There are temperature cycles. There is no significant temperature increase in the Late 20th Century Warming and the post-1998 El Niño cooling event is shown.*

[1837] Ineson, S. and Scaife, A. A. 2009: The role of the stratosphere in the European climate response to El Niño. *Nature Geoscience* 2: 32-36.

[1838] Saji, N. H., Goswami, B. N., Vinayachandran, P. H. and Yamagata, T. A. 1999: A dipole mode in the tropical Indian Ocean. *Nature* 401: 360-363.

[1839] Webster, P. J., Moore, M. D., Loschnigg, J. P. and Leben, R. R. 1999: Coupled ocean-atmosphere dynamics in the Indian Ocean during 1997-98. *Nature* 401: 356-360.

[1840] Pfeiffer, M. and Dullo, W.-C. 2006: Monsoon-induced cooling of the western equatorial Indian Ocean as recorded in coral oxygen isotope records from the Seychelles covering the period of 1840-1994 AD. *Quaternary Science Reviews* 25: 993-1009.

[1841] Goswami, B. N., Venugopal, V., Sengupta, D., Madhusoodanan, M. S. and Xavier, P. K. 2006: Increasing trend of extreme rain events over India in a warming environment. *Science* 314: 1442-1445.

[1842] Abram, N. L., Gagan, M. K., Cole, J. E., Hantoro, W. S. and Mudelsee, M 2008: Recent intensification of tropical climate variability in the Indian Ocean. *Nature Geoscience* 1: 849-853.

[1843] Charles, C. D., Cobb, K. M., Moore, M. D. and Fairbanks, R. G. 2003: Monsoon-tropical ocean interaction in a network of coral records spanning the 20th Century. *Marine Geology* 201: 207-222.

There are many competing and contradictory theories for the origin of El Niño events. One suggestion was that an anomalously warm spot in the eastern Pacific could weaken the east-west temperature difference, causing trade winds to weaken and warm water to push to the west.[1844] The origin of the warm spot was unexplained. Increased trade winds could build up a bulge of warm water in the western Pacific, weakening winds could then cause this water to move eastwards.[1845,1846,1847] No such build-up was measured with recent El Niño events. Variations on this theory are that warm equatorial waters are dispersed to high latitudes during El Niño events and the equatorial area takes a variable amount of time to recharge.[1848] In the western Pacific, intense weather events could create a weakening of westward currents stimulating an El Niño event.[1849]

Another theory suggests that tropical volcanoes may eject dust that decreases the input of solar radiation to trigger El Niño.[1850] No correlation is recorded. Others suggest that El Niño events correlate with the Sun's cycles[1851,1852,1853] and, if this is the case, then there is an opportunity for long-range prediction of El Niño events. Random behaviour of the equatorial Pacific Ocean (possibly triggered by weather or terrestrial volcanic events) may initiate an event.[1854] A relationship may exist between fluctuations in low-level winds,[1855] rain in the western and central equatorial zones, sequences of

---

[1844] Bjerknes, J. 1969: Atmospheric teleconnections from the equatorial Pacific. *Monthly Weather Review* 97: 163-172.

[1845] Wyrtki, K. 1975: El Niño – The dynamic response of the equatorial Pacific Ocean to atmospheric forcing. *Journal of Physical Oceanography* 5: 572-584.

[1846] Cane, M. A. and Zebiak, S. E. 1985: A theory for El Niño and the Southern Oscillation. *Science* 228: 1085-1087.

[1847] Yamagata, T. and Masumoto, Y. 1989: A simple ocean-atmosphere coupled model for the origin of warm El Niño Southern Oscillation event. *Philosophical Transactions of the Royal Society of London*. A 329: 225-236.

[1848] Recharge oscillator theory.

[1849] Zhang, R.-H, Rothstein, L. M. and Busalacchi, A. J. 1998: Origin of upper-ocean warming and El Niño change on decadal scales in the tropical Pacific Ocean. *Nature* 391: 879-883.

[1850] Adams, B. J., Mann, M. E. and Ammann, C. M. 2003: Proxy evidence for an El Niño-like response to volcanic forcing. *Nature* 426: 274-278.

[1851] Drijfhout, S. S., Haarsma, R. J., Opsteegh, J. D. and Selten, F. M. 1999: Solar-induced versus internal variability in a coupled climate model. *Geophysical Research Letters* 26: 205-208.

[1852] Vereteneko, S. V., Dergachev, V. A. and Dmitriyev, P. B. 2005: Long-term variations of the surface pressure in the North Atlantic and possible association with solar activity and galactic cosmic rays. *Advances in Space Research* 35: 484-490.

[1853] Weng, H. J. 2005: The influence of the 11 yr solar cycle on the interannual-centennial climate variability. *Journal of Atmospheric and Solar-Terrestrial Physics* 8-9: 793-805.

[1854] Eckert, C. and Latif, M. 1997: Predictability of a stochastically forced hybrid coupled model of El Niño. *Journal of Climate* 10: 1488-1504.

[1855] Madden-Julian Oscillation.

eastward waves[1856] and El Niño.[1857] The interference between downwelling eastward-trending waves and upwelling waves[1858] may carry warm surface water eastwards.[1859] This may have produced the 2006 event but does not explain other El Niño events.

The processes that cause an El Niño event are still being investigated. Weather events tend to be somewhat chaotic and El Niño may be one of these.[1860] There is no evidence to suggest that these events are related to global warming and there is no reliable method of predicting them.[1861] There are, however, some intriguing correlations.[1862] The timing of earthquake swarms (resulting from increased seismic activity on the East Pacific Rise) correlates strongly with El Niño cycles.[1863,1864] Some astronomers have demonstrated a relationship between solar activity and these cycles.[1865]

Seismic activity in the East Pacific Rise results from rocks being pushed apart by a mass of rising molten rock. This creates earthquakes. Some 85% of the world's volcanoes are submarine, unseen, non-explosive and ignored by the IPCC. The East Pacific Rise is one of the most active spreading centres on Earth, with thousands of submarine vents exhaling hot water and eruptions focusing heat in one small part of the ocean.[1866] Vent water can be up to 420°C. Clustered earthquake swarms 10–33 km deep occur over several days to weeks. Such earthquakes are shallow. Following earthquake clusters, submarine hot spring venting rates and the rise of magma are increased.

Sea floor basaltic lavas comprise molten rock, at least 100 times the

---

[1856] Kelvin waves.

[1857] Eisenman, I., Yu, L and Tziperman, E. 2005: Westerly wind bursts: ENSO's tail rather than the dog? *Journal of Climate* 18: 5224-5238.

[1858] Rosby waves.

[1859] Hackert, E., Ballabrera-Poy, J., Busalacchi, A. J., Zhang, R.-H. and Murtugudde, R. 2007: Role of the initial ocean state for the 2006 El Niño. *Geophysical Research Letters* 34: L09605, doi: 10.1029/2007GL029452.

[1860] Tziperman, E., Stone, L., Cane, M. A. and Jarosh, H. 1994: El Niño chaos: Overlapping of resonances between the seasonal cycle and the Pacific Ocean-Atmosphere oscillator. *Science* 264: 72-74.

[1861] Graham, N. E., Michaelsen, J. and Barnett, T. P. 1987: An investigation of the El Niño-Southern Oscillation cycle with statistical models.1. Predictor field characteristics. *Journal of Geophysical Research. C. Oceans* 92: 14251-14271.

[1862] Leybourne, B. A. and Adams, M. B. 2001: El Niño tectonic modulation in the Pacific Basin. *Oceans* 4: 2400-2406

[1863] Walker, D. A. 1995: More evidence indicates link between El Niños and seismicity. *EOS* 76: 33.

[1864] McPhaden, M. J. 1999: Genesis and evolution of the 1997-98 El Niño. *Science* 283: 950.

[1865] Landscheidt, T. 2000: Solar activity controls El Niño and La Niña. *ESA Special Publication* 463: 135.

[1866] Morgan, W. J. 1972: Deep mantle convection plumes and plate motions. *Bulletin of the American Petroleum Geologists* 56: 203-213.

volume of water is required to cool a unit volume of molten rock. Minimum estimates are that at least 100 cubic kilometres of lava are erupted onto the sea floor each year, about three times as much lava as a large terrestrial eruption (e.g. Santorini, Krakatoa). This means that every year at least 10,000 cubic kilometres of seawater is required to cool the annual budget of new lava. The mid-ocean ridge submarine volcanic chains are 64,000 km long so the energy released from cooling such a volume of magma is monstrous.[1867]

El Niño may originate in the east-west trending active volcanic Galapagos Ridge. This area has an unusual tectonic setting with the ocean ridge plunging beneath the active continental volcanic zone and with equatorial ocean current trends almost coincident with the Ridge. If El Niño derives from plate tectonics, then the interaction of the Earth's fluid processes such as convection of heat from molten rock, ocean currents and atmospheric circulation are clearly important. Eruption centres, such as the East Pacific Rise and Galapagos Islands, are the locus for transfer of massive amounts of heat to seawater with the normal rate of submarine volcanic activity, let alone a major eruption. Circulating seawater in cooling seafloor volcanoes modifies the cooling rock by adding water, sodium and other material to the rock. The chemical reactions during modification of rock create heat. The seawater also leaches material such as metals from the cooling rock. The modified seawater is vented to the surface as a hot spring. These submarine hot springs are most typically along the medial rifts of the mid ocean ridges and exhale fine black particles of metal sulphide minerals if they are very hot (black smokers) or fine white particles of sulphate minerals if they are warm (white smokers). Smokers are located by measuring heat, trace gases and hot spring particles in the seawater.

Mid ocean ridges are areas on the Earth's crust where there is a huge heat flow from deep in the Earth to the surface and sea surface temperatures have a tectonic signature.[1868] The transfer of volcanic heat via circulating seawater is a major global heat transfer process whereby localised volumes of the ocean are quickly heated. Mid ocean ridge earthquake swarms herald the rise of molten rock that needs to be cooled. These immediately precede El Niño events, which suggests not only a correlation but also causation.[1869,1870]

The US Geological Survey National Earthquake Information Center (NEIC) provides data from 1600 AD[1871] and the International Earth

[1867] At least 4.5 x $10^{21}$ joules.

[1868] ftp://ftp.ncep.noaa.gov/pub/cmb/sst

[1869] Shaw, H. S. and Moore, J. G. 1988: Magmatic heat and the El Niño cycle. *EOS* 69: 1553.

[1870] Walker, D. A. 1999: Seismic predictors of El Niños revisited. *EOS* 80: 25-26.

[1871] http://neic.usgs.gov/neis/epic/epic.html

Rotation Service (IERS) provides historical Earth rotation speeds.[1872] A plot of El Niño from the Japanese Meteorological Agency[1873] against increased frequency of earthquakes and increase in the Earth's rotation speed shows a striking correlation. The exception was the 1982–1983 El Niño event which occurred with a higher frequency of earthquakes and a decrease in the Earth's rotation. In 1982, 203 earthquakes were centred in the eastern Pacific at 3°S 177°E. The area had previously been seismically quiet for 400 years. El Niño ocean warming started in the mid equatorial region (5°N to 5°S, 150°E to 160°E) and then moved eastwards. In the 1990s, earthquakes in the eastern Pacific were more common than in the 1980s. During the 1990s there was a relative slowdown in the Earth's rotation and, as a result, the frequency of El Niño in the 1990s was higher than in the 1980s.

However, the sea surface warming in the eastern Pacific requires another factor. That is the weakening of the trade winds and the ocean current in the equatorial Pacific. Slowing of the Earth's rotation can produce such weakening. When a huge amount of molten rock is erupted along the mid ocean ridge or elsewhere, the rotational inertia of the Earth increases and the rotation speed is reduced. When the eruption has stopped, the rotational inertia decreases and the rotation speed gradually increases.

Elsewhere, sea surface temperature anomalies are shown to correlate with, but lag behind earthquake clusters. This lag is due to slow heat movement in rocks because they are good insulators. Heat transfer from 10 km depth took about 2 months in the Adriatic and Mediterranean seas, but in the eastern Pacific the figure was less, probably because of a far thinner crust and a much greater volume of near-surface molten rock.[1874]

Although there are hints that earthquakes may influence weather, there has been little work done on the geophysical conditions for the formation of a storm,[1875] but we can speculate as follows: Molten rock rises because it is lighter than the surrounding solid rock. This molten rock decreases the Earth's gravity at the site of the rise and this changes atmospheric pressure. Very little change in gravity (0.3 to 0.4 microgals) will produce an atmospheric change of one millibar. Superconducting gravimeters[1876] show shifts in gravity of 6

---

[1872] www.iers.org/iers/earth

[1873] www.coaps.fsu.edu/pub/JMA-SST-Index

[1874] Johnson, H. P., Hutnak, M., Dziak, R. P., Fox, C. G., Urcuyo, I., Cowen, J. P., Nabelek, J. and Fisher, C. 2000: Earthquake-induced changes in a hydrothermal system on the Juan de Fuca mid-ocean ridge. *Nature* 407: 174-177.

[1875] Wood, M. D. and King, N. E. 1977: Relation between earthquakes, weather and soil tilt. *Science* 197: 154-156.

[1876] Sun, H., Xu, H., Ducarme, B. and Hinderer, J. 2008: Comprehensive comparison and analysis of the tidal gravity observations obtained with superconducting gravimeters at stations in China, Belgium and France. *Chinese Science Bulletin* 44: 750-755.

microgals with typical weather patterns and maximum shifts of 45 microgals. The small regional atmospheric fluxes of the Southern Oscillation, which are typically 4 to 6 millibars, may be produced by only a 2 microgal change in gravity.[1877]

Submarine volcanic activity is very localised in a global sense. Although solar energy is orders of magnitude greater than geothermal energy, it is the focusing of geothermal energy that may initiate El Niño. A body of ascending molten light rock produces gravity changes in the order of milligals, orders of magnitude higher than the microgal changes associated with weather. The high-pressure zone of the Southern Oscillation is over the most dynamic mid ocean ridge on Earth, the East Pacific Rise. The low-pressure zone of the Southern Oscillation is centred just north of Darwin over the Banda Sea. The Banda Sea is the junction of the Pacific, Australian and Eurasian tectonic plates. The mantle has risen to within 21 km of the surface in the Banda arc, 14 km from the surface in the Weber Deep and 7 km from the surface in the North Banda Sea. It has been suggested that this huge body of shallow hot material affects the gravity in the area thereby affecting the atmospheric pressure flux of the Southern Oscillation (which modulates El Niño). Should we be looking into the Earth for an explanation of El Niño-La Niña events?

Storms are formed by high horizontal gradients of atmospheric pressure. Both earthquakes and masses of rising molten rock create huge localised changes in atmospheric pressure. This may significantly change ocean circulation patterns thereby, along with atmospheric pressure, affecting weather, hurricane formation and ocean-atmosphere circulation. The heat wave of Europe in 2003 may have resulted from a preceding burst of geothermal flow that produced the Adriatic, Aegean and Algerian earthquake events in the same year. The whole show may be controlled by events at the core-mantle boundary that generate electromagnetic, gravity and orbital surges.[1878]

## Water cycle

Current climate change models just ignore well-documented, multi-year, alternating sequences in the hydrometeorological processes that have been documented during the past 100 years. The linking of a 21-year sequence of

[1877] Widmer-Schnidrig, R. 2003: What can superconducting gravimeters contribute to normal-mode seismology. *Bulletin of the Seismological Society of America* 93: 1370-1380.

[1878] Leybourne, B., Orr, W., Haas, A., Gregori, G. P., Smoot, C and Bhat, I. 2006: Tectonic forcing function of climate – revisited: Four elements of coupled climate evidence of an electromagnetic driver for global climate. *New Concepts in Global Tectonics* 40.

rainfall, river flow and air temperature to sunspot activity is well recorded.[1879] In Australia, 40 major floods were recorded from 1900 to 1982. Of these, 24 occurred during the first cycle of a double sunspot cycle and 16 in the second cycle, again showing the very strong relationship between solar activity and climate.[1880]

The global system is principally an interaction between the solid Earth, oceans, air and life with some extraterrestrial input. The cycle of carbon throughout the history of time is controlled by water and buffered by life, the rocks, major plate tectonic recycling processes, and conversion of rocks to soil.[1881] Although there might be some argument about the rates of weathering, the story is the same.[1882] There have been some exciting periods of time on Earth when there was far more volcanicity, huge amounts of $CO_2$ were pumped into the atmosphere, increased amount of carbon was buried in sediments and the global average surface temperature was some 5–6°C warmer than at present.[1883] Silicate weathering to form soils and in submarine settings and deposition of carbonate were again the buffer to stop a runaway greenhouse.[1884] The Earth will just keep doing what it has always done, whether humans live here or not.

The carbon cycle dominates opinions on global warming at the expense of the water cycle. On one scale, water does a slow lap of the planet and is recycled by masses of wet ocean floor being plunged deep into the Earth. This water, together with some of the original water from the planet, is released to the surface by volcanoes.

On another scale, the movement of water on the surface of the planet, the hydrosphere and the atmosphere represent the largest movement of mass and energy in the Earth's outer spheres. The isotopic composition of river water combined with information on plant transpiration and soil evaporation shows that the larger water cycle controls the smaller carbon cycle.[1885] Other

---

[1879] Alexander, W. 2007: Locally-developed climate model verified. *Water Wheel* 6: 1, 27-29.

[1880] Francou, J. and Rodier, J. A. 1984: *World Catalogue of Maximum Observed Floods.* International Association of Hydrological Series.

[1881] Franck, S., Kossacki, K. and Bounama, C. 2007: Modelling the global carbon cycle for the past and future evolution of the earth system. *Chemical Geology* 159: 305-317.

[1882] Caldeira, K. and Kasting, J. F. 1992: The life span of the biosphere revisited. *Nature* 360: 721-723.

[1883] Tajika, E. 1999: Carbon cycle and climate change during the Cretaceous inferred from a biogeochemical carbon cycle model. *Island Arc* 8: 292-303.

[1884] Nakamori, T. 2001: Global carbonate accumulation rates from Cretaceous to Present and their implications for the carbon cycle model. *Island Arc* 10: 1-8.

[1885] Ferguson, P. R. and Veizer, J. 2007: Coupling of water and carbon fluxes via the terrestrial biosphere and its significance to the Earth's climate system. *Journal of Geophysical Research* 112: D24S06, doi:10.1029/2007/JD00843.

factors influencing vegetation such as $CO_2$, nutrients and temperature are not the main drivers of plant growth.

As water vapour is the main greenhouse gas, huge amounts of energy are locked into oceans and large amounts of energy are transferred in the melting and evaporation of water. Compared to water, $CO_2$ is a trace gas in the atmosphere and is a minor component in atmosphere and hydrosphere systems. The carbon cycle is essentially driven by solar energy via the water cycle.

Contrary to popular belief, the carbon cycle does not control climate. It is the water cycle that does and water vapour is the main greenhouse gas in the atmosphere. Terrestrial water vapour fluxes represent one of the largest movements of mass and energy in the Earth's outer sphere, yet the relative contributions of non-biological water vapour fluxes and those that are regulated solely by the physiology of plants are not well known.

## The rock that made a fool out of humans

In 1831, Admiral Sir James Robert George Graham had the Union Jack hoisted on a volcanic landmass that had suddenly appeared near Sicily. It was called Graham Bank and was claimed by England. It was also claimed by the Kingdom of the Two Sicilies who called it Isola Ferdinandea, the French (L'Isle Julia), and other powers who variously named it Nerita, Hotham, Scicca and Corrao. In the subsequent dispute over ownership, France and the Kingdom of the Two Sicilies almost came to war. England and the Two Kingdoms of Sicily had a diplomatic row. During the intense diplomatic dispute, the island quietly slipped back underwater.

In 1987, US warplanes thought the dark mass 8 metres below sea level was a Libyan submarine and attacked it with depth charges. In February 2000 when the volcano again stirred, Domenico Macalusa, a surgeon, diver and the Honorary Inspector of Sicilian Cultural Relics, took action. He persuaded Charles and Camilla, the last two surviving relatives of the Bourbon Kings of the Two Sicilies, to fund the bolting of a 150 kg marble plaque to the volcano at some 20 metres below sea level. The plaque pre-empted ownership if the volcano ever again rose above sea level. It was placed underwater in September 2001. By November 2002, the plaque had been smashed into 12 pieces by person or persons unknown.

This rock is worth nothing, is of no use as a territorial possession and is of no scientific interest and yet the French and Bourbons nearly came to war 170 years ago and the English and Italians are still in dispute. Graham Bank serves to show that whatever futile political decisions we humans make, the land rises and falls, sea level rises and falls, and climates change as they have done since the dawn of time.

The forces of natural processes are far greater than human ego.

# Chapter 7

# AIR

Question: Do thermometer measurements show the planet is warming?
Answer: No.

Q: Do other temperature measurements show the planet is warming?
Answer: No.

Question: Is atmospheric $CO_2$ of human origin increasing?
Answer: Possibly.

Question: Is atmospheric $CO_2$ approaching a dangerous level?
Answer: No.

Question: Do higher sea temperatures cause more hurricanes?
Answer: No.

Question: Do clouds influence climate?
Answer: Yes.

*There is no such thing as the greenhouse effect. The atmosphere behaves neither as a greenhouse nor as an insulating blanket preventing heat escaping from Earth. Competing forces of evaporation, convection, precipitation and radiation create an energy balance in the atmosphere.*

*Historical thermometer measurements are flawed, contain bias, have a low order of accuracy, do not cover the planet's surface equally and are a combination of land surface measurements, the effects of urbanisation and the effects of changing land use patterns. The data quality is not research quality hence no conclusions about future trends can be made.*

*Satellite and balloon measurements provide a more accurate data set. These show that there is no global warming. Temperature proxies give general climate trends and are not accurate enough for global temperature predictions.*

*The atmosphere contains 800 billion tonnes of carbon, whereas the oceans have 39,000 billion tonnes and surface rocks have 65,000,000 billion tonnes. Deep crustal and mantle rocks contain orders of magnitude more carbon. Atmospheric carbon only occupies 0.001% of the total carbon in the upper crust, oceans, life and atmosphere. Soil contains more carbon that the total carbon in life and air combined. Each year, 18% of the atmospheric $CO_2$ is exchanged with life and the oceans, and $CO_2$ produced today stays in the atmosphere for only 4 to 5 years. Many of the major sources and sinks of $CO_2$ are ignored by the IPCC.*

*Historical $CO_2$ measurements show that in the 19th and 20th Centuries, there were times when the atmospheric $CO_2$ content was far higher than at present. Historical measurements of $CO_2$ have not been correlated or integrated with modern measurements and long-term trends cannot be deduced. Most modern measurements are rejected and the remaining edited $CO_2$ measurements show seasonal trends and an increase in measured $CO_2$ over the last 50 years. The seasonal trends show that $CO_2$ is rapidly scrubbed from the atmosphere in the Northern Hemisphere growing season.*

*Hurricanes were more common in past times and hurricane damage is related to population demographics and expensive coastal real estate and not sea surface temperature. Hurricanes are creating increasing damage only because humans have created more structures near coastlines for hurricanes to destroy.*

*Clouds are not factored into climate models and yet clouds have a significant influence on climate. The formation of low-level clouds by cosmic radiation has not been considered in climate models despite experiments and observations showing that the Sun, cosmic rays and clouds are probably major drivers of climate.*

## The greenhouse effect

Everyone knows what the greenhouse effect is. Well ... do they? Ask someone to explain how the greenhouse effect works. There is an extremely high probability that they have no idea. What really is the greenhouse effect?

The use of the term "greenhouse effect" is a complete misnomer. Greenhouses or glasshouses are used for increasing plant growth, especially in colder climates. A greenhouse eliminates convective cooling, the major process of heat transfer in the atmosphere, and protects plants from frost. Because there is no convective cooling, the temperature in the glasshouse stays warm. There are no magical properties of glass. Transparent plastic sheeting and polycarbonate are also used in greenhouses. Plant growth is also stimulated by pumping in $CO_2$ such that the air in a greenhouse has at least three times the $CO_2$ content of the air outside.

The Earth has an average surface temperature of about 15°C. The tropics are some 10°C warmer. In the atmosphere, $CO_2$ is a highly effective trap of energy in the infra-red wavelength band of 14 to 16.5 microns. Blocking the escape of heat radiation with wavelengths in this range reduces the radiating efficiency of the Earth by 15%. If the atmosphere had no $CO_2$, far more heat would be lost from Earth and the average surface temperature would be -3°C. The efficiency of the $CO_2$ trap is essentially insensitive to the amount of $CO_2$ in the atmosphere. All the $CO_2$ does is slows down heat loss. Atmospheric $CO_2$ does not trap heat, as insulation does. If the current atmospheric $CO_2$ content of 380 ppmv were doubled to 760 ppmv, there would be a minuscule impact on the radiation balance and the temperature. An increase in air temperature of 0.5°C is likely. This is hardly catastrophic. Furthermore, the effects of the additional $CO_2$ would be completely masked by other climate drivers such as the Sun and the Earth's orbit and there would be great benefits derived from accelerated plant growth. If indeed humans are slightly changing climate, then addressing climate change is not just like a home heating system where the thermostat can be changed to the desired temperature.

The first IPCC Report tried to set the scene by stating that the Earth radiates 240 watts per square metre of energy back into space. This equates to an emission temperature at the surface of -19°C yet the Earth's average surface temperature is +15°C.[1886] The IPCC then states:

> The reason the earth's surface is this warm is the presence of greenhouse gases, which act as a partial blanket for the long-wave radiation coming from the earth's surface. This blanketing is known as the natural greenhouse effect.

The explanation is wrong. It implies that the atmosphere comprises layers that do not mix. Furthermore, greenhouse gases do not act like an insulator or a blanket. Air itself is a good insulator. Its main gases are nitrogen and oxygen, which are excellent insulators against the conduction of heat. However, conduction is not a process of the atmosphere. The trace amounts of $CO_2$ in the atmosphere have no effect whatsoever on the insulating properties of air.

The IPCC again makes an incorrect explanation of the greenhouse effect.

---

[1886] Equivalent to an altitude of 5 km.

Much of the thermal radiation emitted by the land and the ocean is absorbed by the atmosphere, including clouds, and re-radiated back to earth. This is called the greenhouse effect.

The IPCC's global energy budget has the Earth's surface emitting 390 watts per square metre and the energy radiated back to the surface as 324 watts per square metre of infra-red radiation. The IPCC do not explain how the loss of infra-red radiation of 66 watts per square metre to the atmosphere can actually create warming. If radiation is lost, there should be cooling.

In 1896, Svante Arrhenius suggested that the burning of coal produces an increase in atmospheric $CO_2$ and the absorption of infra-red radiation by $CO_2$ could warm the planet's atmosphere. However, Arrhenius was not aware of the carbon cycle, considered only infra-red radiation, and did not evaluate the movement of heat and air in the atmosphere by convection. The process of transferring heat from the Earth's surface to the atmosphere has long been known.[1887]

The surface of the Earth is heated by incoming radiation. What we humans feel on the surface of the Earth is short wavelength radiation, a wavelength that is not absorbed by greenhouse gases. Because of the molecular structures of water vapour, $CO_2$, methane and a few other gases, they absorb energy of a specific but longer wavelength. However, the wavelengths absorbed are only in very tight bands so not all the long-wave radiation (infra-red) is absorbed. This wavelength coincidentally matches those of the energy radiated away from the heated surface of Earth. The molecules in water vapour, $CO_2$ and methane not only absorb this energy but they also re-radiate much of it back into the atmosphere and back into space. The middle and high latitudes radiate more infra-red energy to space than solar radiation absorbed. To maintain energy balance, we need to transport energy from the tropics to higher latitudes.

There are two processes at work. Solar radiation warms the Earth's surface, mainly in the tropical oceans. Because the evaporation of water from the surface of the oceans requires heat, the surface of the warm oceans evaporates water and this water vapour (with its latent heat) is transferred to the atmosphere. The ocean surface then cools. The heat transferred from the oceans to the atmosphere is now held in the atmospheric boundary layer. The second process is one where infra-red radiation to space is emitted from greenhouse gases in the upper atmosphere. Energy is radiated from the upper atmosphere at a rate and wavelength controlled by temperature. This leads to cooling of the upper atmosphere.

---

[1887] Riehl, H. and Malkus, J. S. 1958: On the heat balance in the equatorial trough zone. *Geophysics* 6: 503-538.

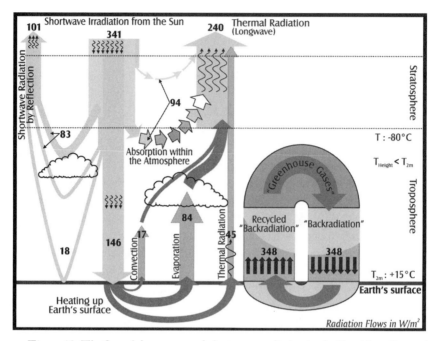

*Figure 43: The flow of long wave and short wave radiation in the Earth's surface and atmosphere showing that the atmosphere is at equilibrium.*

Energy in the atmosphere is then not balanced and, because air is such a good insulator, energy will not be conducted through the atmosphere. The only way to maintain energy balance between the lower and upper atmosphere is to have convection. However, the atmosphere is layered, temperature decreases with altitude[1888] and the rate of temperature decrease must be sufficient to allow air to rise buoyantly in updraughts. Atmospheric boundary layer air rises buoyantly in updrafts in tropical convection clouds. This converts heat and latent heat to potential energy. Away from the updrafts, turbulent motions of the atmosphere mix energy downward and convert potential energy to heat. This process distributes heat and latent heat from the tropics throughout the lower atmosphere.

When infra-red radiation is absorbed by the molecules of greenhouse gases (e.g. $H_2O$, $CO_2$), its energy is transformed into thermal expansion of air. This causes convection of expanded lighter warmer air and the resultant redistribution of air and energy. Most of the heat in the dense lower atmosphere is transferred by convection (67%) and not radiation (8%) with water condensation providing the rest (25%). When air heats, it expands, becomes lighter and rises. Denser cooler air in the upper layers of

---

[1888] 6.5°C per kilometre.

the lower atmosphere descends and replaces the warmer air from the lower layers. This system of multiple cells of air convection acts like a continuous surface cooler and the cooling effect greatly surpasses the warming effect from radiation.[1889] This is where models used by the IPCC greatly understate the increase of evaporative cooling with temperature.[1890]

The climate system is balanced when the total amount of solar radiation absorbed by the infra-red radiation to space and the rate of heat and latent energy distribution by convection is offset by the total radiation loss. This balance is the so-called greenhouse effect when the atmospheric average surface temperature of +15°C is considerable greater than the radiation temperature of the surface of the Earth (-19°C). The role of greenhouse gases is to cool the atmosphere through radiating energy to space.[1891] The dominant gases in the atmosphere such as nitrogen and oxygen cannot do this. The next most dominant gas, water vapour, can. In fact, water is the main greenhouse gas and more than 98% of the effect of atmospheric greenhouse gases is due to water vapour. The next most dominant gas, argon, cannot. The traces of $CO_2$ and methane can. This radiation, together with solar surface warming, generates convective instability and the shifting around of massive amounts of energy, especially in the tropics, from the surface to the upper atmosphere and polar regions. The total radiative loss from the atmosphere is more than 107 watts per square metre and the radiative forcing from doubling the current atmospheric $CO_2$ is only 4 watts per square metre. The greenhouse gases in the atmosphere and clouds emit infra-red radiation because of their temperature and because the characteristic radiation emitted is this long-wave radiation (14–16.5μ). Unless energy was added to the atmosphere, the greenhouse gases and clouds would keep emitting energy until there was no more energy to emit. The planet would then be frozen and sterile. Thank God for greenhouse gases.

An increased $CO_2$ content of the atmosphere reduces infra-red radiation to space in the infra-red absorbing bands of $CO_2$. Increased $CO_2$ also increases the downward radiation at the surface. These are opposing effects and tend to cancel. Changes in the total radiation loss will change the convective overturning in response. Increased radiation to space will prompt more convective overturning and distribution of more energy from the atmospheric boundary layer. Increased convective overturning dries the lower atmosphere, increasing the infra-red emissions to space in the water

[1889] Hadley Cells.

[1890] Held, I. M. and Sodon, B. J. 2006: Robust responses of the hydrological cycle to global warming. *Energy and Environment* 18: 951-983.

[1891] Chilingar, G. V., Khilyuk, L. F. and Sorokhtin, O. G. 2008: Cooling of atmosphere due to $CO_2$ emission. *Energy Sources* 30: 1-9.

vapour bands, and this compensates for the reduction in emission of the $CO_2$ bands.

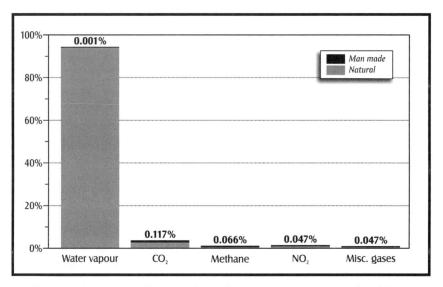

*Figure 44: Atmospheric greenhouse gases showing the proportion of greenhouse gases derived from natural and human activities. About 98% of the greenhouse effect in the atmosphere is due to water vapour and very little of the effect of $CO_2$ is due to human activity.*

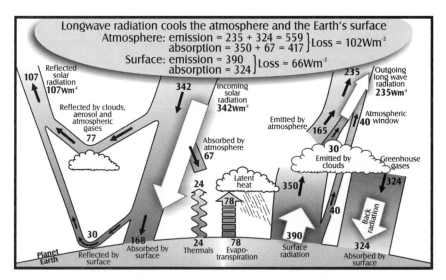

*Figure 45: Additions and losses of long wavelength radiation to the Earth showing role of clouds and latent heat.*

An increase in atmospheric $CO_2$ will increase the back radiation that is absorbed at the Earth's surface and will change the energy balance. The increase in back radiation will tend to warm the surface. This will increase the rate of energy loss by heat exchange with the overlying boundary layer, evaporation of latent energy and emission of radiation. One kilogram of air at -15°C can hold 4 grams of water as vapour, whereas at +35°C, one kilogram of air can hold 40 grams of water as vapour. What we don't know is whether this extra water vapour in warm air stays up there to amplify the $CO_2$ warming or returns to Earth quickly. Whatever happens, the water vapour operates as a greenhouse gas and the evaporation of water and condensation of water vapour involve a transfer of huge amounts of energy.

There will be a new energy balance as the increased energy loss equates with the $CO_2$ forcing of the back radiation. If the current $CO_2$ content is doubled, the surface temperature responds by rising 0.3°C. However, there is an amplification due to water vapour feedback and the temperature rise will be in the order of 0.5°C. The warmer surface temperature will increase the infra-red radiation to space to compensate for the reduction in the $CO_2$ bands. The warm surface temperature will stimulate convective overturning, resulting in an increased radiation to space in the water vapour bands. This radiation from space from the water vapour bands will dampen the effects of the increased $CO_2$ and the constraining of the surface temperature by the increased evaporation of latent energy at the surface.

This is why in past times when the atmospheric $CO_2$ concentration has more than 25 times higher than the present value, there has been no runaway greenhouse or "tipping points". If there were no water on Earth, then there would have been a runaway greenhouse billions of years ago.

The IPCC greenhouse gas models have the troposphere warming and changes in stratospheric circulation as a result of humans adding greenhouse gases to the atmosphere.[1892] A good way of measuring the amount of circulation is to measure the age of stratospheric air[1893,1894] because the age decreases with rising levels of greenhouse gases in the atmosphere.[1895]

---

[1892] Solomon, S., Qin, D., Manning, M., Marquis, M., Marquis, M., Averyt, K., Tignor, M. M. B., Miller, Jnr, H. L. and Zhenlin, C. (eds) 2007: *Climate change 2007: The physical science basis – Contribution of Working Group 1 to the Fourth Assessment Report of the Intergovernmental Panel on Climate Change.* Cambridge University Press.

[1893] Li, S. and Waugh, D. W. 1999: Sensitivity of mean age and long-lived tracers to transport parameters in a two-dimensional model. *Journal of Geophysical Research* 104: 30559-30569.

[1894] Waugh, D. W. and Hall, T. M. 2002: Age of stratospheric air: Theory, observations and models. *Reviews in Geophysics* 40: 1-10.

[1895] Austin, J. and Li, F. 2006: On the relationship between the strength of the Brewer-Dobson circulation and the age of stratospheric air. *Geophysical Research Letters* 33: L17807.

Balloon measurements over the last 30 years show that the stratosphere air is unchanged.[1896]

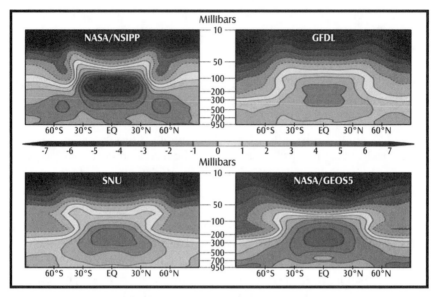

*Figure 46: Four computer predictions used by the IPCC for the effects of global warming by doubling atmospheric $CO_2$ as a function of latitude and pressure in the upper troposphere-lower stratosphere. All models predicted an atmospheric equatorial warming (shown with increasing darker tones).[1897]*

The greatest impact of $CO_2$ is in the first 100 ppmv in the atmosphere. After this concentration, the source of infra-red radiation to space is from the active $CO_2$ radiation bands in the stratosphere. There is every reason to suspect that the Earth is near an upper temperature limit given its present distribution of land and ocean and the strength of solar irradiance. The Earth's surface is heated by way of solar radiation and back infra-red radiation emanating from clouds, greenhouse gases, aerosols, soil, vegetation and rocks. It is cooled by conduction, evaporation and infra-red emission. Solar radiation and conduction are near constant and the Earth's surface temperature will vary according to increasing back infra-red radiation (radiation forcing from water vapour and $CO_2$) being offset by surface infra-red emission and latent heat of evaporation.

---

[1896] Engel, A., Möbius, T., Bösnich, H., Schmidt, U., Heinz, R., Levin, I., Atlas, E., Aoki, S., Nakazawa, T., Sugawara, S., Moore, F., Hurst, D., Elkins, J., Schauffler, S., Andrews, A. and Boering, K. 2009: Age of stratospheric air unchanged within uncertainties over the past 30 years. *Nature Geoscience* 2: 28-31.

[1897] Lee, M.I., Saurez, M. J., Kang, I.-S., Held, I. M. and Kim, D. 2008: A moist benchmark calculation for atmospheric general circulation models. *Journal of Climate* 21: 4934-4954.

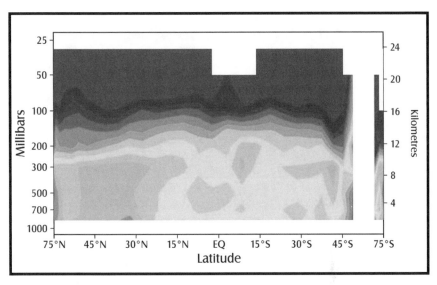

*Figure 47: Actual measurements of temperature in the upper troposphere-lower stratosphere from radiosonde balloon measurements. Not one of the global warming predictions in Figure 46 could be validated yet these predictions are still used by the IPCC.[1898]*

At global average surface temperature of 15°C, the rate of increase of surface infra-red emission with temperature is about 5 watts per square metre per °C and the rate of increase of latent energy from evaporation is of similar magnitude. This means that back infra-red radiation from doubling of $CO_2$ concentration must be at least 1 watt per square metre to sustain a 1°C rise and more than 30 watts per square metre to sustain a 3°C rise. Using the most accurate line-by-line radiation calculations, the increase in back infra-red radiation due to doubling of $CO_2$, increasing atmosphere temperature by 3°C and holding relative humidity constant (the full positive feedback of 3°C) only produces an increase in back infra-red radiation of 18 watts per square metre, well short of the 30 watts per square metre necessary to sustain a 3°C increase at equilibrium surface temperature. The rapidly increasing surface infra-red emission and latent heat loss with temperature are a barrier to significant surface temperature increase unless there is a change in the solar radiation input, either directly or through a change to cloudiness and ice.

A doubling of $CO_2$ is used in the Stern Review as the benchmark for climate sensitivity. This represents a forcing of about 3.7 watts per square metre. Since greenhouse gases of human origin have been estimated at 2.7 watts per

---

[1898] HadAT2 radiosonde observations, CCSP (2006), p. 116 (Figure 5.7E).

square metre, then we are already three quarters of the way to an effective doubling of $CO_2$. However, we have experienced much less warming than such forcing would suggest. There is a poor correlation between temperature and $CO_2$ in the 20th Century, with temperature increasing from 1905 to 1918 before there were substantial human greenhouse gas emissions. The rapid World War II and post-war emissions were during the period from 1940 to 1976 when temperature dropped. The correlation of $CO_2$ with temperature from 1976 to 1998 is good, but after 1998 there is again a poor correlation. The Stern Review assumes, against all empirical evidence and reasoning, that future increments of $CO_2$ will have substantially greater effects than those in the past.

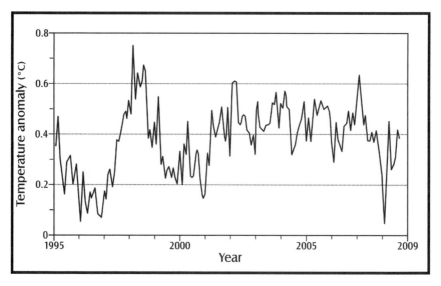

*Figure 48: Globally averaged surface temperature showing a decrease in surface temperature from a peak during the 1998 El Niño event. A similar trend is seen in the lower troposphere.*

Carbon dioxide in the atmosphere operates like a curtain on a window. If you want to keep out light, add a curtain. A second curtain makes little difference, a third curtain makes even less difference and a fourth curtain is totally ineffectual. $CO_2$ operates the same way. Once there is about 400 ppmv $CO_2$ in the atmosphere, the doubling or tripling of the $CO_2$ content has little effect on atmospheric temperature because $CO_2$ had adsorbed all the infra-red energy it can adsorb.

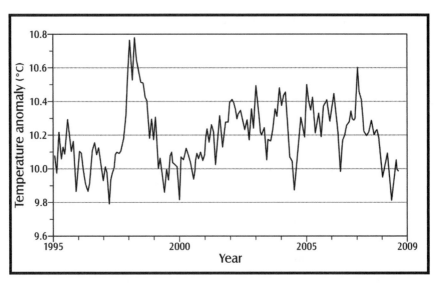

*Figure 49: Globally averaged lower troposphere temperature showin a decrease in temperature from a peak during the 1998 El Niño event. A similar trend is seen in surface temperature.*

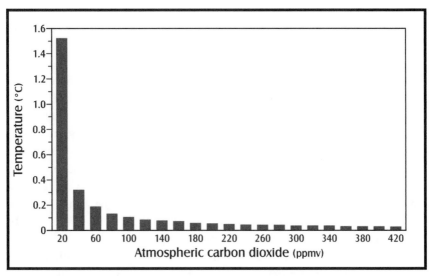

*Figure 50: The first 20 ppmv of $CO_2$ operating as a greenhouse gas in the atmosphere has the greatest effect on temperature. After about 200 ppmv, $CO_2$ has done its job as a greenhouse gas and has adsorbed almost all the infra-red energy it can absorb. Once the atmosphere is at the present $CO_2$ content of 385 ppmv, a doubling or quadrupling of the atmospheric $CO_2$ content will have very little effect on atmospheric temperature. This is why in former times when the atmospheric $CO_2$ content was up to 25 times the current content, there was no runaway greenhouse or "tipping point".*

## The measurement of temperature

*Measurement and order of accuracy*

Data collection in science involves observation, measurement and experiment. We need to be certain that this data can be validated repeatedly and we need to understand that every measurement has an order of accuracy. For example, the distance of a marathon race is 42.195 km but, because of the uncertainties, errors and equipment used for measurement, this number is really 42.195 ± 0.002 km which means the marathon length is somewhere between 42.193 and 42.197 km. So too with temperature measurements. We commonly read that the Earth has warmed by 0.7°C during the 20th Century. This is not a scientific measurement as there is no order of accuracy. If the order of accuracy is large, then the figure is meaningless. If the Earth has warmed by 0.7 ± 0.8°C then the figure is meaningless; if it is 0.7 ± 0.4°C then the figure may have some scientific meaning. If a figure is quoted for global temperature, the normal scientific questions are: What was the measuring technique? What are the limitations of the measuring technique? What is the order of accuracy of the technique? How was the data analysed? This is the normal business of science.

It is worthwhile to dwell on how temperature is measured, the inaccuracies of measurement and the order of accuracy of measurement. Until recently, surface temperature was only measured using a thermometer. It is this historical data set that is used to show global temperature change. However, these measurements are limited because not all thermometers have been calibrated against a standard thermometer, hence what one thermometer records as 31°C another might record as 33°C when the temperature is really 32°C. If the thermometer is calibrated at 15°C, then far higher and lower temperatures are prone to errors. Furthermore, most thermometers are graduated to the nearest degree, hence one observer may record the temperature as 31°C and another might record it as 32°C. Both measurements are correct because if a thermometer is graduated to the nearest degree, then any measurement is ± 0.5°C. Therefore, if most of the data that has been used to establish that the global temperature has changed by 0.7°C over the 20th Century was derived from thermometers, then this figure is really 0.7 ± 0.5°C. In effect, the accuracy of historical temperature measurements is close the instrumental order of accuracy and must be treated with caution. Many other errors can creep into measurement.

In the 20th Century, temperature measurements were made using a Stevenson Screen. This is an outdoor whitewashed louvred wooden box on stilts. The box houses thermometers. The Stevenson Screen must be away from back reflection (walls, bricks, concrete, asphalt, rocks, soil), at a fixed height and sited above and surrounded by mown grass. Some weather stations used today are actually on top of concrete buildings, adjacent to buildings, next to the street or in carparks. Temperature measured 1 metre above the ground surface can be more than 1°C different from temperature 2 metres above the ground surface. Not all Stevenson Screens measure temperature at exactly the same height above the ground.

Important properties of the Stevenson Screen are the size of the box and the thickness of the wooden slats. The aim of the Stevenson Screen is to stop direct sunlight hitting the instruments and to have the temperature inside the box as close as possible to that of the airflow. The automation of weather stations using different ways of measuring temperature has resulted in a change in the size and materials of the Stevenson Screen. Slow-response thermometers have been replaced by much higher-response electronic measuring instruments, and the large wooden Stevenson Screen boxes have been replaced by smaller ventilated boxes made of plastics. In hot climates, there is a greater heating of the outer plastic surface and conduction of heat through to the inner surface. The potential to give an artificially enhanced temperature on hot windless days is much greater in the newer plastic screens than for the older and larger wooden screens.

One can argue that it is human vanity to imagine that our relatively small inhabited percentage of global surface has the ability to alter the climate of the whole planet, as some 98.6% of the surface of the planet is essentially uninhabited. There are many measuring stations that have huge back reflections, mown grass does not occur in many parts of the world, and whitewash paint is reflective of infra-red radiation. Most climate stations in eastern Colorado did not meet the requirements of the World Meteorological Association's requirements for correct siting[1899] and this was found to be common in the USA.[1900] It has now been shown that about half of the US weather measuring stations are incorrectly positioned, do not fulfil the siting requirements of the US government and introduce a warming bias.[1901]

Weather stations are on the land (29% of Earth's area), clustered in the

[1899] Peilke, Snr, R. A., Davey, C., Niyogi, D., Fall, S., Steinweg-Woods, J., Hubbard, K., Lin, X., Cai, M., Lim, Y.-K., Li, H., Nielson-Gammon, J., Gallo, K., Hale, R., Mahmod, R., Foster, S., McNider, R. T. and Blanken, P. 2007: Unresolved issues with the assessment of multi-decadal global land surface temperature trends. *Journal of Geophysical Research* 112: D24508, doi: 10.1029/2006JD008229.

[1900] Peilke, Snr, R. A., Nielson-Gammon, J., Davey, C., Angel, J., Bliss, O., Doesken, N., Cai, M., Fall, S., Niyogi, D., Gallo, K., Hale, R., Hubbard, K. G., Lin, X., Li, H. and Raman, S. 2007: Documentation of uncertainties and biases associated with surface temperature measurement sites for climate change assessment. *Bulletin of the American Meteorological Society* 88: 913-928.

[1901] http://surfacestations.org

industrialised countries. There are no Stevenson Screens on the surface of the oceans, hence we have no Stevenson Screen measurements of global temperatures for 71% of the surface of the Earth. Furthermore, 3% of the Earth's surface is covered by ice and, of the remaining 26%, less than 2% is habitable if we remove swamps, deserts, lakes and mountain ranges. Humans live on only 1.4% of the surface of the Earth, which is hardly representative of the surface of the planet. Earth's metropolitan areas would fit into an area less than the size of Spain.

In order to understand historical temperature measurements, all we have to work with are the thermometer measurements made by those long gone. These results have bias, the instruments were not calibrated with other instruments, different stations have different orders of accuracy, the results may have systematic errors, the Stevenson Screen may have been incorrectly positioned, measurements have been made at different heights above the ground, the measuring station may have been moved many times, and industry and urbanisation may have encroached on what was originally a measuring station in a rural setting. Global temperature from thermometer measurements from 1860 are used to determine modern temperature trends. Some thermometers used mercury, others used coloured alcohol. More modern stations have changed to thermocouples. The change to the HO-83 hygro-thermometer created a discontinuity of about 0.5°C and records have not been adjusted.[1902,1903]

Before 1860, various proxies can be used. However, the reliability of proxies and the calibration of proxy temperature measurements with thermometer measurements are very unreliable. We only have some 150 years of somewhat uncertain and sparse thermometer measurements upon which predictions are made about climate changes that span hundreds of years. There are 30 years of measurement by satellites and slightly more from radiosonde balloon measurements. Satellite measurements cover the entire globe, there is no concentration of measurement in industrialised Western counties and the oceans are covered equally as well as the lands. These are clearly far more reliable measurements than those from thermometers. Radiosonde balloons give a three-dimensional profile of temperature.

Weather stations show bias. This is due to the positioning of far more weather stations in populated areas of developed countries than in Third World countries, the Arctic tundra, deserts, Antarctica and oceans. There

[1902] Karl, T. R., Diaz, H. F. and Kukla, G. 1988: Urbanization: its detection and effect in the United States climate record. *Journal of Climate* 1: 1099-1123.

[1903] Karl, T. R., Derr, V. E., Easterling, D. R., Folland, C. K., Hofman, D. J., Levitus, S., Nicholls, N., Parker, D. E. and Withee, G. W. 1995: Critical issues for long-term climate monitoring. *Climate Change* 31: 185-221.

have been enormous changes in the numbers of weather stations used for compilation of the average surface temperature. In 1900 there were 1500, in 1980 there were 6000, in 1998 there were 2700 and now there are about 2000. Most of the stations that disappeared were rural stations, especially in the former Soviet Union. The biggest shutdown of measuring stations was around 1990 and the data for the USA shows a sudden increase in temperature at that time.[1904] A larger percentage of stations after 1990 were in urban areas. The averages for average surface temperature are taken from grid boxes of squares 5° latitude x 5° longitude on a Mercator projection map. Out of a possible total of 2592 grid boxes, there were 300 available in 1900 and 850 in 1980. This means that historical temperature measurements by thermometer did not evenly cover the globe and for most places on Earth there was no temperature data in a grid box.

There is a high density of grid boxes in the USA and Western Europe and vast gaps in Africa, South America, the Middle East, India, Greenland, Siberia and Antarctica. There are very few measuring stations in the oceans. More grid boxes are in populated Western countries than elsewhere. This bias is exacerbated because many weather stations that were once in rural areas are now in urban areas. The back reflection from buildings, roads and pavements creates a temperature rise, as does domestic heating, factories, vehicles, machinery, power lines and shopping centres. This is called the urban heat island effect. When the average global surface temperature is cited, what does it really mean?

Human bias occurs with all measurements. The hottest town in South Australia has the Stevenson Screen next to a 2-metre galvanised iron fence and above bare soil. The back reflection of radiation is huge and the measurements are biased upwards. No wonder the measurements show Marree is hot. A friend of mine was once the policeman in the desert hamlet of Marla in far north South Australia. One of his jobs was to take daily temperature measurements, the first of which was at 3 a.m. The thermometer in the Stevenson Screen was calibrated to the mid range yet the summer temperature was far higher than the calibration temperature and the police officer reading the Stevenson Screen instruments had no training from the Bureau of Meteorology. At those times when a nocturnal embrace with Bacchus led to a late rising, the 3 a.m. and morning temperature was estimated a few hours later. In Siberia, the Soviet Union provided a diesel fuel and vodka subsidy when the temperature was below -15°C. When the temperature was -13°C, I am sure the observers' thirst for vodka could easily be accommodated with

---

[1904] GHCN2 data.

a simple slip of the pen. After the collapse of the Soviet Union, the Russian Federation closed many isolated rural measuring stations, withdrew this subsidy and there was an event of apparent warming in Siberia. Many of the weather measuring stations from the former Soviet Union and Russia have many months of missing data between 1971 and 2001[1905] and yet they form part of the global data that is used to calculate temperature change. In many cases missing data is estimated by using information from stations nearby. This is invalid in remote areas where stations may be hundreds of kilometres apart or where months of data are missing. There is a computer program ("filnet")[1906] for estimating missing values.[1907]

All this may appear to be nitpicking. However, this is the way scientific data is evaluated. If there are claims that the global temperature has risen by 0.7°C over the last century, then we need to know if the measurements of temperature are accurate, can be validated and can be repeated. Measurement errors are ± 0.5°C, errors due to siting of a Stevenson Screen may be ± 0.3°C, errors due to wood or plastic may be ± 0.1°C and errors due to the urban heat island effect may be ± 0.4°C. The total errors are ± 1.3°C. Therefore, over the last century global temperatures have therefore risen by 0.7 ± 1.3°C. This is a meaningless figure. The only valid scientific conclusion is that temperature may have increased, been static or decreased over the 20th Century.

In science, when a controversial or novel discovery or observation is made, the scientific community awaits validation by other independent groups. The literature is cluttered with unusual claims that cannot be validated (e.g. cold fusion). The IPCC models suggest that as a result of human-induced global warming, there should be greater warming in the middle of the lower atmosphere at the tropics. Temperature change in the lower troposphere in the tropics during 1979–2004 was examined using 58 radiosonde stations and the microwave satellite data sets of the University of Alabama in Huntsville. The data shows that between 1979 and 2004, the temperature change was 0.05 ± 0.07°.[1908] This means that temperature has not changed in the tropics during a period when global warming should have been warming

[1905] McKitrick, R. R. and Michaels, P. J. 2004: A test of corrections for extraneous signals in gridded surface temperature data. *Climate Research* 26: 159-173.

[1906] Created by USHCN.

[1907] McKitrick, R. R. and Michaels, P. J. 2007: Quantifying the influence of anthropogenic surface processes and in homogeneities on gridded global climate data. *Journal of Geophysical Research* 112: D24509, doi: 10.1029/JD008465.

[1908] Christy, J. R., Norris, W. B., Spencer, R. W. and Hnilo, J. J. 2007: Tropospheric temperature change since 1979 from tropical radiosonde and satellite measurements. *Journal of Geophysical Research* 112, D06102, 1029/2005JD006881.

the tropics. The IPCC model failed the test and hence is invalid.[1909] Again, the IPCC computer models have no bearing on reality. In fact, satellites and radiosondes show that there is no global warming.[1910]

Aerosols have a cooling effect on the atmosphere. It has been argued that modern air pollution has produced aerosols, which has created cooling, and the global warming from $CO_2$ could have been even greater.[1911,1912,1913] It is argued that aerosols have a short life in the atmosphere[1914] and that global dimming is also delayed.[1915] This is fundamentally flawed. The 20th Century temperature measurements, despite limitations, showed a decrease from 1940 to 1976, an increase from 1976 to 1998 and a decrease from 1998 onwards. If these temperature changes were not from natural phenomena, then there should have been global dimming from pollution from 1940 to 1976 and 1998 onwards. There was not. Furthermore, to suddenly turn off the cooling effects from aerosols in both 1976 and 1998 taxes credulity.

The ocean heat content change is probably a more reliable tool to diagnose global warming. By using ocean heat content change, there is a negative feedback of radiation, i.e. there is no global warming. However, the ocean heat content data does not consider heat added below 700 metres water depth and is unavailable over time periods of many decades.[1916] Huge but unknown amounts of heat are added to ocean waters from cooling molten rocks, hence the ocean heat content may reflect tectonic and atmospheric heating.

[1909] Lindzen, R. S. 2007: Taking greenhouse warming seriously. *Energy and Environment* 18: 937-950.

[1910] Keller, C. F. 2007: Global warming 2007. An update to global warming. The balance of evidence and its policy implications. *Scientific World Journal* 7: 381-399.

[1911] Bellouin, N. O., Boucher, J., Haywood, J. and Reddy, M. S. 2005: Global estimate of aerosol directi radiative forcing from satellite measurements. *Nature* 311: 1720-1721.

[1912] Pinker, R. T., Zhang, B. and Dutton, E, G. 2005: Do satellites detect trends in surface solar radiation. *Science* 308: 850-854.

[1913] Wild, M., Gilgen, H., Roesch, A., Ohmura, A., Long, C. N., Dutton, E. G., Forgan, B., Kallis, A., Russak, V. and Tsvetkov, A. 2005: From dimming to brightening: Decadal changes in solar radiation at Earth's surface. *Science* 308: 847-850.

[1914] Andreae, M. O., Jones, C. D. and Cox, P. M. 2005: Strong present-day aerosol cooling implies a hot future. *Nature* 435: 1187-1190.

[1915] Delworth, T. L., Ramaswamy, V. and Stenchikov, G. L. 2005: The impacts of aerosols on simulated ocean temperature and heat content in the twentieth century. *Geophysical Research Letters* 32: L24709, doi 10.1029/2005GL024457.

[1916] Pielke Snr, R. A., Davey, C., Niyogi, D., Fall, S., Steinweg-Woods, J., Hubbard, K., Lin, X., Cai, M., Lim, Y-K., Nielson-Gammon, J., Gallo, K., Hale, R., Mahmood, R., McNider, R. T. and Blanken, P. 2007: Unresolved issues with the assessment of multi-decadal global land surface temperature trends. *Journal of Geophysical Research* 112: D24508, doi: 10.1029/2006JD008229.

*Treatment of data*

A number of centres compile meteorological data from ground stations and satellites. These are the Hadley Centre at the University of East Anglia, which is a branch of the UK Meteorological Office, the Goddard Institute of Space Studies (GISS), part of NASA, the National Oceanographic and Atmospheric Administration (NOAA) which in turn is part of the US Department of Commerce, the University of Alabama in Huntsville (UAH) and Remote Sensing Systems (RSS) in Santa Rosa, California. The last two groups only use satellite data whereas the first three groups integrate historical and modern thermometer data with radiosonde and satellite data. The data shows essentially the same trend although some centres such as GISS show consistently higher temperatures than the others. The question is: Do the ground-based data and satellite groups measure the same temperature? There is a weak correlation between the land-based stations and a strong correlation between the satellite data. This is not surprising, as land-based measurements are based on a number of uncertainties.

How do we look at the primary data collected on the ground, from radiosonde balloons and from satellites? Has this data undergone correction for various errors? Do we use a 12-month, 6-month, 3-month or 1-month average? In each case, there are different results. What data is rejected? What data is amended, corrected or adjusted? Once we have the data, what statistics do we use to present the data? Because of the massive volume of data, the simple graphs that Hadley, GISS, UAH etc produce are the end result of a large amount of data collection, amendment, correction and reduction.

There is little reconciliation of the surface temperature measurements with those determined from balloons and satellite.[1917] In fact, satellites and radiosondes show that there is no global warming.[1918] Maybe the land surface data is measuring something else, such as population growth and an increase in the *per capita* use of energy.

The Medieval Warming and Little Ice Age are a real nuisance because they show great variations in temperature which are unrelated to human activity. The solution for those who wish to show there is human-induced global warming is simply to remove these periods of history from the record, an intellectual triumph by Michael Mann and colleagues who created the "hockey stick" (i.e. Mann-made global warming *ex nihilo*). Antarctica is also

---

[1917] Christy, J. R., Norris, W. R., Spencer, R. W. and Hnilo, J. J. 2007: Troposphere temperature change since 1979 from tropical radiosonde and satellite measurements. *Journal of Geophysical Research* 112: D06102, doi.10:1029/2005JD006881.

[1918] Keller, C. F. 2007: Global warming 2007. An update to global warming. The balance of evidence and its policy implications. *Scientific World Journal* 7: 381-399.

a nuisance because the ice sheet is expanding and temperature refuses to rise, no matter how much $CO_2$ humans pump into the atmosphere. If it can be shown that the temperature of Antarctica is rising, then this little spot of bother can also be solved. However, temperature measurements show that Antarctica is not warming, so it was time to torture that data and get it to confess. And confess it did, and we had another event of Mann-made warming.

A great media fanfare accompanied the announcement in January 2009 that Antarctica is warming.[1919] The international alarmist movement and uncritical scientifically-illiterate media were quick to act, and showed horror images and presented apocalyptic implications of global warming. One had to be immediately suspicious of this work because of the people who published this study (co-author Michael Mann) and the use of tortuous statistical techniques rather than the use of ground temperature measurements. The procedure used to show Antarctica was warming was a "statistical-climate-field reconstruction technique to obtain a 50-year-long complete estimate of monthly Antarctic temperature anomalies", a long-winded euphemism meaning data was cooked to get the required result. Furthermore, there was missing data from some measuring stations and temperature between measuring stations was interpolated because of the large distances between the stations. One measuring station (Harry) was buried in snow for years and then resited in 2005. The data that Steig used in his modelling which he claimed came from Harry was old data from another station on the Ross Ice Shelf (Gill) with new data from Harry added to it producing the abrupt warming. The data is worthless.

Although the abstract of the paper claimed that temperature "reconstructions" were able to "demonstrate" warming of 0.17°C per decade over the past 50 years in West Antarctica, 0.10°C over East Antarctica and 0.12°C over the whole continent, the devil is in the detail. In the body of the paper was hidden an analysis showing that when the statistical study was compared with detrended data, there was no warming in Antarctica. There is an overstating of temperature trends[1920,1921] and, as a result of adjusting data, Antarctic cooling disappears.

In a current affairs radio program (*AM* on ABC radio), one scientist (Barry Brook) argued that this work showed that the end is nigh, whereas

[1919] Steig, E. J., Schneider, D. P., Rutherford, S. D., Mann, M. E., Cosimo, J. C. and Shindall, D. T. 2009: Warming of the Antarctic ice-sheet surface since the 1957 International Geophysical Year. *Nature* 457: doi 30.1038/nature07669.

[1920] http://www.osdpd.nasa.gov/PSB/EPS/SST/data/anomnight.1.15.2009.gif

[1921] http://arctic.atmos.uiuc.edu/cryosphere/IMAGES/current.anom.south.jpg

another scientist (William Kininmonth) argued that the Antarctic sea surface temperature and spread of sea ice show the opposite trend and that this was not discussed in the paper on Antarctic warming. In the transcript of the *AM* interview on the ABC's website, the comments of Kininmonth were omitted.

*Urban heat island effect*

Concrete, bitumen, buildings, pavements, air conditioners, heating, cars and basic city infrastructure all add heat to urban areas. Most ground-based temperature measurements are made in urban areas. The greater the population, the greater the addition of heat.[1922] Temperature measurements at city locations are contaminated data and can either be ignored or attempts can be made to "correct" the primary data. If the uncorrected primary data is doubtful, then any conclusion about global temperature must also be in doubt.

Many measuring stations used to be in rural areas or at airports, well away from cities. The spread of urbanisation, the clustering of businesses around airports and the increase in air traffic has contaminated these temperature measurements. Furthermore, the very hot gases from jet engines and turboprop aeroplanes adds heat to the airport precinct. Some measuring stations have been moved to reduce these problems. This creates another problem. Can a correlation of measurements from the old site be made with those from the new site? Over the period of historical measurements, the time of day of temperature measurements has also changed. Many long-term weather records come from in or near cities which have warmed as they grew. Even if a weather station is located in a rural area, changing land use patterns change temperature. These changes are as great as the temperature changes that we are told derive from human emissions of greenhouse gases.[1923] Many poor countries have sparse weather station records and few resources to ensure data quality. Less than one third of the weather stations operating in the 1970s now remain in operation. For some time there have been reservations about temperature measurements at land-based weather stations.[1924]

---

[1922] Landsberg, H. E. 1981: *The urban climate.* Academic Press.

[1923] Pielke, R. A. Snr, Marland, G., Betts, R. A., Chase, T. N., Eastman, J. L., Niles, J. O., Niyogi, D. and Running, S. W. 2002: The influence of land-use change and landscape dynamics on the climate system: Relevance to climate-change policy beyond the radiative effect of greenhouse gases. *Philosophical Transactions of the Royal Society of London* A360: 1705-1719.

[1924] Gall, R. K., Young, R., Schotland, R. and Schmitz, J. 1992: The recent maximum temperature anomalies in Tuscon: Are they real or an instrumental problem? *Journal of Climate* 5: 657-665.

The Climate Change Science Program states:[1925]

> Urban areas are among the most rapidly changing environments
> on Earth. As cities grow, they affect local climates. The urban
> heat island effect has raised average urban air temperatures by 2
> to 5°F more than surrounding areas over the past 100 years, and
> up to 20°F more at night.

Towns with a population of just 1000 people could have a warming of
2°C relative to rural areas.[1926] The village of Barrow (Alaska) with 4600
inhabitants showed a warming of 3.4°F in winter compared to surrounding
rural areas.[1927] In southeastern Australia, populated towns and cities show the
urban heat island effect.[1928] In Europe[1929] and Central America,[1930] the same
phenomenon has been recognised. In a study of some 140 weather stations
in Australia, the 100 or so stations in cities and towns showed a temperature
increase whereas those outside built-up areas showed no temperature
increase.[1931] Long-term analysis of urbanisation shows that there is a strong
bias to the surface-based temperature record.[1932] The global databases do not
consider the area of a city, do not make "adjustments" for urbanisation until
the population exceeds 100,000 and do not recognise that many cities of
more than 100,000 people are an amalgam of many towns each with under
100,000 inhabitants. Because towns with as few as 1000 inhabitants create an
urban heat island effect, there are no urban heat island corrections for most
population centres in the world and the urban heat island "adjustments"
can only be a minimum figure. More and more of the world's population
live in cities, and cities grow around airports where we commonly measure
temperature.

Some scientists have recognised that surface temperature measurements
are contaminated and they make adjustments to fix this problem. Some of

---

[1925] CCSP, 2006: temperature trends in the lower atmosphere: Steps for understanding and
reconciling differences. *US Climate Science program* http://www.climatescience.gov/Library/
sap/sap1-1/public-review-draft-sap1-1prd-all.pdf

[1926] Oke, T. R. 1973: City size and urban heat island. *Atmospheric Environment* 7: 769-779.

[1927] Hinkel, K., Nelson, F., Klene, A. and Bell, J. 2003: The urban heat island in winter at
Barrow, Alaska. *International Journal of Climatology* 23: 1889-1905.

[1928] Torok, S., Morris, C., Skinner, C. and Plummer N. 2001: Urban heat island features of
southeast Australian towns. *Australian Meteorological Magazine* 50: 1-13.

[1929] Block, A., Keuler, K and Schaller, E. 2004: Impacts of anthropogenic heat on regional
climate patterns. *Geophysical Research Letters* 31: L12211, doi: 10.1029/2004GL019852.

[1930] Velazquez-Lozada, A. V., Gonzalez, J. E. and Winter, A. 2006: Urban heat island effect
analysis for San Juan, Puerto Rico. *Atmospheric Environment* 40: 1731-1741.

[1931] Gladstone, J. D. 1992: *Viticulture and the environment.* Winetitles Australia.

[1932] Karl, T. R., Diaz, H. F. and Kukla, G. 1988: Urbanization: its detection and effect in the
United States climate record. *Journal of Climate* 1: 1099-1123.

these fudge factors are, for example, the "urbanisation adjustment", the "time of observation bias adjustment" and the "homogeneity adjustment". In an "urbanisation adjustment", rural and urbanised measuring stations are not defined. What is rural to a city person may be urban to a rural person. We do not know if these adjustments are correct or can be validated. Annual mean surface temperatures for continental USA are subjected to a comprehensive correction procedure called "homogeneity adjustment". However, there are two versions used[1933] which give different results from the same data.[1934] After adjustment, there is a modest warming of less than 1°C for the 20th Century.

Statistical studies for land-based data[1935] and sea surface temperatures[1936] show that the global surface temperature is biased upwards. In New York there have been temperature measurements since 1869. Central Park measurements were taken at the Arsenal Building (between 63rd and 64th streets) from 1909 to 1919, and from 1920 onwards measurements were taken at the Belvedere Castle on Transverse Road (near 79th and 81st streets). Reconciling of these different data sets, population growth, increasing per capita energy use and increased traffic has created a very high degree of uncertainty. Some 20% of the historical data has now been modified 16 times over the last 3 years and most changes produced an upward trend in temperature (e.g. August 2006).

When the same methodology is applied to China, the temperature fluctuates with peaks at 1943 and 1998.[1937] There was no evidence of global warming. In China, the urban warming at the city measuring stations can account for about 65–80% of the overall warming in 1961–2000, and 40–61% of the overall warming in 1980–2000.[1938] In areas of high industrialisation, aerosols may also have an effect on temperature. In Europe over the period 1961–2004, thirteen different weather stations in different climate and geographic areas show that there is an increase in sea level pressure over

---

[1933] USHCN Version 1 and GHCN Version 2.

[1934] McKitrick, R. R. and Michaels, P. J. 2007: Quantifying the influence of anthropogenic surface processes and inhomogeneities on gridded global climate data. *Journal of Geophysical Research* 112: D24509, doi: 10.1029/JD008465.

[1935] McKitrick, R. R. and Michaels, P. J. 2004: A test of corrections for extraneous signals in gridded surface temperature data. *Climate Research* 26: 159-173 and Erratum *Climate Research* 27: 265-268.

[1936] Christy, J. R., Parker, D. E., Brown, S. J., Macadam, I., Stendel, M. and Norris, W. 2001: Differential trends in tropical sea surface and atmosphere temperatures since 1979. *Geophysical Research Letters* 28: 183-186.

[1937] Zhou, Z., Ding, Y., Luo, Y. and Wang, S. 2005: Recent studies on attributions of climate change in China. *Acta Meteorologica Sinica* 19: 389-400.

[1938] Ren, G. Y., Chu, Z. Y., Chen, Z. H. and Ren, Y. Y. 2007: Implications of temporal change in urban heat island intensity observed at Beijing and Wuhan stations. *Geophysical Research Letters* 34: L05711, 1029/2006GL027927.

the weekends and a consequent decrease in anticyclonic conditions during the central weekdays. Weekend weather is wetter and cooler than weekday weather. These weekly changes in atmospheric circulation may be related to anthropogenic aerosols.[1939]

The IPCC documents and lead authors have ignored changes to temperature by changes in land use and urbanisation. In the IPCC 2007 Report, lead authors of Chapter 3 (Jones and Trenberth) argue that the urban heat island effect is no more than 0.05°C. This is the figure used in a 1990 paper by Jones[1940] who could not be unaware of a 2003 paper in *Nature* that gives an urban heat island effect of at least 0.25°C.[1941]

*Conclusions from temperature data*

It has been suggested that the average air temperature at the Earth's surface has increased by 0.06°C per decade during the 20th Century[1942] and by 0.19°C per decade from 1979 to 1998.[1943] The cooling from 1940 to the mid 1970s and the record El Niño of 1998 make the second figure somewhat higher than the average per decade for the century. However, Antarctic meteorological data demonstrates a net cooling on the Antarctic continent between 1996 and 2000, particularly during summer and autumn. The authors[1944] of this *Nature* paper somewhat embarrassingly state: "continental Antarctic cooling, especially the seasonality of cooling, poses challenges to models of climate and ecosystem change". This data only poses challenges if your mind is closed.

Satellites provide a more comprehensive coverage of atmospheric temperature than thermometers, which are sparsely distributed over the Earth's surface. Measurements of global temperatures from satellites have shown that there is no net warming of the upper tropical troposphere[1945]

---

[1939] Sanchez-Lorenzo, A., Calbó, J., Martin-Vide, J., Garcia-Manuel, A., García-Soriano, G. and Beck, C. 2008: Winter "weekend" effect in southern Europe and its connections with periodicities in atmospheric dynamics. *Geophysical Research Letters* 35: doi:10.1029/2008GL034160.

[1940] Jones, P. D., Kelly, P. M., Goodess, C. M. and Karl, T. 1990: The effect of urban warming on the Northern Hemisphere temperature average. *Journal of Climate* 2: 285-290.

[1941] Kalnay, E. and Cai, M. 2003: Impacts of urbanization and land-use change on climate. *Nature* 423: 528-531.

[1942] Houghton, J. T., Jenkins, G. J. and Ephraums, J. J. (eds) 2001: *Climate change 2001: The IPCC Scientific Assessment*. Intergovernmental Panel on Climate Change. Cambridge University Press.

[1943] National Research Council 2000: *Reconciling observations of global temperature change*. National Academy Press, Washington DC.

[1944] Doran, P. T., Priscu, J. C., Lyons, W. B., Walsh, J. E., Fountain, A. G., McKnight, D. M., Moorhead, D. L., Virginia, R. A., Wall, D. H., Clow, G. D., Fritsen, C. H., McKay, C. P. and Parsons, A. N. 2002: Antarctic climate cooling and terrestrial ecosystem response. *Nature* 418: 292.

[1945] A part of the atmosphere stretching from 12 to 16 km in altitude.

despite what the various climate models predict. Satellite measurements were supported by balloon radiosonde measurements.[1946,1947,1948] When the temperature data from tropical radiosonde balloons was rejected and temperature trends were inferred from troposphere wind data, it was concluded that the tropical troposphere is "very likely increasing as global surface temperatures rise".[1949] It seems that the IPCC models work if measured data is rejected and inferences and assumptions are used. It has been argued that satellite temperature measurements show bias hence will not detect the modelled troposphere temperature increase.[1950] Despite the complexity of temperature trends with altitude, ozone depletion in the tropical lower troposphere was attributed to troposphere warming.[1951] Most troposphere research shows that few conclusions can be made. Work is in progress, and inferring temperature rather than using measured temperature does not assist with understanding the atmosphere.[1952]

Global mean values of temperature conceal the actual meaningful regional and local temperature trends. Physical, mathematical and observational grounds are employed to show that there is no physically meaningful global temperature for the Earth in the context of the issue of global warming. Statistical studies can show that for a given set of information, a given temperature can be interpreted as both warming and cooling.[1953] Nature is not obliged to respect our empirical studies of climate, statistics and concepts.

The European heat wave in the 2003 summer was touted as evidence of global warming. Some unsporting scientists[1954] analysed the data and stated that there is: "not strong support for the idea that regional heat or cold

---

[1946] Spencer, R. W. and Christy, J. R. 1990: Precise monitoring of global temperature trends from satellites *Science* 247: 1558-1562

[1947] National Research Council 2000: *Reconciling observations of global temperature change.* National Academy Press, Washington, D. C.

[1948] Karl, T. R., Hassol, S. J., Miller, C. D. and Murray, W. L. (eds) 2006: *Temperature trends in the lower atmosphere: Steps for understanding and reconciling differences.* US Climate Change Science Program, Washington, D.C.

[1949] Allen, R. J. and Sherwood, S. C. 2008: Warming maximum in the tropical upper troposphere deduced from thermal winds. *Nature Geoscience* 1: 399-403.

[1950] Fu, Q., Johanson, C. M., Warren, S. G. and Seidel, D. J. 2004: Contribution of stratospheric cooling to satellite-inferred tropospheric temperature trends. *Nature* 429: 55-58

[1951] Foster, P. M., Bodeker, G., Schofield, R., Solomon, S. and Thompson, D. 2007: Effects of ozone cooling. *Geophysical Research Letters* 34: L23813.

[1952] Haimberger, L. 2006: Homogenization of radiosonde temperature time series using innovation statistics *Journal of Climate* 20: 1377-1403.

[1953] Essex, C., McKitrick, R. and Andresen, B. 2007: Does a global temperature exist? *Journal of Non-Equilibrium Thermodynamics* 32: 1-27.

[1954] Chase, T. N., Wolter, K., Pielke Snr, R. A. and Rasool, I. 2006: Was the 2003 European summer heat wave unusual in a global context. *Geophysical Research Letters* 33: L23709, doi.10.1029/2006GL027470.

waves are significantly increasing or decreasing with time during the period considered here (1979–2003)."

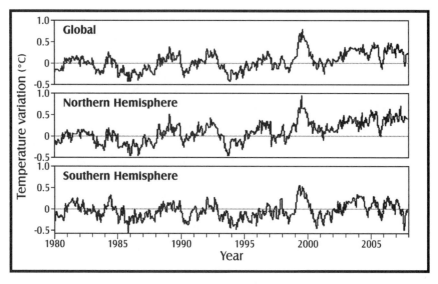

*Figure 51: Global temperatures calculated from satellite measurements. There is slight global warming in the late 20ᵗʰ Century in the Northern Hemisphere, slight global cooling in the late 20ᵗʰ Century in the Southern Hemisphere and no apparent global warming or cooling for the whole planet.*

And the answer to their question: Was the 2003 European summer heat wave unusual in the global context? It is a resounding *no*. The data showed that the heat wave was statistically unusual and there was a deep troposphere phenomenon in 2003. Furthermore, extreme cold and warm anomalies both occur regularly and these occasionally exceed the magnitude of the 2003 anomaly. There is no correlation between global and hemispheric average temperature and the presence of warm or cold regional anomalies and there is no support for the idea that regional heat waves are increasing with time. While Europe was sweltering in the 2003 heat wave, other areas were undergoing unseasonably cold times. A conclusion from this work was that natural variability in the form of El Niño and volcanic eruptions appear to be of much greater importance in causing extreme regional temperature anomalies than a simplistic upward trend in time. This shows that regional features cannot be viewed as a global phenomenon.

Elsewhere, some of the hottest temperatures were not recorded in the 1990s, claimed by the IPCC to be the hottest decade for a century. The hottest temperature ever recorded was on 13 September 1922 in Libya (58°C), the hottest in the USA was 56.7°C (Death Valley, 10 July 1913) and the hottest in Canada was 45°C (5 July 1937). The dust bowl years in the USA (1920s–1930s with a peak in 1934–1936) were the hottest in the USA, Canada and Mexico and it is probable that the 1930s was a far hotter decade than the 1990s.

Summer in southeastern Australia is characterised by heat waves, bushfires and shark attacks and in northern Australia by tropical cyclones and flooding. All of this makes spectacular news copy. The 2008–2009 summer had all of these elements with the longest continuous heat wave for decades. Catastrophic fires in Victoria on 7 February 2009, a 47°C day, were fanned by strong winds. More than 1800 homes were burned down and there was a tragic loss of hundreds of lives. Four small towns were completely wiped off the map. Some fires were deliberately lit. The media and some environmental commentators attributed these summer bushfire conditions to the effects of human-induced global warming. In the same television news segments on the Victorian bushfires were images of cyclone damage and floods in North Queensland and the extreme cold, large snow dumps and blizzards in the USA, Canada, UK and Europe. Professor Geoffrey Blainey[1955] put some sobering perspective on the bushfire tragedy of 7 February 2009:

> In our recorded history, there have been no bushfires as spectacular as February 1851, on the very eve of the first goldrushes. They called it Black Thursday. Half of Victoria seemed to be on fire. A wild northerly was blowing, and it drove a such column of black smoke right across Bass Strait that one town near Devonport was so darkened in mid-afternoon that people actually thought the end of the world had come. The period after World War I was especially dry in Victoria. Five devastating years for bushfires were 1919, 1926, 1932, 1939 and 1944. Those of 1926 have been largely forgotten, but in the town of Gilderoy, only two of 14 forest workers survived. More than 50 Victorians were killed in bushfires in that February. I remember Black Friday, January 13, 1939. The two million hectares burned probably exceeded the extent of bush and forest destroyed this weekend. Some 1300 homes were lost, and 69 sawmills. More than 70 Victorians died that day.

---

[1955] *Herald Sun*, Monday 9th February 2009.

The Southern Hemisphere lower troposphere data[1956] since satellites were launched in 1979 shows that the average temperature anomaly just oscillates around zero. In other words, temperature is unchanged. This lack of warming in the Southern Hemisphere is in contrast to the Northern Hemisphere where there has been about 0.5°C warming since measurements were taken. This warming is deduced from thermometer measurements. However, the satellite temperature measurements show that global warming is not global. In the Northern Hemisphere, there has not been a steady temperature rise from 1979 to 2006. Temperature anomaly oscillates around zero from 1979 to 2000 and then takes an upward step. This warming of about 0.5°C cannot be correlated with a rise in atmospheric $CO_2$. Since 1998, the temperature has fallen. Thermometer measurements show that the temperature rise from 1890 to 1940 preceded the rise in $CO_2$. Temperature decreased from 1940 to 1976 at a time when $CO_2$ was increasing. Temperature changes do not correlate with $CO_2$ changes but correlate well with solar activity.

The Hadley Centre in the UK has shown that warming stopped in 1998. $CO_2$ emissions did not. Even Rajendra Pachauri, the IPCC chairman, has stated that the IPCC must now review its calculations. Maybe the calculation revision could look at the rounding-up of the no-feedbacks climate sensitivity parameter from 0.265[1957] to 0.300[1958,1959] in their 2007 report. In this revision, maybe the order of accuracy could be given and maybe the uncertainties about the level of scientific understanding for this key variable could be aired. The rounding-up of 13% directly causes a 50% increase in the IPCC's central climate sensitivity, already inflated by an undeclared 70% increase since 1995 in the IPCC's central estimate of the feedback factor. In the calculation of the water-vapour feedback, the IPCC ignores two thirds of the cooling effect of evaporation.[1960] The IPCC assumes that the energy reflected back from clouds must be negative despite data showing that it is positive.[1961] The models that the IPCC relies on for the interaction between water vapour,

[1956] http://www.atmos.uah.edu/data/msu/t2lt/tltglhmam_6.0p

[1957] Hansen, J., Lacis, A., Rind, D., Russell, G., Stone, P., Fung, I., Ruedy, R. and Lerner, J. 1984: Climate sensitivity: Analysis of feedback mechanisms. In: *Climate Processes and Climate Sensitivity* (eds Hansen, J. E. and Takahashi, T.). American Geophysical Union Monograph 29: 130-163.

[1958] Bony, S., Colman, R., Kattsov, V. M., Allan, R. P., Bretherton, C. S., Dufresne, J-L., Hall, A., Hallegate, S., Holland, M., Iingram, W., Randall, D. A., Sodon, B. J., Tselioudis, G. and Webb, M. 2006: How well do we understand and evaluate climate change feedback processes? *Journal of Climate* 19: 3445-3482.

[1959] Colman, R. 2003: A comparison of feedbacks in general circulation models. *Climate Dynamics* 20: 865-873.

[1960] Wentz, F. J., Ricciardulli, L., Hilburn, K. and Mears, C. 2007: How much more rain will global warming bring? *Science* 317: 233-235.

[1961] Spencer, R. W., Braswell, W. D., Christy, J. R. and Hnilo, J. 2007: Cloud radiation budget changes association with tropical intraseasonal oscillations. *Geophysical Research Letters* 34: L15707, doi:10.1029/2007GL029698.

$CO_2$ and the radiant energy flux in the tropical mid-troposphere use are flawed[1962] with the consequence that the IPCC's climate sensitivity estimates are at least a threefold exaggeration.[1963] It is little wonder that many scientists have great disquiet about the IPCC.

Rather than make predictions about data before the data is collected, the long-term data from the Armagh Observatory (Northern Ireland)[1964] shows that maximum and minimum temperatures have risen in line with global averages but minima have risen faster than maxima, thereby reducing the daily temperature range. The total number of hours of bright sunshine has fallen since 1885 at the four sites studied. This is consistent with both a rise in cloudiness and the fall in the daily temperature range. There were significant cycles of temperature over periods of 7–8 years, 20–23 years and 30–33 years in the seasonal and annual meteorological series from Armagh. Some of these cycles are linked to the North Atlantic Oscillation. There is a big difference between sensationalist predictions for the media from the Meteorological Office in Britain and the analysis of a large amount of data collected over more than 100 years. One gets news headlines, the other is ignored.

The last 30 years of weather are not in accord with what greenhouse models predict and can be far better explained by natural processes, such as solar variability.[1965] Furthermore, the satellite and balloon data show that atmospheric warming trends do not exceed surface warming trends, whereas greenhouse models demand that the atmospheric trend be 2–3 times greater. Such results are in conflict with those of the IPCC and many research papers based on essentially the same set of data. Climate change by natural processes cannot be affected or modified by controlling $CO_2$ emissions.

There are innumerable examples to show that "global warming" is not global. For example, analysis of data from 1958–2001 from eight measuring stations in southern Greenland shows that there has been a significant cooling of 1.29°C over the 44-year period.[1966] Sea surface temperatures in

---

[1962] Douglass, D. H., Christy, J. R., Pearson, B. J. and Singer, E. F., 2007: A comparison of tropical temperature trends with model predictions *International Journal of Climatology of the Royal Meteorological Society* 28: 1693-1701.

[1963] Lindzen, R. S. 2007: Taking greenhouse warming seriously. *Energy and Environment* 18: 937-950.

[1964] Butler, C. J., Garcia-Surez, A. and Palle, E. 2007: Trends and cycles in long Irish meteorological series. *Proceedings of the Royal Irish Academy* 107B: 157-165.

[1965] Douglass, D. H., Christy, J. R., Pearson, B. J. and Singer, E. F., 2007: A comparison of tropical temperature trends with model predictions. *International Journal of Climatology of the Royal Meteorological Society* 28: 1693-1701.

[1966] Hanna, E. and Cappelen, J. 2003: Recent cooling in coastal southern Greenland and relation with the North Atlantic Oscillation. *Geophysical Research Letters* 30: 1132: doi:10.1029/2002/GL015797.

the Labrador Sea also show cooling. This cooling is related to changes in the North Atlantic Oscillation over the past few decades and has probably significantly affected the Greenland Ice Sheet. Measurement of temperature since 1784 from 13 stations along the southern and western coasts of Greenland verifies the ice proxy temperature calculations.[1967] The coldest year was 1863 and the coldest decade was 1810–1820. This was the decade when Tambora erupted. The warmest year on record was 1941 while the 1930s and 1940s are the warmest decades. The last two decades of the 20th Century were colder than any of the previous six decades. No net warming has occurred since the warm period of the 1930s and 1940s.

On a scale of 500 years, the planet is warming. But warming compared to what? We were in the grip of the Little Ice Age 500 years ago, so thankfully it is now warmer. On a scale of 5000 years, there have been many periods of warming and cooling. On a scale of 5 million years, there were numerous periods of intense cold and many short periods of warmth. The average global temperature over the last 2.67 million years is less than the current global temperature. Why? Because we are living in the Pleistocene glaciation which has not yet run its full course.

*Temperature proxies*

There are a number of ways climate can be determined for the times before systematic thermometer measurements started to be made in 1860. As with any scientific method, all of these proxy methods have limitations.

Coal, peat, fossil pollen, spores, lichen, wood and leaves can be used to reconstruct climate. Many plants have altitude limitations hence can be used to construct old tree lines, some have temperature limitations and all depend on $CO_2$. Fossil leaves allow a reconstruction of the $CO_2$ content of an ancient atmosphere. If the fossil plant is similar to a modern plant then measurement of the leaf breathing holes[1968] allows a calculation of the atmospheric $CO_2$ content. For example, during the Middle Cretaceous the atmospheric $CO_2$ content was calculated at 560–960 ppmv followed by a time when it was 620–1200 ppmv.[1969] The current atmospheric $CO_2$ content is 385 ppmv. The distribution of fossilised wood, leaves, pollen, spores and lichen in coal is used to create the palaeoenvironment. Coal deposition varies from upper delta

[1967] Vinther, B. M., Andersen, K. K., Jones, P. D., Briffa, K. R. and Cappelen, J. 2006: Extending Greenland temperature records into the late eighteenth century. *Journal of Geophysical Research* 111: 10.1029/2005JD006810.

[1968] Stomata

[1969] Haworth, M., Hesselbo, S. P., McElwain, J. C., Robinson, S. A. and Brunt, J. W. 2005: Mid-Cretaceous $p$CO$_2$ based on stomata of the extinct conifer *Pseudofrenelopsis* (Cheirolepidiaceae). *Geology* 33: 749-752.

plain to coastal swamps tucked in behind dunes and the vegetation reflects these different environments.[1970] Temperature measurements deduced from plants are not very accurate because even Arctic plants like it warm. The main restrictions on plant habitat are altitude, rainfall, wind, nutrients and the $CO_2$ content of the atmosphere. Temperature is not the only variable.

In many lake sediments, there is a story trapped in mud. Layers of mud in lakes with pollen are normally about 0.03 mm thick with the occasional layer thicker than 0.1 mm. Pollen, spores and leaves have a waxy coating that protects them from decay. This plant material shows changes in regional vegetation over time and by integrated results from pollen from a great diversity of locations can show changes over the Northern Hemisphere (e.g. Massif Central of France,[1971] southern Italy[1972] and Cascade Range of northwestern USA[1973]) and the climate of the Northern Hemisphere in some time past can be deduced. In terrestrial and freshwater-brackish environments, beetles (which represent three quarters of all animals) have habitats dependent upon local climate factors and swiftly migrate to more congenial localities if there is a slight climate change. They are a wonderful temperature proxy. They can also tell us about the Sun. Tree ring studies in Fennoscandia, the Kola Peninsula and Northern Siberia show cyclical $C^{14}$ values. The main periods of solar cycles at 22, 33 and 80–90 years were recognised, the 11-year cycle was very faint.[1974]

Tree rings are one of the popular proxies to determine ancient climates. How accurate are they really? The European heat wave of 2003 provided the opportunity to test tree ring use as a proxy.[1975] The heat was expected to have a strong impact on tree growth, especially where trees occur at their ecological limits. In the dry inner Alpine climate of the Tyrol Mountains in Austria, Scots pine and Norwegian spruce tree ring growth was suppressed by up to 35%, the growth was very much dependent upon the position of the tree in

[1970] Diessel, C. F. K. 1992: *Coal-bearing depositional systems*. Springer-Verlag.

[1971] Thouveny, N., de Beaulieu, J. L., Bonifay, E., Creer, K. M., Guiot, J., Icole, M., Johnsen, S., Jouzel, J., Reille, M., Williams, T. and Williamson, D. 1994: Climate variations in Europe over the last 140 kyr deduced from rock magnetism. *Nature* 371: 503-506.

[1972] Allen, J. R. M., Brandt, U., Brauer, A., Hubberten, H-W., Huntley, D., Keller, J., Krami, M., Mackensen, A., Mingram, J., Negendank, J. F. W., Nowaczyk, N. R., Oberhänsli, H., Watts, W. A., Wulf, S. and Zolitschka, B. 1999: Rapid environmental changes in southern Europe during the last glacial period. *Nature* 400: 740-743.

[1973] Whitlock, C. and Bartlein, P. J. 1997: Vegetation and climate change in northwestern America during the last 125 kyr. *Nature* 423: 57-61.

[1974] Kasatkina, E., Shumilov, O., Lukina, N. V., Krapiec, M. and Jacoby, G. 2007: Stardust component in tree rings. *Dendrochronologica* 24: 131-135.

[1975] Pichler, P. and Oberhuber, W. 2007: Radial growth response of coniferous trees in an inner Alpine environment to heat-wave of 2003. *Forest Ecology and Management* 242: 688-699.

the alpine environment, and growth was related to rainfall. Temperature had no effect on tree ring growth. Amazonian forests have changes of up to 25% in leaf area, depending upon seasonality of solar radiation.[1976] Because tree ring growth varies within site and canopy position, the use of tree rings as proxies must be treated with caution.

A study of 39 tree species across 50 degrees of latitude from tropical Colombia to boreal Canada shows that the internal temperature of trees remains constant at 21.4°C over the growing season.[1977] For decades, the oxygen isotope ratios in tree rings have been used to determine air temperature on the assumption that leaf temperature was equal to the air temperature. This was done to use tree rings to determine historical climates. However, this technique may be flawed. Leaves are internally cooler in warm climates and warmer in cold climates. Leaves are kept cool by the constant evaporation of water and reducing the exposure to the Sun through leaf angles and leaf reflective qualities. Warmth is gained by decreasing evaporation and increasing the number of leaves per branch. The leaf area of the Canadian black spruce is the same as the Caribbean pine. These tricks are probably evolutionary adaptations to help trees gain maximum nutrients through optimal photosynthesis.

Using larch trees, the mean summer temperature record of the European Alps from 755 to 2004 AD showed high temperature in the early 10th, early 13th and early 20th Centuries. Prolonged cooling was measured from 1350 to 1700. This is in accord with other studies and reflects the Medieval Warming, the Little Ice Age and the Late 20th Century Warming. The coldest decade of the record was 1810–1820, coincidental with the cooling induced by the eruption of Tambora. The warmest decade of the record was the 1940s. Other information suggests that there was a global cooling from 1940 to 1976, contrary to actual temperature measurements. This example suggests that extreme caution must be exercised when using temperature proxies. The warm summers coincided with periods of high solar activity, the cold summers coincided with periods of reduced solar activity. The final sentence of this work[1978] gives the picture as it is: "the twentieth-century contribution

[1976] Myneni, R. B., Yang, W., Nemani, R. R., Huete, A. R., Dickinson, R. E., Knyzazikhin, Y., Didan, K., Fu, R., Juárez, R. I. N., Saatchi, S. S., Hashimoto, H., Ichii, K., Shabanov, N. V., Tan, B., Ratana, P., Privette, J. L., Morisette, J. T., Vermote, E. F., Roy, D. P., Wolfe, R. E., Friedl, M. A., Running, S. W., Votava, P., El-Saleous, N., Devadiga, S., Su, Y. and Salomonson, V. V. 2007: Large season swings in leaf area of Amazonian rainforests. *Proceedings of the National Academy of Sciences* 104: 4820-4823.

[1977] Helliker, B. R. and Richter, S. L. 2008: Subtropical to boreal convergence of tree-leaf temperatures. *Nature* 454: 511-514.

[1978] Buntgen, U., Frank, D. C., Nievergelt, D. and Esper, J. 2006: Summer temperature variations in the European Alps, A.D. 755-2004. *Journal of Climate* 19: 5606-5623.

of anthropogenic greenhouse gases and aerosol remains insecure".

Periods in the past were certainly far warmer than in the Late 20th Century Warming and this is in accord with variations in vegetation, glacier expansion, tree lines, tree ring width, and lake and river histories. The data from Austria is in accord with data from the other side of the world in Tibet[1979] and Canada.[1980]  In Tibet, it was warmer than today in 1150–1380 AD. From 1430 to the 19th Century, there was a series of cold intervals with glacial advance in 1580–1590, from the end of the 18th Century to the beginning of the 19th Century and 1860–1880. In Canada, pollen spores show that between 700 and 150 years ago the Little Ice Age had its grip on Canada.

Dust continually falls to Earth. It derives from deserts, volcanoes, industrial activities and extraterrestrial sources. If this dust drops on a beach or on the land, there is little chance of it being preserved to tell a long-term story. However, dust that falls in the deep oceans, lakes and ice does not get reworked by currents, burrowing animals, plants and human activities and provides a better record of past events. Very finely layered sediments that have formed slowly are the best places to look for dust and if a radioactive technique of determining time can be used, the history of dust can be put together as a climate proxy. More dust from deserts is deposited during times of global cooling when winds are stronger. Polar winds are stronger and ice shows an increase in sea spray during colder times. Elsewhere we have looked at volcanic dust, extraterrestrial dust and dust from ancient industry such as Roman mining in Iberia. Dust is a useful proxy but cannot be used to give extremely accurate temperatures of past times.

The process of weathering involves water, air and micro-organisms. This process converts rock to soil. During weathering rocks are oxidised, leached, hydrated, carbonated, reduced in density and volume and increased in porosity and permeability. However, there are some minerals that are very resistant to weathering and remain in soil. These minerals are titanium minerals (rutile, ilmenite) and zircon.[1981]  Weathered material is removed by erosion and the eroded material is deposited as sediment. The best sediments to study are those in basins where there is no oxygen and therefore no burrowing organisms turning over the sediment and destroying the evidence. Some

---

[1979] Bräuning, A. 2006: Tree-ring evidence of "Little Ice Age" glacier advances in southern Tibet. *The Holocene* 16: 369-380.

[1980] Hallett, D. J. and Hills, L. V. 2006: Holocene vegetation dynamics, fire history, lake level and climate change in the Kootenay Valley, southeastern British Columbia, Canada. *Journal of Paleolimatology* 35: 351-371.

[1981] Rutile $TiO_2$, ilmenite $FeTiO_3$, zircon $ZrSiO_4$.

sediments have alternating layers of light (plankton-rich) and dark (mineral-rich). These represent the seasonal fluctuations in surface waters from trade wind-induced upwelling and regional precipitation and can show interannual climate changes for 15,000 years and decadal fluctuations for 500,000 years.[1982] This sediment in basins eventually becomes sedimentary rock.

During periods of high rainfall, the erosion rate is high, whereas in drought, the erosion rate is low. During a period of low rainfall, clays are washed out from soils. The titanium minerals remain behind in the soil and the sediment deposited from erosion is low in titanium. In a period of high rainfall, the soil is removed as a sheet and the deposited sediment is high in titanium. Therefore titanium in sediments, especially lake sediments, can provide an indication of past climate. On both sides of the Pacific the titanium in lake sediments has identified climate oscillations,[1983] the Roman Warming, the Dark Ages, the Medieval Warming, the Little Ice Age and the Late 20th Century Warming.[1984,1985,1986,1987] Furthermore, global climate change has produced socially damaging droughts that have lasted for years to decades.[1988]

Coral, which thrives in low latitude shallow warm waters, is a good indicator of sea surface temperature. This can be deduced from oxygen isotope studies. Coral proxies have been dealt with earlier to show sea level change and sea surface temperature change. Corals can show events that only occurred yesterday. For example, there has been a correlation of sea surface temperature with low latitude volcanic activity for the last 450 years. This coral isotope data was integrated with tree ring and ice core data to show that

---

[1982] Peterson, L. G., Haug, G. H., Hughen, K. A. and Rohl, U. 2000: Rapid changes in the hydrologic cycle of the tropical Atlantic during the last glacial. *Science* 290: 1947-1951.

[1983] Mayewski, P. A., Rohling, E. E., Stager, J. C., Karlen, W., Maasch, K. A., Meeker, L. D., Meyerson, E. A., Gasse, F., van Kreveld, S., Holmgren, K., Lee-Thorp, J., Rosqvist, G., Rack, F., Staubwasser, M., Schneider, R. R. and Steig, E. J. 2004: Holocene climate variability. *Quaternary Research* 62: 243-255.

[1984] Haug, G. H., Hughen, K. A., Sigman, D. M., Peterson, L. C. and Rohl, U. 2001: Southward migration of the intertropical convergence zone through the Holocene. *Science* 293: 1304-1308.

[1985] Almeida-Lenero, L., Hooghiemstra, H., Cleef, A. M. and van Geel, B. 2005: Holocene climatics and environmental change from pollen records of Lakes Zempoala and Quila, central Mexican highlands. *Reviews of Palaeobotany and Palynology* 136: 69-32.

[1986] Hodell, D. A., Brenner, M. and Curtis, J. H. 2005: Terminal classic drought in the northern Maya lowlands inferred from multiple sediment cores in Lake Chichancanab (Mexico). *Quaternary Science Reviews* 24: 1413.

[1987] Yancheva, G., Nowaczyk, N. R., Mingram, J., Dulski, P., Schettler, G., Negendank, J. F. W., Liu, J., Sigman, D. M., Peterson, L. C. and Haug, G. H. 2007: Influence of the intertropical convergence zone on the East Asian monsoon. *Nature* 445: 74-77.

[1988] Haug, G. H., Gunther, D., Peterson, L. C., Sigman, D. M., Hughen, K. A. and Aesschlimann, B. 2003: Climate and the collapse of the Maya civilisation. *Science* 299: 1731-1735.

there was a sustained cold period following the eruption of Tambora in 1815 and another cold period, presumably associated with an unknown tropical volcanic eruption.[1989]

Stalagmites and stalactites grow in caves and are linked to the weather as they form from rainwater which has passed through soil and limestone to form calcium carbonate precipitates. As the chemistry of this water and the climate of the cave change throughout the year, different types of calcium carbonate are deposited, creating layers or rings. The width of the ring or how fast the stalagmite has grown in any one year can be linked directly to the climate. Because stalagmites grow for long periods of time, they give a more accurate picture about long-term climate variability than studies of tree rings. They are buffered from any wild swings in weather that might occur above ground. Stalagmites grow for thousands of years and, if they are in the right type of cave, they can preserve a signal of temperature and precipitation levels over the time they grew. Furthermore, during weathering, water dissolves uranium and so the uranium, thorium and radioactive decay products in stalactites can be used to measure when each layer was deposited. However, it is not that simple. If rainfall is below critical level or snow becomes locked in ice and permafrost then the stalactite gives an incomplete record.

The carbon ($C^{12}$, $C^{13}$ and $C^{14}$) and oxygen ($O^{16}$ and $O^{18}$) isotopes give clues about water temperature, atmospheric $CO_2$ and the contribution of humans to atmospheric $CO_2$. The strontium, magnesium, calcium and organic material in stalactites can also be used to determine ancient environments. Stalagmite ring width, rather like tree rings, has also been used to determine past climates. However, attempts to use only stalagmite ring width to ascertain climate variation shows that there is no relationship between stalagmite ring width and tree rings in the same area.[1990] It is really only the oxygen isotopes that can give an accurate determination of past temperatures and if this is combined with a dating technique (e.g. uranium-thorium), the timing of temperature changes can be calculated.

Three stalagmites in the Spannagel Cave some 2500 metres above sea level in the Tux Valley of the Tyrol in Austria near the Hintertux Glacier are nearly 10,000 years old.[1991] The Medieval Warming and Little Ice Age

[1989] D'Arrigo, R., Wilson, R. and Tudhope, A. 2009: The impact of volcanic forcing on tropical temperatures during the past four centuries. *Nature Geoscience* 2: 51-56.

[1990] Polyak, V. J. and Amerson, Y. 2001: Late Holocene climate and cultural changes in Southwestern United States. *Science* 294: 148-151.

[1991] Vollweiler, N., Scholz, D., Mühlinghaus, C., Mangini, A. and Spötl, C. 2006: A precisely dated climate record for the last 9 kyr from three high alpine stalagmites, Spannagel Cave, Austria. *Geophysical Research Letters* 33: L20703, doi: 101029/2006GL027662.

are some of the strong climate features recorded there. There were warm periods at 7500–5900 years ago (Holocene Climate Optimum), 3800–3600 and 1200–700 years ago (Medieval Warming) and cool periods at 7900–7500, 5900–5000, 3500–3000 and 600–150 years ago (Little Ice Age). During the Roman Warming (2250–1700 years ago), some of the Alpine passes were ice-free in both summer and winter. A pronounced warming peak at 2200 years ago coincides with Hannibal's crossing of the Alps in 218 BC. A study of layered stalagmites which show annual layers over a 500-year period in China, Italy and Scotland shows the Little Ice Age and subsequent warming in the Northern Hemisphere of 0.65°C.[1992] Some stalactites have been undisturbed for tens of thousands of years and can give an even longer-term record of temperature and vegetation responses to climate change.[1993]

Deep boreholes on the land can provide a lot of information. Not only can we see the rocks beneath our feet, we can measure the flow-up heat from deep in the Earth. Furthermore, during warm times heat is conducted downwards and detailed temperature measurements in the first few metres of a borehole can provide indications about past warm and cold times. Again, like the other proxies, it does not give the exact temperature or the exact timing of a climate change.

The great advances in understanding past climates came with the Ocean Drilling Program. Deep drill holes through ocean floor sediments and sea floor basalts gave information that was fundamental to the construction of the plate tectonics theory. Additional information produced the first glimpses of submarine hot spring activity and the associated strange life forms and a look back at global climates. Ocean floor sediment is especially good material to study because the rate of sediment formation is slow, there are few disturbances as the sediments do not form on slopes, they have not had burrowing organisms turning them over and because microfossils of floating and bottom dwelling organisms are well preserved. These organisms, mainly foraminifera, are very sensitive to temperature, light, salinity and water depth. They have a shell, normally composed of calcium carbonate. Some are so sensitive that if the temperature changes by just 1°C, they either speciate or die. Large volcanic eruptions blanket the Earth with dust which falls into the ocean basins and forms thin layers in ocean floor sediments. The chemistry of these ash layers can be used to identify the source volcano, as

---

[1992] Smith, C., Baker, A., Fairchild, I. J., Frisia, S. and Borsato, A. 2006: Reconstructing hemisphere-scale climates from multiple stalagmite records. *International Journal of Climatology* 26: 1417-1424.

[1993] Dorale, J. A., Edwards, L. R., Ito, E. and González, L. A. 1998: Climate and vegetation history of the Midcontinent from 75 to 25 ka: a speleothem record from Crevice Cave, Missouri, USA. *Science* 282: 1871-1874.

can the radioactive dating of the ash. Furthermore, the irregular flips in the Earth's magnetic field can be used as a crude dating technique, as can the life and death of various short-lived rapidly evolving floating micro-organisms. Extraterrestrial dust can show when there was increased meteor and asteroid activity, and desert dust can show when times were windier.

Fossil foraminifera give a picture of the past environment, and the measurement of $O^{16}$ and $O^{18}$ in the shells can be used to calculate the temperature of ocean water. All this information can be integrated to give a more complete picture of sea temperature, salinity, wind, extraterrestrial activity, currents, volcanic activity and changing environmental patterns at very tightly constrained time intervals. In the North Atlantic, armadas of icebergs melted and dropped material into these seafloor sediments and this gives information on the break-up of ice sheets during and after the last glaciation. This information can be combined with similar information from lakes and ice to give an overview of climate change.

What we see from these studies is that climate cycles have been with us for a long time. Most proxy studies use a diversity of proxies which give a more coherent broad integrated picture of climate events. For example, floating organisms, oxygen isotopes, organic chemicals and pollen have shown that there was a cooling of 2°C of the surface of the Adriatic Sea 6000 years ago and another cooling of 2–3°C some 3000 years ago. These cooling events lasted for several hundred years, occurring at the same time as other cooling events in the Aegean Sea and Greenland.[1994]

Drill holes into the Greenland and Antarctic ice sheets were also revolutionary and gave a great leap forward in our understanding of the planet over the last few hundred thousand years.[1995,1996,1997,1998] Isotopes of oxygen, hydrogen, carbon, argon, helium, beryllium and chlorine can give an accurately dated history of temperature, solar activity, extraterrestrial dust

[1994] Sangiorgi, F., Capotondi, L., Combourieu Nebout, N., Vigliotti, L., Brinkhuis, H., Giunta, S., Lotter, A. F., Morigi, C., Negri, A. and Reichart, G.-J. 2003: Holocene sea-surface temperature variations in the southern Adriatic Sea inferred from a multiproxy approach. *Journal of Quaternary Science* 18: 723-732.

[1995] Barnola, J. M. D., Raynaud, D., Korotkevich, Y. S. and Lorius, 1987: Vostok ice core provides a 160,000-year record of atmospheric $CO_2$. *Nature* 329: 408-414.

[1996] Chappellaz, J., Barnola, J.-M., Raynaud, D., Korotkevich, Y. S. and Lorius, C. 1990: Atmospheric $CH_4$ record over the last climatic cycles revealed by the Vostok ice core. *Nature* 345: 127-131.

[1997] Jouzel, J., Lorius, C., Petit, J. R., Genthon, C., Barkov, N. I., Kotlyakov, V. M. and Petrov, V. M. 1987: Vostok ice core: a continuous isotopic temperature record over the last climatic cycle (160,000 years). *Nature* 329: 403-407.

[1998] Lorius, C., Jouzel, J., Ritz, C., Merlivat, N. E., Barkov, N. E. and Korotkevich, A. 1985: 150,000-year climatic record from Antarctic ice. *Nature* 316: 591-595.

and climate change. Furthermore, volcanic dust and acid can give time layers in the ice and a history of volcanic activity and nitric acid can give clues about supernoval explosions. Terrestrial dust and sea spray give a picture of windiness. Air bubbles that were trapped in snow only to be later trapped in ice can be used to look at the chemistry of ancient air. However, with all these proxy techniques there are orders of accuracy, limitations and uncertainties.

Ice core contains volcanic ash which is useful for identifying the specific volcano and dating. Climate sensitivity cannot be inferred from the temperature record alone even if the forcing is known. Equilibrium climate sensitivities can be reasonably determined only if both the forcing and the change in heat storage are known very accurately.[1999] Annual layers of snowfall in ice cores can be counted as easily as tree rings, allowing precise dating of events such as volcanic eruptions. Summer snow crystals are large and the snow has a higher acidity than winter snow. In some core, even the seasons can be identified because spring winds blow around more dust and sea spray which settles into ice. Wet seasons, with heavy snow, and dry seasons, with huge dust storms, can easily be identified.

Ice cores also record human activity. The use and testing of atomic bombs is well recorded, as is the decrease in radioactivity followed by the atomic test ban treaty. Chernobyl is a radioactive spike at 1987–1988.[2000] The burning of coal and petroleum is shown by an increase in atmospheric nitrous oxide. The increase in $CO_2$[2001] and methane[2002] is also recorded, however it cannot be determined whether this is from human activity. As discussed elsewhere in this book, $CO_2$ has a great diversity of sources and sinks and it is not possible to ascribe a $CO_2$ increase to human activity. So too with methane. Nitrates have short-lived extraordinarily high concentrations before the Industrial Revolution, possibly due to supernovae, and the background nitrate concentration begins to increase after the Industrial Revolution.[2003] Sulphates also have short-lived high concentrations before the Industrial Revolution due to volcanic eruptions and then increased after the Industrial Revolution. Since the various clean air acts of the 1990s, both the nitrate and

[1999] Boer, G. J., Stowasser, M. and Hamilton, K. 2007: Inferring climate sensitivity from volcanic events. *Climate Dynamics* 28: 481-502.

[2000] Dibb, J., Mayewski, P.A., Buck, C. F. and Drummey, S. M. 1990: Beta radiation from snow. *Nature* 344: 6270, 25.

[2001] Keeling, C. K., Adams, J. A., Ekdahl, C. A. and Guenther, P. R. 1976: Atmospheric carbon dioxide variations at the South Pole. *Tellus.* 28, 552-564.

[2002] Etheridge, D. M., Pearman, G. I. and Fraser, P. J. 1992: Changes in trophospheric methane between 1841 and 1978 from a high accumulation rate Antarctic ice core. *Tellus* 44: 282-294.

[2003] Machida, T., Nakazawa, T., Fuji, Y., Aoke, S. and Watanabe, O. 1995: Increase in atmospheric nitrous oxide concentrations during the last 250 years. *Geophysical Research Letters* 22: 2921-2924.

sulphate contents of polar ice have stabilised.

There was a suggestion from a study of the Vostok (Antarctica) ice core that $CO_2$ had forced climate changes throughout the past ice ages. However, this has been challenged because of substantial mismatches. These mismatches were corrected using a model that incorporates variations of climate in the source regions for water vapour.[2004] Measurements of $CO_2$, methane and temperature in ice cores show a sawtooth pattern.[2005] At the scale of measurement, it looked as if there was a close correlation between temperature and $CO_2$.[2006] Furthermore, temperature rose and then suddenly declined. There was no "tipping point" and the temperature-$CO_2$ plots clearly showed that the rise of temperature was stopped by something other than $CO_2$.[2007] To analyse thousands of metres of ice core chemically is a difficult, costly, slow process. Some early sampling was using layers separated by 800 years of snow deposition.[2008] This missed a key feature of the core. It was only with later, far more detailed studies of ice cores that another story emerged. New high resolution studies over the last 450,000 years of Vostok core show that at all times of cold to warm transitions, temperature rise is followed by a rise in $CO_2$ some 800 years later.[2009] Hence the rise of $CO_2$ in past climates is a response to warming, not the cause.

All scientific measurements have their limitations. So too with ice core measurements. A common proxy of temperature is to measure the proportion of heavy ($O^{18}$) to light ($O^{16}$) oxygen in trapped air in ice and in ice itself. Detailed measurements of the lower part of the Greenland GRIP ice core show that the chemical signature of air trapped in ice lags some 1000–4000 years behind the chemical signature of the ice that trapped the air. The chemical signature of Greenland ice core older than 100,000 years does not correlate with Vostok ice core and shows climate changes that are far too

[2004] Cuffey, K. M. and Vimeux, F. 2001: Covariation of carbon dioxide and temperature from the Vostok ice core after deuterium-excess correction. *Nature* 412: 523-527.

[2005] Dahl-Jensen, D. K., Mosegaard, K., Gundestrup, N., Clow, G. D., Johnsen, S. J., Hansen, A. W. and Balling, N. 1998: Past temperatures from the Greenland ice sheet. *Science* 282: 268-271.

[2006] Mudelsee, M. 2001: Phase relations among atmospheric $CO_2$ content, temperature and global ice volume over the past 420 la. *Quaternary Science Reviews* 20: 583-589.

[2007] Wunsch, C. 2003: Greenland-Antarctic phase relations and millennial time-scale climate fluctuations in the Greenland ice-cores. *Quaternary Science Reviews* 22: 1631-1646.

[2008] Barnola, J. M., Pimienta, P., Raynaud, D. and Korotkevich, Y. S. 1991: $CO_2$-climate relationship as deduced from the Vostok ice core: a re-examination on new measurements and on a re-evaluation of the air dating. *Tellus* B 43: 83-90.

[2009] Caillon, N., Severinghaus, J. P., Jouzel, J., Barnola, J-M., Kang, J. and Lipenkov, V. Y. 2003: Timing of atmospheric $CO_2$ and Antarctic temperature changes across Termination III. *Science* 299: 1728-1731.

fast for detailed comparisons.[2010] Other work with the GRIP core shows that gas measurement in core older than 110,000 years has limitations.[2011] Rapid chemical variations occur on the scale of 20 cm in the core. These do not reflect rapid climate changes and do not appear to form by gas diffusion after snow-air mixing or after coring. These variations place limits on interpretation of the oxygen isotope proxy on small scales. This is a fundamental scientific limitation: the scales of measurement.

The oldest parts of the Greenland ice core (older than 110,000–120,000 years) have been disturbed because the ice has been folded by creep near its contact with bedrock. This shows that ice is plastic, that there is a frictional drag on ice grinding over land and that this folding may release gas trapped in ice. There is a major field of study in geology that deals with fluid inclusions in rocks.[2012] Every time rocks are recrystallised, the fluid inclusions are broken open, liquid and gas is released and new fluid inclusions form. These new secondary fluid inclusions are not the same composition as the primary fluid inclusions and tell us nothing about past fluid processes or gases. So too with ice, which is a common rock.[2013] Every time ice recrystallises during creep, the primary fluid inclusions are broken open, the ice recrystallises and secondary fluid inclusions are trapped. Extreme caution must be taken if we are to make interpretations about ancient climate from air trapped in ice.

However, GRIP core shows that the last interglacial (130,000–116,000 years ago) was about 5°C warmer than now and that there was abrupt warming about 115,000 years ago[2014] which is not seen in Antarctic ice cores. Again, we see that global climate is not global and that scales of measurement are important. If we claim that the Late 20th Century Warming is unprecedented then it is only over a scale of 150 years. Over a scale of 1000 years it is not unprecedented as Greenland was about 5°C warmer in the Medieval Warming than now. Over a scale of 120,000 years, Greenland was also about 5°C warmer than now in the last interglacial.

Again, the obvious question remains unanswered: If human addition of

[2010] Fuchs, A. and Leuenberger, M. C. 1996: $\partial^{18}O$ of atmospheric oxygen measured on the GRIP ice core document stratigraphic disturbances in the lowest 10% of the core. *Geophysical Research Letters* 33: 1049-1052.

[2011] Caillon, N., Landais, A., Chappellaz, J., Steffensen, J., Delmotte, M., Jouzel, J., Masson-Delmotte, V. and Raynaud, D. 1993: New gas measurements in the GRIP ice core: Rapid interglacial climate variability ruled out and proposal for a reconstruction of the last interglacial in Greenland. *American Geophysical Union*, Fall Meeting 2003, Abstract PP32A-0276.

[2012] Roedder, E. 1984: *Fluid inclusions*. Mineralogical Society of America.

[2013] Goldstein, R. H. 2001: Fluid inclusions in sedimentary and diagenetic systems. *Lithos* 55: 159-193.

[2014] North Greenland Ice Core Project Members, 2004: High-resolution record of Northern Hemisphere climate extending into the last interglacial period. *Nature* 431: 147-151.

$CO_2$ to the modern atmosphere is producing climate change, then how can previous pre-industrialisation warmings be explained? Might not the Late 20th Century Warming have its origins in the same processes that gave us the Medieval Warming, the Roman Warming and countless other interglacial warmings?

The chemical fingerprints in ice core are smoothed, probably as a result of diffusion of air and water in ice.[2015,2016,2017] This amount of diffusion is probably related to the temperature and snow accumulation rate at the site. This diffusion therefore places a limit on the amount of detail that can be extracted from looking at the chemical fingerprints in ice. Various experiments have been conducted to try to understand the large diffusion lengths.[2018] However, as in all science, there are other explanations. One of these is that the stronger than expected smoothing over the last 100,000 years can be explained by warmer firn temperatures thereby resulting in longer firn diffusion lengths.[2019] Some 12,000–10,000 years ago, the Greenland temperature was about 5°C warmer than now and, if warmth and not diffusion gave smoothing and longer diffusion lengths, then Greenland would have been even warmer.[2020] This is in accord with other calculations.

There are systematic errors in ice core studies, and many of the techniques used (e.g. air inclusions in ice) must be treated with great caution.[2021] These errors are such that one Greenland ice core (GRIP) cannot be correlated with another (GRIP2) over the last 100,000-year period. Correction of ice core ages to orbital features of the Earth allow GRIP and GRIP2 ice cores to be correlated. The unanswered question remains: what other errors occur in ice core on small scales or in core that formed in times of low accumulation of snow?

---

[2015] Nye, J. F. 1998: Diffusion of isotopes in the annual layers of ice sheets. *Journal of Glaciology* 44: 467-468.

[2016] Johnsen, S. J., Clausen H. B., Cuffey, K. M., Hoffmann, G., Schwander, J. and Creyts, T. 2000: Diffusion of stable isotopes in polar firn and ice. The isotopic effect in firn diffusions. In: *Physics of Ice Core Records* (ed. Hondoh, T.) 121-140, Hokkaido University Press.

[2017] Rempel, A. W. and Wettlaufer, J. S. 2003: Isotopic diffusion in polycrystalline ice. *Journal of Glaciology* 49: 397-406.

[2018] Johnsen, S. J. and Andersen, U. 1997: Isotopic diffusion in Greenland ice and firn. Evidence from crystal boundary diffusion. *Eos Transactions, AGU Fall Meeting*, 78:F7 Poster U21A-4.

[2019] Vinther, B. M., Johnsen, S. J. and Clausen, H. B. 2005: Central Greenland late Holocene temperatures. *EGU*, Abstract, session CL21, Vienna, Austria.

[2020] Johnsen, S. J., Vinther, B. M., Clausen, H. B., Creyts, T. T., Seierstad, I. and Sveinbjornsdottir, A. 2005: The GRIP ice core isotopic excess diffusion explained. *Geophysical Research Abstracts* 7: 10540.

[2021] Hinnov, L. 2003: Evidence for a systematic error in GRIP ice-flow chronology. *XVI INQUA Congress*, Paper 87-14.

# Hurricanes

Because sea surface temperatures vary less than land surface temperatures through the seasonal cycles, any place on land where the wind blows off the ocean will experience a maritime climate (e.g. Pacific Northwest, UK) whereas winds blowing across a continent provide blazing summers and bitterly cold winters (e.g. eastern North America, central and northeast Asia). In winter, cold westerly continental winds in Florida cool the Gulf Stream by evaporation and heat transfer. Much of this heat is acquired by storms in the atmosphere and is carried off eastern USA and Canada, mitigating what would otherwise be a very cold climate. The Gulf Stream heads northeast and becomes the North Atlantic Drift, and further downstream the Norwegian current. After spawning many Atlantic storms, it loses most of the remaining heat in the Nordic seas.

Earth's rotation and the mountains in North America[2022,2023] contribute substantially to differences in temperature across the Atlantic Ocean.[2024,2025] The southward and then northward deflection creates an irregularity in the west-to-east flow of air across North America and far downwind to the east. Such deflections are on a massive scale. Southward deflections occur across central and eastern North America. This brings Arctic air south and the cool winters of the east coast of the USA and Canada. The northward return of air occurs across the eastern Atlantic Ocean and Western Europe, bringing mild subtropical air north and warm winters on the far eastern side of the North Atlantic. Furthermore, at low latitude the oceans gain heat from the Sun and lose heat by evaporation and the excess heat is moved polewards by ocean currents at atmospheric circulation.[2026]

There have been suggestions that increased sea surface temperature has increased the frequency of hurricanes, especially around the Gulf of Mexico in the southern USA.[2027] The increased sea surface temperature is attributed to global warming resulting from addition of $CO_2$ to the atmosphere from

[2022] Nigam, S., Held, I. M. and Lyons, S. W. 1988: Linear simulation of stationary eddies in a GCM. Part II: Mountain model. *Journal of Atmospheric Sciences* 45: 1433-1452.

[2023] Hoskins, B. J. and Valdes, P. J. 1990: On the existence of storm tracks. *Journal of Atmospheric Sciences* 47: 1854-1864.

[2024] Lau, N.-C. 1979: The observed structure of tropospheric stationary waves and the local balances of vorticity and heat. *Journal of Atmospheric Sciences* 36: 996-1016.

[2025] Manabe, S. and Stouffer, R. J. 1988: Two stable equilibria of a coupled ocean-atmosphere model. *Journal of Climate* 1: 841-866.

[2026] Trenberth, K. E. and Caron, J. M. 2001: Estimates of meridional atmosphere and ocean heat transports. *Journal of Climate* 14: 3433-3443.

[2027] Webster, P. J., Holland, G. J., Curry, J. A. and Chang, H.-R. 2005: Changes in tropical cyclone number, duration, and intensity in a warming environment. *Science* 309: 1844-1846.

human activities.[2028] This is a fallacy. If the oceans become warmer, there are not necessarily more hurricanes. If there has been more $CO_2$ in the atmosphere, this does not necessarily mean it is of human origin. Other factors involved are the difference between the upper and lower layers of the atmosphere (wind shear) and dry air. These factors are far more critical than the ocean being warmer.

This view that the intensity, frequency and longevity of tropical cyclones has increased as a result of the modern global warming increasing the sea surface temperature[2029,2030] is contrary to other studies that show global warming has not led to changes in the intensity of tropical cyclones.[2031,2032,2033] Tropical cyclones, at least in the Atlantic Basin, result from complex interactions of many factors, including atmospheric stability, sea surface temperature and vertical wind shear.[2034] A study of global cyclone activity has shown there has been no significant change in cyclone activity over the past 20 years and that there has been a slight increase in global Category 4 to 5 hurricanes from 1986–1995 to 1996–2005. This change is probably not related to changes in sea surface temperature and may well be due to improved observational technology. The conclusion from this work and that of others[2035,2036] is that there is little correlation between sea surface temperatures and hurricane development in the Atlantic Basin and that vertical wind shear is a far more fundamental component of hurricane development and maintenance. It appears that if anthropogenic global warming has contributed to increased hurricane activity in the Atlantic Basin or anywhere else in the world, then it is not measurable.

---

[2028] Mann, M. and Emanuel, K. 2006: Atlantic hurricane trends linked to climate change. *EOS* 87: 233-241.

[2029] Emanual, K., 2005: Increasing destructiveness of tropical cyclones over the past 30 years. *Nature* 436: 686-688.

[2030] Webster, P. J., Holland, G. J., Curry, J. A. and Chang, H.-R. 2006: Changes in tropical cyclone number, duration, and intensity in a warming environment. *Science* 309: 1844-1846.

[2031] Klotzbach, P. J. 2006: Trends in global cyclone activity over the last twenty years (1986-2005). *Geophysical Research Letters* 33: L010805, doi:10.1029/2006GL025881.

[2032] Pielke, Jr., R. A., Landsea, C., Mayfield, M., Laver, J. and Pasch, R. 2006: Reply to "Hurricanes and global warming-potential linkages and consequences. *Bulletin of the American Meteorological Society* 87: 628-631.

[2033] Hoyos, C. D., Agudelo, P. A., Webster, P. J. and Curry, J. A. 2006: Deconvolution of the factors contributing to the increase in global hurricane intensity. *Science* 312: 94-97.

[2034] Klotzbach, P. J. 2006: Trends in global tropical cyclone activity over the past twenty years (1986-2005). *Geophysical Research Letters* 33: L010805, doi: 101029/2006GL025757.

[2035] Shapiro, L. J. and Goldenberg, S. B. 1998: Atlantic sea temperatures and tropical cyclone formation. *Journal of Climate* 11: 578-590.

[2036] Klotzbach, P. J. and Gray, W. M. 2006: *Extended range forecast of Atlantic seasonal hurricane activity and US landfall strike probability for 2007.* http://hurricane.atmos.colostate.edu/Forecasts/2006/dec2006.pdf

The 2007 hurricane season predictions were for 16 named storms, nine hurricanes, five intense hurricanes and a 74% chance of a storm hitting the east coast of USA. These predictions were all above the historical average and numerous news stories cited global warming as the culprit. By December 2007 when the hurricane season ended, there had been six hurricanes (the historical average) and two intense hurricanes (below average). Not one hit USA (below average). The accumulated cyclone energy was almost a third of the prediction and two thirds of the historical average. The 2007 hurricane season was a fizzer and all disaster predictions from global warming were wrong.

Similar wild speculations were made by the UK Meteorological Office for temperature in 2007.[2037] The official government press release screamed:

> 2007 is likely to be the warmest year on record globally, beating the current record set in 1998, say climate change experts at the Met Office. Global temperature for 2007 is expected to be 0.54°C above the long term (1961–1990) average of 14°C.

This prediction was made on 4 January 2007. The media screamed tales of forthcoming disaster. Actual measurements for 2007 show that it was one of the coldest years this century and the coldest since 1995.

There were four very destructive hurricanes over a short period in the 2004 hurricane season in North America. The question raised was: Is this due to human-induced global warming? By analysing the history of hurricanes, it was shown that global tropical cyclone activity, as measured by the accumulated cyclone energy index, decreased from 1990 to 2005.[2038] Hurricane Katrina was in 2005. The 2004 season was due to a "consequence of multidecadal fluctuations in the strength of the Atlantic multidecadal mode and strength of the Atlantic Ocean thermohaline circulation".

This means that global warming played no part whatsoever. The increased coastal population and investment in real estate along the US coastline means that hurricane-spawned damage and destruction in the coming decades will be far greater than in the past.

Other scientific studies question whether the global tropical cyclone database is sufficiently reliable to ascertain long-term trends in cyclone intensity, particularly in the frequency of extreme tropical cyclones.[2039] This caution contrasts with the unquestioning mantra that human-induced

---

[2037] http://www.metoffice.gov.uk/corporate/pressoffice/2007/pr20070104.html

[2038] Klotzbach, P. J. and Gray, W. M., 2006: Causes of the unusually destructive 2004 Atlantic basin hurricane season. *Bulletin of the American Meteorological Society* 87: 1325-1333.

[2039] Landsea, C. W., Harper, B. A., Hoarau, K. and Knaff, J. A. 2006: Can we detect trends in extreme tropical cyclones? *Science* 313: 452-454.

global warming increases the frequency and strength of hurricanes. One of the problems in science is a change in the methods of measurement thereby resulting in comparisons that may be invalid. Aircraft used to make measurements in cyclones over an area of a few dozen square kilometres. Now data is collected from geostationary satellites that can provide high-resolution images, overhead views of tropical cyclones and more accurate intensity results. It is only in the Atlantic and northwest Pacific that regular aircraft measurements have been taken and measurements show that since 1960, there are no significant trends in cyclone activity. Extreme tropical cyclones and overall tropical cyclone activity have been globally unchanged from 1986 to 2005.

Global warming appears to decrease the number of hurricanes in the Atlantic Ocean. If the planet warms, then it has been calculated that by the end of the 21st Century, the number of hurricanes in the Atlantic will fall by 18%, the number of hurricanes reaching the USA and its neighbours will drop by 30%, the biggest storms (wind velocity of greater than 100 kilometres per hour) will fall by 8% and tropical storms (wind velocity 39 to 73 km per hour) will decrease by 27%.[2040] This work suggested a rise in the number of hurricanes between 1980 and 2006 and concluded:

> Our results do not support the notion of large increasing trends in either tropical storm or hurricane frequency driven by increases in atmospheric greenhouse-gas concentrations.

This confirms other studies that analysed a century of tornadoes,[2041] long-term records of Atlantic seasonal hurricanes,[2042] statistics and position of Atlantic hurricanes,[2043] statistics of extreme weather events[2044,2045] and storm activity along the east coast of the USA.[2046,2047] This study contradicts

[2040] Knutson, T. R., Sirutis, J. J., Garner, S. T., Vecchi, G. A. and Held, I. M. 2008: Simulated reduction in Atlantic hurricane frequency under twenty-first-century warming conditions. *Nature Geoscience* 1: 359-364.

[2041] Brooks, H. E. and Sowell, C. A. 2001: Normalized damage from major tornadoes in the United States: 1890-1999. *Weather and Forecasting* 16: 168-176.

[2042] Gray, W. and Klotzbach, P. J. 2003: Extended range forecast of Atlantic seasonal hurricane activity and U.S. landfalling strike probability for 2004. *Colorado State University, December 2003.*

[2043] Elsner, J. B., Liu, K. and Kocher, B. 2000: Spatial variations in major US hurricane activity: Statistics and a physical mechanism. *Journal of Climate* 13: 2293-2305.

[2044] Changnon, S. A. and Changnon, D. 2000: Long-term fluctuations in hail incidence in the United States. *Journal of Climate* 13: 658-664.

[2045] Changnon, S. and Changnon, D. 2001: Long-term fluctuations in thunderstorm activity in the United States. *Climate Change* 50: 489-503.

[2046] Zhang, K., Douglas, B. C. and Leatherman, S. P. 2000: Twentieth-Century storm activity along the U.S. East Coast. *Journal of Climate* 13: 1748-1761.

[2047] Hirsch, M. E., Degaetano, A. T. and Colucci, S. J. 2001: An East Coast winter storm climatology. *Journal of Climate* 14: 882-899.

previous work.[2048,2049,2050]

The only extraordinary aspect of hurricanes in the Northern Hemisphere is that there are not as many in the 21st Century as in the 20th and 19th Centuries even though the number of storms has not changed.[2051] In a study of hurricanes from 1851 to 2007, it was shown that hurricanes occur in 60-year cycles. The period from 1851 gave the opportunity to study two complete cycles. There was no increase in hurricane activity connected to increased human $CO_2$ emissions.[2052] Corals from the Caribbean were used as a proxy record of wind shear and sea surface temperatures since 1730.[2053] Major hurricanes decreased from the 1760s, reaching an all-time low in the 1970s and 1980s. Since then, the number of hurricanes has started to increase again with several very active hurricane seasons. Although it has been well known that the frequency of tropical cyclones has almost doubled globally since the 1970s,[2054] in the context of the past three centuries the current frequency and intensity of hurricanes is just business as normal.

Hurricane Katrina in August 2005 was especially damaging because New Orleans is sinking due to compaction of Mississippi River sediments and extraction of groundwater and petroleum.[2055] New Orleans sank rapidly by about 1 metre in the three years before Katrina struck.[2056] New Orleans was already below the level of the Mississippi River. This sinking was exacerbated by the weight of buildings, traffic, extraction of underground water and extraction of oil from the Texas Gulf and the proximity to a major river. The normal risks of cyclones and hurricanes are amplified because increasingly wealthy people in countries like the USA seek expensive waterfront areas or

[2048] Emanuel, K. A. 2005: Increasing destructiveness of tropical cyclones over the past 30 years. *Nature* 436: 686-688.

[2049] Webster, P. J., Holland, G. J., Curry, J. A. and Chang, H.-R. 2005: Changes in tropical cyclone number, duration, and intensity in a warming environment. *Science* 309: 1844-1846.

[2050] Mann, M. and Emanuel, K. 2006: Atlantic hurricane trends linked to climate change. *EOS* 87: 233-241.

[2051] Bengtsson, L., Hodges, K. I., Esch, M., Keenlyside, N., Kornblueh, L., Luo, J.-J. and Yamagata, T. 2007: How may tropical cyclones change in a warmer climate? *Tellus A* 59: 539-561.

[2052] Chylek, P and Lesins, G. 2008: Multi-decadal variability in Atlantic hurricane activity 1851-2007. *Journal of Geophysical Research* 113: doi 10.1029/2008JD010036.

[2053] Nyberg, J., Winter, A. and Malmgren, B. A. 2005: Reconstruction of major hurricane activity. *EOS* 86: 52.

[2054] Webster, P. J., Holland, G. J., Curry, J. A. and Chang, H. R. 2005: Changes in tropical cyclone number, duration and intensity in a warming environment. *Science* 309: 1844-1846.

[2055] Ericson, J. P., Vorosmarty, C. J., Dingham, S. L., Ward, L. G. and Meybeck, M. 2006: Effective sea-level rise in deltas: Causes of change and human dimension impacts. *Global and Planetary Change* 50: 63-82.

[2056] Dixon, T. H., Amelung, F., Ferretti, A., Novali, F., Rocca, F., Dokka, R., Selia, G., Kim, S.-W., Wdowinski, S. and Whitman, D. 2006: Space geodesy: subsidence and flooding in New Orleans. *Nature* 441: 587-588.

warm climates for living and playing. Expensive houses are built below sea level, preparation for the inevitable hurricanes is not made and damage is more costly.[2057]

The storms have not changed. What has changed is that we humans have migrated to coastlines, the population has grown and we have put expensive structures in the path of storms. There is no correlation between the incidence and severity of hurricanes at atmospheric concentrations of $CO_2$. However, insurance payouts have increased greatly because increased affluence has allowed Americans to migrate to their warmer southeastern states.

Suggestions that global warming has produced an increasing number of hurricanes seem to be constrained to the USA. Furthermore, these hurricanes seem to occur in cyclical patterns and these cycles are unrelated to warming or cooling trends. Other areas prone to hurricanes (cyclones) do not seem to have had the media scare campaigns (e.g. Queensland, Australia) despite a long history of cyclones.[2058]

Cyclones in India show a similar history.[2059] Over the last 115 years, the highest number of cyclones (10) hit India in 1893, 1926 and 1930. Over the last 20 years, the highest number of cyclones (6) that hit Indian shores was in 1992 and 1998. Highest rainfall in India was recorded in 1917 (1457.3 cm) with 1918 recording 913 cm. Over the last 20 years, the highest rainfall was in 1988 (1288 cm) and the lowest was in 2000 (939 cm). The heavy rainfall in Mumbai on 27 July 2005 is used as evidence of climate change, yet over the same 24-hour period, the nearby localities of Colaba (7.3 cm) and Santa Cruz (94.4 cm) received rainfall that differed by an order of magnitude. There appears to be little evidence that global warming has affected the cyclone history of India.

Hurricanes have a little trick up their sleeves. Tropical hurricanes increase erosion on the land. The eroded material is transported down river systems into the oceans. The largest proportion of carbon is transported down river systems into the oceans during tropical hurricanes. The carbon transfer from land to the oceans is modulated by the frequency, intensity and duration of tropical hurricanes.[2060] This addition of carbon to the oceans never appears in any IPCC models.

---

[2057] Emanuel, K. A. 2005: Increasing destructiveness of tropical cyclones over the past 30 years. *Nature* 436: 686-688.

[2058] Nott, J. and Hayne, M. 2001: High frequency of super-cyclones along the Great Barrier Reef over the past 5,000 years. *Nature* 413: 508-512.

[2059] Valdiya, K. S. 2005: *Coping with natural hazards: Indian context*. Orient Longman.

[2060] Hilton, R. G., Galy, A., Hovius, N., Chen, M-C, Horng, M-J. and Chen, H. 2008: Tropical-cyclone-driven erosion of the terrestrial biosphere from mountains. *Nature Geoscience* 1: 759-762.

## Carbon dioxide

Carbon dioxide puts the fizz into soft drinks, the holes in bread, the dry ice in a cooler, the bubbles in beer and sparkling wines and the gas in fire extinguishers. It does not form polluting smog or acid rain. Carbon dioxide is plant food. Without carbon dioxide, there would be no life on Earth.

### The global carbon cycle

Carbon is more basic to life than sex. You heard it here first! Carbon dioxide is a colourless odourless non-poisonous gas. It is plant food, and it drives the whole food chain. All life is based on and contains carbon. Every cell in every living organism on the planet is based on carbon. Bacteria, algae and plants remove $CO_2$ from the air and water and store it in their tissues. Together with water vapour, $CO_2$ keeps our planet warm such that it is not covered in ice, too hot or devoid of liquid water. If there were no mechanism for recycling carbon back into the air and water, then these reservoirs would have been depleted long ago.

During the history of the planet, $CO_2$ levels have continuously fluctuated. During periods of high $CO_2$ in the air, life underwent massive expansion and diversification, whereas in periods of low $CO_2$, like today, plant life is not as energetic. The $CO_2$ content of air has hardly ever been as low as today and ecosystems suffer because of this. Early in the Earth's history, the $CO_2$ content of air was tens to hundreds of times higher than today and, over time, this $CO_2$ has been stored as carbon compounds in rocks, oil, gas, coal and carbonate rocks.

We are what we eat and, no matter what we eat, the source of the carbon ultimately derives from plants. And where did the plants get their carbon? From atmospheric $CO_2$ either directly from the atmosphere or from water. Water has dissolved $CO_2$ that derives from the atmosphere. Plants use sunlight, chlorophyll and $CO_2$. These are the essential processes of photosynthesis ticking along. We humans ingest and exhale carbon compounds, all used in keeping the human body alive. The planet has flourished at times when $CO_2$ was much higher than now and, at present, the $CO_2$ content of the atmosphere is relatively low. Because of this, commercial horticulture grows crops in an atmosphere than contains three times the current atmospheric $CO_2$ content to accelerate plant growth. The most common compounds in the Solar System are carbon compounds – there are almost 10 million different carbon compounds known. To call for lowering the carbon footprint is asinine. To refer to "carbon pollution" is ascientific political spin. To tax, ration and control the basic element for life is a micro-management of human freedom.

Carbon dioxide is only a trace gas in the Earth's atmosphere. Its current concentration is 385 ppmv (parts per million by volume) or 582 ppm by mass. A $CO_2$ measuring station at Mauna Loa (Hawaii) has measured an increase from 325 ppmv (1970) to 380 ppmv at the beginning of the 21st Century. As with all science, and especially the assumed pre-industrial value of 280 ppmv used by the IPCC, the $CO_2$ measuring methods and the reduction of measured data from Hawaii are questioned. A small seasonal variation is recorded, resulting from the consumption of $CO_2$ by Northern Hemisphere biota in summer and the release of $CO_2$ in winter. In urban areas, $CO_2$ is generally higher and indoors $CO_2$ can be over 3000 ppmv.

Using the total mass of the atmosphere,[2061] the amount of $CO_2$ in the atmosphere is estimated at 3000 billion tonnes,[2062] equivalent to 800 billion tonnes of carbon. The atmosphere is one of the smallest reservoirs of $CO_2$ on the planet. The world's oceans contain about 39,000 billion tonnes of carbon; soils, vegetation and humus contain about 2000 billion tonnes of carbon, and carbonate rocks such as limestone contain 65,000,000 billion tonnes of carbon.[2063] There is more carbon in soil than the total amount of carbon in the atmosphere and living matter.[2064] Arctic soils especially are a huge sink of carbon.[2065] The atmosphere contains only 0.001% of the total carbon present in the atmosphere-ocean-upper crust system. This figure is probably an underestimate.

There is an unknown but large amount of carbon as carbon-based compounds in sedimentary rocks such as black shales; as carbon in both known and uneconomic fossil fuels such as lignite, black coal, oil shale, oil, gas; as carbonate cementing grains together in sedimentary rocks; as carbonate alteration products of igneous rocks; as carbonate alteration products associated with the reaction of hot fluids with rocks; and as $CO_2$ gas in inclusions in mineral grains.

The figures used only deal the upper crust of the Earth and not the ultimate source of $CO_2$, the lower crust and mantle. It is $CO_2$ especially from the mantle and lower crust that has been degassed into the atmosphere since

[2061] Trenberth, K. E., Christy, J. R. and Olson, J. G. 1988: Global atmospheric mass, surface pressure, and water vapour variations. *Journal of Geophysical Research* 93D: 10925.

[2062] Atmosphere is $5.14 \times 10^{15}$ tonnes hence $CO_2$ mass is ~$3 \times 10^{12}$ tonnes.

[2063] Oelkers, E. H. and Cole, D. R. 2008: Carbon dioxide sequestration: A solution to a global problem. *Elements* 4: 305-310.

[2064] Batjes, N. H. 1996: Total carbon and nitrogen in the soils of the world. *European Journal of Soil Science* 47: 151-163.

[2065] Ping, C-L., Michaelson, G. J., Jorgenson, M. T., Kimble, J. M., Epstein, H., Romanovsky, V. E. and Walker, D. A. 2008: High stocks of soil organic carbon in the North American Arctic region. *Nature Geoscience* 1: 615-619.

the beginning of time through volcanoes. This process is still taking place and the $CO_2$ deep in the Earth far exceeds any of the surface reservoirs.[2066]

There are some hidden sources of $CO_2$ and methane. They lie in fluid inclusions in minerals. All minerals contain inclusions of solids, liquids and gas. The solids are most commonly salts, the liquids are most commonly saline water and less commonly liquid $CO_2$, and the gases are most commonly $CO_2$ and methane.[2067] Some rocks are especially rich in fluid inclusions such as quartz veins in faults. The reason why the glassy mineral quartz is milky is due to myriads of fluid inclusions. Every time there is an earthquake, a fault zone moves and fluid inclusions in quartz are opened.[2068] They release saline water, $CO_2$ and methane and the quartz recrystallises.[2069] Many fault zones after earthquakes release warm water, and $CO_2$ and methane are vented from these faults to the atmosphere.[2070] Considering that there are more than 10,000 earthquakes each year,[2071] it is surprising that this source of $CO_2$ is not even considered by the IPCC.

It is estimated that each year the atmosphere exchanges 90 billion tonnes of carbon with the surface ocean and 110 billion tonnes with vegetation, showing that the residence time of $CO_2$ in the atmosphere is less than 4 years.[2072] A quarter of all human emissions of $CO_2$ are naturally sequestered in soil each year.[2073] Terrestrial carbon of biological origin in freshwater ecosystems outgasses large quantities of $CO_2$ to the atmosphere.[2074] This means that the natural geological processes in the carbon cycle are extraordinarily rapid. It has been estimated that each year at least half the human emissions of $CO_2$ are locked up in the oceans and soils. Volcanoes produce more $CO_2$ than the world's cars and industries combined. Animals produce 25 times as much $CO_2$ as cars and industry.

---

[2066] Holland, H. D. 1978: *The chemistry of the atmosphere and oceans*. Wiley.

[2067] Shepherd, T. J., Rankin, A. H. and Alderton, D. H. M. 1985: *A practical guide to fluid inclusion studies*. Blackie.

[2068] Ramsay, J. G. 1980: The crack-seal mechanism of rock deformation. *Nature* 284: 135-139.

[2069] Sibson, R. H., Moore, J. M. M. and Rankin, A. H. 1975: Seismic pumping – a hydrothermal fluid transport mechanism. *Journal of the Geological Society of London* 131: 653-659.

[2070] Sibson, R. H. 1986: Brecciation processes in fault zone: Inferences from earthquake rupturing. *Pure and Applied Geophysics* 124: 159-175.

[2071] US Geological Survey National Earthquake Information Center, http://neic.usgs.gov/neis/epic/epic.html

[2072] Houghton, R. A. 2007: Balancing the global carbon budget. *Annual Review of Earth and Planetary Sciences* 35: 313-347.

[2073] Lal, R. 2003: Global potential of global carbon sequestration to mitigate the greenhouse effect. *Critical Reviews in Plant Science* 22: 151-184.

[2074] Battin, T. J., Kaplan, L. A., Findlay, S., Hopkinson, C. S., Marti, E., Packman, A. I., Newbold, J. D. and Sabater, F. 2008: Biophysical controls on organic carbon fluxes in fluvial networks. *Nature Geoscience* 1: 95-100.

Tropical peat lands are one of the largest surface reserves of terrestrial organic carbon.[2075] They can be up to 20 metres thick and drainage and forest clearing have made them susceptible to fire.[2076] The widespread fires through the forest peat lands and vegetation of Indonesia during the 1997–1998 El Niño event released 0.81–2.57 billion tonnes of carbon into the atmosphere. This is equivalent to 13–40% of the mean annual carbon emission from fossil fuels.[2077] Satellite studies have shown that the Mediterranean shrub lands are very sensitive to the increased $CO_2$ in the Late 20th Century Warming.[2078] Rising $CO_2$ provides protection from drought by increasing plant water use efficiency.[2079] Together with greater rainfall, this has led to an increase in vegetation cover over the past decades.

Carbon has two stable isotopes, carbon 12 ($C^{12}$) and carbon 13 ($C^{13}$). During photosynthesis, there is a preferential uptake of $C^{12}$. Therefore bacteria, plants and coals are enriched in $C^{12}$ compared to $C^{13}$. Release of $CO_2$ into the atmosphere by burning coal and hydrocarbons, decomposition of plants and farming changes the ratio of $C^{12}$ to $C^{13}$ in the atmosphere. So does the release of methane from bogs, swamps, bacteria in the gut of animals and leakage of methane from rocks. If we have an idea of the ratios of $C^{12}$ to $C^{13}$ before, during and after industrialisation, we can calculate how much $CO_2$ released into the atmosphere is from human activities. There has been a progressive increase in the $C^{12}$ to $C^{13}$ ratio in the modern atmosphere, showing that the biological contribution of carbon to the atmosphere is increasing. This can be from a great diversity of sources, including the burning of fossil fuels.

The cosmogenic isotope of carbon (carbon 14, $C^{14}$) derives from cosmic bombardment of nitrogen in the upper atmosphere and from nuclear bombs. Some 2200 nuclear devices have been triggered on Earth since 1945. In the upper atmosphere, an active Sun blows away cosmic radiation and less $C^{14}$ is produced. The $C^{14}$ content of the atmosphere fluctuates as a result of solar activity. However, because $C^{14}$ has such a short half-life,[2080] none is

---

[2075] Sorensen, A. K. W. 1993: Indonesian peat swamp forests and their role as a carbon sink. *Chemosphere* 27: 1065-1082.

[2076] Siegert, R., Rücker, G., Hinrichs, A. and Hoffmann, A. 2001: Increased fire impacts in logged over forests during El Niño driven fires. *Nature* 414: 437-440.

[2077] Page, S. E., Siegert, F., Rieley, J. O., Boehm, H-D. V., Jaya, A. and Limin, S. 2002: The amount of carbon released from peat and forest fires in Indonesia in 1997. *Nature* 420: 61-65.

[2078] Osborne, C. P., Mitchell, P. L., Sheehy, J. E. and Woodward, F. I. 2000: Modelling the recent historical impacts of atmospheric $CO_2$ and climate change on Mediterranean vegetation. *Global Change Biology* 6: 445-458.

[2079] Osborne, C. P. and Woodward, F. I. 2001: Biological mechanisms underlying recent increases in the NDVI of Mediterranean shrublands. *International Journal of Remote Sensing* 22: 1895-1907.

[2080] 5,730 years

preserved in coals which are hundreds of millions of years old. Burning of coal produces $CO_2$ with enriched $C^{12}$ and no $C^{14}$. The $C^{14}$ proportion of total carbon in the atmosphere is decreasing, suggesting that there is an increased biological contribution of $CO_2$ to the atmosphere. Isotope chemistry tells us what we know: human activities such as deforestation, coal burning, animal husbandry and cropping add $CO_2$ to the atmosphere. Biological carbon also enters the atmosphere with methane leaks from continental shelf sediments, coal seams and gas leaks and this gives the same isotope signature as $CO_2$ derived from burning coal and oil.

We can measure the amount of $CO_2$ added to the atmosphere by measuring the amount of fossil fuels burned.[2081] Most of the $CO_2$ is released in the Northern Hemisphere because of its greater population and industry,[2082] although the exact amount of material burned in the Third World is hard to calculate. Once fossil fuel is burned, the $CO_2$ enters the atmosphere[2083] and later becomes dissolved in the oceans.[2084,2085,2086] The carbon isotope content of corals,[2087] tree rings and ice cores[2088] is independent from the total amount of $CO_2$ in the atmosphere. If we humans burn fossil fuels and biomass, then the atmospheric oxygen content should be decreasing. The oxygen/nitrogen ratio shows that this is the case[2089] and is not consistent with outgassing of $CO_2$ from oceanic or crustal sources. A cautionary word: we don't really know what the crustal outgassing rate is, as it is variable over time and we currently live in a period of volcanic quiescence. If this is the case, the

[2081] Raupach, M. R., Marland, G., Ciais, P., Le Quere, C., Canadell, J. G., Klepper, G. and Field, C. B. 2007: Global and regional drivers of accelerating $CO_2$ emissions. *Proceedings of the National Academy of Sciences* 104: 10288-10293.

[2082] Marland, G. and Boden, T. 1993: The magnitude and distribution of fossil fuel related carbon releases. In: *The Global Carbon Cycle* (ed. M. Heimann), 117-138, Springer-Verlag.

[2083] Keeling, C. D. 1993: Global observations of atmospheric $CO_2$. In: *The Global Carbon Cycle* (ed. M. Heimann), 1-31, Springer-Verlag.

[2084] Quay, P. D., Tilbrook, B. and Wong, C. S. 1992: Oceanic uptake of fossil fuel $CO_2$: Carbon-13 evidence. *Science* 256: 74-79.

[2085] King, A. L. and Howard, W. R. 2004: Planktonic foraminiferal $\partial^{13}C$ records from Southern Ocean sediment traps: New estimates of the oceanic Suess effect. *Global Biogeochemical Cycles* 18: GB2007.

[2086] Quay, P., Sonnerup, R., Stutsman, J., Maurewr, J., Körtzinger, A., Padin, X. A. and Robinson, C. 2007: Anthropogenic $CO_2$ accumulation rates in the North Atlantic Ocean from changes in the $^{13}C/^{12}C$ of dissolved inorganic carbon. *Global Biogeochemical Cycles* 21: GB1009.

[2087] Druffel, E. R. M. 1997: Geochemistry of corals: Proxies of past ocean chemistry, ocean circulation, and climate. *Proceedings of the National Academy of Sciences* 94: 8354-8361.

[2088] Francey, R. J., Allison, C. E., Etheridge, D. M., Trudinger, C. M., Enting, I. G., Leuenberger, M., Langenfelds, R. L., Michel, E. and Steele, L. P. 1999: A 1000-year high precision record of $\partial^{13}C$ in atmospheric $CO_2$. *Tellus B* 51: 170-193.

[2089] Keeling, R. F., Piper, S. C. and Heimann, M. 1996: Global and hemispheric sinks deduced from changes in atmospheric $O_2$ concentration. *Nature* 381: 218-221.

more highly populated and more industrialised Northern Hemisphere emits more $CO_2$ than the Southern Hemisphere. This is exactly what is measured. Notwithstanding, various lines of evidence suggest that some of the increase in atmospheric $CO_2$ measured over the last 150 years is of human origin.

## Measurement of $CO_2$

The measurement of $CO_2$ in the atmosphere is fraught with difficulty. There is a 180-year record of atmospheric $CO_2$ measurement by the same method. It has been measured with an accuracy of 1–3% from 1812 until 1961 by a chemical method.[2090] Between 1812 and 1961, there have been more than 90,000 measurements of atmospheric $CO_2$ by the Pettenkofer method. These showed peaks in atmospheric $CO_2$ in 1825, 1857 and 1942. In 1942, the atmospheric $CO_2$ content (400 ppmv) was higher than now.[2091] A plot of the $CO_2$ measured by these methods shows that for much of the 19th Century and from 1935 to 1950, the atmospheric $CO_2$ was higher than at present and varied considerably. There are great variations in $CO_2$. A simple home experiment indoors can show that in a week, $CO_2$ can change by 75 ppmv. A variable $CO_2$ content is exactly as expected, a smooth $CO_2$ curve rings alarm bells.

In 1959, the measurement method was changed to infra-red spectroscopy with the establishment of the Mauna Loa (Hawaii) station, and measurements were compared with a reference gas sample. Compared to the Pettenkofer method, infra-red spectroscopy is simple, cheap and quick. The infra-red technique has never been validated against the Pettenkofer method. The raw data from Mauna Loa is "edited" by an operator who deletes what may be considered poor data. Some 82% of the raw infra-red $CO_2$ measurement data is "edited" leaving just 18% of the raw data measurements for statistical analysis.[2092,2093] With such a savage editing of raw data, whatever trend one wants to show can be shown. In publications, large natural variations in $CO_2$ were removed from the data by editing in order to make an upward-trending curve showing an increasing human contribution of $CO_2$.

The early Mauna Loa and South Pole $CO_2$ measurements were considerably below measurements made at the same time in northwestern Europe from 21

---

[2090] Pettenkofer method.

[2091] Beck, E. 2007: 180 years of atmospheric $CO_2$ gas analysis by chemical methods. *Energy and Environment* 18: 259-282.

[2092] Pales, J. C. and Keeling, C. D. 1965: The concentration of atmospheric carbon dioxide in Hawaii. *Journal of Geophysical Research* 70: 6053-6076.

[2093] Backastow, R., Keeling, C. D. and Whorp, T. P. 1985: Seasonal amplitude increase in atmospheric $CO_2$ concentration at Mauna Loa, Hawaii, 1959-1982. *Journal of Geophysical Research* 90: 10529-10540.

measuring station using the Pettenkofer method.[2094] During the period these 21 stations were operating (1955–1960), there was no recorded increase in atmospheric $CO_2$.[2095] There is a poor correlation between temperature and the greatly fluctuating atmospheric $CO_2$ content measured by the Pettenkofer method.

The Pettenkofer method measurements in northwestern Europe showed that $CO_2$ varied between 270 and 380 ppmv, with annual means of 315–331 ppmv. There was no tendency for rising or falling $CO_2$ levels at any one of the measuring stations over the 5-year period. Furthermore, these measurements were taken in industrial areas during post- World War II reconstruction and increasing atmospheric $CO_2$ would have been expected. While these measurements were being undertaken in northwestern Europe, a measuring station was established on top Mauna Loa in order to be far away from $CO_2$-emitting industrial areas. The volcano Mauna Loa emits large quantities of $CO_2$, as do other Hawaiian volcanoes.[2096] During a volcanic eruption, the observatory was evacuated for a few months and there was a gap in the data record which represented the period of no measurement. There are now no gaps in the Mauna Loa data set.[2097]

The annual mean $CO_2$ atmospheric content reported at Mauna Loa for 1959 was 315.93 ppmv. This was 15 ppmv lower than the 1959 measurements for measuring stations in northwestern Europe. Measured $CO_2$ at Mauna Loa increased steadily to 351.45 ppmv in early 1989.[2098] The 1989 value is the same as the European measurements 35 years earlier by the Pettenkofer method, which suggests problems with both the measurement methods and the statistical treatment of data. In fact, when the historical chemical measurements are compared with the spectroscopic measurements of air trapped in ice and modern air, there is no correlation. Furthermore,

---

[2094] Bischof, W. 1960: Periodical variations of the atmospheric $CO_2$-content in Scandinavia. *Tellus* 12: 216-226.

[2095] Bischof, W. 1962: Variations in concentration of carbon dioxide in free atmosphere. *Tellus* 14: 87-90.

[2096] Ryan, S. 1995: Quiescent outgassing of Mauna Loa Volcano 1958-1994. In: *Mauna Loa revealed: structure, composition, history and hazards* (eds Rhodes, J. M. and Lockwood, J. P.), *American Geophysical Union Monograph* 92: 92-115

[2097] Jaworowski, Z., Segalstad, T. V. and Hisdal, V. 1992: Atmospheric $CO_2$ and global warming: a critical review; 2nd Revised Edition. *Norsk Polarinstitutt Meddelelser* 119.

[2098] Keeling, C. D., Backastow, R., Carter, A. F., Piper, S. C., Whorf, T. P., Heimann, M., Mook, W. G. and Roeloffzen, H. 1989: A three-dimensional model of atmospheric $CO_2$ transport based on observed winds. 1: Analysis of observational data. In: *Aspects of climate variability in the Pacific and the Western Americas* (ed. Peterson, D. H.) *American Geophysical Union Monograph* 55: 165-236.

measurement at Mauna Loa is by infra-red analysis[2099,2100] and some of ice core measurements of $CO_2$ in trapped air were by gas chromatography.[2101]

The Mauna Loa results change daily and seasonally. Night-time decomposition of plants and photosynthesis during sunlight change the data, as does traffic and industry. Downslope winds transport $CO_2$ from distant volcanoes and increase the $CO_2$ content. Upslope winds during afternoon hours record lower $CO_2$ because of photosynthetic depletion in sugar cane field and forests. The raw data is an average of four samples from hour to hour. In 2004, there were a possible 8784 measurements. Due to instrumental errors, 1102 samples have no data, 1085 were not used because of upslope winds, 655 had large variability within one hour but were used in the official figures and 866 had large hour-by-hour variability and were not used.[2102]

The Mauna Loa $CO_2$ measurements show variations at sub-annual frequencies associated with variations in carbon sources, carbon sinks and atmospheric transport.[2103] Air that arrives during the April–June period favours a lower $CO_2$ concentration. Seasonal changes derive from Northern Hemisphere deciduous plants that take up $CO_2$ in spring and summer and release it in autumn and winter due to the decay of dead plant material. Every April, the Northern Hemisphere reduction of atmospheric $CO_2$ shows that Nature reacts quickly to $CO_2$ in the atmosphere and can remove large amounts in a very short time. This is not news. For millennia farmers have called this time the growing season.

There may be errors in sampling and analytical procedure.[2104] Measuring stations are now located around the world and in isolated coastal or island areas to measure $CO_2$ in air without contamination from life or industrial activity to establish the background $CO_2$ content of the atmosphere. The problem with these measurements is that land-derived air blowing across the sea loses about 10 ppm of its $CO_2$ as the $CO_2$ dissolves in the oceans.

[2099] Keeling, R. F., Piper, S. C. and Heimann, M. 1996: Global and hemispheric sinks deduced from changes in atmospheric $O_2$ concentration. *Nature* 381: 218-221.

[2100] Keeling, C. D. and Whorf, T. P. 2005: Atmospheric $CO_2$ records from sites in the SIO air sampling network. In: *Trends: A compendium of data on global change. Carbon Dioxide Analysis Center*, Oak Ridge National Laboratory, TN.

[2101] MacFarling Meure, C., Etheridge, D., Trudinger, C., Steele, P., Langenfelds, R., van Ommen, T., Smith, A. and Elkins, J. 2006: Law Dome $CO_2$, $CH_4$ and $N_2O$ ice core records extended to 2000 years BP. *Geophysical Research Letters* 33: L14810.

[2102] ftp://ftp.cmdl.noaa.gov/ccg/co2/in-situ

[2103] Litner, B. R., Buermann, W., Koven, C. D. and Fung, I. Y.,2006: Seasonal circulation and Mauna Loa $CO_2$ variability. *Journal of Geophysical Research* 111, D13104, 10.1029/2005JD006535.

[2104] Jaworowski, Z., Segalstad, T. V. and Hisdal, V. 1992: Atmospheric $CO_2$ and global warming: a critical review; 2nd Revised Edition. *Norsk Polarinstitutt Meddelelser* 119.

If the ocean is colder, more $CO_2$ is lost. A greater problem is that the infra-red absorption spectrum of $CO_2$ overlaps with that for $H_2O$ vapour, ozone, methane, dinitrogen oxide and CFCs.[2105] Some infra-red equipment has a cold trap to remove water vapour. However, $CO_2$ dissolves in cold water and some $CO_2$ is also removed. These other gases are detected and measured as $CO_2$. Gases such as CFCs, although at parts per billion in the atmosphere, have such a high infra-red absorption that they register as parts per million $CO_2$. Unless all these other atmospheric gases are measured at the same time as $CO_2$, then the analyses by infra-red techniques must be treated with great caution. If the Pettenkofer method was used concurrently with infra-red measurement for validation, then there could be more confidence in the infra-red results. The infra-red $CO_2$ figures are now at the level recorded by the Pettenkofer method 50 years ago. Do we really have absolute proof that $CO_2$ has risen over the last 50 years?

The IPCC's Third Assessment Report of 2001 argued that only infra-red $CO_2$ measurements can be relied upon and prior measurements can be disregarded.[2106] The atmospheric $CO_2$ measurements since 1812 do not show a steadily increasing atmospheric $CO_2$ as shown by the Mauna Loa measurements. The IPCC chose to ignore the 90,000 precise $CO_2$ measurements compiled despite the fact that there is an overlap in time between the Pettenkofer method and the infra-red method measurements at Mauna Loa. If a large body of validated historical data is to be ignored, then a well reasoned argument needs to be given. There was no explanation. Just silence.

A pre-IPCC paper used carefully selected Pettenkofer method data. Any values more than 10% above or below a baseline of 270 ppmv were rejected.[2107] The rejected data included a large number of the high values determined by chemical methods. The lowest figure measured since 1812, the 270 ppmv figure, is taken as a pre-industrialisation yardstick. The IPCC want it both ways. They are prepared to use the lowest determination by the Pettenkofer method as a yardstick yet do not acknowledge Pettenkofer method measurements showing $CO_2$ concentrations far higher than now many times since 1812.

---

[2105] Briegleb, B. P. 1992: Longwave band model for thermal radiation in climate studies. *Journal of Geophysical Research* 97: 11475-11485.

[2106] IPCC, 2001: *Climate Change 2001. The scientific basis. Contributions of working group 1 to the Third Assessment Report of the Intergovernmental Panel on Climate Change* (eds Houghton, J. T., Jenkins, G. J. and Ephraums, J. J.), Cambridge University Press.

[2107] Callender, G. S. 1938: The artificial production of carbon dioxide and its influence on temperature. *Quarterly Journal of the Royal Meteorological Society* 66: 395-400.

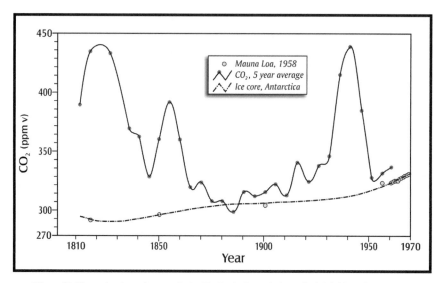

*Figure 52: Determinations of atmospheric CO$_2$ by the Pettenkofer method (solid line of 5 year averages) between 1812 and 1961, deductions of atmospheric CO$_2$ from Antarctic ice core (gas chromatography) and edited measurements of atmospheric CO$_2$ from Mauna Loa (infra-red spectroscopy, 1958 and onwards). One method of measurement shows great variability in atmospheric CO$_2$ yet another method does not. The high values of CO$_2$ by the Pettenkofer method have been rejected by the IPCC yet the lowest value is used by the IPCC as the baseline pre-industrial value for atmospheric CO$_2$.*

The Mauna Loa observatory has the longest continuous measurement of atmospheric CO$_2$ at the one site by infra-red spectroscopy. It is considered to be a precise measurement of middle tropospheric CO$_2$ because of the minimal influence of human or vegetation effects, the minimal influence of volcanism and because there has been nearly 50 years of measurement using the same technique.[2108] The Mauna Loa record shows a 19.4% increase in mean annual concentration of CO$_2$ from 315.98 ppmv of dry air in 1959 to 377.38 ppmv in 2004. The biggest single jump of 2.87 ppmv was in the El Niño year 1997–1998, in accord with other increases associated with El Niño events.[2109]

Ocean degassing is a major source of atmospheric CO$_2$. Cold seawater surface dissolves CO$_2$; when the seawater is at the tropics it releases CO$_2$.

[2108] Keeling, C. D., Bacastow, R. B., Bainbridge, A. E., Ekdahl, C. A., Guenther, P. R., Waterman, L. S. and China, J. F. S. 1976: Atmospheric carbon dioxide variations at Mauna Loa Observatory, Hawaii. *Tellus* 28: 538-551.

[2109] Bacastow, R. B., Adams, J. A., Keeling, C. D., Moss, D. J., Whorf, T. P. and Wong, T. S. 1980: Atmospheric carbon dioxide, the Southern Oscillation, and the weak 1975 El Niño. *Science* 210: 66-68.

It is only the surface seawater that is releasing $CO_2$. Degassing of the oceans is taking place in the tropical waters around Hawaii. Deep ocean water is undersaturated in $CO_2$ and can dissolve huge amounts of $CO_2$ from submarine volcanoes. At sites of upwelling, this $CO_2$ is released to the atmosphere. Increased solar activity warms the surface of the oceans, especially in the tropics. This increases the amount of water vapour and $CO_2$ in the atmosphere and reduces the uptake of $CO_2$ by the oceans. As a result, some of the $CO_2$ that has been attributed to human activity is of solar origin.[2110,2111] The release of more water vapour and $CO_2$ into the atmosphere may have contributed to the Late 20th Century Warming in addition to warming from increased solar output and variations in the Sun's electromagnetic and gravitational fields.

If each $CO_2$ molecule in the atmosphere has a short lifetime, it means that the $CO_2$ molecules will be removed fast from the atmosphere to be adsorbed in another reservoir. But how fast? Because atmospheric $CO_2$ is increasing, it was argued that $CO_2$ has not been dissolved in the sea and must have an atmospheric lifetime of several hundred years.[2112] The IPCC suggests that the lifetime is 50–200 years.[2113] The IPCC lifetime has been criticised because lifetime is not defined[2114] and because the IPCC has not factored in numerous known sinks of $CO_2$.[2115,2116] The $CO_2$ atmospheric lifetime can be calculated by measuring the amount of $C^{12}$, $C^{13}$ and $C^{14}$ in atmospheric $CO_2$. From this, the $CO_2$ lifetime can be calculated. This can be double-checked by measuring the amount of the inert but radioactive gas radon ($Rn^{222}$), the solubility of $CO_2$ and various complicated carbon isotope calculations.

[2110] Scafetta, N. and West, B. J. 2006a: Phenomenological solar signature in 400 years of reconstructed Northern Hemisphere temperature record. *Geophysical Research Letters* 33: L17718 doi:10.1029/2006GL027142.

[2111] Scafetta, N. and West, B. L. 2006b: Phenomenological solar contribution to the 1900-2000 global surface warming. *Geophysical Research Letters* 33: L05708 doi:10.1029/2005GL025539.

[2112] Rodhe, H. 1992: Modeling biogeochemical cycles. In: *Global biogeochemical cycles* (eds Butcher, S. S., Charlson, R. J., Orians, G. H. and Wolfe, G. V.), 55-72, Academic Press.

[2113] Houghton, J. T., Jenkins, G. J. and Ephraums, J. J. (eds) 1990: *Climate Change. The IPCC Assessment. Intergovernmental Panel on Climate Change.* Cambridge University Press.

[2114] O'Neill, B. C., Gaffin, S. R., Tubiello, F. N. and Oppenheimer, M. 1994: Reservoir timescales for anthropogenic $CO_2$ in the atmosphere. *Tellus* 46B: 378-389.

[2115] Jaworowski, Z., Segalstad, T. V. and Hisdal, V. 1992: Atmospheric $CO_2$ and global warming: a critical review; 2nd Revised Edition. *Norsk Polarinstitutt Meddelelser* 119.

[2116] Segalstad, T. V. 1996: The distribution of $CO_2$ between atmosphere, hydrosphere and lithosphere; minimal influence from anthropogenic $CO_2$ on the global "Greenhouse Effect". *The Global Warming Debate. The Report of the European Science and Environment Forum.* (ed. Emsley, J.), 41-50, Bourne Press.

Calculations[2117] of the lifetime of atmospheric $CO_2$ based on natural $C^{14}$ give lifetime values of 3 to 25 years (18 separate studies), dilution of atmosphere from fossil fuel burning a lifetime of 2 to 7 years (two separate studies), atomic bomb $C^{14}$ lifetime value of 2 to more than 10 years (12 separate studies), measurements of $Rn^{222}$ give $CO_2$ atmospheric lifetime of 7.8 to 13.2 years (three separate studies), $CO_2$ solubility gives an atmospheric lifetime of 5.4 years[2118] and $C^{12}$ to $C^{13}$ mass balance value for the lifetime as 5.4 years.[2119] There is very little disagreement. The lifetime of $CO_2$ in the atmosphere is about 5 years, a number previously acknowledged by the former IPCC chairman Bert Bolin.[2120] The short atmospheric lifetime of $CO_2$ means that about 18% of the atmospheric $CO_2$ pool is exchanged each year. Those that argue that the measured increase of $CO_2$ in the atmosphere at Mauna Loa must be due to man's burning of fossil fuels have modelled that the lifetime of this $CO_2$ in the atmosphere is 50–200 years.[2121]

There is a considerable difference in the atmospheric $CO_2$ lifetime between the 37 independent measurements and calculations using six different methods and the IPCC computer model. This discrepancy has not been explained by the IPCC. Why is this important? If the $CO_2$ atmospheric lifetime were 5 years, then the amount of the total atmospheric $CO_2$ derived from fossil fuel burning would be 1.2%,[2122] not the 21% assumed by the IPCC. In order to make the measurements of the atmospheric $CO_2$ lifetime agree with the IPCC assumption, it would be necessary to mix all the $CO_2$ derived from the world's fossil fuel burning with a different $CO_2$ reservoir that was five times larger than the atmosphere.[2123]

There have been attempts to explain this discrepancy. These involve

---

[2117] Sundquist, E. T. 1985: Geological perspectives on carbon dioxide and the carbon cycle. In: *The carbon cycle and atmospheric CO2: natural variations Archean to present* (eds Sundquist, E. T. and Broecker, W. S.). *American Geophysical Union Monograph* 32: 5-59.

[2118] Murray, J. W. 1992: The oceans. In: *Global Biogeochemical Cycles* (eds Butcher, S. S., Charlson, R. J., Orians, G. H. and Wolfe, G. V.), 175-211, Academic Press.

[2119] Segelstad, T. V. 1992: The amount of non-fossil-fuel $CO_2$ in the atmosphere. *American Geophysical Union, Chapman Conference on Climate, Volcanism and Global Change, March 23-27, 1992, Hilo, Hawaii, Abstracts* 25.

[2120] Bolin, B. and Eriksson, E. 1959: Changes in the carbon dioxide content in the atmosphere and sea due to fossil fuel combustion. In: *The atmosphere and sea in motion. Scientific contributions to the Rossby Memorial Volume.* The Rockefeller Institute Press: 130-142.

[2121] IPCC, 2001: *Climate Change 2001. The scientific basis. Contributions of working group 1 to the Third Assessment Report of the Intergovernmental Panel on Climate Change* (eds Houghton, J. T., Jenkins, G. J. and Ephraums, J. J.), Cambridge University Press.

[2122] Revelle, R. and Suess, H. 1957: Carbon dioxide exchange between atmosphere and ocean and the question of an increase in atmospheric $CO_2$ during past decades. *Tellus* 9: 18-27.

[2123] Broecker, W. S., Takahashi, T., Simpson, H. J. and Peng, T.-H. 1979: Fate of fossil fuel carbon dioxide and the global carbon balance. *Science* 206: 409-418.

speculations about unmeasured carbon isotope behaviour[2124] and ignore chemical and isotopic experiments that show equilibrium between $CO_2$ and water within a few hours.[2125,2126] Furthermore, the proportion of $C^{12}$ and $C^{13}$ measured in atmospheric $CO_2$ is substantially different from that used in the IPCC model.[2127] The IPCC model on how long $CO_2$ resides in the atmosphere and the amount of fossil fuel $CO_2$ in the atmosphere is not supported by radioactive or stable carbon isotope evidence and hence the basic assumptions made by the IPCC are incorrect. Maybe the IPCC's troubles derive from concerns that ice cores do not give reliable data for past atmospheres, including the pre-Industrial Revolution atmosphere. Maybe the IPCC's troubles derive from the fact that recent atmospheric $CO_2$ measurements have been by a non-validated instrumental method[2128] where results were visually selected and "edited", deviating from unselected measurements of constant $CO_2$ levels by the highly accurate wet chemical methods at 19 stations in Europe.[2129]

The contribution of fossil fuel $CO_2$ to the atmosphere, and the lifetime of this $CO_2$ in the atmosphere, are the cornerstone of the whole IPCC *raison d'être*. There is now no *raison d'être*. For the atmospheric $CO_2$ budget, marine adsorption and degassing, $CO_2$ from geological sources (volcanic degassing, metamorphism, mountain building, faulting), $CO_2$ loss from weathering and $CO_2$ adsorption by micro-organisms must be far more important than assumed by the IPCC. The total of the $CO_2$ released from fossil fuel burning and biogenic releases (4% atmospheric $CO_2$) is far less important than the 21% of atmospheric $CO_2$ assumed by the IPCC.

The $CO_2$ content of the atmosphere is ultimately determined by geological processes. Over the last 4567 million years, the Earth has degassed about

---

[2124] Oeschger, H. and Siegenthaler, U. 1978: The dynamics of the carbon cycle as revealed by isotope studies. In: *Carbon dioxide, climate and society* (ed. Williams, J.), 45-61, Pergamon Press.

[2125] Inoue, H. and Sugimura, Y. 1985: Carbon isotopic fractionation during the $CO_2$ exchange process between air and seawater under equilibrium and kinetic conditions. *Geochimica et Cosmochimica Acta* 49: 2453-2460.

[2126] Dreybrodt, W., Lauckner, J., Zaihua, L., Svensson, U. and Buhmann, D. 1996: The kinetics of the reaction $CO_2 + H_2O \rightarrow H^+ + HCO_3^-$ as one of the rate limiting steps in the system $H_2O$-$CO_2$-$CaCO_3$. *Geochimica et Cosmochimica Acta* 60: 3375-3381.

[2127] Keeling, C. D., Bacastow, R. B., Carter, A. F., Piper, S. C., Whorf, T. P., Heimann, M., Mook, W. G. and Roeloffzen, H. 1989: A three-dimensional model of atmospheric $CO_2$ transport based on observed winds: 1. Analysis of observational data. In: *Aspects of climate variability in the Pacific and the Western Americas* (ed. Peterson, D. H.). *American Geophysical Union Monograph* 55: 165-236.

[2128] Jaworowski, Z., Segalstad, T. V. and Hisdal, V. 1992: Atmospheric $CO_2$ and global warming: a critical review; 2nd Revised Edition. *Norsk Polarinstitutt Meddelelser* 119.

[2129] Bischof, W. 1960: Periodical variations of the atmospheric $CO_2$-content in Scandinavia. *Tellus* 12: 216-226.

half of its estimated $CO_2$ by geological processes.[2130] This $CO_2$ has not been lost to space, it is stored in rocks (such as limestone) and life. The balance between degassing $CO_2$ from the Earth's interior and weathering (i.e. atmosphere-biosphere-hydrosphere-lithosphere system), carbonate sedimentation and biological carbon determines the atmospheric $CO_2$ content. Seawater alkalinity is controlled by $CO_2$ uptake during weathering, and inorganic and biological carbonate deposition in the oceans are major $CO_2$ sinks. These sinks have kept the oceans alkaline for billions of years. Warmer climates accelerate weathering thereby accelerating the uptake of atmospheric $CO_2$. The atmospheric $CO_2$ content is controlled by climate and it appears unlikely that atmospheric $CO_2$ drives climate.[2131] But we knew this anyway from the 800-year lag of increased $CO_2$ following temperature rise as measured in ice cores.

The key assertions in the global warming claim is that $CO_2$ has increased approximately 35% over the last 150 years. This assertion is challenged. The release of $CO_2$ from cooling molten rocks is unmeasured (e.g. Kamchatka), the release of $CO_2$ from bacteria in the first 4 kilometres of the Earth's crust is unknown, the release of $CO_2$ from uplift of mountain ranges and alps is unknown and the release of $CO_2$ from submarine volcanoes is unknown. Most of this $CO_2$ has a biological signature and hence cannot be differentiated from $CO_2$ derived from the burning of fossil fuels.

If cold ocean water dissolves atmospheric $CO_2$ or dissolves volcanic $CO_2$ deep in the oceans, this $CO_2$ is not released for hundreds to thousands of years later. If atmospheric $CO_2$ has increased over the last 150 years, both the lag processes and all processes of release of $CO_2$ into the atmosphere need to be critically evaluated. Cold $CO_2$-bearing water is carried to the tropics by deep currents. When this water rises in upwelling or is heated in the tropics, it releases $CO_2$. The release of $CO_2$ into the atmosphere from the oceans tells us about processes that took place hundreds to thousands of years ago and provides no information about modern processes.

So, what does the IPCC say about $CO_2$? There are many papers that use the same data set as the IPCC and derive a different conclusion. For example, studies in China concluded that global warming is not related only to $CO_2$ and that the effect of $CO_2$ on temperature is grossly exaggerated.[2132] An assertion by the IPCC is that global atmospheric temperature has risen

---

[2130] Holland, H. 1984: *The chemical evolution of the atmosphere and oceans*. Princeton University Press.

[2131] Kondratyev, K. Y. 1988: *Climate shocks: natural and anthropogenic*. John Wiley.

[2132] Zhen-Shan, L. and Xian, S. 2007: Multi-scale analysis of global temperature changes and trend of a drop in temperature in the next 20 years. *Meteorology and Atmospheric Physics* 95: 115-121.

by 0.7°C since 1850. The bulk of the global temperature rise was before massive industrialisation (1850 to 1940), then during the post-World War II economic boom temperature decreased while the industrial emissions of $CO_2$ greatly increased, and from 1976 to 1998 temperature increased. It has been static since 1998. Furthermore, there are uncertainties in thermometer measurements and the thermometer measurements are not in accord with balloon and satellite temperature measurements. In addition, the Little Ice Age ended in 1850, so naturally temperatures rose. The imputation is that this temperature rise is due to human activity. The post-1850 temperature rise cannot be related to post-1850 industrialisation and must be related to natural processes. The post-1940 increase in atmospheric $CO_2$ may well be due to the release of $CO_2$ from the oceans lagging behind the solar-driven heating from 1850 to 1940 or even the Medieval Warming or Roman Warming.

Another assertion of the IPCC is that $CO_2$ is a greenhouse gas. This assertion does not acknowledge that $H_2O$ is the main greenhouse gas, that $CO_2$ derived from human activity produces 0.1% of global warming and that there is a maximum threshold for $CO_2$, after which an increase in $CO_2$ has very little effect on atmosphere warming. The current $CO_2$ content of the atmosphere is the lowest it has been for thousands of millions of years, and life (including human life) has thrived at times when $CO_2$ has been significantly higher.[2133]

The IPCC 2007 report stated that the $CO_2$ radiative forcing had increased 20% during the last 10 years. Radiative forcing puts a number on increases in radiative energy in the atmosphere and hence the temperature. In 1995, there was 360 ppmv of $CO_2$ in the atmosphere, whereas in 2005 it was 378 ppmv, some 5% higher. However, each additional molecule of $CO_2$ in the atmosphere causes a smaller radiant energy increase than its predecessor and the real increase in radiative forcing was 1%. The IPCC have exaggerated the effect of $CO_2$ 20-fold.

During times of ice ages such as 140,000 years ago, the $CO_2$ content of the atmosphere was higher than the pre-industrial revolution figure of 270 ppmv.[2134] It is clear that $CO_2$ is not the only factor that controls air temperature, otherwise we could not have ice age conditions with a high atmospheric $CO_2$ content. The transition from a global ice age to global warming at about 250 Ma was characterised by huge rises (to 2000 ppmv) and

[2133] deFreitas, C. R. 2002: Are observed changes in the concentration of carbon dioxide in the atmosphere really dangerous? *Bulletin of Canadian Petroleum Geology* 50: 297-327.

[2134] Lorius, C., Jouzel, J., Raynaud, D., Hansen, J. and Le Treut, H. 1990: The ice-core record: climate sensitivity and future greenhouse warming. *Nature* 347: 139-145.

falls (to 280 ppmv) in the amount of atmospheric $CO_2$.[2135] During this time plant and animal life thrived. If $CO_2$ was not recycled and humans burned all the known fossil fuels on Earth, then the atmospheric $CO_2$ content would be 2000 ppmv.

Over geological time, a correlation between $CO_2$ and temperature over coarse (10 million-year time scales) and fine (million-year time scales) resolutions shows that $CO_2$ operates together with many factors including solar luminosity, tectonics and palaeogeography. Nearly 500 proxy records of $CO_2$ over the last 500 million years show that all cool events are associated with $CO_2$ levels below 1000 ppmv and when $CO_2$ is less than 500 ppmv, then this is the recipe for widespread continental glaciation.[2136] This is contrary to other studies that show that atmospheric $CO_2$ in previous cold times was more than 4000 ppmv in the Ordovician-Silurian (450–420 Ma) glaciation and about 2000 ppmv in the Jurassic-Cretaceous (151–132 Ma) glaciation.[2137]

It was argued recently that the current glaciation in Greenland was driven by a low atmospheric $CO_2$ content.[2138] Various trigger mechanisms for the glaciation were tested (orbital, closing the Panama seaway, permanent El Niño, Himalayan uplift) using modelling software.[2139] This is the same software that attempts to predict future climates on the basis that $CO_2$ drives climate. This result is not surprising as the modelling is programmed to produce such a result. What is not discussed is how the atmospheric $CO_2$ fell and the role of extraterrestrial and solar activity.

The current content of $CO_2$ in the atmosphere is 385 ppmv. This is commensurate with the modern oscillating climate comprising 90,000 years of continental glaciation and 10,000 years of interglacial warmth. The variation in $CO_2$ shows that a climate sensitivity of greater than 1.5°C has probably been the most robust feature of the Earth's climate system over the last 420 million years.[2140]

The driver of climate affects the climate equilibrium and the temperature changes accordingly. As a result of a higher temperature, more $CO_2$ is outgassed

[2135] Montanez, I. P., Tabor, N. J., Niemeier, D., DiMichele, W. A., Frank, T. D., Fielding, C. R., Isbell, J. L., Birgenheier, L. P. and Rygel, M. C. 2007: $CO_2$-forced climate and vegetation instability during Late Paleozoic deglaciation. *Science* 315: 87-91.

[2136] Royer, D. L. 2006: $CO_2$-forced climate thresholds during the Phanerozoic. *Geochimica et Cosmochemica Acta* 70: 5665-5675.

[2137] Berner, R. A. and Kothavala, Z. 2001: Geocarb III: A revised model of atmospheric $CO_2$ over Phanerozoic time. *American Journal of Science* 301: 182-204.

[2138] Lunt, D. J., Forster, G. L., Haywood, A. M. and Stone, E. J. 2008: Late Pliocene Greenland glaciation controlled by a decline in atmospheric $CO_2$ levels. *Nature* 454: 1102-1105.

[2139] HadCM3.

[2140] Royer, D. L., Berner, R. A. and Park, J. 2007: Climate sensitivity constrained by $CO_2$ concentrations over the past 420 million years. *Nature* 446: 530-532.

from the oceans. These matters are complicated and equilibrium calculations are difficult because of the large number of unknown factors. For example, at low temperatures such as during glaciation, the stronger winds bring more dust into the ocean basins. This dust contains iron and other elements and this stimulates an increase in biological activity, especially in those parts of the ocean that contain limited nutrients. This biological activity removes $CO_2$ from the atmosphere, yet there has not been a runaway glaciation. In the scheme of things, $CO_2$ plays second fiddle in global climate.

Photosynthesis is rapid and there are annual cycles where more $CO_2$ is consumed in the warm sunny summer months than in winter. There are also lags in the return of $CO_2$ to the atmosphere if organic material produced by photosynthesis is buried in sediments, peat or limey rocks. This prevents conversion of carbon compounds to $CO_2$ and methane that are released to the atmosphere. Burial of carbon compounds was accelerated at about 400 Ma after the evolution of terrestrial vascular plants. Forests grew quickly, there was a removal of $CO_2$ from the atmosphere and carbon was not recycled as atmospheric $CO_2$ because it was buried as coals, carbonaceous sediments, limey sediments and limestone reefs. There were times, such as the Carboniferous, when there was an explosion of plant life on Earth. There was a massive removal of $CO_2$ from the atmosphere, further oxygenation and storage of recycled carbon in Northern Hemisphere coals. The removal of $CO_2$ from the atmosphere was immediately before the Permo-Carboniferous glaciation and may well have been one of the factors that set the stage for an ice age. At that time, the supercontinent Gondwana was at the South Pole and the Permo-Carboniferous ice sheets were at high latitudes.

If the Permo-Carboniferous glaciation was influenced by a low atmospheric $CO_2$ content, then what about the Pleistocene glaciation which we now enjoy? During the break-up of the supercontinent Gondwana over the last 100 million years, India drifted northwards and collided with Asia 50 million years ago. The Himalayas were pushed up. India is still pushing against Asia and the Himalayas are still rising. Accumulation of high-altitude snow and ice increased the amount of reflected solar energy. The area of the Himalayas is about half that of the USA so the feedback effect of reflected solar energy is large. The size and height of the Tibetan Plateau changed global wind patterns, resulting in regional changes to climate. The great difference between winter and summer temperatures shatters rocks. This produces pieces of rock with a large surface area for attack by rain and micro-organisms. The huge Tibetan Plateau heats in summer, the air above it heats and rises and cooler moist air from the tropical Indian Ocean is dragged up to the plateau. This results in monsoonal rains that attack the rocks to form soils, a process that removes $H_2O$, oxygen and $CO_2$ from the atmosphere and adds salts and bicarbonate to the oceans. Soils are stripped from the steep slopes in periods of high rainfall and the process starts all over again. The thick pile of sediments at the Ganges Delta in the Bay of Bengal shows that this process of $CO_2$ removal from the atmosphere has been taking place for 50 million years. Some 15 million years after this process started, ice appeared on Antarctica.[2141]

The measurement of $CO_2$ by two different methods is unresolved. The residency of $CO_2$ in the atmosphere is far less than that used by the IPCC and this greatly affects the estimation of the amount of $CO_2$ produced by humans. Not all sources and sinks of $CO_2$ are considered by the IPCC. The calculation of the transfer of $CO_2$ between the atmosphere and the oceans uses a body of incomplete data. By using various general circulation models, computer models are compared and for some strange reason agree with other computer models which used the same data set.[2142] There is clearly a lot more to learn about $CO_2$.

A final word on measurement. The methods of measurement and data treatment for sea surface temperature, air temperature and $CO_2$ have errors, bias, lack of correlation between different methods, lack of validation, selective culling of data and uncertainties. If you want to measure something to obtain a predestined conclusion, it is easy. For example, if you want to show that the Moon is made of green cheese, then the seismic velocity of a Gjetost (a Norwegian green cheese) is the same as lunar rock 10017. Because seismic velocity is a measure of the composition of a material, the Moon must be made of green cheese. In fact, to be more precise, Norwegian green cheese. Science proves it.

*Methane and other greenhouse gases*

Methane ($CH_4$) is emitted from cattle, rice paddies, termites, decomposing life, hydrates, petroleum leaks, rocks, mid ocean ridges and sedimentation. One can only speculate as to why political activists concentrate their attention on $CO_2$ rather than methane. It may be because $CO_2$ is linked to industrial growth whereas methane is considered more "natural" and emitted by less developed nations.[2143]

---

[2141] Raymo, M. E. and Ruddiman, W. F. 1992: Tectonic forcing of late Cenozoic climate. *Nature* 359: 117-122.

[2142] Le Quere, C., Aumont, O., Bopp, L., Bousquet, P., Ciais, P., Francey, R., Heimann, M., Keeling, C. D., Keeling, R. F., Kheshgi, H., Peylin, P., Piper, S. C., Prentice, I. C. and Rayner, P. J. 2003: Two decades of ocean $CO_2$ sink and variability. *Tellus B* 55: 649-656.

[2143] Singer, S. F. 2008: *Nature, not human activity, rules the climate*. Heartland Institute.

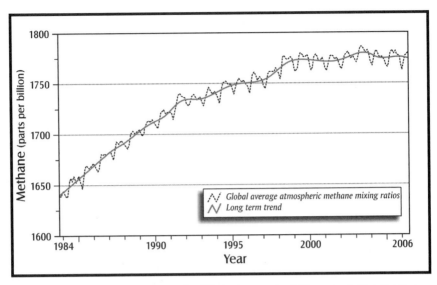

*Figure 53: Changes in atmospheric methane[2144] showing annual and 25-year trends. Note that the long-term trend has flattened since about 1998 despite an increase in methane-emitting domesticated animals hence there are unaccounted sources of methane unrelated to human activities.*

Methane, although in trace amounts in the atmosphere, is weight for weight a far more powerful greenhouse gas than $CO_2$ or water vapour. However, methane is highly reactive, hence has a limited life in the atmosphere. Some publications show global methane content decreasing, others show it increasing.[2145] The reasons for the contrast and changes in methane content of the atmosphere are unknown. If we look at emissions of $CO_2$ and methane from humans and their livestock, it appears that livestock are responsible for more emissions than all the cars, planes and other forms of transport put together.[2146] Burning fuel to produce fertiliser to grow feed, produce meat and to transport meat and clearing of vegetation for grazing produces 9% of all human and livestock $CO_2$ emissions, and flatulence and manure fumes produce massive quantities of methane. Methane from cattle, termites and other animals is generated by bacteria in the gut. Bacteria in bogs, swamps, soil and rocks also emit huge quantities of methane. Because

---

[2144] Carbon Cycle Co-operative air sample network data.

[2145] Rigby, M., Prinn, R. G., Fraser, P. J., Simmonds, P. G., Langenfelds, R. L., Huang, J., Cunnold, D. M., Steele, L. P., Krummel, P. B., Weiss, R. F., O'Doherty, S., Salameh, P. K., Wang, H. J., Harth, C. M., Mühle, J. and Porter, L. W. 2008: Renewed growth of atmospheric methane. *Geophysical Research Letters* 35: L22805, doi: 10.1029/2008GL036037.

[2146] UN Food and Agricultural Organisation 2006: Livestock's Long Shadow.

of the short life of ruminants, termites and gut bacteria, biomass burning, ruminants and bacteria put $C^{14}$ back into the atmosphere, hence calculations of the amount of carbon from fossil fuel burning and the residence time of $CO_2$ in the atmosphere can only be maximum figures.

Livestock produce about 100 different gases which are regarded as pollutants, overgrazing turns 20% of pastures into desert, and some 500 litres of water are needed to produce 1 litre of milk. Wastes from feedlots and fertilisers over-nourish waterways, causing weeds to choke all other life, pesticides, antibiotics and hormones can enter drinking water, and dead zones in coral reefs can be caused by waste material from cattle farming. The average cow releases 150 to 250 litres a day of methane, a greenhouse gas 20 times more intense than $CO_2$. If we drive a diesel car with an efficiency of 5.5 litres per 100 km, then the daily greenhouse effect of one cow is equivalent to driving 10,000 km in this car. And then there are termites, an even greater producer of methane.

We are not sure where the world's methane comes from. For example, Lake Kivu in the East African Rift gives a clue to the difficulties with methane. Huge amounts of $CO_2$ and methane are emitted from Rift Valley volcanoes and hot springs. These emissions are sporadic, so measurement of gas emissions over years or decades is unable to give the total methane losses over time. If $CO_2$ released from a volcano reacts with hydrogen, another common volcanic gas, then methane is formed. Some 70% of the methane in Lake Kivu derives from this process, the rest from bacterial fermentation of acetate in sediments.[2147] The reduction of methane in the Earth's atmosphere is decreasing yet the number of methane-emitting animals such as cattle, sheep and goats is increasing. To use methane in any climate model is dangerous because we know so little about it.

Wetlands are a large source of methane. Northern Hemisphere sub-Arctic wetlands are a significant source but the relative proportion of methane from high and low latitude wetlands is unknown.[2148] Hourly methane measurements from autumn into winter freezing of soils shows that emissions fall after the growing season, and when the ground is frozen there are sporadic large bursts of methane.[2149] Craters in sediments on the sea floor, especially in

[2147] Tietze, K., Geyh, M., Müller, H., Schröder, L., Stahl, W. and Wehner, H. 1980: The genesis of methane in Lake Kivu (Central Africa). *International Journal of Earth Sciences* 69: 452-472.

[2148] Mikaloff Fletcher, S. E., Tans, P. P., Bruhwiler, L. M., Miller, J. B. and Heimann, M. 2004: $CH_4$ sources estimated from atmospheric observations of $CH_4$ and $C^{13}/C^{12}$ isotopic ratios: 1. Inverse modeling of source processes. *Global Biogeochemical Cycles* 18: GB4004.

[2149] Mastepanov, M., Sigsgaard, C., Dlugokencky, E. J., Houweling, S., Ström, L., Tamstorf, M. P. and Christensen, T. R. 2008: Large tundra methane burst during onset of freezing. *Nature* 456: 628-631.

high latitude areas, are probably due to the decomposition of methane hydrate and the catastrophic release of methane and $H_2O$. Carbon in the methane hydrates derives from bacteria and it has a characteristic proportion of $C^{12}$ to $C^{13}$. Methane hydrate changes to methane beneath the sea floor and the methane is oxidised in seawater to $CO_2$ and $H_2O$. The carbon in the $CO_2$ retains this distinct proportion of $C^{12}$ to $C^{13}$ of microscopic marine life. When this marine life dies, it falls to the sea floor to be fossilised in sediments. The $C^{12}:C^{13}$ signature is retained. Continental shelf sediments show a series of distinctive $C^{12}:C^{13}$ spikes showing that methane has been suddenly injected into seawater many times. This process may well take place today. If there is a significant injection of methane into the ocean, the water is charged with gas and loses density. Some rather odd instantaneous sinkings of North Sea fishing trawlers may result from methane outbursts creating a sudden loss of buoyancy.

There are huge emissions of methane from life.[2150] Ruminants contain methanogenic bacteria in their gut which produce large volumes of methane from fermentation.[2151] Some methane emissions are from domesticated animals,[2152] especially cattle.[2153] Fermentation also takes place in dung, producing more methane.[2154] Many terrestrial arthropods such as termites, cockroaches, millipedes and scarab beetles have a hind gut that is colonised by methanogenic bacteria. These animals emit a huge but unknown amount of methane to the atmosphere.[2155] There have been some attempts to calculate the amount of methane emitted by termites.[2156] The bacteria that emit methane are one of the oldest life forms on Earth, are a unique biological group[2157] and exist as a major life form in the rocks beneath our feet.

Nature just keeps on ignoring humans. The 1987 Montreal Protocol

---

[2150] Lerner, J., Matthews, E. and Fung, I. 1988: Methane emissions from animals: A global high-resolution database. *Global Biogeochemical Cycles* 2: 139-156.

[2151] Bryant, M. P. 1979: Microbial methane production – theoretical aspects. *Journal of Animal Science* 48: 193-201.

[2152] Johnson, D. E. and Ward, G. M. 1996: Estimates of animal methane emissions. *Environmental Monitoring and Assessment* 42: 133-141.

[2153] Johnson, K. A. and Johnson, D. E. 1995: Methane emissions from cattle. *Journal of Animal Science* 73: 2483-2492.

[2154] Varel, V. H., Isaacson, H, R. and Bryant, M. P. 1977: Thermophilic methane production from cattle waste. *Applied Environmental Microbiology* 33: 298-307.

[2155] Hackstein, J. H. and Stumm, C. K. 1994: Methane production in terrestrial arthropods. *Proceedings of the National Academy of Sciences* 91: 5441-5445.

[2156] Zimmerman, P. R., Greenberg, J. P., Wandiga, S. O. and Crutzen, P. J. 1982: Termites: A potentially large source of atmospheric methane, carbon dioxide, and molecular hydrogen. *Science* 218: 563-565.

[2157] Balch, W. E., Fox, G. E., Magrum, L. J., Woese, C. R. and Wolfe, R. S. 1979: Methanogens: reevaluation of a unique biological group. *Microbiological and Molecular Biological Reviews* 43: 260-296.

stopped the production and consumption of most ozone-destroying chemicals. Many will linger in the atmosphere for decades. The method of breakdown and time scale for breakdown depends upon the wavelength of light the molecule can adsorb. One of the critical molecules, dichlorine peroxide, appears to break down far slower than was though.[2158,2159,2160] This means it is not possible to generate enough breakdown to explain the observed loss of ozone. Everything used to fit nicely. Now it is back to the drawing board. This is the nature of science.

## Clouds

Climate science is baffled by clouds. Not enough is understood yet about clouds to state whether they have a cooling or a warming effect on the Earth. A change in the cloud cover by 1% can produce changes as great as the most exaggerated claims that are meant to derive from humans adding $CO_2$ to the atmosphere. Climate models do not do clouds well and don't consider fog and mist, which have the same effect as clouds. The IPCC notes that water vapour and especially clouds are major sources of uncertainty in their models. And yet the IPCC dismisses the role of cosmic radiation creating low-level clouds, and cosmic rays are one of the mechanisms of generating low-level clouds. Concurrently, the IPCC promotes conclusions based on complex computer models that use dubious assumptions, incomplete data and a poor understanding of how the planet operates.

The French-US satellite (Calipso) and the NASA CloudSat are currently looking at and through clouds to try to increase the accuracy of weather models. Climate models regard clouds as passive and atmospheric $CO_2$ as active. As for using cloudless models or models with an incomplete understanding of clouds to predict climate 100 years into the future, forget it. The influence of clouds is easy to test. Walk outside. It is cooler on a cloudy day because of the reflection of incoming solar energy back into space. On a cloudy tropical night, it is warm because heat radiated from the Earth back to space is captured by water vapour. At the same latitude in the desert on a cloudless night, it is very cold because of the lack of water vapour in the atmosphere to capture the back-radiated heat. High altitude thin clouds can warm the Earth. For example, the feathery cirrus clouds are at about -40°C

[2158] Pope, F. D., Hansen, J. C., Bayes, K. D., Friedl, R. R. and Sander, S. P. 2007: Ultraviolet absorption spectrum of chlorine peroxide, ClOOCl. *Journal of Physical Chemistry* A111: 4322-4332.

[2159] Molina, L. T and Molina, M. J. 1987: Production of $Cl_2O_2$ from the self-reaction of the ClO radical. *Journal of Physical Chemistry* 91: 433-436.

[2160] Farman, J. C., Gardiner, B. G. and Shanklin, J. D. 1985: Large losses of total ozone in Antarctica reveal seasonal ClOx/Nox interaction. *Nature* 315: 207-210.

and radiate less heat into space than they trap beneath them.

Most clouds cool the Earth and reduce the amount of incoming sunshine to the Earth by about 8%. If clouds were removed, the Earth would warm by 10°C. Thick clouds at middle altitudes cover about 7% of the Earth at any one time whereas low-level clouds cover about 30% of the Earth. They stop sunshine, radiate radiation back into space and account for some 60% of the Earth's cooling. Of these clouds, the strato-cumulus clouds that occur mainly over the oceans account for about 20% of the cover of the Earth. The daily television weather forecasters uses satellite cloud information and, since 1966, satellite images have been able to produce more accurate, higher quality and more extensive weather forecasts. Cloud measurements show constant changes in clouds with large changes during El Niño events.

Except for water vapour, all of the greenhouse gases are trace constituents of the atmosphere. Water vapour tends to follow temperature change rather than cause it. At higher temperatures there is more evaporation and higher water vapour concentrations. At lower temperatures, the opposite occurs. Water vapour is an amplifier rather than a trigger. This can easily be put to a test. In a humid coastal area with a clear sky it is warm in the day and mild at night. At the same latitude inland where the air is dry, it will be hot in the day and cold at night. Ice ages and cold polar areas are characterised by cold dry air.

An argument commonly heard is that heat-related clear-sky drought and increased heavy rains are evidence of human-induced climate change. Both are related to clouds. The International Satellite Cloud Climatology Project (ISCCP) has a multi-decadal record of cloudiness[2161] which shows that computer climate models that incorporate cloudiness are flawed.

A heat vent over the warmest part of the Pacific Ocean, the planet's warmest spot, was discovered in 2001 by analysing nearly two years of daily geostationary satellite data on cloud cover and sea surface temperatures. The vent opens to release heat when the sea surface temperature rises, resulting in a decrease in high clouds above the western tropical Pacific Ocean when sea surface temperatures are higher.[2162] This work validates earlier work[2163]

---

[2161] Evan, A. T., Heidinger, A. K. and Vimont, D. J., 2007: Arguments against a physical long-term trend in global ISCCP cloud amounts. *Geophysical Research Letters* 34: LO4701.

[2162] Lindzen, R. S., Chou, M.-D. and Hou, A. Y. 2001: Does the Earth have an adaptive infrared iris? *Bulletin of the American Meteorological Society* 82: 417-432.

[2163] Sud, Y. C., Walker, G. K. and Lau, K. M. 1999: Mechanism regulating sea-surface temperatures and deep convection in the tropics. *Geophysical Research Letters* 26: 1019-1022.

and was confirmed in later studies.[2164,2165,2166] As with all new ideas, the heat vent idea stimulated vibrant discussion.[2167,2168,2169] The decrease in high clouds is because more of the cloud droplets form raindrops and few are left in the clouds to form ice crystals. These ice crystal cirrus clouds are poor sun shields and good insulators hence a decrease in cloud area allows more heat energy to leave the atmosphere. The amount of heat vented to space in the 1980s and 1990s is equivalent to the model-based speculations of atmospheric heat gain if there was an instantaneous doubling of atmospheric $CO_2$.

Nature continues to surprise and belittle us and every new discovery shows that science is never settled.

---

[2164] Fu, Q., Baker, M. and Hartman, D. L. 2002: Tropical cirrus and water vapour: an effective Earth infra-red iris feedback? *Atmospheric Chemistry and Physics* 2: 31-37.

[2165] Chen, J., Carlson, B. E. and Del Genio, A. D. 2002: Evidence for the strengthening of the tropical general circulation in the 1990s. *Science* 295: 838-841.

[2166] Weilecki, B. A., Wong, T., Allan, R. P., Slingo, A., Kiehl, J. T., Sodon, B. J., Gordon, C. T., Miller, A. J., Yang, S-K., Randall, D. A., Robertson, F., Susskind, J. and Jacobowitz, H. 2002: Evidence for large decadal variability in the tropical mean radiative energy budget. *Science* 295: 841-844.

[2167] Hartmann, D. L. and Michelsen, M. L. 2002: No evidence for iris? *Bulletin of the American Meteorological Society* 83: 249-254.

[2168] Lindzen, R. S., Chou, M.-D. and Hou, A. Y. 2002: Comments on "No evidence for iris". *Bulletin of the American Meteorological Society* 83: 1345-1348.

[2169] Chou, M.-D., Lindzen, R. S. and Hou, A. Y. 2002: Comments on "The iris hypothesis: A negative or positive cloud feedback? *Journal of Climate* 15: 2713-2715.

# Chapter 8

# ET MOI

## Ubi dubium ibi libertas[2170]

We are facing the greatest global threat in my three score and two years. It is not global warming. It is the threat from policy responses to perceived global warming and the demonising of dissent. These policies also threaten freedoms and the nature of science and religion. Policy changes have the ability to reduce base load energy supplies of electricity that underpin employment and the standard of living.

There are calls for trials and imprisonment of those scientists who, on scientific evidence, do not agree that human emissions have changed climate. Such scientists are called deniers and are compared to Holocaust deniers[2171] yet their scientific doubts are not addressed. Some environmental advocates of human-induced global warming are promoting Nuremberg-type trials for those who dissent.[2172] Those who have enjoyed the stifling benefits of totalitarian communist systems see the green movements in democratic countries as destroying hard-earned freedoms.[2173]

The world has been warming, slightly and intermittently, since the Little Ice Age. It has also been cooling. This is not surprising and is consistent with what we measure from the past. Sea level, the ice sheets and life on Earth have also changed, albeit slightly. This is also consistent with what we measure from the past. Natural changes have no respect for politics, treaties or emissions policies. If there are indeed human-induced climate changes, then we are unable to separate them from natural variability. How many years must the planet cool before we acknowledge that the planet is not warming?

---

[2170] Latin proverb: Where there is doubt, there is freedom.

[2171] *"Almost everywhere, climate change denial now looks as stupid and as unacceptable as Holocaust denial."* George Monbiot, environmental writer, *The Guardian* 21st September 2006.

[2172] *"When we've finally gotten serious about global warming, when the impacts are really hurting us and we're in a full worldwide scramble to minimise the damage, we should have war crimes trials for these bastards – some sort of climate Nuremberg."* Gristmill, the environmental blog writing in support of George Monbiot on http://gristmill.grist.org, 19th September 2006.

[2173] Klaus, V. 2008: *Blue planet in green shackles.* Competitive Enterprise Institute.

The global warming observed during the last 150 years is just one frame in the movie about the history of the Earth.

Western countries have been frightened senseless about unstoppable global warming related to human additions of $CO_2$ to the atmosphere. There is widespread fear, sometimes manifested as medical afflictions. Many in Western societies believe they are doomed. There has been an uncritical, unthinking acceptance by the community of the media barrage about catastrophic climate change. For many, critical thinking is an anathema.[2174] There is now a realisation that it is not easy to cut emissions and maintain a stable society, that wind and solar electricity generation are unreliable, far too expensive and add little base load power, that biofuel production is greatly damaging to the environment and creates food shortages, and that growing economies like China and India will emit large amounts of $CO_2$ into the atmosphere despite decisions by other countries. The potential loss of industrial competitiveness is such that many countries are now stating that they'll not sign a post-Kyoto agreement unless China, India and the USA also sign.

Disillusionment is setting in because the dogma stated that as $CO_2$ rises, so does temperature. There has been a rise in $CO_2$ yet cooling is taking place. Dangerous global warming has not occurred and politicians are starting to realise that carbon taxes and emissions trading catastrophes are on a far shorter electoral time-frame than projected warming. The big losers are climate scientists whose advice is no longer being followed. They have overplayed their hand and decision-makers no longer trust their advice.

Politicians need wriggle room for unexpected changed circumstances. Some climate scientists have been telling politicians that there are no options and all the eggs must be put in the one basket. If politicians believe that in a few decades the major industrial economies can be changed to low carbon economies, then there is a harsh reality. Politicians face this reality at every election, climate scientists do not. Politics is the art of compromise, doing deals and counting heads. Extreme green movements leave no room for compromise and use blackmail. When the situation and facts change, politics changes. When the facts change, extreme green ideology, politics and pressure do not change. Green ideology and political pressure take place in a science-free zone. Politicians have embraced a premature scientific hypothesis because they are bombarded by environmental pressure groups and one set of scientists who have everything to gain. This has been done at the expense of integrated interdisciplinary scientific evidence.

---

[2174] *"Many people would rather die than think. In fact they do."* (Bertrand Russell)

The founder of Greenpeace, Dr Patrick Moore,[2175] has stated that the green movements have been taken over by neo-Marxists promoting anti-trade, anti-globalisation and anti-civilisation. The average citizen fears global warming, is concerned about the future for the next generation and feels helpless. They want to do something. They want governments to do something, which is why governments have bowed to green pressure. Well-meaning citizens have joined groups such as Greenpeace in order to try to do something concrete, not knowing that the leaders of such groups have scientifically misled and deceived politicians and their advisers. Why is it that the leaders of the green movements do not have to fulfil the same ethical, governance and due diligence provisions of the law as company directors?

The UK Labour government since 1997 has made climate change a top priority yet it has not been able to reduce $CO_2$ emissions over the last decade. Labour Party ideology and reality are poles apart. The UK government is now retreating from a position where they created greater expenses for their core voters, yet the political class is in denial. The UK and many other democratic governments have placed themselves in a difficult position because the civil service has become politicised, governments receive advice which suits their ideology instead of dispassionate professional advice. This is a failure of governance and is not a process that any corporation would undertake in multi-billion-dollar decision-making process.

If governments had read the fine print of the crucial Chapter 5 of IPCC AR4 (*Humans Responsible for Climate Change*), they would have realised that it is based on the opinions of just *five* independent scientists. Governments are planning to structurally change their nations' economies where most people will suffer from increased taxes and costs and few will prosper, based on the opinion of the fabulous five whose computer models have not been able to accurately predict the cooling that has occurred since 1998. Common sense was abandoned in the face of unending pressure from green groups.

Some of the more astute politicians have realised that there has been no unprecedented global warming, that climates always change, that there are cyclical climate changes, that there is little relationship between $CO_2$ and climate, that there is no consensus of scientists and that massive structural changes to the economy could be economic and electoral suicide. Even more astute politicians would follow the money and look at the interests of the green groups and scientists advocating human-induced global warming. These politicians should be asking questions. How do you explain that global temperatures according to the IPCC have not increased since 1998 and that

[2175] http://www.greenspirit.com.index.cfm

there has been no significant warming since 1995? Are you aware that even the IPCC does not consider climate models to be "predictions" or "forecasts" but merely "emission scenarios"? Are you aware of the numerous studies from science and history that show that in the Medieval Warming it was warmer than today and this was a time with no cars or industrialisation? How do you explain that $CO_2$ levels have been much higher in the Earth's history and have not coincided with extinctions and warm periods? Why has Greenland cooled since the 1940s? Why was the Arctic warmer than now in the 1920s and 1930s? Why has Antarctic sea ice expanded to record levels in recent years? Why has Arctic sea ice expanded since 2008? Can you explain why Arctic sea ice reductions in 2006 and 2007 were unrelated to man-made emissions of $CO_2$? If high $CO_2$ creates global warming, why has $CO_2$ been far higher than at present during at least four glaciations? There are hundreds of questions that have yet to be asked.

The IPCC has also painted itself into a corner with its narrow focus because it is a political organisation. The IPCC has huge UN and government backing and yet it ignores scientific areas that examine other climate forcings. The IPCC has essentially ignored the role of natural climate variability. In other areas of science, if the hypothesis could not be proven then one would search for an alternative explanation. It appears that natural variability of climate has not been considered as an alternative explanation of the observed changes.

In the time since the IPCC was formed 20 years ago there has been no demonstration of human-induced global warming, and the IPCC (and Kyoto Protocol) should now be abandoned. They have shown ephemeral correlations of weather patterns and $CO_2$ and, when selected inputs into computer models are used, approximately the correct answers are produced until 1997. The models suggested constant warming until the end of time. However, neither the post-1998 cooling nor El Niño events were predicted. This alone shows that the computer models are only sophisticated computer games with input based on the programmer's predilections. Climate predictions are not evidence and certainly are not suitable for environmental or political planning. The refusal of climate to follow the IPCC computer models is the grim reality of Nature. Not only have the computer models not worked, they are, as Freeman Dyson observes, "full of fudge factors and do not begin to describe the real world."[2176]

The extensive reliance by global warmers on computer models impresses those with little scientific training. However, the significant manipulation of

---

[2176] http://www.abc.net.au/rn/inconversation/stories/2008/2444172.htm

the source data and the lack of use of many known variables exacerbate uncertainties and can produce the predestined outcome before the model can be run. This is a common flaw of mathematical modelling. Models with simulations, projections and predictions prove nothing. All a model shows is something about the model itself, normally its limitations. Data collection in science is derived from observation, measurement and experiment, not from modelling. We can't make Nature conform to virtual computer models. Climate catastrophes regularly occur and, no matter what models are used, they will still occur.

Models tell us more about the self-regulating undisclosed interests of groups than they do about present or future climate. One has only to look at previous failed models with their frightening scenarios.

One of the early climate modellers, Edward Lorenz (1917–2008), was the first to advocate the idea of chaos theory.[2177] Lorenz showed that very small changes in input to his climate models produced unexpected but reproducible results. He concluded: "the prediction of the sufficiently distant future is impossible by any method, unless the present conditions are known exactly". Nothing has changed.

Five simulations of global climate were undertaken for the period 1860–2000 using the same general circulation models that are used by the IPCC. Each simulation had slightly different initial conditions but otherwise was the same. Very small differences in the initial conditions of climate resulted in large variations in later climate.[2178] Because we do not know exactly the initial conditions, the variations could arise from internal variations in the computer system or unknown external variations. What we do know is that natural processes are dynamic, non-linear and chaotic whereas computer simulations are generally based on simple non-linear systems. These are some of the reasons why climate change projections from computer models have little to do with the real world. The very next year after this publication of the five simulations in *Science*, the report of the IPCC claimed that, based on computer model simulations, climate has only limited variability and hence was not dynamic, non-linear and chaotic.

Complex computer models produce what are called "predictions", as if they were factual. They are models that try to simulate future climate based on incomplete information, attempts to deal with non-linear variables in a dynamic system, and with far too many assumptions. Climate computer

---

[2177] Lorenz, E. 1963: Deterministic nonperiodic flow. *Journal of Atmospheric Science* 20: 130-141.

[2178] Delworth, T. L. and Knutson, T. R. 2000: Simulations of early 20th century global warming. *Science* 287: 2246-2250.

models amount to a religious statement by the modeller stating his belief as to how his small part of the world should operate. Global climate models deal with atmosphere-ocean-land interactions, while variables such as the Sun, the cosmos, bacteria, history and the geological processes that take place beneath our feet are just ignored.

Although global climate models have advanced and are now able to make weather predictions a week in advance, they are still not able to make seasonal or yearly forecasts. When the variable of time is added to forecasts, they become far less reliable. How can we trust global climate forecasts 100 years into the future when the same models can't demonstrate shorter-range forecasts? To make matters worse, the modellers don't seem to have much confidence in their own models.

Australia's CSIRO runs this disclaimer with its climate predictions:

> The projections are based on results from computer models that involve simplification of real physical processes that are not fully understood. Accordingly, no responsibility will be accepted by CSIRO for the accuracy of the projections inferred from this brochure or for any person's interpretations, deductions, conclusions or actions in reliance on this information.

You don't need me to express reservations about climate models. The climate modellers have done it themselves.

The scientific evidence presented in this book shows that other factors besides $CO_2$ drive climate. This conflicting data brings uncomfortable feelings amongst advocates of human-induced global warming, so they dismiss this data in a defiant act of cognitive dissonance. However, the data remains and the apocalyptic prophesies fail.

Global warmers are uplifted by believing that they have a mission to save the world. Unfortunately, Nature does not appreciate them. For many environmentalists travelling the path of certain salvation, it is ideologically impossible to acknowledge that the planet is dynamic and that past natural changes are far greater than anything measured in modern times. The weather dominates daily life. For the last few decades, global warming has replaced the weather. When we did not fry, climate change replaced global warming. And when the climate stubbornly did not change, the language has been downgraded to carbon pollution, carbon footprint and carbon-free economy. Meanwhile, the planet has been doing what the planet always does – change.

You cannot have a conversation about the biggest policy argument of the last few decades and then not allow one side to debate. Yet this is what has happened. This has allowed the advice given by green activists to government ministers to devolve into farce. Politicians are exposing themselves to ridicule. At a recent G8 summit, German Chancellor Angela Merkel attempted to convince world leaders to play God by restricting $CO_2$ emissions to a level that would magically limit the rise in global temperatures to 2°C. She can't really be that stupid. Maybe she is?

Another example. In 2008 Australia's Penny Wong, whose official title is Minister for Climate Change and Water, published a green paper which contained a short opening sentence[2179] with seven scientific errors. One error for every three words must be some sort of record. The same minister has demonised element number 6 in the periodic table of elements. Is her knowledge of science so weak that she does not know that carbon is fundamental for all life on Earth? "Carbon pollution" implies pollution by the element carbon, not the compound carbon dioxide. These absurdities flow endlessly with talk of carbon pollution and carbon footprints. Unless this minister lives on air, she pollutes her body every day with the carbon-based food she eats. If this minister were really serious about carbon pollution, she would stop eating.

The same minister is responsible for the Murray-Darling Basin. In Australia, a prolonged drought and political mismanagement by four state governments and the federal government have led to the collapse of the food industry in the Murray-Darling Basin, the bread basket of Australia, and a shortage of potable water for a state capital city. No matter how much the community pays in carbon taxes, it will not make it rain in the Murray-Darling Basin. The Murray-Darling Basin has died the death of a thousand government plans. The same governments who cannot solve a regional problem now claim they can change global climate!

## The vernalisation of science

In the Soviet Union, Trofim Lysenko was an insignificant agricultural scientist. He claimed he could save the starving USSR by modifying biology in the same way that the communists wanted to modify human behaviour. He claimed that treating seeds before cultivation (vernalisation) would affect their behaviour. Lysenko was an adept propagandist who boasted he could triple or quadruple grain production. The government wanted to increase food production. He was made head of the Soviet Lenin All Union Institute of Agricultural Sciences, and argued that conventional genetics was the basis

---

[2179] "Carbon pollution is causing climate change, resulting in higher temperatures, more droughts, rising sea levels and more extreme weather."

of fascist eugenics. Opposition to Lysenko was not tolerated. Mendelian geneticists were demonised as "fly-lovers and people haters", and between 1934 and 1940 numerous geneticists were shot or exiled to Siberia to enjoy starvation. The great Vavilov was sent to Siberia, starved to death in 1943, and Lysenko took Vavilov's position as Director of the Lenin Academy of Agricultural Sciences. Genetics research stopped. In 1948 genetics was officially labelled as "bourgeois pseudoscience".

After Stalin's death, Khrushchev supported Lysenko. After Khrushchev's departure in 1964, the Academy of Sciences made a devastating critique of Lysenko. Meanwhile, tens of millions of people had starved in the Soviet Union because Lysenko's agricultural policies did not produce enough food. The ban on genetics was lifted in 1965 but the USSR had lost 30 years of advances in agriculture.[2180] This could still be seen in August 1977 in the Soviet Union when I drove from Soviet Karelia into Finnish Karelia. The soil and climate are identical. Finnish crops were tall with full heads of grain, Soviet grain was short, sparse and with shrunken heads.

There are close parallels between Lysenko and the global warming movement. Climate change is promoted in seven-second grabs, so like Lysenkoism it is much easier to understand than the complexities of real science. Carbon dioxide is now the equivalent of Mendelian genetics. In June 1988, James Hansen gave evidence to a US Senate committee. Hansen's star rose and he then became climate adviser to the US President, to Al Gore and many others such as Lehman Brothers, the great financiers, who saw carbon emissions trading as a new unregulated global financial instrument which they could control. Lehman Brothers has now gone broke.

Gore was a director of Lehman Brothers. Gore founded his own "green" corporation, Generation Investment Management. He is a board member of a renewable energy company. In many legal jurisdictions, if Gore made speeches about climate change and did not declare his interests, he would be committing a criminal offence. The whole gravy train gained momentum with the establishment of a single-issue group (IPCC), propaganda via Al Gore's fictional Hollywood blockbuster movie *An Inconvenient Truth* and Mann's infamous "hockey stick", various partisan economic reports (e.g. Stern, Garnaut) for populist political leaders and an uncritical media looking for horror stories.

Carbon and emissions trading schemes are a God-given opportunity to plunder. New legislation on pie-in-the-sky emissions will result in the public paying more for everything. Trading schemes will be based on a mythical commodity. Such schemes have not stood the test of time and will require constant amendment. The opportunities for fraud are breathtaking. There will be great profits to make out of chaos and many businesses such as (the late) Lehman Brothers positioned themselves to make a killing. Governments just cannot resist an opportunity to raise more taxes, to increase the bureaucracy and impose more regulations.

It is only the courts of the UK that have addressed the scores of scientific errors in Gore's movie.[2181] Viscount Monckton of Brenchley, an expert witness in the UK legal case condemning *An Inconvenient Truth*, showed from science that the movie has 35 errors which distort or exaggerate, all in the direction of unjustified alarmism. The statistical likelihood that all 35 errors would accidentally trend in one direction is 1 in 34 billion. The judge in this litigation listed nine major discrepancies which all differ from the scientific data (shown here in parentheses). Gore claimed sea level would rise 20 feet (Greenland and Antarctica may add 2.5 inches to sea level from melt water over 100 years), that whole Pacific populations will have to be moved (Pacific Ocean sea level has hardly changed), thermohaline circulation will stop (thermohaline circulation may slow), $CO_2$ increases create temperature increase ($CO_2$ increase follows temperature increase), global warming is melting the Kilimanjaro glaciers (changes due to long-term climate shifts and deforestation), global warming has decreased the water level in Lake Chad (over-extraction of water and farming has dried Lake Chad), Hurricane Katrina was due to global warming (isolated low-frequency events cannot be attributed to global warming), polar bears die while trying to find ice (high winds killed four polar bears in an area where sea ice was growing) and global warming bleached corals (the 1998 El Niño bleached corals).

In the UK, a high school book version of *An Inconvenient Truth* has been circulated. It is replete with a good dose of schoolboy howlers. For example, the photograph of the mosquito is that of a parasitic wasp and the photograph of the tsetse fly is indeed a photograph of a tsetse fly, but with two legs missing. And the list goes on and on. The Norwegian parliament awarded the Nobel Peace Prize jointly to Al Gore for *An Inconvenient Truth* and the IPCC. The Norwegian parliament had previously awarded the Nobel

---

[2181] (a) Floods in 18 countries, plus Mexico: Four errors (b) The Arctic ice cap will be gone in 5 to 7 years: Six errors (c) Forest fires are causing devastation: Five errors (d) Many cities are short of water: Four errors (e) More severe storms: Six errors (f) West Antarctica has lost an area of ice the size of California: Four errors (g) Deserts are growing: Three errors (h) Sea level is rising: Eight errors (i) $CO_2$ is global warming pollution: Seven errors (j) Venus has experienced a runaway greenhouse and the European Union states the Earth is a sister planet of Venus: Four errors (k) The IPPC reports are unanimous: Five errors (l) Svante Arrhenius made 10,000 calculations 116 years ago, demonstrating that temperature would rise "many degrees" in response to $CO_2$ doubling: Four errors.

Peace Prize to Yasser Arafat.

The IPCC claims that its reports are written by 2500 scientists. In fact, they are written by 35, who are controlled by an even smaller number. A new Lysenko has arisen in climate circles – no prizes for guessing who it is. Green political groups quickly followed the climate change political line because it gave them enormous power and, in many places, the balance of political power. No politician could argue against global warming for fear of being painted as anti-environment. All droughts, floods, storms, malaria and even cooling were blamed on global warming, and massive bureaucracies and research institutes have been established to deal with climate change.

The parallels with Lysenkoism are that both movements first started through political organisations. They both claim that the science is settled and that there is nothing to debate. They disregard or deny evidence from other areas of science, they disregard or deny examples where predictions are wrong, they demonise and victimise the opposition and they relate science to ideology. Neither evidence nor reason is necessary. Both movements have a huge propaganda machine, create huge bureaucracies where many people have careers based on the ruling ideology, underpinned by slogans and simplistic solutions imposed on the people by administrative fiat. After great human and economic cost, Lysenkoism was replaced by real science because real science just does not go away.

Fads in science (e.g. Freudian science) and scientific dictators (e.g. Lysenko) are common. These reduce science to a single issue, ignore the integrated interdisciplinary nature of science, ignore uncertainties and commonly are married to one concept and one procedure. With climate change, the science is driven by computer models and ideology. Computers are used in science to analyse scientific data. Computer models are not data. If computer models torture the data enough, the data will confess to anything. Computer models do not require the rigours of observational science where data is invariably collected outdoors in horrendous weather conditions.

Models do not require the rigours of scientific methodology and the constant nagging but healthy uncertainties of observations, measurements and experiments. Computer models are constructed by the master caste in air-conditioned offices far removed from reality. The modellers claim that they can model all the complex processes on the planet to predict climate hundreds of years into the future, yet they cannot predict the biggest surface transfer of energy on Earth: El Niño. The public is bamboozled by computer models which are meant to predict the future. They can't.[2182] In many ways,

---

[2182] Pilkey, O. H. and Pilkey-Jarvis, L. 2007: *Useless arithmetic: Why environmental scientists can't predict the future.* Columbia University Press.

the growth of the global warming industry has replaced the collection of primary field data, measurement and experiment. The simplicity of a model may reflect the beauty of science but it may also reflect our ignorance of the complexity and subtlety of natural processes. This may reflect human arrogance as much as the elegance of science.

A radio show based on H.G. Wells' *The War of the Worlds* was aired in the USA as a Halloween special on 30 October 1938. Orson Welles, convincingly playing the role of a news reporter, declared: "The Martians are coming". Many listeners, unable to distinguish between science and science fiction, believed him and panicked. Climate change is an issue fraught with exactly this kind of confusion. Not only has widespread panic occurred, but governments across the world are being pressured into adopting questionable policies to address climate change which is not of human origin. Policy debate has proceeded without the normal level of scientific proof. Even though climate change is poorly understood, restrictive laws have already been passed in order to prevent alleged human-induced global warming.

It has been known for hundreds of years that Nature hides many secrets.[2183] These secrets guarantee that computer models will be incorrect. Global warming adherents tell us the science is settled. This means there is nothing more to learn about Nature. If the science of human-induced global warming is settled, then there is no need for climate research funding, climate institutes and government bureaucracies. If the matter is settled then it is not science. Scientific knowledge is doubling every seven years and it is not possible for anything to be settled. As this book shows, the global warming movement ignores unfashionable rigorous sciences such as geology, astronomy and solar physics. If science is selectively used and large bodies of validated science are dismissed, then the global warming movement is not underpinned by science.

We live in a time when the methodology of science is suspended. Reactions to human-induced global warming based on incomplete science can only be extraordinarily costly, will distort energy policy and will make the poor poorer.

The real problems of pollution can not be ignored. However, in the case of the effect of $CO_2$ on climate, the correct solution to the non-problem of $CO_2$ is to have the courage to thoughtfully do nothing.

---

[2183] *"In Nature's infinite book of secrecy, A little I can read."* (Soothsayer in *Antony and Cleopatra* 1.2, 9-10; William Shakespeare [1564-1616 AD]).

## Tip of the iceberg

The public debate over global warming will never be settled by reason and evidence. The complexities of climate are too great for casual understanding based on seven-second television grabs. The unknowns and uncertainties provide too much scope for subjective preference. Few people arrive at belief by diligent effort to understand and evaluate. Most choose belief because it is emotionally satisfying. Evidence that supports the established belief is then readily accepted. Evidence to the contrary is ignored or dismissed. "Belief" is a word of religion and politics. It is not a word of science.

Human-induced global warming is a popular belief because it offers the satisfaction of righteousness without actually having to do anything. Subscribing costs nothing, it provides the immediate reward of moral superiority, and there is the bonus of seeing "polluters" having to pay for their sins. This makes an attractive package and the media provides an abundance of pseudo-evidence to support it. Trying to counter this with considered thoughtful scientific argument which can't be conveyed in a seven-second sound bite cannot compete in popular appeal.

However, despite its appeal and momentum the human-induced global warming juggernaut is highly vulnerable in two critical respects. One is that almost all of the purported "scientific evidence" as well as the catastrophic consequences being predicted are directly dependent upon warming, so an ongoing cooling trend in climate will eventually be impossible to either ignore or explain. The other is that a transition from the dominant fossil fuel energy industry will not be a relatively painless change because there is no known method or technology by which industrial economies can survive on expensive, unreliable, clean, green, renewable energy.

The era of cheap abundant energy may be ending, not because of supply or demand but because green movements are trying to force society to survive on high-cost unreliable energy supplies. Renewable energy is expensive, diffuse and many decades of development away from becoming more than a minor contributor to meeting demands. For example, all of the electricity generated by 2000 wind turbines in the UK is still considerably less than that produced by a single medium-sized conventional power station. There are nearly 50 nuclear, gas or coal-fired power plants in Britain today. Because of wind's intermittency, wind turbines generate electricity at less than a fifth of rated capacity.

Cheap abundant energy is fundamental to all economies. With energy shortages, almost everything will become more expensive and we will simply be poorer. Many things we now take for granted could be unaffordable or even unavailable. Proposed carbon taxes, emissions trading and controls will, at best, only exacerbate the difficulties. Governments and a few selected financial institutions will reap huge amounts of money from the public. Emissions trading will become a weapon of mass taxation. This money will not and cannot change climate. No amount of money will change cosmic rays, the behaviour of the Sun, the Earth's orbit, ocean currents and plate tectonics. Emissions trading will enrich a few and make most people poorer.

Whatever the belief of environmentalists, political pressure about human-induced global warming can only create social disruption because the world demand for energy will not disappear. To reduce *per capita* energy consumption can only reduce the standard of living. This will create social unrest and instability in governments. At present the only efficient sources of energy are fossil fuels, hydroelectricity and nuclear fission. Until the costs of energy from other sources are substantially reduced and the energy used to build power-generating plants is less than the lifetime of energy the plant will produce, then it is political suicide to impose other energy sources onto communities.

We are already seeing a surge in food prices, energy prices and the general cost of living. Governments seem to forget that many big political changes in the past have been driven by high food prices. At 200 ppmv $CO_2$ in the atmosphere, plant growth almost stops. The green revolution is possible because atmospheric $CO_2$ is now at 385 ppmv and, if $CO_2$ were at 1000 ppmv, there would be a huge improvement in plant growth and no harm to animals. This we know because we pump $CO_2$ into agricultural glasshouses to create higher yields and faster growth at about 1000 ppmv $CO_2$. Most people on Earth struggle to obtain protein, so to reduce or bury this plant food by $CO_2$ sequestration creates a moral issue.

Most proponents of human-induced global warming are either ignorant of science or have far too much personally invested. They rely on authority, consensus and a maniacal desperation to stamp out contrary views. We are told that the theory of human-induced global warming is robust and that you would be off your head to disagree. Not a day goes by without the media presenting shock-horror stories about global warming. We are continually told that there is a consensus of eminent scientists supporting the theory and that the only scientists who disagree must be in the pay of oil companies. There is no understanding that disagreement could be for rational, dispassionate intellectual reasons. By *ad hominem* attacks on those who disagree, the gains made in the Enlightenment are abandoned in pursuit of a political cause. What is not stated is that the predictions of climate scientists about a human-induced climate catastrophe are somewhat tainted by their own patronage arrangements with politicians, governments, NGOs and research organisations that have invested heavily in a global warming catastrophe. The intertwining of political prestige, patronage and scientific authority is probably why many climate scientists seize the moment and chase fame, political fortune and a lifetime of funding.

Despite the billions of dollars spent on the IPCC, Kyoto, climate research, talkfests and government policies, the world does not appear to be getting warmer. We are told that the 20th Century was the warmest on record yet the IPCC's own data shows that this is not the case. The 20th Century showed warming and cooling. The 20th Century was the first century after the Little Ice Age and it is no surprise that it was warmer. According to the IPCC's own figures, the world has been getting cooler since 1998. The ice core data shows that a $CO_2$ increase follows a temperature rise. The Mann "hockey stick" managed to make the Medieval Warming disappear and made the recent warming look alarming. The human-induced global warming theory suggests that high in the Earth's atmosphere above the equator there should be warming, but weather balloons and satellites show that this is not the case.

There is a good correlation between temperature and $CO_2$ from 1976 to 1998. Correlation does not mean causation. We only have to step back in time to show that there is no relationship between $CO_2$ and temperature. Those supporting the human-induced global warming theory cannot explain why the Minoan Warming, Roman Warming and Medieval Warming were warmer than today, as there was no industrialisation in those times. They cannot explain why temperature increased from 1860 to 1875, then decreased from 1875 to 1890, rose until 1903, fell until 1918 and then rose dramatically until 1941. They cannot explain why the rate and amount of warming at the beginning of the 20th Century were greater than now despite lower $CO_2$ emissions by humans at the time. They cannot explain why there was cooling from 1941 to 1976, the year of the Pacific Decadal Oscillation. Since 1976, temperature rose by 0.4°C until 1998. They cannot explain why there has been cooling since 1998. With the theory in tatters, it is no wonder that they defend their political dogma with religious zeal.

This book is not advocating that we pollute our airways, waterways and land. This book does not advocate that we be wasteful with our energy, water, land, food and mineral resources. This book does not advocate that we starve a large number of people in the Third World countries because of a political ideology. However, that colourless, odourless, tasteless, non-toxic gas called carbon dioxide is not a pollutant. It is a trace gas in the atmosphere. It underpins all life on Earth. It is plant food. All plants require $CO_2$ for photosynthesis, and plants (especially photosynthetic micro-organisms) are the pillars supporting all life on Earth. The slight increase in $CO_2$ over the last 35 years from 325 ppmv to 385 ppmv has increased grain yields.

Almost every time a global warming story is featured on television, there is a background image of a cooling tower on a coal-fired power station. The visible plume is not $CO_2$, it is $H_2O$. The plume of minute water droplets is the visual stimulus to support the popular view that $CO_2$ is a pollutant of human origin. Not only is this a misleading image, but without the power station the television program could not be broadcast. Whenever a story of sea level change or melting of glacial ice is given airtime on television, we see shots of icebergs calving off the Antarctic Ice Sheet. We do not see footage of snow falling in the Antarctic highlands thousands of metres above sea level. We do not see information that shows that the loss of ice by calving and the gain by snowfall are in balance. We do not see footage to show that satellite observations are telling us the polar ice sheets are thickening. We do not see information to show that the calving of ice is the end result of a process that started thousands of years ago, well before humans used fossil fuels.

## Scientific consensus

When science was born, the consensus at that time was driven by religion, politics, prejudice, mysticism and self-interested power. From Galileo to Newton and through the centuries, science debunked the consensus by experiment, calculation, observation, measurement, repeated validation, falsification and reason. Appeals to consensus are not new. The methodology of science allows problems to be solved, whereas the science of the global warmers is designed to confirm a political opinion. There is a consensus regarding the science of global warming but only amongst ascientific environmental activists.

Scientific fact now no longer seems to be necessary. Human-induced global warming is one such example, where one camp attempts to demolish the basic principles of science and install a new order based on political and sociological collectivism. Science is becoming a belief system wherein the belief with the greatest number of followers becomes the established fact and received knowledge. This belief is sustained by consensus and authority. With this new authoritarian science based on consensus and espoused by

UN's IPCC and other agencies as authorities, it appears that true science does not matter any more. If Mann's "hockey stick" chart showing rising global temperature is based on fraud and invalid statistical methods, it just does not matter because we still have a consensus.

If one body of science shows that the polar ice sheets are expanding or that polar bears are now thriving, then this is ignored because this data is not within the consensus. When astronomers have the temerity to show that climate is driven by solar activities rather than $CO_2$ emissions, they are dismissed as dinosaurs undertaking the methods of old-fashioned science. These astronomers are not members of the "consensus club" and are not heard. Once there is consensus, anything that questions the popular paradigm can be dismissed without reason. The history of science shows that today's popular paradigm is tomorrow's discredited theory. A good example is the popular idea that an asteroid impact 65 million years ago in Mexico led to the extinction of dinosaurs.

Science where a majority of votes by climate scientists determines a scientific truth is politics, not science. And that is exactly what human-induced global warming is: politics. After the consensus method fails one too many times, there will be a quiet advance to real science. In the interim, we have to live with the carping of ascientific unelected political pressure groups who behave like scalded cats should anyone have the temerity to argue that global warming is not a man-made phenomenon. Some scientists have placed science on the platform of religious dogma. When the received truth (i.e. IPCC) is challenged, those scientists who dissent suffer from the equivalent of the Congregation for the Doctrine of the Faith. Global warming fundamentalists exclude or suppress any information that challenges or contradicts their beliefs and use state power to pursue their objectives, sometimes contrary to the will of the majority. Those that dissent are labelled insane and derided, venal or in the pay of big business, and there is no debate of facts in what is honest disagreement in science.

The IPCC reports read like documents orchestrated to produce the outcome that human activities cause global warming. The media, whose survival depends upon sensationalism, lap up the doom-and-gloom story from the *Summary for Policymakers* of these reports and do not read the science in the body of the reports. In all IPCC reports, the *Summary for Policymakers* is written for policy makers, politicians, media outlets and environmental activists and is not written by scientists. It does not reflect the science in the body of the reports. In the IPCC reports, the *Summary for Policymakers* is the only thing that journalists use to get their information. These summaries have generated a sense of certainty, omniscience and orthodoxy. The summaries do not reflect the uncertainties that are in the main body of the science. In fact, the uncertainties of the science in the body of the reports are themselves uncertain.

Climate change is viewed as a problem. The problem is portrayed as a simple problem derived from a single cause (human emissions of $CO_2$), and a problem that can be solved by one small group of experts from a single discipline. In fact, there are a whole series of very complex questions in many larger topics that are not necessarily related. There are many highly qualified people on all sides of these debates with greatly differing informed opinions. This is exactly what would be expected in any scientific debate. It does not happen with the IPCC, which has gained a consensus from like-minded people all in a small area of science. All these people depend upon each other for peer review publications and research grants.[2184]

There was a statistical study to show that the 20th Century was unusually warm.[2185] This was another attempt to validate the "hockey stick" and it sent the media into overdrive. What was not reported by the media was another paper showing that appropriate statistical tests that link climate proxy records to observational data were not utilised and, as a result, the unusual warmth of the 20th Century disappeared.[2186] Both papers were in the same scientific journal. That is the nature of science. There is no consensus.

Russian scientists deny that the Kyoto Protocol or IPCC reports reflect a consensus view of the scientific community. As Western nations step up pressure on India and China to curb the emission of greenhouse gases, Russian scientists reject the very idea that carbon dioxide may be responsible for global warming. Russian research on Antarctic ice cores shows that temperature rise triggers an atmospheric $CO_2$ rise with a lag of 500 to 600 years. Russian scientists suggest climate models are inaccurate, since scientific understanding of many natural climate factors is still poor and climate cannot be properly modelled. Oleg Sorokhtin of the Russian Academy of Sciences Institute of Ocean Studies suggested that global climate depends predominantly upon many factors such as solar activity, precession of the Earth's axis, changes in ocean currents, fluctuations in saltiness of oceans, and surface water, whereas industrial emissions do not play any significant role. He argues that high $CO_2$ concentrations are good for life on Earth and

---

[2184] Schulte, K.-M. 2008: Scientific consensus on climate? *Energy and Environment* 19: 281-286.

[2185] Osborn, T. J. and Briffa, K. R. 2006: The spatial extent of 20th-century warmth in the context of the past 1200 years. *Science* 311: 841-844.

[2186] Bürger, G. 2007: Comment on "The spatial extent of 20th-century warmth in the context of the past 1200 years. *Science* 316: 1844.

that the Earth's atmosphere has built-in regulatory mechanisms that moderate climate changes. When temperatures rise, ocean evaporation increases, denser clouds stop solar rays and surface temperatures decline.

Under the previous leadership of Lord Robert May, the Royal Society of London became an advocate of climate alarmism. Another previous leader of the Royal Society of London used his authority to state that heavier-than-air machines could never fly and that we knew all there was to know about physics.[2187] Authority does not validate a scientific conclusion. It takes evidence to do that. Under Lord May, the Royal Society tried to enlist other academies in joining them in an alarmist manifesto. The US Academy of Science, while sharing some of the views, decided not to sign up. The Russian Academy of Science has taken the opposing view. The 32,000 American scientists who signed the Oregon Petition expressed serious doubt about the major conclusions of the IPCC. Many surveys now involve little time and effort as they utilise the click of a computer mouse. The process for the Oregon Petition involved filling in a printed document, finding a stamp and envelope and posting it to the Oregon Institute of Science and Medicine.[2188]

The American Physical Society stated:

> There is a considerable presence within the scientific community of people who do not agree with the IPCC conclusion that anthropogenic $CO_2$ emissions are very probably likely to be primarily responsible for the global warming that has occurred since the Industrial Revolution.

When is a consensus not a consensus? The social scientist Naomi Oreskes claimed in the scientific journal *Science* that a search of the ISI Web of Knowledge Database for the years 1993–2003 under the key words "global climate change" produced 928 articles, all of which had abstracts supporting the consensus view.[2189] Another social scientist, Benny Peiser, tried to validate this claim, checked Oreskes' procedure and found that only 905 of the 928 articles actually had abstracts, and that only 13 of the 905 explicitly supported the consensus view.[2190] Some papers opposed the consensus view. Referees and editors of *Science* could have done their job and easily checked Oreskes' claim, as did Peiser. They did not. Claims of consensus relieve policy bureaucrats, environmental advocates and politicians of the need to validate claims or have any knowledge of science and are used to intimidate

---

[2187] Lord Kelvin.

[2188] www.petitionproject.org

[2189] Oreskes, N. 2004: The scientific consensus on climate change. *Science* 306: 1686.

[2190] http"//www.staff.livjm.ac.uk/spsbpeis/Oreskes-abstracts.htm

those who beg to differ.

The IPCC's repeated assertion that there is a scientific consensus behind its reports and policy prescriptions reflects its own unscientific foundation. Science only progresses by continuously questioning existing orthodoxy. The Royal Society published a populist pamphlet[2191] in which they relied heavily upon the IPCC as the reliable source of scientific information. The Society adopted the IPCC claim that warming is almost certainly human in origin. The Society presented no independent evidence to support this claim. The Society did not critically evaluate the IPCC claims. The Society did not state that there are many eminent scientists who have arrived at a contrary conclusion based on evidence. The Royal Society is dependent upon funds from the UK government, and to support the government of the day was politically astute. The Royal Society purports to speak on behalf of a consensus of scientists, but direct polling of scientists with an interest in climate shows that about 30% of them are sceptical of human-induced global warming.

So much for consensus.

The peer review process of scientific journals is probably the best process we have for advancing science. However, it is highly flawed.[2192] A paper can be accepted, accepted with revision or rejected. An editor can influence the acceptance or rejection by the choice of reviewers who may support the submitted paper that supports their own work. Reviewers can greatly influence whether a paper is published or rejected. If a reviewer receives criticism in the submitted manuscript or if the conclusions in the manuscript are contrary to work published by the reviewer, the paper may be rejected. Reviewers normally do not ask for the primary data or repeat calculations.

The same process applies with research grant applications. Many outstanding papers and research grant applications are not supported because they do not concur with the popular paradigm. Many papers and grants are not supported because they are weak. If it were not for a few sceptical journalists being objective, then the public would probably never know that there is a considerable body of science that does not support the hypothesis that human activities are creating global warming.

Much good science does not get published in peer-reviewed scientific journals because it does not announce something breathtakingly new. For example, Flinders University in South Australia established an array of tidal measuring stations in the Pacific Ocean. The results show that sea level was neither rising nor falling. It was static. This does not appear to be a great

---

[2191] The Royal Society of London, 2007: *Climate change controversies: a simple guide.*

[2192] Miller, D. W. 2007: The government grant system: Inhibitor of truth and innovation. *Journal of Information Ethics* 16: 59-69.

advance in science, and it is hard to market a publication to a journal editor on the basis that nothing has happened, so this work was not published in a major peer-reviewed journal. Although this information, a null result, may well be useful for later work on evaluating the relationship between climate change and sea level, it is not easily accessible. The results from this research would not win research grants.

Science is primarily supported by government money. Environmentalists commonly claim that there is bias in science when it is supported by industry funds. However, governments like certain beliefs that are popular and help re-election. If one argues that climate change has a solar origin or that low doses of radiation are harmless, then it will be hard to obtain funding from a government. In effect, the non-scientific public are driving the directions of scientific research. A culture of believers has now developed and they survive by playing safe to acquire research funding. Research grant committees want a timetable of achievements in Year 1, Year 2 and Year 3 of the proposed research project. This, in effect, implies that there will be no research breakthrough.

Some 50 or 100 years ago, great science breakthroughs were common events. Not so today. The research grant system today would not fund a James Watson, Francis Crick, Richard Feynman, Jonas Salk or Linus Pauling let alone a Charles Darwin, Albert Einstein or Marie Curie. No scientific journal today would have published a paper submitted by an unknown patent clerk on a fundamental, breathtaking new concept of physics. Thomas Kuhn, the philosopher of science, argued that science progresses in revolutionary bursts in which the dominant paradigm is overturned. However, in the world of modern research grants, those that have established themselves with the dominant paradigm are those who peer review the scientific papers and grant applications. Science will be damaged in the eyes of the public when a popular dominant paradigm is shown to be invalid.

I would be interested to learn how scientific research in solar physics, astronomy and plate tectonics funded by big business could arrive at a different conclusion from the same science funded by government. No matter where the funding comes from, to my knowledge there are neither right-wing nor left-wing cosmic rays. Mind you, I could be wrong. Cosmic rays enjoy a secular apolitical state. They have no ideology. If the science is validated, then the source of the funding matters not, because evidence is evidence. Cosmic rays don't care who funds their measurement, they just keep hitting Earth. Spurious arguments about the source of the funding for research show how environmentalists operate, not how science operates.

Meanwhile, the contrary scientific arguments are not addressed.

Most contrarians have spent a lifetime on science and now no longer rely on research funds for professional advancement. They do not have to toe the party line for professional advancement. Scepticism is fundamental to science. Unless dogma and orthodoxy are challenged, we retreat into a world of superstition and authoritarianism. The suppression of contrary ideas is probably more dangerous to society than global warming, as the 2009 EU President, Vaclav Klaus, has argued.[2193]

Government money was used for the production of a book in 2000 by the German federal government's Department for Geological and Mineral Resources Research in Hannover (BGR). It was also supported by Lower Saxony's state body for primary industry research (NLfB) and the Hannover Institute for Geology (GGA). This compilation of the existing knowledge by two eminent scientists was the knowledge of climate derived from history, archaeology and geology and looked back in time to try to ascertain whether the amount and rate of modern changes is anything extraordinary.[2194] It showed that there was nothing extraordinary about modern times, that $CO_2$ did not drive climate and that factors such as the Sun, the Earth's orbit and tectonics were far more important drivers of climate. The book created a storm. It did not fit the German government's party line, as their political survival was dependent upon Die Grünen and the socialists. The minister of the environment strongly criticised the book and distanced herself from a publication by her own independent dispassionate scientists. In her eyes, it was clear that government money had been wasted. The book, despite being a best-seller, was never reprinted. I treasure my copy of what is effectively a banned book.

Maybe the governments can learn from big business. No corporation would make trillion-dollar decisions without a comprehensive and expensive due diligence, a process that involves transparent validation, re-evaluation, repetition of the primary data and recalculation. This has not been done by governments, many of whom claim that the risks are too high to do nothing about global warming. What they do not say is that the risk of making trillion-dollar decisions without an adequate due diligence is suicide. Already, the painfully long suicide notes have been written (e.g. Stern Review).

Like most people, politicians and the media want to be popular and say things that they believe make most people comfortable. If this is supported by

[2193] Klaus, V. 2008: *Blue planet in green shackles*. Competitive Enterprise Institute.

[2194] Berner, U. and Streif, H. 2000: *Klimafakten: Die Rückblick – Ein Schlüssel für die Zunkunft*. E.Schweizerbart'sche Verlagsbuchhandlung.

scientific consensus, then all the better. Popular media stories boost advertising revenues, catastrophes sell, advertisers want to appear environmentally conscious and contrary views don't sell. Unpopular politicians had better have a night job.

What would the IPCC and warmers accept as disconfirming evidence for their current popular paradigm? I have asked warmers many times. It is not possible to get an answer.

I am often asked the rhetorical question: "You are claiming that the IPCC, the Royal Society and other academies and the IPCC are all wrong?"

I answer that consensus is not a scientific fact, it is a political process. I also argue that the massive failure of the world's financial systems shows how world leaders, politicians, academics, markets, business leaders and the media were collectively fooled.

Are governments able to change climate? If governments could change the galactic path of the Solar System, the variable energy emitted by the Sun, changes in cosmic ray flux, the wobbles in the Earth's orbit, the behaviour of bacteria and plate tectonics, then I would be convinced that they could change climate.

Governments should never forget King Canute.

## The end is nigh

A growing number of scientists are recognising that climate, environmental and economic modelling of an inherently unpredictable future is futile and illogical.[2195] Long-distance predictions have a monumental rate of failure and those predictions made using computer modelling are no different. In fact, the dire predictions by climate groups have damaged science.[2196] Such predictions probably tell us more about the group behaviour of the climate modelling community than about global warming. But then again, predictions of the future are not really new.

We live in a technological world. This technology is underpinned by science. The average punter understands neither the science nor the technology used in everyday life. Carl Sagan argued that science is the candle in the darkness and is opposed to the new Dark Age which is underpinned by irrationality and superstition.[2197] By corrupting science, we step back into irrationality and superstition. This irrationality of destructive delusions costs

---

[2195] Pilkey, O.H. and Pilkey-Jarvis, L. 2007: *Useless arithmetic: why environmental scientists can't predict the future.* Columbia University Press.

[2196] Pearce, F. 2008: Poor forecasting undermines climate debate. *New Scientist*, 1st May 2008, 8-9.

[2197] Sagan, C. 1996: *The demon-haunted world.* Headline Book Publishing.

communities dearly.[2198] Technology appears to produce political problems, the politicians and the public expect science to provide answers to problems and the answers are expected to be unequivocal. Nevertheless, we are told, the world is going to end, we are all going to die slowly, we are going to be fried in a hot greenhouse world and, what's more, we are going to die poor. And it's all our fault. Folks, it's time for indulgences. Or is it?

There is a pretty dismal history of experts making predictions about the end of the planet and other such frightening catastrophes. Most predictions, including those of the climate zealots, have religious overtones. Pessimistic predictions attract interest and there is always a crowd ready to listen to dire apocalyptic predictions.[2199]

The New Testament tells us (Matthew 16:28) that the world will end before the death of the last Apostle.[2200] The world didn't end. In 992 AD, the scholar Bernard of Thuringen confidently announced that, from his calculations, the world had only 32 years left. The world did end for Bernard, who died before the 32 years elapsed. The Last Judgement was to take place 1000 years after the birth of Christ. As the world was to end, it was not necessary to exert energy and effort planting crops in what were subsistence cultures. Many didn't plant crops. In 1000 AD the world ended for many because there was famine. The astrologer John of Toledo circulated pamphlets in 1179 AD showing that the world would end at 4.15 pm (GMT) on 23 September (Julian calendar) when the planets were in Libra. This was taken so seriously that in Constantinople the Byzantine Emperor walled up his windows and the Archbishop of Canterbury called for a day of atonement. Walling up windows worked. The world did not end.

The early 16th Century was a great time for end-of-world predictions. Despite numerous failed predictions, the population was only too willing to believe the next prediction. The best prediction of the lot was by astrologers who suggested that a biblical-type deluge would end the world in 1523. Some 20,000 Londoners left for higher ground as they preferred to perish outdoors rather than in the comfort of their own homes. Others, like the Prior of St Bartholomew's, stocked up on food and water. The world didn't end and the astrologers claimed that their calculation was a mere 100 years wrong and the world was going to end anyway in 1623. It did for all those that were alive

---

[2198] Booker, C. and North, R. 2007: *Scared to death: From BSE to global warming. Why scares are costing us the Earth.* Continuum.

[2199] A good example is the poem *Said Hanrahan*, by John O'Brien (1878-1952) from *Around the Boree Log and Other Verses* (1921).

[2200] A good summary of end of world ideas, some of which are used herein, is James Randi's 1990 book *The mask of Nostradamus. A biography of the world's most famous prophet.* James Scribner's Sons, New York.

in 1523.

This prediction sounds very much like the predictions we hear from the climate zealots who make predictions so far in advance that they will not be around to be stoned by angry mobs when their predictions fail. Despite the failure of the 1523 end of the world, astrologers were at it again in 1524 when the planets would be aligned in Pisces. Of course, the end was to be a global flood. This planetary conjunction was Mercury, Venus, Mars, Jupiter and Saturn along with the Sun. Neptune, unknown then, was also in Pisces but Uranus, Pluto (unknown then) and the Moon were not. Who could fail to believe an astrologer called Nicolaus Peranzonus de Monte Sainte Marie, one of the main promoters of the end of the world? Others, like Georg Tannstetter of the University of Vienna, argued that the world would not end. The cacophony of hysteria was so great that Tannstetter was not heard. The same applies today. The global flood was predicted for 20 February 1524, there was frantic boat building activity and many scaled-down replicas of Noah's Ark were built. Many in port towns retreated to boats.

In Germany, Count von Iggleheim built a three-storey ark. He retreated into the ark on the designated deluge day and an angry crowd gathered outside because a rich man was to go through the eye of a needle and they were to perish. On the deemed deluge day, it rained lightly, the crowd behaved as a crowd and hundreds were killed in a stampede. It certainly was the end of the world for the Count, who was stoned to death by the crowd. Records show that 1524 was a drought year in Europe!

The delightfully-named Frederick Nausea, Bishop of Vienna, predicted in 1532 that the end was nigh because he had seen all sorts of strange things such as bloody crosses in the sky with a comet, black bread had fallen from the sky, there were three suns, and a burning castle had been seen in the sky. The world did not end.

The mathematician and biblical scholar Stifelius of Lochau (Germany) calculated that the world would end at 8 am on 3 October 1533. This created great fear in the people of Lochau. The world did not end. Fortunately, the citizens of Lochau came to their senses and gave Stifelius a well-deserved flogging, stripped him of his ecclesiastical title and ran him out of town. In Strasbourg (France), the Anabaptist Melchior Hoffmann announced that the world would be immolated in 1533. Only 144,000 were to live. The rich forgave their debts and gave away their earthly goods in order to be among the chosen few. The year 1533 was one with very few house fires, principally because there was great caution about the dangers of fire and the resultant fiery end. The world did not end in 1533. However, recalculations showed that it was now to end in 1534. Over 100 credulous punters were baptised in Amsterdam as a precaution. The world also did not end in 1534.

Calculations by the astrologer Pierre Turrell in Dijon (France) showed that the world would end in 1537, 1544, 1801 or 1814. Such calculations are the computer models of today. Turrell was smarter than the average astrologer and predicted the end of the world would occur well after his expected lifetime. This is exactly what the climate alarmists are doing. Another astrologer, Cyprian Leowitz, calculated that the world would end in 1584. He must have had great confidence in his calculations because he issued astronomical tables showing celestial events until the year 1614 in case the world did not end. It didn't. The year 1588 was another end-of-the-world year according to Johann Müller. Müller, a self-professed sage who used the name Regiomontanus, was also smart enough to predict the end of the world well after his expected lifetime.

In the 17th Century, 1648 was an end-of-the-world year according to Rabbi Sabbati Zevi of Smyrna (now Izmir, Turkey). In a fit of humility, Zevi claimed he was the Messiah and persuaded the citizens of Smyrna to give up work and prepare for their return to Jerusalem. Zevi was arrested for sedition by the Sultan and, while in prison in Constantinople, was converted to Islam. The end of the world did not happen in 1648. In 1578, Helisaeus Roeslin of Alsace calculated that the world would end during a solar eclipse on 12 August 1654. This was a pretty safe bet as the physician Roeslin would have expected to be pushing up daisies in 1654. However, the eclipse occurred on 11 August 1654. Notwithstanding, people stayed indoors and the churches were filled on 12 August.

It was business as usual in the 18th Century with Cardinal Nicholas de Cusa declaring that the end of the world would be in 1704. Although he was a Cardinal, his prediction was not supported by the Vatican, the end did not come. The Swiss Bernoulli family produced eight outstanding mathematicians in three generations. Jacques Bernoulli is well known for discovering the mathematical series now called Bernoulli Numbers and less well known for his prediction that the world would end from a cometary impact on 19 May 1719. Who could dispute such calculations if they were done by a Bernoulli? Both Bernoulli Numbers and the world survived.

The English had their own William Whiston, who predicted that the world would end on 13 October 1736. It didn't. Emmanuel Swedenborg, known for his scholarly concordance, claimed that he frequently consulted with the angels who revealed to him that 1757 would be the end. It was not. The English sect leader Joanna Southcott claimed that the world would end in 1774 and that she was pregnant with the New Messiah. The world didn't end and Joanna did not deliver a bundle of joy.

England has sporadic earthquakes. The 8 February 1761 earthquake was followed 28 days later on 8 March by another earthquake. William Bell persuaded Londoners that the next earthquake would be 28 days later on 5 April. This, rather like climate predictions, was a linear projection based on two points. Many left towns, mainly by boat. On 6 April Londoners came to their senses and threw Bell into Bedlam, the institution for the mentally disturbed.

John Turner, prophet and follower of Joanna Southcott, predicted that D-Day was 14 October 1820. It was not. The dates of 3 April 1843, 7 July 1843, March 21 1844 and 22 October 22 1844 were predicted by William Miller as end-of-the-world dates. Surely just one date was enough? The end was to be preceded by a midnight cry in 1831, and a spectacular meteor shower in 1833 only strengthened Miller's prophecies. On each appointed date, Millerites would gather on hilltops awaiting the end. Up until Miller's death in 1849, the credulous still believed that Miller could predict the end of the world. Egyptologists got into the act and, from measurements of the Great Pyramid at Giza, some claimed that the world would end in 1881. Remeasuring gave a more accurate date of 1936 and even more detailed remeasuring gave the date at 1953. Richard Head, an example of nominative determinism, published a book in 1684 called *The Life and Death of Mother Shipton*. A reprint in 1862, replete with forged rhymes attributed to Mother Shipton, predicted the end in 1881. It was then claimed that the end was to be in 1891. Anyone for 1981 or 1991?

The 20th Century was no different. Despite the horrors of two world wars, John Ballou Newbrough predicted that 1947 was the year. The US and other governments were going to be crushed and Europe again would have massive depopulation from war. It didn't. There have been numerous late 20th Century predictions of population and environmental catastrophes[2201] in the style of Thomas Malthus (1766–1834), all of which have been spectacularly wrong because they omitted to consider advances in science and technology. In 1980s, we had a few choices of dates for the end of the world. When Saturn and Jupiter were almost in conjunction in the sign of Libra on 31 December 1980, the world was to end. It didn't. The planets were aligned on 10 March 1982. A 1974 book, *The Jupiter Effect*, predicted that there would be earthquakes on that day. The problem is that there are earthquakes every day, whether the planets are aligned or not. Earthquakes in 1980 were touted as

---

[2201] Ehrlich, P. 1968: *The population bomb*. Ballantine.

the premature result of "the Jupiter effect". In fact, anything that occurred on planet Earth at that time was due to "the Jupiter effect". Just as today any extraordinary weather phenomenon is promoted as evidence of global warming.

At some unspecified time in the 1980s, Jeane Dixon predicted that a comet would destroy the Earth. It didn't. Since one day for God represents 1000 years for man and God creatively toiled for six days, then man should toil for 6000 years and then take a rest. A long rest. A permanent rest. By this calculation the world didn't end in 1996 although, while in an embrace with Bacchus, I might have missed the end. Quatrain 10-72 of Nostradamus tells us that July 1999 was the time. It wasn't. Millennium cults had a field day in 1999 and 2000. The world didn't end. Computers did not fail with the Y2K bug. Aeroplanes did not fall out of the sky. The world just kept on doing what the world does.

In the late 20th and early 21st Centuries, the world was going to end with a nuclear holocaust, Chernobyl, fluoride, a new ice age, DDT, soil loss, planetary alignments, mega-famines, AIDS, peak oil and the end of oil, the second coming, 9/11, GM foods, breast implants, acid rain, ozone holes, Y2K bug, avian flu, SARS, mad cow disease, acid oceans, asteroid impacts, Cuban crisis, global warming, inflation, financial booms, financial crashes, political assassinations, wars and goodness knows what else. At my age, I have experienced all of these end-of-the-world scenarios, I have experienced three climate changes and have seen better health, greater longevity, greater wealth, better education, better transport and less famine. All of them brought to the planet by science, technology and capitalism. More people die in winter than in summer, there is more depopulation in global cooling events than in global warming events, and yet we are the first generation on planet Earth to fear warmth.

For millennia, people have been predicting the end of the world. These predictions have been based on religion, science and mathematics. They are normally blessed with moral overtones. If just one of these predictions were correct, then we would not be here. Apocalyptic predictions have a 100% failure rate. It is really very hard indeed to be 100% incorrect. Those making apocalyptic predictions have no interest in improving life on Earth, they just demonstrate a complete denial of reality. We fragile humans probably need to fear the unknown as a fundamental biological survival mechanism making us alert to dangers.

Climate zealots warn us of a future catastrophe and that we must pay penance and change our ways. They use a narrow body of science and some mathematics and the message is given with religious vigour. There is no reasoned argument presented, hence reason cannot be used to evaluate contrary data and change conclusions. One is reminded of the words of Jonathan Swift (1667–1745): "It is useless to attempt to reason a man out of a thing he never reasoned into."

Politicians and the public are frightened witless. There is also a universal fear that the hypothesis that human activity causes global warming will be debated and that this hypothesis will be shown to have poor foundations. Attempts to restrict free speech and calls for censorship of alternative views are made by climate zealots. Such actions have characterised salvationist cults down through the ages.

I suggest that we give the climate zealots the same treatment given to previous prophets of doom such as Count von Iggleheim, Stifelius, Rabbi Zevi and William Bell. The next time someone comes to your front door and tells you that the world is going to end, sool the dog onto them. History is on your side.

There is no use for an honest scientist who says "I don't know". Yet uncertainty is the crux of science whereas certainty underpins religious beliefs. The politicians and the public prefer to hear scientists give confident black-and-white answers and make confident predictions. Uncertainty and predictions that all is well are far less likely to attract attention than those that say we're doomed.

It is hereby declared that the end of the world is cancelled. History is on my side.

## Religion, environmentalism and romanticism

Any system that allows the questioning of beliefs is an enlightened system. The truth can only be determined by having, without fear, vibrant critical and analytical discussion, by embracing rather than fearing uncertainty and by not suppressing evidence that may be contrary to one's treasured beliefs. Some 150 years ago, John Stuart Mill stated:

> Complete liberty of contradicting and disproving our opinion is the very condition which justifies us in assuming its truth for purposes of action; and on no other terms can a being with human faculties have any rational assurance of being right.

This does not happen with the populist global warming movement. In schools today we teach scientific "facts" the same way as theological "facts" were taught centuries ago. Global warming has become the secular religion of today. In contrast, those in the knowledge business pursue facts and objective truths, they are rooted in reality and, on the basis of new validated information, constantly change their conclusions.

Dogma, suppression of alternative ideas and reliance on authority are characteristics of fundamentalist religions. Of great concern is that there are data errors in the system designed to provide accurate temperature data. It is this data that is the cornerstone of the climate models. However, it appears that data and the scientific process are not required in the new secular religion.

Human-induced global warming is an unproven scientific hypothesis yet it has become an article of scientific dogma. The peer review process in climatology research is controlled by the secular equivalent of the Collegium Romanum, the IPCC. They in turn are answerable to the Inquisition, the global warming fundamentalists, who in today's world cannot yet resort to instruments of torture.

Despite our comfortable materialistic lives in the Western world, there are many who ask: Is that all? They want a meaning to life and yearn for a spiritual life. Some follow the traditional religions, others embrace paranormal beliefs, superstitions and irrationality and many follow a variety of spiritual paths. Established religion in Western societies has taken a giant backward step, there is a huge spiritual gap and many people want something to believe in. In fact, many will believe almost anything just to fill the yawning spiritual vacuum.

A new religion has been invented to fulfil this need: extreme environmentalism. It is an urban atheistic religion disconnected from the environment. The rise in environmentalism parallels in time and place the decline of Christianity and socialism and incorporates many of the characteristics of Christianity and socialism. Just as the Roman Empire discovered, when the masses have embraced a new religion, the state must follow. Environmentalism is an urban religion disconnected from Nature, or rural life, or the realities of food and mineral production. This environmental religion is terrified of doubt, scepticism and uncertainty yet claims to be underpinned by science. It is a fundamentalist religion with a fear of Nature. It has its own high priests such as Al Gore and its holy writ, such as the IPCC reports and the Kyoto Protocol. Instant theological gratification occurs with the various future scenarios in the IPCC summaries. Like many religious followers, few have ever thoughtfully read and understood the holy books.

Like many fundamentalist religions, it attracts believers by announcing apocalyptic calamities unless we change our ways. Fear is bankable. The new

environmental religion exploits fear and those who genuinely believe that
something must be done to save the planet are attracted to this new religion.
The converts speak and interact like traditional evangelicals upon conversion
("How were you converted?" i.e. How have you established your credentials?).
Its credo is repeated endlessly and a new language has been invented that
separates believers from non-believers. Environmental evangelism has a ritual
and language that has replaced substance. Logic, questioning or contrary data
are not permitted. Heretics are inquisitorially destroyed. In 1600, Giordano
Bruno was burned at the stake for supporting the Copernican theory of a
Sun-centred universe. In 1632, Galileo was accused of heresy for supporting
the Copernican view and was forced to retract. Now the Royal Society has
issued an edict excommunicating all those who argue that the Sun and not
human activity causes variations in the Earth's climate. Again, it is the Sun
that appears to be causing all the trouble.

A new class of high priest resorts to the traditional methods of
enforcement. In order to establish the essential fear-provoking scenario they
have nominated in the role of original sin one particular element, one element
out of the 92 natural elements in the periodic table. To the rational mind it
is a bizarre choice, yet one that conforms to the long-established principles
of the founding of authoritative religions. Why is it bizarre? If you are of a
mind to seek out mystery, magic and miracles look no further than the sixth
member of the periodic table of elements.

Like other fundamentalist apocalyptic religions, it states that now is the
most important time in history and people are told that humanity is facing
the greatest crisis ever. We must make great sacrifices. Now. This new
environmental religion is underpinned by Judeo-Christian thinking: If the
world has changed, then we humans are to blame. This New Age religion
tries to remystify the world, a world that its adherents neither experience nor
try to understand.

The environmental religion produces widespread fear and a longing for
simple all-encompassing narratives. It offers an alternative and static account
of a natural world with which adherents have little contact. Environmentalism
is not a connection with the natural world. The greens' construction of self-
understanding parallels New England (USA) Puritanism, where the structure
of what is being said is of import, not what is being said. Environmentalism
is an ascientific disconnection from the natural world arising from a modern
urban lifestyle where the necessities of life come from shops. An urban life
shields consumers from the effects of their lifestyle, they become blameless
and enjoy the fruits of an affluent consumer life while producing little.

Vicarious experiences with Nature are learned and romanticised via television or the internet. The urban environmentalist does not experience flood, drought, forest fires, subsistence farming, crop failures, dust storms, insect plagues and changing seasons. The urban environmentalist does not know about or appreciate seasonal foods because of rapid international trade.

Environmentalism develops a righteous cause, politicians promise to save this and that and the media are provided with a never-ending source of drama with all the essential ingredients. Unholy alliances are forged between environmental groups with no accountability, politicians, bureaucrats, academics and the media. Minority groups (such as farmers and miners) who provide the basic necessities for urban life are sitting ducks for cheap shots by environmental groups and politicians.

The doomsdayers promote their new religion with seven-second television grabs. A disunity between religion and science is created. Observational science is field-based, computer modelling is urban-based. The very way in which science is undertaken encourages disunity and exacerbates the divide between the urban atheistic religion of environmentalism and rural reality. The science that derived from the Enlightenment and which bathes in doubt, scepticism and uncertainty is willingly thrown overboard. Contrary facts are just ignored, enthusiastic reporting by non-scientists is undertaken and new science is reported with alarmist implications, yet there is no reporting of contrary information. Non-scientific journalists and public celebrities write polemics that encourage public alarm.

Rabid environmentalism embraces the hallmarks of fundamentalist Christianity. The end of the world is nigh. Judgement Day is at hand. Repent. Environmentalism embraces the Fall – the loss of harmony between man and Nature caused by our materialistic society. It romantically searches for the lost Eden of the past. In fact, long ago there was only bitter cold, struggle, starvation, malnutrition, death and unemployment. There was no harmony with Nature. The environmentalists' lost Eden never existed. Humans have burned and eaten the environment since Adam was a boy. We are the first generation to try to act otherwise.

For billions of people around the world, these are the best times to be alive. From Beijing to Bratislava to Benin, more of us are living longer, healthier and more comfortable lives than at any time in history. Fewer of us are suffering from poverty, hunger or illiteracy. Pestilence, famine, death and war, the Four Horsemen of the Apocalypse, are in retreat, thanks to the liberating forces of capitalism, science and technology. If indeed temperature is rising, it is good news. Previous warm spells (e.g. Medieval Warming,

Roman Warming, Minoan Warming) are associated with prosperity and the advance of civilisation. By contrast, cold has been the deadly killer.

Both environmentalism and fundamentalist religions foster a sense of moral superiority in the believer. They create a sense of guilt. Our wickedness has damaged our inheritance and, although it is almost too late, immediate reform can transform the future. The fact that we Westerners can live a comfortable life is exploited as guilt. There is no necessity to tell the truth, as I have found in various battles with creationists.[2202] I see little difference between the US-based creationist movements and extreme environmentalism. In Western Europe, extreme environmentalism has superseded Christianity as a modern urban atheistic religion of the middle classes, and converts witness their faith through NGOs, the Worldwide Fund for Nature, Greenpeace, World Business Council for Sustainable Development, talk fests at Rio, Nairobi and Bali and various UN activities.

History shows that it was warmer in the Minoan Warming, the Roman Warming, the Medieval Warming and the Late 20th Century Warming. Why is the Late 20th Century Warming anything special? Both history and science are dismissed because they are not in accord with the dogma chanted *ad nauseam* by the high priests. It is very dangerous for society to dismiss history because it is not in accord with dogma. Science is a celebration of doubt, there is always argument about data and about the interpretation of data.

The environmental religion embraces anti-human totalitarianism. Some environmentalists consider their ideas and arguments to be an indisputable truth and use sophisticated methods of media manipulation and public relations campaigns to exert pressure on policy makers to achieve their goals. Their argument is based on the spreading of fear and panic by declaring the future of the world to be under serious threat. In such an atmosphere they continue pushing policy makers to adopt illiberal measures, impose arbitrary limits, regulations, prohibitions and restrictions on everyday human activities and make people subject to omnipotent bureaucratic decision making.

Global warming hysteria has a chilling effect on free speech. Those scientists who, on the basis of evidence, have a contrary view are written off as deniers and compared to Holocaust deniers. If the arguments to support human-induced global warming were undeniably strong, then such hysteria would not be necessary. From Torquemada to McCarthy, we have had calls for censorship on the basis of protecting the public from dangerous seditious contrary ideas. Torquemada wanted to save humanity from religious heresy. McCarthy wanted to save Americans from communism. Our modern censors

---

[2202] Plimer, I. R. 1994: *Telling lies for God.* Random House.

want to save the planet. From what?

A former spokesman of the Royal Society joined 37 other signatories demanding that a TV company make changes to the Martin Durkin film *The Great Global Warming Swindle*. It was alleged that this film had a "long catalogue of fundamental and profound mistakes" and that these "major misrepresentations" should be removed before the film was distributed to the public as a DVD. The same self-chosen 38 did not write to Al Gore demanding corrections to his film *An Inconvenient Truth*. Furthermore, there are thousands of DVDs on sale that make scientific claims to support crackpot ideas. Have the chosen 38, in their spirit of public interest, taken it upon themselves to complain about such anti-scientific ideas?

Gore admits that there are errors in his film but, notwithstanding, his film will stimulate debate about climate change amongst school children. This is allowable by the self-appointed chosen 38 because, despite elementary scientific errors, it presents the correct moral outlook. Because Durkin's film has what has been deemed an incorrect moral outlook, it is vilified. School children do not have the breadth and depth of scientific knowledge to demonstrate the scores of scientific errors in Gore's movie. The Gore film has not stimulated debate amongst school children. It has been accepted as fact.

The filling of a spiritual vacuum by environmentalism creates an ever greater spiritual vacuum. The environmental religion based on climate change catastrophism is itself a catastrophe that we inflict upon ourselves at huge intellectual, moral, spiritual and economic cost. This has been long recognised by religious leaders. For example, Australia's Cardinal George Pell wrote:

> … pagan emptiness and fears about nature have led to hysteria and extreme claims about global warming. In the past, pagans sacrificed animals and even humans in vain attempts to placate capricious and cruel gods. Today they demand a reduction in carbon dioxide emissions.

In science, we are in awe of Nature. In religion, we are in awe of God. Yet the new environmental religion is in awe of nothing. It is spiritually vacuous and negative. Christianity has a long tradition of using music for worship. This music, especially from the time of Bach and onwards, underpins all Western music. The environmental religion has no music, no traditions, no scholarship, no nothing. The new environmental religion has no big questions. It has no unknowns. When environmentalists recognise the religious aspects of their stance, then real discussion with scientists becomes possible. Until then, they are no different from creationists who claim that their stance is scientific when their very foundations are religious, dogmatic and fraudulent. Religion (*religare*) can mean to bind fast. The contradictory religion of environmentalism has given people a purpose in life and binds disparate (and desperate) groups. Despite ignoring all the contrary science, this new religion provides some of the stitches that hold the fabric of society together.

The laboratory for religious life and practice is experience. Religion is not about pie in the sky when we die, it is about the present. Religion tries to make sense of what's happening to us now and gives us the mechanisms whereby we can have hope for a meaningful life, in spite of crippling disappointments. Religion gives us the mechanism to cope with failure. Environmentalism cannot provide for any of these needs.

The global warming movement has joined disparate groups in a common cause. It has unified a collective of prejudices with a perceived moral high ground. The environmental romantics hate industry, love Nature, idealise peasant life, believe capitalism is wicked, think people in modern society lead depraved shallow lives and have forgotten the true value of things, don't like cars or supermarkets and hate the average person taking cheap long-haul flights to warm areas for holidays. We humans normally seek a warmer climate for our holidays. Maybe warming is good for us?

The environmental romantics have a loathing and fear of population increase, seek to return to the past and promote pagan superstitions. Well before the crutch of global warming appeared, the environmental romantics hated the modern world despite the fact that in industrial societies we live longer, we are healthier, the air and water are getting cleaner, the area of forests is expanding and we have far greater freedoms than in past times. It is the energy-intensive communication systems of the modern world that allow the environmental romantics to spread the word.

The world of the environmental romantics does not exist. Their world is one that would result in no electricity or potable water, and warm food and heating would be from the burning of dung, wood or poor quality coal in an unventilated small hovel. These fires would emit carcinogenic smoke, create respiratory diseases (especially in women), destroy forests and create widespread atmospheric pollution. There would be neither light nor heat in many hovels. The death toll would be horrendous. Sustainable living is viewed by the environmental romantics as a virtue. In effect, sustainable living is such that, with the slightest change in weather, climate or politics, there is disease, mass famine and death. Sustainability creates a miserable existence, poverty, disease, depopulation and ignorance.

Self-denial and a return to the past led to the 600-year Dark Ages after the fall of the Roman Empire and an age of darkness in the Islamic Empire starting in 1566 AD. By trying to work with Nature, one faces famine and death. Being creative and riding the waves of change is the only way we humans have survived. Mother Nature does not build Gardens of Eden for the eco-conscious. She has given us at least 50 mass extinctions, she is the mother of catastrophe and constantly changes the unwritten rules.

The tradition of Judeo-Christian religions is strongly coupled with philosophical optimism. God did not put us here on Earth to moan and groan. Our human heritage has equipped us well to deal with the challenges from ice ages and cave bears to disease and overpopulation. The whole species co-operated to eliminate smallpox and the women of Mexico united to reduce their average family size from 7 to 2.5 in 50 years. Science has helped us to understand challenges and to defeat them. Science is the only way to tackle modern environmental problems such as toxic pollutants in our air, water and soil and over-exploitation of the planet's resources. These problems and their solutions can only be understood with apolitical dispassionate science. Non-scientific ideological-based extreme environmentalism inhibits solutions to problems.

We are one of the most pathetic species on Earth. We are born as hairless, immobile, long-term suckling creatures with neither fangs nor claws. It takes years before we are independent enough to feed ourselves. Even cockroaches and mice can sprint faster than we can. Yet we live on the ice, at altitude in mountains, at sea level, in the tropics and in deserts. How did we do it? We did not work with Nature, we conquered Nature. Despite natural disasters. And before we get too arrogant, the ultimate survivors and conquerors of Nature on planet Earth are bacteria. They are the largest biomass on Earth, have been on Earth for 4000 million years and can survive in a diversity of extremely hostile environments. They have survived the five major mass extinctions of life on Earth.

Photosynthesis for the survival and growth of plants requires $CO_2$. Animals require plants. A significant reduction in atmospheric $CO_2$ would slow or even stop plant growth. The end result of green environmental political policies is not going green but going brown by reducing vegetation. This will destroy subsistence agriculture, and create famine and widespread suffering.

Some 40% of the people on Earth have neither reticulated electricity nor potable water. This is a matter of life and death. These 40% live in impoverished nations. The cheapest electricity is generated from dirty coal and the denial of cheap electricity to 40% of humans is immoral. In the

Western world, we take clean water and electricity for granted. This is a demonstration that capitalism, science and technology have delivered on a truly spectacular scale. However, the policies of the Western-based green movements keep impoverished people in misery, exacerbate curable disease, create food shortages, destroy economies, kill people and offer no practical solutions to the plight of those without potable water and electricity.

The romantic environmentalists just don't care, or prefer not to care, as long as they adhere to their political faith. Once the ideology has been shown to be spectacularly wrong, the environmental romantics just move onto another cause underpinned by another flawed ideology. This suspension of disbelief, critical facilities and scepticism and the reliance on faith has long been known as a human fragility.[2203] Environmental romantic policies are already starting to bite. Many Western cities are now having water shortages and irregular electricity supply because no new dams and coal- or nuclear-fired power stations have been built as a result of decades of green political pressure.

One hears, "If Al Gore's film has inaccuracies, isn't it worth it just to get the message across?" It's professionally unacceptable, unethical and untruthful. That's why. In comfortable Western countries, a little fiction may do little harm and may even result in some good. However, in developing countries diversion of resources to address the myth of human-induced climate change promoted by Gore is catastrophic. Funds could otherwise be used to help with pressing issues such as clean water supplies, reticulated electricity, health services and education.

Global warming hysteria is big business. Just follow the money. Various green movements claim that those who do not accept the hypothesis that humans are causing climate change have this view because they are supported by the petroleum and coal industries. A US Senate report shows that the greens are the best-funded quarter of the advocacy industry. Between 1998 and 2005, the 50 biggest green movements in the USA attracted revenue of $22.5 billion.[2204] This is the GDP of a few impoverished African countries. Such funds could provide massive improvements in the health of millions of people and would have a far greater environmental impact on the planet than advocacy.

---

[2203] *"That willing suspension of disbelief for the moment, which constitutes poetic faith….."* (*Biographia Literaria*, Chapter XIV [1817], Samual Taylor Coleridge [1772-1834]).

[2204]    http://epw.senate.gov/public/index.cfm?FuseAction=Minority,Blogs&ContentReco rd_id=38d98c0a-23ad-48ac-d9f7facb61a7.

Bjørn Lomborg provides a faint glimmer of boring objectivity.[2205,2206] He has undertaken cost-benefit analyses of global challenges such as disease, pollution, conflict, terrorism, climate change, water and so on. Climate change does not even make his top ten list. If $1 billion is spent on the important problems, some 600,000 children will be prevented from dying and about two billion people will be saved from malnutrition. A $2 billion expenditure on climate change would only stop warming by about two minutes at the end of the 21st Century. If $60 million per annum is spent on micro-nutrient supplements for malnourished children, yearly benefits through improved health, fewer deaths and increased earnings would be worth more than $1 billion. Deworming and community-based nutrition, ending trade tariffs and clean water are not glamorous. The most spectacular pictures of glaciers calving, hurricanes and cute animals such as a polar bear allegedly stranded on ice do not show the most pressing issues for humanity. If the emotion and hysteria of the climate change argument were removed and there were a quest to find the objective truth about the state of the planet, then more pressing issues might be on the agenda. The green movement seems to be quite happy to turn a blind eye to pressing issues such as medicine and nutrition.

Since the inception of the Kyoto Protocol, some $10 billion a month has been spent to avert a speculated 0.5°C temperature rise by 2050. These funds would already have provided all of the Third World with potable water, reticulated electricity and would have reduced global atmospheric pollution.

And yet many environmentalists cant about morality and ethics!

## The Kyoto Protocol

On 2 February 2007, amid great fanfare, the IPCC released the *Summary for Policymakers* for the Fourth Assessment Report. The press release stated:

> The 2007 IPCC report, compiled by several hundred climate scientists, has unequivocally concluded that our climate is warming rapidly, and that we are now at least 90% certain that this is mostly due to human activities. The amount of carbon dioxide in our atmosphere now far exceeds the natural range of the past 650,000 years, and it is rising very quickly due to human activity. If this trend is not halted soon, many millions of people will be at risk from extreme events such as heat waves, drought,

[2205] Lomborg, B. 2001: *The Skeptical Environmentalist.* Cambridge University Press.
[2206] Lomborg, B. 2007: *Cool It! The skeptical environmentalist's guide to global warming.* Published online 31st August 2007.

floods and storms, our coasts and cities will be threatened by rising sea levels, and many ecosystems, plants and animal species will be in serious danger of extinction.

The Kyoto Protocol is based on global temperature changes and atmospheric $CO_2$ changes. A plot of various temperature estimates (Hadley and satellite) against $CO_2$ concentration (Mauna Loa) since the Kyoto Protocol shows that there is no correlation between temperature and $CO_2$, yet a correlation between temperature and $CO_2$ was the entire basis for the Kyoto Protocol. In fact, the temperature-$CO_2$ plot shows cooling despite an increase in atmospheric $CO_2$. Furthermore, the $CO_2$ rise is linear yet $CO_2$ emissions from rapidly growing economies like China and India are certainly not linear. This immediately rings alarm bells about the reliability of $CO_2$ measurements. Oceans, soils and plants already absorb at least half of the human $CO_2$ emissions and human emissions are dwarfed by the balanced natural system. For example, termite methane emissions are 20 times more potent than human $CO_2$ emissions, and massive volcanic eruptions (e.g. Pinatubo) emit the equivalent of a year's human $CO_2$ emissions in a few days.

The Kyoto Protocol classifies the Gulf States as developing countries, hence they are under no obligation to reduce $CO_2$ emissions. They can produce a megawatt-hour of electricity using Australian coal for $US17.49 (mid 2008 dollars). Using home-grown petroleum products, they can generate a megawatt-hour of electricity using their own gas ($US41.34) or oil ($US79.50). It is far cheaper for the Gulf States to import Australian coal for electricity generation and sell petroleum. The sun-drenched Gulf States generate a combined 36 megawatts of solar electricity. In 2007, gloomy Germany produced 1300 megawatts. The Gulf States have used their developing country capacity to generate the cheapest power whereas the Germans have used the most expensive and unreliable method for generating electricity. When science and economics are ruled by ideology, bizarre events occur.

## Oh dear, oh dear, oh dear

The hypothesis that human activity can create global warming is extraordinary because it is contrary to validated knowledge from solar physics, astronomy, history, archaeology and geology. Extraordinary claims require extraordinary evidence. The requirement for extraordinary evidence has tempted some scientists to create evidence by a diversity of dubious methods.

Irving Langmuir[2207] argued that there is good science, pathological science and pseudoscience. Langmuir argued that there are cases where there is no dishonesty involved but where people are tricked into false results by a lack of understanding about how humans can be led astray by subjective effects, wishful thinking and threshold interactions. Pathological science attracts much media attention, even for decades, and then is quietly forgotten. Langmuir's Rules are:

(i) The maximum effect that is observed is produced by a causative agent of barely detectable intensity.

(ii) All observations are near the threshold of optical visibility.

(iii) There are claims of great accuracy.

(iv) Fantastic theories contrary to experience are created.

(v) Criticisms are met by *ad hoc* excuses thought up on the spur of the moment. There is always an instantaneous answer to criticism.

(vi) The ratio of supporters to critics rises to near 50% and then decreases gradually to oblivion. During this process, only supporters can reproduce the effects, critics cannot.

The science supporting human-induced global warming fails on at least three of Langmuir's Rules.

In my experience of more than 40 years of science, scientific research is difficult, takes time, one's favourite theories may require abandonment on the basis of new validated evidence, and significant new discoveries contrary to the established validated body of evidence are rare. Scientists hold no shining moral beacon. Many findings are promoted with over-enthusiasm and delusion is not uncommon. As with any other discipline, fraud, creating data *ex nihilo*, stupidity and mistakes occur. Charles Babbage (1792–1871)[2208] suggested that there are three forms of scientific dishonesty:

(i) Trimming (the smoothing of irregularities to make data look extremely accurate).

(ii) Bias (retention of data that fits the theory and discarding data that does not fit the theory), and

(iii) Forging (inventing some or all of the data).

Some science supporting human-induced global warming (most notably the "hockey stick" of Mann) fulfils at least two of these criteria. Creation "science" fulfils them all. Fraud in science is undertaken, like other fraud, for

---

[2207] Langmuir, I. 1989: Pathological science: scientific studies based on non-existent phenomena. *Physics Today* 36: 47 (transcribed and edited by R. N. Hall).

[2208] Feder, K. 1996: *Frauds, myths, and mysteries: Science and pseudoscience in archaeology*. Mayfield Publishing.

financial gain, fame, nationalism, racial pride, religion and a Rousseauvian desire for a romantic past. Some pseudoscience, such as creation "science", is the fruit of an unsound mind.

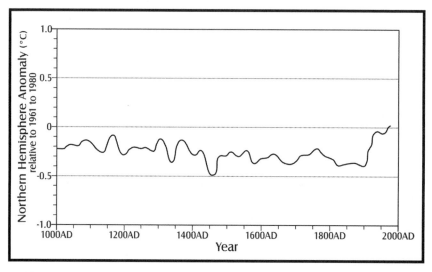

*Figure 54: Mann's "hockey stick" which fulfils two of the three criteria of Babbage for scientific dishonesty. The "hockey stick" manages to ignore the hundreds of scientific studies that show the Medieval Warming (900-1300 AD) was far warmer than at present and the Little Ice Age (1300-1850 AD) was far colder than at present. The "hockey stick" attempts to show that 20th Century temperatures have risen at an unprecedented rate.*

It just suits an awful lot of people to believe that human activity has a significant effect on temperature. These are environmental movements under various guises (e.g. Greenpeace, WWF, FoE), the Green political parties, politicians (who wish to use the hypothesis to justify various actions or a lack of action), the media (whose survival is based on attention-grabbing stories), some scientists (who see it as a treasure chest to plunder for research and to increase reputations) and the IPCC (whose *raison d'être* would disappear if no human influence was found).

Scientists, like other professions, do not have a mortgage on ethics, morality and honesty. However, the validation, repeatability and coherence processes of science are a long-term audit of science. This process has detected mistakes, fraud, cultural bias and incompetency.[2209] I do not think that scientists manipulate their results to get on the gravy train of research

---

[2209] Gould, S. J. 1996: *The mismeasure of man*. Norton.

funding. However, there are limited funds for research and if there is a popular subject of research, then it is funded better than other areas. If there is a war on cancer, then a higher proportion of biological research becomes cancer research. If we are frightened to death about the effects of global warming creating more insect-borne disease, then it makes good electoral sense to make more funding available for insect-borne disease research.

Although Occam's Razor[2210] is commonly used in science and is a general rule of thumb, sometimes Nature plays cruel tricks and the simplest explanation is not necessarily the explanation supported by data. Hypotheses must be tested. The hypothesis that human activity has produced global warming fails on a number of counts. Once there is an initial hypothesis, then specific testable predictions are made. Predictions made from the hypothesis that humans have produced global warming have failed.

Our Western society is based on science. It is this science that has given us technology. This technology has been used by capitalism to create wealth. Notwithstanding, there are a large number of people in Western countries who believe weird things. Michael Schermer has identified some 25 fallacies that lead people to believe weird things.[2211] The theory that human activity has produced global warming follows at least six of these 25 fallacies.

Flawed scientific papers that accept the popular paradigm are easily published. Papers critical of the popular paradigm and the replies to criticism are more difficult to publish. For example, it was claimed that the summer temperature in Burgundy (France) could be derived from 1370 AD to the present from grape harvests.[2212] The paper claimed that 2003 was the warmest summer since 1370 and this bold assertion certainly would have helped publication and media publicity. After a tortuous effort to obtain the authors' primary data, and complaints to *Nature*, it was statistically shown by Douglas Keenan[2213] that there was no basis for this claim. Keenan writes on his website:

> What is important here is not the truth or falsity of the assertion of Chuine *et al.* about Burgundy temperatures. Rather, what is important is that a paper on what is arguably the world's most important scientific topic [global warming] was published in the world's most prestigious scientific journal with essentially no checking of the work prior to publication.

[2210] *Entia non sunt multiplicanda.*

[2211] Schermer, M. 1997: *Why people believe weird things: Pseudoscience, superstition, and other confusions of our time.* W. H. Freeman and Company.

[2212] Chuine, I., Yiou, N., Viovy, N., Seguin, B., Daux, V. and LeRoy Ladurie, E. 2004: Grape ripening as a past climate indicator. *Nature* 432: 289-290.

[2213] Keenan, D. J. 2007: Grape harvest dates are poor indicators of summer warmth. *Theoretical and Applied Climatology* 87: 255-256.

This is not the first time that editors and reviewers of *Nature* had not checked the work of authors making astounding claims (e.g. Mann *et al.*).

The Stern Review looks very impressive. A reader of the first 21 pages devoted to "The Science of Climate Change" could not fail to be impressed by the 74 references and the opening to the Chapter:

> An overwhelming body of scientific evidence now clearly indicates that climate change is a serious and urgent issue. The earth's climate is rapidly changing, mainly as a result of increases in greenhouse gases caused by human activity.

The science, of course, says nothing of the sort. Science does not say climate change is a serious and urgent issue. Politicians and environmental activists make such claims. The science does not say that the Earth's climate is rapidly changing. Climate is changing, as it always does, but the current changes are far slower than previous changes. Even the IPCC does not agree with Stern's assertions. The IPCC[2214] states that: "most of the warming over the last 50 years is likely to have been due to an increase in greenhouse gas concentrations".

Nothing is written about humans or rapidly changing climate yet we are warned by Stern that temperature will rise 0.7°C over the next 150 years. This is exactly what has happened over the last 150 years of post-Little Ice Age climate. This is not a rapid rise and the fluctuations in temperature over the last 150 years are no guarantee that the temperature will rise. It may fall, as it has done in the past.

The Stern report distorts IPCC statements by claiming: "The IPCC concluded in 2001 that there is new and stronger evidence that most of the warming observed over at least the past 50 years is attributable to human activities."

The IPCC statement actually was: "There is new and stronger evidence that most of the warming observed over the past 50 years is attributable to human activities." Stern has exaggerated by adding the words "at least" and somehow omits to mention that for half of the past 50 years the climate was actually cooling. Furthermore, Stern has exaggerated the models for temperature increase by the IPCC. He has tried to include methane gas in his scientific summary, a greenhouse gas that Stern argues has been increasing for the last 50 years. It has not. The Stern report makes no attempt to explain why temperature has been decreasing since 1998. Stern claims that:

> If annual greenhouse gas emissions remained at the current

---

[2214] IPCC, 2001: *Climate Change 2001. The scientific basis. Contributions of working group 1 to the Third Assessment Report of the Intergovernmental Panel on Climate Change* (eds Houghton, J. T., Jenkins, G. J. and Ephraums, J. J.), Cambridge University Press.

level, concentrations would be more than treble pre-industrial levels by 2100, committing the world to 3 to 10°C warming level based on the latest climate projects.

Wrong. There is no established relationship between greenhouse gas emissions and greenhouse concentrations in the atmosphere. Reducing emissions does not necessarily mean a reduction in atmospheric concentrations, unless humans are the only force driving all natural processes on planet Earth.

The Stern Review ignores one of the rare notes of caution in an IPCC *Summary to Policymakers*, which suggests that observed changes may be explained by the greenhouse effect "but do not exclude the possibility that other forcings have contributed".[2215] The Stern Review disregards such caution and claims that: "more than a decade of research and discussion ... has reached the conclusion that there is no other plausible explanation for the observed warming for at least the past 50 years."[2216]

This is untrue. There have been many plausible explanations advanced in the scientific literature,[2217,2218,2219] one of which states:[2220]

> ... the global warming observed during the last 150 years is just a short episode in the geologic history. The current global warming is most likely a combined effect of increased solar and tectonic activities and cannot be attributed to the increased anthropogenic impact on the atmosphere. Humans may be responsible for less than 0.01°C (of approximately 0.56°C total average atmospheric heating during the last century).

The Stern Review ignored any contrary scientific research. Stern does not even understand that models trained to emulate the last 150 years of

---

[2215] *Summary for Policymakers*, p. 10, in: IPCC, 2001: *Climate Change 2001. The scientific basis. Contributions of working group 1 to the Third Assessment Report of the Intergovernmental Panel on Climate Change* (eds Houghton, J. T., Jenkins, G. J. and Ephraums, J. J.), Cambridge University Press.

[2216] Stern Review, page 3.

[2217] Kalmay, E. and Cai, M. 2003: Impact of urbanization and land use change on climate. *Nature* 423: 528-531.

[2218] de Laat, A. T. J. and Maurellis, A. N. 2004: Industrial $CO_2$ emissions as a proxy for anthropogenic influence on lower troposphere temperature trends. *Geophysical Research Letters* 31, L0524, doi: 10.1029/2003GL019024.

[2219] Hale, R. C., Gallo, K. P., Owen, T. W. and Loveland, T. R. 2006: Land use/land cover change effects on temperature trends at U.S. Climate Normals stations. *Geophysical Research Letters* 33: L11703.

[2220] Khilyuk, L. F. and Chilingar, G. V. 2006: On global forces of nature driving the Earth's climate. Are humans involved? *Environmental Geology* 50: 899-910.

climate using data from many sources[2221,2222,2223] are only successful at predicting warming and failed spectacularly at predicting the cooling from 1998 onwards.

The first page of Stern's science has basic errors of fact, exaggeration, misquotation, opinion, science created *ex nihilo* and fulfilment of pre-ordained dogma. And it gets worse. For example, diagrams are redrafted with different base lines and the omission of the order of accuracy and the discussion on Mann's "hockey stick" just happens to omit the scientific demolition of the "hockey stick". These little difficulties in Mann's work are brushed under the carpet because even though temperatures are little different from the past, temperature is only one of the "lines of evidence for human-induced climate change". If temperature essentially does not change, then it is hard to see that climate is changing. Stern states that these lines of evidence rest on "the laws of physics and chemistry". What laws? It appears that if the facts do not agree with the pre-ordained conclusion then they must be dismissed.

The Review is underpinned by the preconceived idea that climate is exclusively influenced by greenhouse gases. The Stern Review, an economic analysis, does not even acknowledge that another major review in the UK on the economics of climate change[2224] was conducted shortly before the Stern Review was commissioned. This does not give confidence.

Leading economists worldwide have totally demolished the Stern Review[2225,2226,2227,2228,2229,2230,2231] and economists and scientists have shown

[2221] Klyashtorin, L. B. and Lyubushin, A. A. 2003: On the coherence between dynamics of the world fuel consumption and global temperature anomaly. *Energy and Environment* 14: 733-782.

[2222] Loehle, C. 2004: Climate change: detection and attribution trends from long-term geologic data. *Ecological Modelling* 171: 433-450.

[2223] Kotov, S. R. 2001: Near-term climate prediction using ice-core data from Greenland. In: *Geological perspectives on global climate change* (eds Gerhard, L. C. et al), Studies in Geology, American Association of Petroleum Geologists 47: 305-316.

[2224] House of Lords Select Committee on Economic Affairs, 2nd report of Session 2005-2006. *The Economics of Climate Change*. Volume 1: *Report*. Volume II: *Evidence*.

[2225] Byatt, I., Castles, I., Goklany, I. M., Henderson, D., Lawson, N., McKitrick, R., Morris, J., Peacock, A., Robinson, C. and Skidelsky, R. (2006): The Stern Review: A dual critique. Part II: Economic aspects. *World Economics* 7(4) 199-229.

[2226] Tol, R. S. J. and Yohe, G. W. 2006: A review of the Stern Review. *World Economics* 7 (4): 233-250.

[2227] Henderson, D. 2006: Report, response and review: An argument in Britain on climate change issues. *Energy and Environment* 17.

[2228] Mendelson, R. O. 2006: A critique of the Stern Report (sic). *Regulation* 29(4).

[2229] Dasgupta, P. 2007: The Stern review's economics of climate change. *National Institute Economic Review* 199.

[2230] Beckerman, W. and Hepburn, C. 2007: Ethics of the discount rate in the Stern Review on the economics of climate change. *World Economics* 8 (1).

[2231] Henderson, D. 2007: Governments and climate change issues: A case for rethinking. *World Economics* 8 (2): 183-228.

that both the economics and science of the Stern Review are highly questionable.[2232,2233] It is little wonder that the Stern Review has been dismissed as a rather shabby political document unrelated to science which reached its pre-ordained conclusions.

The Stern Review, as well as the IPCC, implicitly assume that without some regulatory intervention future use of fossil fuel would continue unconstrained by supply and price. None of these reports considers any possibility of shortages created by green politics and what this might mean. None of these reports predicted the global financial meltdown. If Stern, an economist, cannot make a prediction in his own area of expertise, then what hope is there of his predictions about climate change?

The Stern Review fails to refer to any science that contradicts the claim that the science of greenhouse gas-induced global warming is settled. It also fails to note that the computer models are trained to emulate the last 150 years of temperature measurement but have failed in forward projections. It fails to acknowledge that studies of past climate over different time scales all have suggested that there is both warming and cooling and that Earth is in an interglacial hence the next major climate change will be cooling.[2234,2235,2236]

The Stern Review emphasises "significant melting and an acceleration of ice flows near the coast" and the possibility of "irreversible" melting of the Greenland ice sheet. Of the four papers relied on to make these statements, two show a net gain in the Greenland ice sheet,[2237,2238] the third indicates an ice loss[2239] and the fourth uses meteorological models to show no significant trend.[2240] This selective use of data does Stern no credit. Papers that show

[2232] Carter, R. M., de Freitas, C. R., Goklany, I. M., Holland, D. and Lindzen, R. S. 2006: The Stern review: A dual critique. Part I: The science. *World Economics* 7(4): 167-198.

[2233] Carter, R. M., de Freitas, C. R., Goklany, I. M., Holland, D. and Lindzen, R. S. 2007: Climate science and the Stern Review. *World Economics* 8(2).

[2234] Khiluk, L. F. and Chilingar, G. V. 2006: On global forces of nature driving the Earth's climate. Are humans involved? *Environmental Geology* 50: 899-910.

[2235] Klyashtorin, L. B. and Lyubushin, A. A. 2003: On the coherence between dynamics of the world fuel consumption and global temperature anomaly. *Energy and Environment* 14: 733-782.

[2236] Loehle, C. 2004: Climate change: detection and attribution of trends from long-term geologic data. *Ecological Modelling* 171: 433-450.

[2237] Zwally, H. J., Giovanetto, M. B., Li, H., Cornejo, H. J., Beckley, M. A., Brenner, A. C., Saba, J. L. and Yi, D. 2005: Mass changes of the Greenland and Antarctic ice sheets and shelves and contributions to sea-level rise: 1992-2002. *Journal of Glaciology* 51: 490-527.

[2238] Johannessen, O. M., Khvorostovsky, K., Miles, M. W. and Bobylev, L. P. 2005: Recent ice-sheet growth in the interior of Greenland, www.scienceexpress.org, 20th October 2005.

[2239] Rigot, E. and Kanagaratnam, P. 2005: Changes in the velocity structure of the Greenland ice sheet. *Science* 311: 986-990.

[2240] Hanna, E., Huybrechts, P., Janssens, I., Cappelin, J., Steffan, K. and Stephens, A. 2005: Runoff and mass balance of the Greenland ice sheet. *Journal of Geophysical Research* 110: doi 10.1029/2004JD005641.

that Greenland is colder than it was in 1940[2241] and that temperature is little
changed from the 1780s[2242] were just ignored. The Review also does not state
that Greenland was as warm or warmer in the 1930s[2243] and that the Antarctic
ice sheet is growing.[2244] The Stern Review also did not acknowledge the
uncertainties and knowledge gaps in science, it relied on studies for which
there was no disclosure of primary data and methods, it relied on advocacy
documents and not scientific documents, and it uncritically accepted models
to explain the causes of climate.

The Stern Review does not recognise that the societies with the cleanest
air and water, the greatest abundance and highest nutrition of foods, and
the most aesthetically and socially pleasing and hygienic environments, are
communities that maximise their use of technology and industry.

The Stern Review gave the UK government exactly what it wanted. The
British taxpayer did not get value for money and should be angry. Very
angry.

In Australia in 2008, the Garnaut Report on climate change and emissions
trading commissioned by the Labor government was released. Professor
Garnaut regularly speaks in public about carbon pollution derived from
industry. He has a hobby cattle farm and must know that his stock are a
massive emitter of methane, a potent greenhouse gas. Since 1995, he has
been chairman of Lihir Gold Ltd, a company that dumps millions of tonnes
of waste and tailings in the ocean adjacent to coral reefs. He is also a director
of Ok Tedi Mining Ltd. In 1999, the Ok Tedi tailings dam failed, some 80
Mt of tailings contaminated 120 villages over 1300 square kilometres and
affected 50,000 people along the Fly River (Papua New Guinea).

There was one paper[2245] that was used to resolve an issue in the latest
IPCC report. The IPCC argued that effects on temperature by urbanisation
are insignificant (i.e. urban heat island effect). This paper had Wei-Chyung

[2241] Chylek, P., Box, J. E. and Lesins, G. 2004: Global warming and the Greenland ice sheet.
*Climatic Change* 63: 201-221.

[2242] Vinther, B. M., Andersen, K. K., Jones, P. D., Briffa, K. R. and Cappelen, J. 2006:
Extending Greenland temperature records into the late eighteenth century. *Journal of Geophysical Research* 111: doi 10.1029/2005JD006810.

[2243] Polyakov, I. V., Alekseev, G. V., Bekryaev, R. V., Bhatt, U., Colony, R. L., Johnson, M.
A., Karklin, V. P., Makshtas, A. P., Walsh, D. and Yulin, A. V. 2002: Observationally based
assessment of polar amplification of global warming. *Geophysical Research Letters* 29: doi
10.1029/2001GL011111.

[2244] Wingham, D. J., Shepherd, J. A., Muir, A., and Marshall, G. J. 2006: Mass balance of the
Antarctic ice sheet. *Philosophical Transactions of the Royal Society A* 364: 1627-1635.

[2245] Jones, P. D., Groisman, P. Y., Coughlan, M., Plummer, N., Wang, W.-C., and Karl, T. R.
1990: Assessment of urbanization effects in time series of surface air temperature over land.
*Nature* 347: 169-172.

Wang as an author, as did another on the same subject.[2246] Each paper compares temperature measurements from selected meteorological stations in China from 1954–1983. One of the stations relied upon by Wang was on the prevailing upwind side of a city and later moved 25 km to be on the downwind side of a city. Another station was in the centre of a city and then was moved 15 km to the seashore. The papers claim respectively:

> The stations were selected on the basis of station history; we chose those with few, if any, changes in instrumentation, location or observation times... [and]  They were chosen based on station histories: selected stations have relatively few, if any, changes in instrumentation, location or observation time.

These statements are the essential foundation for both papers and both statements are untrue. The authors chose 84 meteorological stations, 49 of which had no station history.[2247] Of the remaining 35 stations, one had five different locations from 1954 to 1983 up to 11 km apart, at least half the stations had substantial moves and several stations had an inconsistent history making reliable data analysis unattainable.[2248]

The lead author in the first paper (P.D. Jones) is one of the two co-ordinating lead authors of the IPCC chapter "Surface and Atmospheric Climate Change". It was the same Jones who co-authored a paper with data from Beijing and Shanghai that showed that station relocation substantially affected the measured temperatures.[2249] This paper is in direct contradiction to the paper Jones co-authored with Wang. Approaches[2250] for a copy of the primary data from Jones (University of East Anglia) were met with silence and then, "Why should I make the data available to you, when your aim is to try to find something wrong with it?" Every attempt to acquire the primary data was stonewalled. The UK Freedom of Information Act was then used, and the university initially refused to release the data. A copy of a draft letter to the UK Information Commissioner's Office finally forced the university to release the data. The primary data was made available in April 2007, some

---

[2246] Wang, W.-C., Zeng., Z and Karl, T. R. 1990: Urban heat islands in China. *Geophysical Research Letters* 17: 2377-2380.

[2247] http://cdiac.esd.ornl.gov/ndps/ndp039.html

[2248] http://www.infomath.org/apprise/a5620/b17.htm

[2249] Yan, Z., Jones, P. D., Davies, T. D., Moberg, A., Bergström, H., Camuffo, D., Cocheo, C., Maugeri, M., Demarée, G. R., Verhoeve, T., Thoen, E., Barriendos, M., Rodríguez, R., Martín-Vide, J. and Yang, C. 2002: Trends of extreme temperatures in Europe and China based on daily observations. *Climatic Change* 53: 355-392.

[2250] Approaches by Dr Warwick Hughes (Australia) wanting to evaluate primary data as part of an urban heat island study and by Professor Hans von Storch (Germany) in an US Academy of Sciences presentation, 2nd March 2006.

17 years after the two papers relied upon by the IPCC were published. Since then, a number of papers have shown that the global warming measured in China is due to urbanisation.[2251,2252]

There was fabrication of data upon which the IPCC relied, there was a lack of integrity in some important work on global warming, and the normal process of transparency and release of scientific data was not used. It is clear that urbanisation effects on temperature measurements are highly significant. The IPCC's claim in 2007 that urbanisation is insignificant is invalid.

It took nearly eight years and direct action from the US House of Representatives before the data and computer programs for the 1998 Mann *et al.* "hockey stick" were released. These showed a lack of robustness, statistical flaws and fraud. The IPCC used the "hockey stick" with great fanfare in the IPCC's 2001 report and highlighted it in the *Summary for Policymakers*.[2253] It did not appear in the IPCC's next report, save for an obtuse reference buried in the scientific part of the report.[2254] There was no explanation. This suggests that the IPCC knew that the "hockey stick" had no validity. If the "hockey stick" was valid, it would have been the only science that suggested human-induced global warming. The "hockey stick" is still used by environmental extremists and some scientists promoting human-induced global warming.

Two prominent meteorologists took issue with the cherry picking of scientific information used in the IPCC.[2255] Where peer-reviewed papers in international journals conflicted with the view that global near-surface temperature is increasing, these papers were just simply ignored. The public would not be aware that such science was not considered, would assume that the IPCC had been objective and would clearly be influenced by a conclusion based on incomplete information. In fact, the public had every reason to be confident that all the competing data and conclusions would be aired because the IPCC's 2007 report stated that it was:[2256] "A comprehensive and rigorous

[2251]Yihui, D., Guoyu, R., Zongci, Z., Ying, X., Yong, L., Qiaoping, L. and Jin, Z. 2007: Detection, causes and projection of climate change over China: overview of recent progress. *Advances in Atmospheric Sciences* 24: 954-971.

[2252] He, J. F., Liu, J. Y., Zhuang, D. F., Zhang, W. and Liu, M. L. 2007: Assessing the effect of land use/land cover change on the change of urban heat island intensity. *Theoretical and Applied Climatology* 90: 217-226.

[2253] IPCC, 2001: *Climate Change 2001. The scientific basis. Contribution of working group 1 to the Third Assessment Report of the Intergovernmental Panel on Climate Change* (eds Houghton, J. T., Jenkins, G. J. and Ephraums, J. J.), Cambridge University Press.

[2254] IPCC, 2007: *Climate Change 2007. The physical science basis. Contribution of working group 1 to the Fourth Assessment Report of the Intergovernmental Panel on Climate Change* (eds Solomon et al.) Cambridge University Press.

[2255] http://climatesci.colorado.edu/2007/06/20/documentation-of-ipcc-wg1-bias-by-roger-a-pielke-sr-and-dallas-staley-part-1/

[2256] http://www.ipcc.ch

picture of the global present state of knowledge of climate change."

Such papers were readily available to the IPCC lead authors. Their omission suggests that both the science and the executive summary of IPCC reports had pre-ordained conclusions. The rejected papers were not redundant, they just did not support the assessment presented in the IPCC WG1 Report. Examples of only six of the dozens of omitted papers show the lack of objectivity.[2257,2258,2259,2260,2261,2262] In the crucial chapter (Chapter 9) of the IPCC report, 40 of the 53 authors have either co-authored papers with each other or work in the same establishments as other authors of this chapter.[2263] There is every possibility that they have also acted as peer reviewers for each other. This strongly suggests that the IPCC's claims are those of a small group of climate modellers and not a cross-section of various disciplines that deal with climate, palaeoclimate, atmospheric studies, solar physics, astronomy, archaeology, history and geology. This group of climate modellers has everything to gain and nothing to lose by promoting just one argument. There is one constant: there is no shortage of self-styled climate experts willing to make diabolical predictions and to cast shadows of doom.

Numerous scientific papers contradict the IPCC predictions of increased extreme weather, floods and droughts due to human-induced global warming. All of these scientific studies are ignored by the IPCC. For example, the June 2003 issue of the scientific journal *Natural Hazards* was devoted to ascertaining whether extreme weather is a result of human emission of $CO_2$. The editors concluded that most studies found no such connection. This shows that there is also no causation. River flow data of 44–100 years

[2257] Hansen, J., Ruedy, J., Glascoe, J. and Sato, M. 1999: GISS analysis of surface temperature change. *Journal of Geophysical Research* 104: 30997-31022.

[2258] Chase, T. N., Pielke, R. A. Snr, Knaff, J. A., Kittel, T. F. G. and Eastman, J. L. 2000: A comparison of regional trends in 1979-1997 depth-averaged tropospheric temperatures. *International Journal of Climatology* 20: 503-518.

[2259] Lim, Y. K., Cai, M., Kalnay, E. and Zhou, L. 2005: Observational evidence of sensitivity of surface climate changes to land types and urbanization. *Geophysical Research Letters* 32: L22712, doi: 10.1029/2005GL02424267.

[2260] González, J. E., Luvall, J. C., Rickman, D., Comarazamy, D. E., Picón, A. J., Harmsen, E. W., Parsiani, H., Ramírez, N., Vázquez, R., Williams, R., Waide, R. B. and Tepley, C. A. 2005: Urban heat islands developing in tropical coastal cities. *EOS* 86: 397.

[2261] Hubbard, K. G. and Lin, X. 2006: Re-examination of instrument change effects in the U.S. historical climatology network. *Geophysical Research Letters* 33: L15710, doi:10.1029/2006GL027069.

[2262] Mahmood, R., Foster, S. A. and Logan, D. 2006: The GeoProfile met data, exposure of instruments, and measurement of bias in climate record revisited. *International Journal of Climatology* 26: 1091-1124.

[2263] McLean, J. 2008: *Prejudiced authors, prejudiced findings. Did the UN bias its attribution of "global warming" to humankind.* Science and Public Policy Institute.

duration from the Global Runoff Data Centre (Koblenz, Germany) shows no pattern of increasing or decreasing flooding.[2264] There has been no global change in rainfall since satellite measurements commenced.[2265] In a study of drought in the USA from 1925–2003,[2266] the authors stated: "droughts have, for the most part, become shorter, less frequent, less severe, and cover a smaller portion of the country".

All of the IPCC's and Stern Review's speculated impacts of possible global warming are consistently biased and selective. They all heavily lean towards unwarranted alarm.

The 23-page IPCC *Summary for Policymakers* was rushed out for the 2007 Bali Conference. It is a political document and is unrelated to the science in the body of the IPCC report. The main IPCC report containing the science was published much later, hence the science in the *Summary for Policymakers* could not be validated while the Bali political discussions were in progress. However, this was not necessary. The *Summary for Policymakers* showed cooling for 100 of the last 160 years, during which time greenhouse gases were increasing. This certainly suggests that $CO_2$ of human origin does not drive modern climate. There is absolutely no demonstration of a relationship between the increase in greenhouse gases and temperature.

The *Summary for Policymakers* states that precipitation in southern Africa declined from 1900–2005. This is false. This may have been an attempt to win African votes in Bali. Precipitation increased by 9% over this period. This statement is contrary to a later claim on the same page that heavy precipitation events increased during this period. Despite the fact that climate is driven by the receipt and redistribution of solar energy, there is no attempt to relate solar phenomena with global surface temperature, global average sea levels and snow cover. The driving force for climate on Earth is just omitted. No explanation. No critical analysis. Nowhere can we read that there is a large body of solar physics, astronomical, geological, archaeological and historical evidence to show that there are competing theories for the driving of modern and ancient climate changes. Policymakers are given the opposite scenario and that is that solar activity has caused cooling despite a mountain of data showing that there is a parallel relationship between solar activity and temperature on a multi-decadal scale. Although solar activity is summarily dismissed, there is no attempt to dismiss the well-documented

---

[2264] Svensson, C., Kundzewicz, Z. W. and Maurer, T. 2005: Trend detection in river flow series: 2. Flood and low-flow index series. *Hydrological Sciences Journal* 50: 811-824.

[2265] Smith, T. M., Yin, X. and Gruber, A. 2006: Variations in annual global precipitation (1979-2004), based on the Global Precipitation Project 2.5° analysis. *Geophysical Research Letters* 33: doi 10.1029/2005GL025393.

[2266] Andreadis, K. and Lettenmaier, D. 2006: Trends in 20th century drought over the continental United States. *Geophysical Research Letters* 33: doi 10.1029/2006GL025711.

historical relationships between solar activity and temperature. In the *Summary for Policymakers* it was just omitted. It appears that policymakers are being misled.

Nature does not obey our wishes. Global temperature has not risen in accordance with greenhouse gas emissions for the past 10 years. The global mean temperature has dropped, against all predictions, models and scenarios. This is the most rapid and largest temperature shift in the last 100 years. How many years without a $CO_2$-temperature correlation do we have to suffer? The Earth also cooled between 1940 and 1976, but this is rationalised by supporters of human-induced global warming as being due to an increased quantity of aerosols. What is the excuse for the post-1998 cooling? If it is because it is following a natural event (the extraordinary 1998 El Niño), then why can't the 20th Century warming and cooling also be due to another natural event? The 20th Century followed the six centuries of the Little Ice Age. Surely it is not impossible that after the Little Ice Age, the planet started to warm? Despite the fact that it is cooling, we now have warmers' predictions that climate will continue to warm, with at least half the years in the decade after 2009 predicted to exceed the warmest years currently on record.[2267] Time will test this prediction. Most predictions make astounding claims for the forthcoming centuries, which are safe predictions, as the authors will be long dead and buried before their predictions can be checked.

But, the environmentalists argue, what about the precautionary principle? This principle is a concoction by environmentalists[2268] and is underpinned by the assumption that the planet is not dynamic.[2269] The environmentalists' precautionary principle abandons scientific proof as well as the concept of proof. It legitimises unfounded fears and raises irrational decision making to an art form. Who decides what are threats? The Montreal Protocol used the precautionary principle to attempt to ban chlorofluorocarbons because these gases destroy ozone. However, we use chlorine every day to make water fit to drink yet chlorine also destroys ozone. There is no such thing as the precautionary principle in science. No amount of precaution, whatever that is, is going to stop natural climate change. There is a 100% risk of damage from weather and climate change. This happens every day somewhere on Earth. If we followed the precautionary principle to its logical conclusion,

---

[2267] Smith, D. M., Cusack, S., Colman, A. W., Folland, C. K., Harris, G. R. and Murphy, J. M. 2007: Improved surface temperature prediction for the coming decade from a global climate model. *Science* 317: 796-799.

[2268] Deville, A. and Harding, R. 1997: *Applying the precautionary principle*. Federation Press.

[2269] Goklany, I. M. 2001: *The precautionary principle: A critical appraisal of environmental risk assessment*. Cato Institute.

then we would never get out of bed. On second thoughts, maybe we should get out of bed because more people die in bed than standing up. By getting out of bed we reduce our risk of dying.

The late Dr Roger Revelle, Al Gore's scientific adviser, must be turning in his grave. Before he died, he co-authored a popular paper[2270] stating: "We know too little to take any action based on global warming. If we take any action, it should be an action that we can justify completely without global warming."

Gore's staffers tried to have his name posthumously removed from the paper by claiming Revelle was senile. One of Revelle's co-author's took the matter to court and won. Revelle's name stayed on the paper.[2271] Did this attempted suppression of science get widespread publicity? No, and the Gore gravy train is doing well. The movie has cleared $50 million in takings, he charges $100,000–$150,000 per lecture, is the co-founder and chairman of Generation Investment Management which invests in solar and wind power, and accepted a board position on (the late) Lehman Brothers, an organisation that would have benefited from brokering emissions trading permits. Gore lives in a 20-room shack near Nashville (Tennessee) that consumes only 221,000 kW hours of electricity a year, 20 times the US average. He defends this by stating that he has purchased renewable energy credits to offset his own use. And who did he buy these credits from? You guessed it, his own company, Generation Investment Management.[2272] For a man who made his money from petroleum, he is now making the odd shekel frightening people witless about global warming while he positions himself to make serious money from emissions trading.

The list of scientific misrepresentations is long. Trying to deal with these misrepresentations is somewhat like trying to argue with creationists who misquote, concoct evidence, quote out of context, ignore contrary evidence and create evidence *ex nihilo*. To show that one misrepresentation is wrong takes volumes, as was shown in the eight-year battle to show that the Mann *et al.* (1998) paper was fraudulent. And, no matter what methods Mann *et al.* might use, the Medieval Warming and Little Ice Age just refuse to go away.[2273]

---

[2270] Singer, S. F., Starr, C. and Revelle, R. 1991: What to do about greenhouse warming: look before you leap. *Cosmos* 1: 28-33.

[2271] Singer, S. F. 2003: The Revelle-Gore story: Attempted political suppression of science. In: *Politicizing science: The alchemy of policymaking* (ed. Gough, M.), 283-297, Hoover Institution.

[2272] WorldNewsDaily 2nd March, 2007. "*Gore's carbon offsets paid to a firm he owns.*"

[2273] Soon, W. S., Baliunas, S., Idso, C., Idso, S. and Legates, D. 2003: Reconstructing climatic and environmental changes of the past 1000 years: a reappraisal. *Energy and Environment* 14: 233-296.

An example. A book[2274] by a popular science writer (Gabrielle Walker) and the former chief scientist of the UK (Sir David King) is riddled with creationist-type "science". The discredited "hockey stick" appears. This, the warmists' supreme icon, has been described[2275] as "the most discredited study in the history of science." However, although the "hockey stick" has been discredited in the scientific literature, it is still used in *The Hot Topic*. The "hockey stick" rewriting of history suppressed evidence that in Medieval times temperatures were higher than they are today. This does the warming cause no good. Nor do pictures showing polar bears on floating ice which do not acknowledge that polar bear numbers are increasing and that, as there are polar bears today, they must have survived far warmer times in the recent past such as the Medieval Warming, the Roman Warming, Minoan Warming, the Holocene Maximum and the numerous past interglacials. Walker and Smith must know this. Even if polar bears first appeared in the creationist world at 9 am on 26 October 4004 BC, they still had to survive at least five periods of global warming before they entered the Late 20th Century Warming.

Hurricane Katrina is another warmist icon. What the authors fail to mention is that hurricane activity was more extreme in the 1950s and that one of the reasons for the major damage in New Orleans was the collapse of levees and the subsidence of New Orleans before Katrina. Walker and Smith mention the increasing number of times the Thames Barrier has had to be closed without mentioning that eastern England is sinking, or that this sinking has been known since Roman times, or that the Barrier has been closed to keep in river water rather than to keep the seawater out. Again, an omission of critical data leading to a misrepresentation. Walker and Smith suggest that there were 35,000 premature deaths caused by the 2003 European heat wave and they don't stress that a far larger number of people die from extreme cold in Europe. These errors are not trivial matters of perspective – the very fabric of the warmers' arguments is based on discredited information.

In my own country, the prophetic predictions suggest that Australia, a particularly dry continent, is very exposed to the ravages of global warming. The popular thinking is that higher temperatures will produce increased evaporation, further drying the continent. This view is contrary to the geological data on previous global warmings that showed increased rainfall, filling of inland dry lakes with runoff water and a thriving of life. Long-term desertification and low rainfall have occurred during times of global cooling, especially during glaciation when Australia, although not largely

---

[2274] Walker, G and King, D. 2007: *The Hot Topic: How to tackle global warming and still keep the lights on*. Bloomsbury.
[2275] Melanie Phillips, *The Spectator*, 7th February 2009.

covered by ice, had cold strong winds which reduced vegetation, added sea spray and shifted sand dunes. The popular view is contrary to the 100 years of temperature and rainfall records in Australia that show droughts were in cooler times, in accord with the geological evidence.

This view is also contrary to the fact that increased evaporation from the oceans (about 7% for every 1°C sea surface temperature increase) provides increased water vapour in the atmosphere. This water vapour precipitates as rain. The IPCC computer models significantly underestimated the rate of global precipitation with increased temperature by using 1–3% increased evaporation for every 1°C temperature increase.[2276] Projections of increased drought in Australia are a manifestation of a computer model deficiency and do not reflect what has happened in the 20th Century and over geological time. To make matters worse, the IPCC models just don't do clouds, the engine of weather.

This might appear to be a minor error but it has major ramifications. Evaporation requires energy. The more evaporation, then the cooler the sea surface temperature. Increased atmospheric $CO_2$ enhances the greenhouse effect by increased back radiation. Accordingly, there is a rise in sea surface temperature until the increasing energy loss from the surface balances the back radiation. By underestimating evaporation, the calculated increase in surface temperature is far too high in order to achieve the energy balance. Increased evaporation puts more greenhouse gases into the atmosphere, notably water vapour, the most abundant greenhouse gas. This produces a further increase in temperature. An underestimation of evaporation, as in the current computer models, will amplify effects of greenhouse gases in the atmosphere. The incorrect models show that doubling of $CO_2$ in the atmosphere will produce a temperature increase of 2.5°C (rather than the 0.5°C calculated by other techniques). Such high figures are used by the IPCC and environmental activists. Such a large predicted temperature rise has led to frightening suggestions of a runaway greenhouse, "tipping points" and irreversibility of climate trends. Variations in past climates show that an atmospheric $CO_2$ content of many orders of magnitude higher than the current atmospheric $CO_2$ content have not led to a runaway greenhouse, "tipping points" and irreversibility of climate trends. I wonder if the climate alarmists have ever really thought about the ice core records that show that atmospheric $CO_2$ rises after a temperature increase and that temperature increases to a high point and then starts to decrease. The answers lie in looking at Nature, not at a computer screen.

---

[2276] Wentz, F., Ricciardulli, L., Hilburn, K. and Mears, C. 2007: How much more rain will global warming bring? *Science 317:* 233-235.

Climate models with different parameters for clouds and moisture can change global temperature forecasts, ranging from an increase of 11.5°C to a slight cooling. Another great uncertainty is atmospheric pollution by minute particles derived from industry. Recent dramatic reductions in industrial pollution in Europe may have driven temperature far higher than predicted by global warming models.[2277] The great unknowns are extraterrestrial dust particles and volcanic aerosols. Atmospheric particles reflect some of the Sun's energy back into space.[2278] The size of the effect was far greater than expected and raises a number of questions.[2279] The cooling from 1940 to 1975 was rationalised as due to World War II and post-war industrialisation producing an increase in atmospheric particles. If this was the case, then better industrial practices could have contributed to all the warming from 1976–1998 and then industry started emitting particles again in 1998 to account for the post-1998 cooling.

## What if I'm wrong?

This is not a question asked by warmers. I ask this question of myself because, as a scientist, I am only too well aware that science is never bathed in certainty. What happens if all the solar physics, astronomy, history, archaeology and geology presented in this book is wrong? What if humans are really changing the climate by adding $CO_2$ to the atmosphere?

An address by Viscount Monckton of Brenchley[2280] answers the question I ask. These are Viscount Monckton's points.

Even if global temperature has risen, it has risen in a straight line at a natural 0.5°C per century for 300 years since the Sun recovered from the Maunder Minimum. This was long before industrialisation could have had any influence.[2281]

---

[2277] Ruckstuhl, C., Philipona, R., Behrens, K., Coen, C., Dürr, B., Heimo, A., Mätzler, C., Nyeki, S., Ohmura, A., Vuilleumier, L., Weller, M., Wehrli, C. and Zelenka, A. 2008: Aerosol and cloud effects on solar brightening and the recent rapid warming. *Geophysical Research Letters* 35 L12708, doi:10.1029Gl034228.

[2278] Wild, M., Gilgen, H., Roesch, A., Ohmura, A., Long, C. N., Dutton, E. G., Forgan, B., Kallis, A., Russak, V. and Tsvetkov, A. 2005: From dimming to brightening: Decadal changes in solar radiation at the Earth's surface. *Science* 308: 847-850.

[2279] Bellouin, N., Boucher, O., Haywood, J. and Reddy, M. S. 2005: Global estimate of aerosol direct radiative forcing from satellite measurements. *Nature* 438: 1138-1141.

[2280] The comments in the next few pages derive from an address by Christopher Monckton to the Local Government Association, Bournemouth, 3 July 2008 and are reproduced with permission.

[2281] Akasofu, S.-I. 2008: A suggestion to climate scientists and the Intergovernmental Panel on Climate Change. *EOS* 89: 108.

Even if warming had sped up, the present temperature is 7°C below most of the last 500 million years, 5°C below the last four recent interglacials and up to 3°C below the Minoan, Roman and Medieval Warmings.[2282,2283] We live on a warm wet greenhouse volcanic planet which has had ice for less than 20% of time, and it is unusual for the planet to be so cool.

Even if today's warming were unprecedented, the Sun is the probable cause. It was more active in the past 70 years than in the previous 11,400 years.[2284,2285,2286]

Even if the Sun were not to blame for 20th Century temperature changes, the IPCC has not shown that humans are to blame. $CO_2$ occupies only one-ten-thousandth more of the atmosphere today than it did in 1750 AD.[2287]

Even if $CO_2$ were to blame, no runaway greenhouse catastrophe occurred in the Cambrian Period at 500 Ma when there was more than 20 times today's atmospheric $CO_2$. Temperature was up to 7°C warmer than today.[2288]

Even if $CO_2$ levels had set a record, the major temperature measuring stations have recorded no warming since 1998. For seven years, temperatures have fallen. The January 2007–January 2008 fall was the steepest since 1880.[2289]

Even if the planet were not cooling, the rate of warming is far less than the IPCC speculates and too small to be of interest. There may well be no new warming until 2015, if then.[2290]

Even if warming were harmful, humankind's effect is minuscule. The

---

[2282] Petit, J. R., Jouzel, J., Raynaud, D., Barkov, N. I., Barnola, J. M., Basile, I., Bender, M., Chappellaz, J., Davis, J., Delaygue, G., Delmotte, M., Kotlyakov, V. M., Legrand, M., Lipenkov, V., Lorius, C., Pépin, L., Ritz, C., Saltzman, E. and Stievenard, M. 1999: Climate and atmospheric history of the past 420,000 years from the Vostok ice core, Antarctica. *Nature* 399: 429-436.

[2283] Houghton, J. T., Jenkins, G. J. and Ephraums, J. J. 1990: *Climate Change. The IPCC Scientific Assessment. Working Group 1 Report.* Cambridge University Press.

[2284] Usoskin, I. G., Solanki, S. K., Schüssler, M., Mursula, K. and Alanko, K. 2003: Millennium-scale sunspot number reconstruction: Evidence for an unusually active sun since the 1940s. *Physical Review Letters* 91: 211101-211105.

[2285] Hathaway, D. H., Nandy, D., Wilson, R. M. and Reichmann, E. J. 2004: Evidence that a deep meridional flow sets in the sunspot cycle period. *The Astrophysical Journal* 589: 665-670.

[2286] Solanki, S. K. and Krivova, N. A. 2004: Solar irradiance variations: From current measurements to long-term estimates. *Solar Physics* 224: 197-208.

[2287] Keeling, C. D. and Whorf, T. P. 2004: Atmospheric carbon dioxide record from Mauna Loa. http://cdiac.ornl.gov/trends/co2/sio-mlo.html

[2288] Houghton, J. T., Ding, Y., Griggs, D. J., Noguer, M., van der Linden, P. J., Dai, X., Maskell, K. and Johnson, C. A. 2001: *Climate change: The scientific basis.* Cambridge University Press.

[2289] GISS; Hadley; NCDC; RSS; UAH: all 2008

[2290] Keenlyside, N. S., Latif, M., Jungclaus, J., Kornblueh, L. and Roeckner, E. 2008: Advancing decadal-scale climate prediction in the North Atlantic sector. *Nature* 453: 84-88.

observed changes may be natural.[2291,2292,2293,2294,2295,2296]

Even if the effect of humans on climate were significant, the IPCC's proof of human effect on climate (i.e. tropical mid-troposphere warming at thrice the surface rate) is absent.[2297,2298,2299]

Even if the human fingerprint of global warming were present, climate models cannot predict the future of the complex, chaotic climate unless we know its initial state to an unattainable precision.[2300,2301,2302]

Even if computer models could work, they cannot predict future rates of warming. Temperature response to atmospheric greenhouse-gas enrichment is an input to the computers, not an output from them.[2303]

Even if the IPCC's high "climate sensitivity" to $CO_2$ were correct, disaster would not be likely to follow. The peer-reviewed literature is near-unanimous in not predicting climate catastrophe.[2304]

Even if Al Gore were right that harm might occur, the Armageddon scenario he depicts is not based on any scientific view. Sea level may rise 1

[2291] Houghton, J. T., Ding, Y., Griggs, D. J., Noguer, M., van der Linden, P. J., Dai, X., Maskell, K. and Johnson, C. A. 2001: *Climate change: The scientific basis*. Cambridge University Press.

[2292] Hanna, E., Huybrechts, P., Steffan, K., Cappelen, J., Huff, R., Shuman, C., Irvine-Fynn, T., Wise, S. and Griffiths, M. 2007: Increased runoff from melt from the Greenland ice sheet: a response to global warming. *Journal of Climate* 21: 331-341.

[2293] Lindzen, R. S. 2008: Global warming: The origin and nature of the alleged scientific consensus. *Cato Institute* 15: 2.

[2294] Spencer, R. W. 2007: How serious is the global warming threat? *Society* 44: 45-50.

[2295] Wentz, F. J., Ricciardulli, L., Hilburn, K. and Mears, C. 2007: How much more rain will global warming bring? *Science* 317: 233-235.

[2296] Armstrong, J. S. 2008: Global warming: Forecasts by scientists versus scientific forecasts. *Energy and Environment* 18: 7-8.

[2297] Douglass, D. H., Pearson, B. D. and Singer, S. F. 2004: Altitude dependence of atmospheric temperature trends: Climate models versus observation. *Geophysical Research Letters* 31: arXiv:physics/0407074v1.

[2298] Lindzen, R. S., Chou, M. D. and Hou, A. Y. 2001: Does the Earth have an adaptive infrared iris. *Bulletin of the American Meteorological Society* 83: 417-432.

[2299] Spencer, R. W. 2007: How serious is the global warming threat? *Society* 44: 45-50.

[2300] Lorenz, E. N. 1963: Deterministic nonperiodic flow. *Journal of the Atmospheric Sciences* 20: 130-141.

[2301] Giorgi, F. and Bi, X. 2005: Regional changes in surface climate interannual variability for the 21st century from ensembles of global model simulations. *Geophysical Research Letters* 32: L:13701, doi:10.1029/2005GL023002.

[2302] Houghton, J. T., Ding, Y., Griggs, D. J., Noguer, M., van der Linden, P. J., Dai, X., Maskell, K. and Johnson, C. A. 2001: *Climate change: The scientific basis*. Cambridge University Press.

[2303] Akasofu, S.-I. 2008: A suggestion to climate scientists and the Intergovernmental Panel on Climate Change. *EOS* 89: 108.

[2304] Schulte, K.-M. 2008: Scientific consensus on climate change? *Energy and Environment* 19: 281-286.

foot to 2100, not 20 feet as Gore claims.[2305,2306,2307]

Even if Armageddon were likely, scientifically unsound precautions are already starving millions as biofuels use agricultural land, doubling staple cereal prices in a year.[2308]

Even if precautions were not killing the poor, they would work no better than the precautionary ban on DDT, which killed 40 million children before the UN ended it.[2309]

Even if precautions might work, the strategic harm done to humanity by killing the world's poor and destroying the economic prosperity of the West would outweigh any climate benefit.[2310]

Even if the climatic benefits of mitigation could outweigh the millions of deaths it is causing, adaptation as and if necessary would be far more cost-effective and less harmful.

Even if mitigation were as cost-effective as adaptation, the public sector, which emits twice as much carbon as the private sector, must cut its own emissions by half before it preaches to us.

Human forces are orders of magnitude lower than the natural forces that drive climate.[2311] The global cooling and warming observed during the last 150 years is just a short episode in geologic history and the current global warming is most likely a result of the combined effects of many natural drivers of climate and cannot be attributed to human impact. It is important to get the science correct before policy is even discussed. This has not happened.

How many examples of failed predictions, discredited assumptions, evidence of incorrect data and evidence of malpractice are required before the idea of human-induced climate change loses credibility?

The influence of human activity on planet Earth needs some perspective. In the words of Pope Benedict XVI:

> It is important for assessments in this regard to be carried out

---

[2305] Moerner, N.-A. 1981: Revolution in Cretaceous sea level analysis. *Geology* 9: 344-346.

[2306] Moerner, N.-A., Nevanlinna, H., Dergachev, V., Shumilov, O., Raspopov, O., Abrahamsen, N., Pilipenko, O., Trubikhin, V. and Gooskova, E. 2003: Geomagnetism and climate V: General conclusions. *Geophysical Research Abstracts* 5: 10168.

[2307] Houghton, J. T., Jenkins, G. J. and Ephraums, J. J. 1990: *Climate change: the IPCC scientific assessment*. Cambridge University Press.

[2308] UNFAO, 2008.

[2309] Dr. Arata Kochi, UN malaria program, 2006.

[2310] Henderson, D. 2005: The treatment of economic issues by the Intergovernmental Panel on Climate Change. *Energy and Environment* 16: 321-326.

[2311] Khilyuk, L. F. and Chilingar, G. V. 2006: On global forces of nature driving the Earth's climate. Are humans involved? *Environmental Geology* 50: doi 10.1007/s00254-006-0261-x.

prudently, in dialogue with experts and people of wisdom, uninhibited by ideological pressure to draw hasty conclusions, and above all with the aim of reaching agreement on a model of sustainable development capable of ensuring the well-being of all while respecting environmental balances.

Human stupidity is only exceeded by God's mercy, which is infinite.

# INDEX